Methods in Molecular Biology

Series Editor
John M. Walker
School of Life and Medical Sciences
University of Hertfordshire
Hatfield, Hertfordshire, UK

For further volumes:
http://www.springer.com/series/7651

For over 35 years, biological scientists have come to rely on the research protocols and methodologies in the critically acclaimed *Methods in Molecular Biology* series. The series was the first to introduce the step-by-step protocols approach that has become the standard in all biomedical protocol publishing. Each protocol is provided in readily-reproducible step-by-step fashion, opening with an introductory overview, a list of the materials and reagents needed to complete the experiment, and followed by a detailed procedure that is supported with a helpful notes section offering tips and tricks of the trade as well as troubleshooting advice. These hallmark features were introduced by series editor Dr. John Walker and constitute the key ingredient in each and every volume of the *Methods in Molecular Biology* series. Tested and trusted, comprehensive and reliable, all protocols from the series are indexed in PubMed.

Plasmodesmata

Methods and Protocols

Second Edition

Edited by

Yoselin Benitez-Alfonso

Centre for Plant Sciences, School of Biology, University of Leeds, Leeds, UK

Manfred Heinlein

Institut de Biologie Moléculaire des Plantes, CNRS, Université de Strasbourg, Strasbourg, France

Humana Press

Editors
Yoselin Benitez-Alfonso
Centre for Plant Sciences, School of Biology
University of Leeds
Leeds, UK

Manfred Heinlein
Institut de Biologie Moléculaire des Plantes, CNRS
Université de Strasbourg
Strasbourg, France

ISSN 1064-3745 ISSN 1940-6029 (electronic)
Methods in Molecular Biology
ISBN 978-1-0716-2131-8 ISBN 978-1-0716-2132-5 (eBook)
https://doi.org/10.1007/978-1-0716-2132-5

Cover Illustration Caption: 3D-SIM image of plasmolysed BY2 cells showing ER labelled with RTN6-GFP (green) and cell wall labelled with Calcofluor White (blue). Image produced by Kirsten Knox.

This Humana imprint is published by the registered company Springer Science+Business Media, LLC part of Springer Nature.
The registered company address is: 1 New York Plaza, New York, NY 10004, U.S.A.

Preface

Plasmodesmata, the nanochannels that connect plant cells, are crucial for intercellular signaling by playing a key role in plant development and response to the environment. Research on plasmodesmata structure and regulation is growing as accumulating evidence relates the function of these channels to crop yield as well as resilience to climate change and devastating diseases. Despite their importance, we are still far from being able to translate knowledge on plasmodesmata structure and function into the design of strategies to improve food security. This is partly due to the lack of tools and resources to study these channels and to enable their targeted modification. Nevertheless, the field has made major advances in both methodology and current knowledge.

Thus, following the first edition of *Plasmodesmata: Methods and Protocols* published in 2015, it became time to create a second edition, which compiles new, up-to-date protocols, written by experts in the field of plasmodesmata research. The aim is to facilitate the study of these structures and to instigate new research that will further advance our knowledge on plasmodesmata in labs around the world. Plasmodesmata research was classically driven by cell biologists, but in recent years we acknowledge the significant contribution of other disciplines including biochemistry, molecular biology, and mathematical modeling. Consequently, protocols in the book expand to a broad range of disciplines suitable for studying plasmodesmata from multiple angles.

The book is divided into five parts: the first part contains two reviews addressing the general structural composition and regulation of plasmodesmata as well as discussing their role in plant development and disease. The second part is dedicated to methods to study plasmodesmata architectures and distribution in cell interfaces, including the most recent advances in preparing and imaging the plant material using electron tomography and high-resolution microscopy. The third part focuses on approaches to dissect plasmodesmata composition, specially targeting callose accumulation and membranes lipidomic and proteomic analysis as well as methods to study protein clustering and membrane organization around plasmodesmata. The fourth part collects protocols to quantify changes in plasmodesmata permeability using fluorescent tracers and mobile proteins. This section also contains tools to interrogate plasmodesmata function using vectors for callose induction and a forward genetic screen to identify trafficking regulators. To conclude, we included a fifth part with protocols that contribute to plasmodesmata research but fall outside the previous classifications. These are tools to visualize mobile RNAs, to image cell interfaces in the meristem, and to measure/predict symplasmic connectivity using modeling and computational approaches.

We, the editors, believe this compendium will be useful to all plant scientists, especially those studying cell-to-cell communication pathways. We have worked with the authors to make the protocols accessible to beginners and experts; thus the book will be useful for advanced researchers as well as for the teaching and training of young scientists. We hope that this book will become a major resource in the laboratories.

Leeds, UK — *Yoselin Benitez-Alfonso*
Strasbourg, France — *Manfred Heinlein*

Contents

Contributors

SAM AMSBURY • *School of Biosciences, The University of Sheffield, Sheffield, UK*
ANASTASIA K. ATABEKOVA • *Department of Virology, Biological Faculty, Moscow State University, Moscow, Russia; A.N. Belozersky Institute of Physico-Chemical Biology, Moscow State University, Moscow, Russia*
GEORGE W. BASSEL • *School of Life Sciences, The University of Warwick, Coventry, UK*
EMMANUELLE M. F. BAYER • *Laboratoire de Biogènese Membranaire, UMR 5200, CNRS, Université de Bordeaux, Villenave d'Ornon Cedex, France*
ILYA BELEVICH • *Helsinki Institute of Life Science/Institute of Biotechnology, University of Helsinki, Helsinki, Finland*
YOSELIN BENITEZ-ALFONSO • *Centre for Plant Science, School of Biology, University of Leeds, Leeds, UK*
PATRICK BLÜMKE • *Institute for Developmental Genetics, Heinrich Heine University, Düsseldorf, Germany*
STANLEY W. BOTCHWAY • *Central Laser Facility, Science and Technology Facilities Council (STFC), Rutherford Appleton Laboratory, Research Complex at Harwell, Didcot, UK*
LYSIANE BROCARD • *Bordeaux Imaging Centre, Plant Imaging Platform, UMS 3420, INRA-CNRS-INSERM-University of Bordeaux, Villenave d'Ornon, France*
JACOB O. BRUNKARD • *Laboratory of Genetics, University of Wisconsin-Madison, Madison, WI, USA; Department of Plant and Microbial Biology, University of California at Berkeley, Berkeley, CA, USA; Plant Gene Expression Center, U.S. Department of Agriculture, Agricultural Research Service, Albany, CA, USA*
TESSA M. BURCH-SMITH • *Department of Biochemistry & Cellular and Molecular Biology, University of Tennessee, Knoxville, TN, USA; Donald Danforth Plant Science Center, Saint Louis, MO, USA*
AJEET CHAUDHARY • *Plant Developmental Biology, TUM School of Life Sciences, Technical University of Munich, Freising, Germany; Department of Plant Biology, Carnegie Institution for Science, Stanford, CA, USA*
STÉPHANE CLAVEROL • *Plateforme Proteome, Université de Bordeaux, Bordeaux, France*
YOAN COUDERT • *Laboratoire Reproduction et Développement des Plantes, ENS de Lyon, Lyon, France*
FLORENCE R. DANILA • *Research School of Biology, The Australian National University, Acton, ACT, Australia*
GWENDOLYN V. DAVIS • *School of Life Sciences, The University of Warwick, Coventry, UK*
EVA E. DEINUM • *Wageningen University, Wageningen, The Netherlands*
MIN DIAO • *iHuman Institute, Shanghai Tech University, Shanghai, China*
JOHN R. DUNLAP • *The Joint Institute for Advanced Materials, University of Tennessee, Knoxville, TN, USA*
CHRISTINE FAULKNER • *Crop Genetics, John Innes Centre, Norwich Research Park, Norwich, UK*
JESSICA C. FERNANDEZ • *Department of Biochemistry & Cellular and Molecular Biology, University of Tennessee, Knoxville, TN, USA; Department of Botany, University of Wisconsin, Madison, WI, USA*

LAETITIA FOUILLEN • *Laboratoire de Biogènese Membranaire, UMR 5200, CNRS, Université de Bordeaux, Villenave d'Ornon Cedex, France*
TILMAN FRANKE • *Thermo Fisher Scientific, Planegg, Germany*
MICHAEL FUCHS • *Department of Stem Cell Biology, Centre for Organismal Studies (COS), Heidelberg University, Heidelberg, Germany*
TOMOMICHI FUJITA • *Department of Biological Sciences, Faculty of Science, Hokkaido University, Sapporo, Hokkaido, Japan*
MARIE GLAVIER • *Laboratoire de Biogenèse Membranaire, UMR5200 CNRS, Université de Bordeaux, Villenave d'Ornon, France*
MARIAH GOBBLE • *Plant and Microbial Biology Department, North Carolina State University, Raleigh, NC, USA*
KAMILA GODEL-JĘDRYCHOWSKA • *Faculty of Natural Sciences, Institute of Biology, Biotechnology and Environmental Protection, University of Silesia in Katowice, Katowice, Poland*
MAGALI S. GRISON • *Laboratoire de Biogènese Membranaire, UMR 5200, CNRS, Université de Bordeaux, Villenave d'Ornon Cedex, France*
YAĞMUR HASBIOĞLU • *Max Planck Institute of Molecular Plant Physiology, Potsdam, Germany*
MANFRED HEINLEIN • *Institut de Biologie Moléculaire des Plantes, CNRS, Université de Strasbourg, Strasbourg, France*
WILSON HORNER • *Laboratory of Genetics, University of Wisconsin-Madison, Madison, WI, USA; Department of Plant and Microbial Biology, University of California at Berkeley, Berkeley, CA, USA; Plant Gene Expression Center, U.S. Department of Agriculture, Agricultural Research Service, Albany, CA, USA*
VICKY HOWE • *Institute for Developmental Genetics, Heinrich Heine University, Düsseldorf, Germany*
CAIPING HUANG • *Institut de Biologie Moléculaire des Plantes, CNRS, Université de Strasbourg, Strasbourg, France*
SHANJIN HUANG • *Center for Plant Biology, School of Life Sciences, Tsinghua University, Beijing, China*
DAVID JACKSON • *Cold Spring Harbor Laboratory, Cold Spring Harbor, NY, USA*
PHILIP KIRK • *Centre for Plant Science, School of Biology, University of Leeds, Leeds, UK*
MUNENORI KITAGAWA • *Cold Spring Harbor Laboratory, Cold Spring Harbor, NY, USA*
KIRSTEN KNOX • *The School of Life Sciences, University of Glasgow, Glasgow, UK*
FRIEDRICH KRAGLER • *Max Planck Institute of Molecular Plant Physiology, Potsdam, Germany*
VERENA KRIECHBAUMER • *Endomembrane Structure and Function Research Group, Department of Biological and Medical Sciences, Oxford Brookes University, Oxford, UK*
EWA KURCZYŃSKA • *Faculty of Natural Sciences, Institute of Biology, Biotechnology and Environmental Protection, University of Silesia in Katowice, Katowice, Poland*
EKATERINA A. LAZAREVA • *Department of Virology, Biological Faculty, Moscow State University, Moscow, Russia*
ALEXANDER A. LEZZHOV • *A.N. Belozersky Institute of Physico-Chemical Biology, Moscow State University, Moscow, Russia; Faculty of Bioengineering and Bioinformatics, Moscow State University, Moscow, Russia*
JAN U. LOHMANN • *Department of Stem Cell Biology, Centre for Organismal Studies (COS), Heidelberg University, Heidelberg, Germany*
FRANK MACHIN • *Max Planck Institute of Molecular Plant Physiology, Potsdam, Germany*

JOSEPH F. MCKENNA • *School of Life Sciences, University of Warwick, Coventry, UK; Oxford Brookes University, Gypsy Lane, UK*

SERGEY Y. MOROZOV • *Department of Virology, Biological Faculty, Moscow State University, Moscow, Russia; A.N. Belozersky Institute of Physico-Chemical Biology, Moscow State University, Moscow, Russia*

ARTHUR MULLER • *Laboratoire Reproduction et Développement des Plantes, ENS de Lyon, Lyon, France*

JERÔME MUTTERER • *Institut de Biologie Moléculaire des Plantes, CNRS, Université de Strasbourg, Strasbourg, France*

ANDREA PATERLINI • *Sainsbury Laboratory, University of Cambridge, Cambridge, UK*

EDUARDO J. PEÑA • *Instituto de Biotecnología y Biología Molecular (IBBM), Facultad de Ciencias Exactas-UNLP, CONICET, La Plata, Argentina*

JULES D. PETIT • *Laboratoire de Biogenèse Membranaire, UMR5200 CNRS, Université de Bordeaux, Villenave d'Ornon, France; Laboratoire de Biophysique Moléculaire aux Interfaces, TERRA Research Centre, GX ABT, Université de Liège, Gembloux, Belgium*

BRANDON C. REAGAN • *Department of Biochemistry & Cellular and Molecular Biology, University of Tennessee, Knoxville, TN, USA; Department of Plant Biology, Michigan State University, East Lansing, MI, USA*

SEBASTIAN SAMWALD • *Crop Genetics, John Innes Centre, Norwich Research Park, Norwich, UK*

KAY SCHNEITZ • *Plant Developmental Biology, TUM School of Life Sciences, Technical University of Munich, Freising, Germany*

RÜDIGER SIMON • *Institute for Developmental Genetics, Heinrich Heine University, Düsseldorf, Germany*

RICHARD S. SMITH • *The John Innes Centre, Norwich, UK*

ANDREY G. SOLOVYEV • *Department of Virology, Biological Faculty, Moscow State University, Moscow, Russia; A.N. Belozersky Institute of Physico-Chemical Biology, Moscow State University, Moscow, Russia*

ROSANGELA SOZZANI • *Plant and Microbial Biology Department, North Carolina State University, Raleigh, NC, USA*

ESTEE E. TEE • *Crop Genetics, John Innes Centre, Norwich Research Park, Norwich, UK*

TAKUMI TOMOI • *Biosystems Science Course, Graduate School of Life Science, Hokkaido University, Sapporo, Hokkaido, Japan; Laboratory for Biothermology, National Institute for Basic Biology, Okazaki, Aichi, Japan*

LISA VAN DEN BROECK • *Plant and Microbial Biology Department, North Carolina State University, Raleigh, NC, USA*

JUSTYNA WRÓBEL-MAREK • *Institute of Biology, Biotechnology and Environmental Protection, Faculty of Natural Sciences, University of Silesia in Katowice, Katowice, Poland*

DAWEI YAN • *State Key Laboratory of Crop Stress Adaptation and Improvement, School of Life Sciences, Henan University, Kaifeng, China*

Part I

Overview Articles

Chapter 1

Plasmodesmata Structural Components and Their Role in Signaling and Plant Development

Philip Kirk and Yoselin Benitez-Alfonso

Abstract

Plasmodesmata are plant intercellular channels that mediate the transport of small and large molecules including RNAs and transcription factors (TFs) that regulate plant development. In this review, we present current research on plasmodesmata form and function and discuss the main regulatory pathways. We show the progress made in the development of approaches and tools to dissect the plasmodesmata proteome in diverse plant species and discuss future perspectives and challenges in this field of research.

Key words Plasmodesmata, Symplastic signaling, Membranous domains, Cell walls, Proteomics

1 Introduction

Plants can adapt their development in response to changes in internal and external conditions. New organs emerge and others undergo reprogramming leading to modifications in the plant body that optimize its capacity for growth and survival in adverse environments. In the context of resource acquisition, plants must balance system requirements for specific or multiple resources with information about the local availability of those resources in different parts of the plant system. This level of coordination across the whole plant body requires systemic (shoot-to-root and root-to-shoot) signaling. Most research on the intercellular pathways involved in signaling focuses on the transport of phytohormones and other mobile signals via the apoplastic/extracellular space [1–4]. Due to the nature of the cell walls, transport via diffusion in the apoplast is limited to small-molecular-weight compounds such as water, inorganic nutrients, sugars, and hormones. In comparison, little is known about the symplastic pathway. Large- and small-molecular-weight compounds, such as proteins and RNAs, travel via direct cell-cell connections made possible by plasmodesmata (PD) [5]. Plasmodesmata are membranous pores embedded in

Yoselin Benitez-Alfonso and Manfred Heinlein (eds.), *Plasmodesmata: Methods and Protocols*,
Methods in Molecular Biology, vol. 2457, https://doi.org/10.1007/978-1-0716-2132-5_1,

the cell wall that create discrete cytoplasmic continuous domains between cells, named the symplasm. There is still a great deal to learn about PD composition and function, but there is evidence of the critical role these structures play in controlling the passage of developmental and signaling factors that participate in the regulation of plant development and responses to the environment [6].

The focus of this review is to discuss our understanding of PD structure and function in the regulation of plant development. Most of the information on this topic is extracted from research on the model plants *Arabidopsis thaliana* and tobacco although information is emerging on new systems, such as *Medicago truncatula* and poplar. Based on the research so far, we will discuss PD structures, regulation, and their contributions to symplastic signaling during development. In the last section, we introduce a new bioinformatic tool that systemically analyzes the PD proteomes published so far to generate in silico proteomes for species of interest. The aim is to provide a framework for the identification of new factors that modify PD as a tool to optimize plant development and environmental responses.

2 PD Structural Diversity and Regulation

PD are plasma membrane (PM)-lined pores that traverse cell walls of neighboring cells, containing a portion of appressed endoplasmic reticulum (ER) termed the desmotubule (DT) that provides continuity of ER between cells. The gap between the PM and the DT is occupied by a cytoplasmic sleeve (CS), and it is considered to be the main conduit for symplastic transport. The ability of molecules to cross the CS is dependent on the physical features of both the PD and the mobile molecule. One parameter routinely used to describe PD transport capability is the size exclusion limit (SEL), described as the maximum size of molecules that can pass through the pore [7, 8]. SEL is often estimated by comparing the intercellular diffusion of fluorescent dyes of small molecular weight with that of larger protein-based fluorophores such as GFP, GFP protein fusions, or its multimeric configurations [9–12]. SEL is not the only factor that influences transport through PD, but it is a useful framework to conceptualize simple nonspecific PD transport. A recent review provides a more comprehensive introduction to PD structure [6].

In a plant cell interface, PD can appear in different forms tightly linked to their permeability or transport capacity. Primary PD are laid down during cytokinesis when cell plate formation entraps strands of the ER (which will form the DT) between Golgi-derived vesicles [13, 14]. Secondary PD are formed in existing cell walls or in cells forming graft junctions and at host/parasite interfaces [15, 16]. The process for secondary PD formation appears to involve local thinning of the cell wall, insertion of the ER, PM fusion between cells, and further modifications to the cell walls [17, 18]. Secondary PD are often clustered together with primary PD, forming pit fields [19]. Structurally, primary and secondary PD closely resemble each other [18]. Both primary and secondary PD can develop simple or branched morphologies. Branched PD are formed later in development by a yet to be discovered mechanism. Although it was assumed that branched PD had a reduced permeability, mutant studies appear to suggest that this is not the case [11]. Other PD morphologies are described as twinning and H-shaped, and some seem to have specialized functions such as "funnel" PD which are shaped to improve the fluid dynamic properties of the pore for phloem unloading [20]. PD can be further classified as type I or II, based on how tightly the DT is associated with the PM, restricting the CS available for transport [21]. Type I lacks a visible CS based on electron tomography and is more abundant in meristematic and rapidly dividing tissues, whereas type II is associated to mature organs.

Each domain (cell walls, PM, DT, and CS) contributes to PD regulation. Lipidomic analysis of PD fractions revealed that the composition of membrane lipids at PD is distinct from other membranous domains [22]. PD contain highly saturated glycerophospholipids and are enriched in sphingolipid glycosyl inositol phospho ceramides (GIPCs) and sterols relative to the PM [22]. Sterols and sphingolipids have a high affinity for one another resulting in the formation of microdomains within membranes referred to as lipid rafts [23]. PD proteins such as remorins (REM) and tetraspanins (TET) are involved in lipid nanodomain specification and organization at PD [24, 25]. Lipid rafts can recruit specific types of proteins to PD [22] and may be crucial in the specification of PD function. This is exemplified by the recent finding that heterologous expression of plant REM in mammalian cells resulted in membraneous channels, reminiscent of PD, that were able to transport macromolecules between cells [26].

Little is known about the importance of the DT on PD regulation. The exact proteomic and lipid composition of the DT is not known as it has not been possible to disentangle it from that of the PM or the rest of the ER. Only a minority of currently identified PD proteins interact with the DT including calnexins, reticulons, and multiple C2 domain and transmembrane region proteins (MCTPs) [27–29]. The DT are tethered to the PM by MCTPs and synaptotagmins (SYT) [29, 30]. The PM and the DT interact at characteristic spoke-like features seen in electron micrographs which may function as membrane contact sites where lipid-associated proteins can be exchanged [6]. These spoke-like patterns were thought to be related to the cytoskeleton. Both actin and myosin were detected in the PD pore, and research shows that PD-localized formins act as actin anchors to PD domains

[31, 32]. Microinjection experiments have demonstrated that molecules can move between cells directly through the lumen of the DT, and this pathway has been postulated as a potential route for the transport of lipids and specific membrane-bound proteins [6, 33]. More work is required to determine the extent and significance of transport directly via the ER-DT continuum and its contribution to transport in relation to the CS, which is often attributed as the main conduit for symplastic transport.

One may assume that transport through PD may be easier if the DT were simply not present [9–12]. However, recent modeling suggests that the presence of a DT increases the transport capacity of the PD for a given SEL [34]. This may seem counterintuitive at first, but computational modeling comparing channels with and without DT for the same particle size (same SEL) demonstrates that effective permeability in channels without a DT is reduced because there is less cross-sectional area available for transport. The DT has also been proposed to control flux through the CS under osmotic stress or wounding events [35]. The pressure wave created under these conditions may flow the DT complex to plug the pore restricting fluid loss and acting much like a safety valve to mitigate damage. These mechanical features have been evaluated in recent models proposing the DT as key structures in mechanosensing and in regulating transport, through physically obstructing PD aperture.

Cell walls are a defining feature of plant cells, and, not surprising, their regulation is essential to determine PD form and function. The next subsection focuses on different cell wall structures around PD and expands on their importance for transport.

3 PD Cell Walls and Their Importance in Regulation

Cell walls are often conceptualized as a highly complex matrix of interacting polysaccharides and proteins [36], including cellulose, hemicelluloses, pectins, arabinogalactan proteins (AGPs), extensins, expansins, and other minor components. Callose, a 1,3-β-glucan, is found in abundance at PD relative to the rest of the primary cell wall where it accounts for just a small proportion of its composition [36]. Callose is an amorphous polymer synthesized and digested directly at the cell wall by membrane-linked enzymes, named callose synthases (CALS) and PD-located beta-1,3 glucanases (PdBGs) [5]. Regulation of callose deposition at PD is the best characterized mechanism for the regulation of symplastic transport. Changing the expression of PdBG1 and PdBG2 affects transport and lateral root development and patterning [37]. Ectopic expression of CalS3 or function-limiting mutations in CalS10 alter the mobility of cell fate-determining transcription factors (TFs) affecting root growth [38, 39]. In line with their

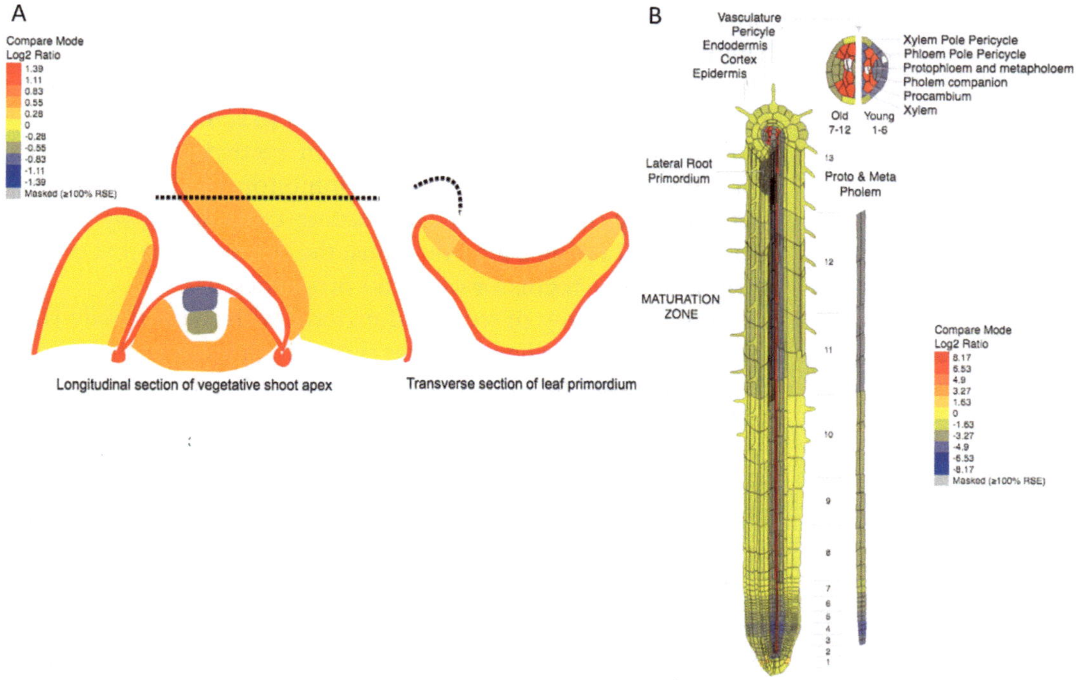

Fig. 1 PdBG1 and CALS3 expression domains in *Arabidopsis thaliana* shoot apex and root. PdBG1 gene expression in relation to CALS3 is represented (scale range red: high PdBG1; blue: high CALS3). Images generated in compare mode with the Shoot Apex eFP (RNA-Seq data) (**a**) and with the Tissue Specific Root eFP (**b**) at bar.utoronto.ca/eplant. Note that the domains of expression in the apical shoot apex and in the root vasculature are largely complementary. (Description of the experiments, references, and the data analysis is provided in http:/bar.utoronto.ca/eplant/)

contrasting functions, CALS and PdBGs show complementary expression profiles in the root and shoot apical meristems (Fig. 1).

The accepted model is that callose physically constricts PD aperture (SEL) which reduces symplastic transport. Callose has been shown to interact with cellulose in vitro, forming elastic and plastic mixtures, depending on concentration. This property might contribute to cell wall resilience by diffusing the stress generated by changes in fluid pressures through PD [40]. Therefore, it is proposed that callose have a dual role in regulating PD, (1) its accumulation restricting pore aperture and (2) improving the properties of cell walls to cope with the mechanical-physical stress imposed by the presence of PD.

Pectins are also distinct at PD relative to the bulk cell wall and may contribute to the PD physical properties or directly in the regulation of symplastic transport. Pectins have been linked to the control of cell wall porosity and are often described as a glue that holds together other cell wall constituents [36]. The role of pectins at PD is unknown, but, as described for callose, pectin composition may contribute to determine the mechanical properties of cell wall around PD. Pectins are a complex class of polysaccharides with an acid backbone mainly formed by homogalacturonan (HG) and rhamnogalacturonan (RGI and RGII) polymers with a diverse set of arabinan, galactan, and arabinogalactan side chains. Methyl ester groups are added to HG during synthesis [41, 42]. These methyl ester groups can be removed by pectin methylesterases (PMEs) which are in turn regulated by PME inhibitors [43]. Non-methyl-esterified polygalacturonans can become cross-linked by binding calcium ions forming a gel [44]; thus, the degree of methyl esterification is a major regulator of pectin properties.

Immunolabeling of sectioned material with the antibody JIM5 (specific to low-esterified HG) has shown high levels of this epitope around pit fields in tomato fruit pericarp [45, 46] and in apple [47]. Similarly, immunolabeling with LM6 (specific to (1-5)-α-L-arabinan side chains in RG-I) showed this epitope enriched in cell walls surrounding PD and in pit fields in tomato and tobacco [19, 46]. PMEs have been identified in PD proteomes [29], and a native PME was found to regulate the transport of the tobacco mosaic virus movement protein [48]. These findings suggest that the degree of pectin esterification regulates PD but the specific mechanism is unknown. Moreover, PME activity regulates apoplastic Ca^{2+} levels during stress responses, and calcium plays a role in signaling changes in symplastic transport [49].

The role of cellulose and other hemicelluloses, such as xyloglucans, in PD regulation has been less studied. Cellulose is a 1,4-β-glucan forming strong linear crystalline structures called cellulose microfibrils [50]. Microfibril arrangement regulates the anisotropic properties of the cell wall to direct, for example, cell expansion [50]. Based on calcofluor white staining and Raman spectroscopy, cell walls around PD appear to contain less cellulose than other domains, but it is not clear if the difference is due to differences in cellulose amount or microfibril organization [19]. Lower cellulose content may make the cell wall more flexible, which might be required for regulated symplastic transport. Furthermore, variations in the composition of pectin and callose, which interact with cellulose, might further determine cellulose properties at PD cell walls [40, 51]. Cell wall proteins, such as expansins, also regulate cellulose properties and have been also identified to target PD cell walls. Overexpression of a PD-specific α-expansin identified in *Nicotiana benthamiana* (NbEXPA1) [52] promotes viral replication and spreading via PD although the underlying mechanism is unclear.

To summarize, although several cell wall components appear to be modified around PD, callose seems to be the major regulator by modifying either PD aperture or the mechanical properties of cell walls to facilitate transport and, perhaps, PD formation.

4 Developmental Signaling Pathways Converge at PD

PD are dynamic structures regulated in response to internal and external cues. High PD connectivity between groups of cells creates symplastic domains where exchange of signals and resources is relatively unrestricted [53]. Low/absence of PD connectivity can effectively restrict communication between neighboring cells creating symplasmically isolated domains. These processes are reversible and are dynamically controlled by developmental and environmental cues.

PD host a wide range of receptor-like proteins (RLP) and receptor-like kinases (RLK) that are key players in intercellular signaling [6]. These PD receptor genes integrate native and foreign apoplastic and symplastic signals with the regulation of symplastic transport [54]. The roles of PD receptors in plant development and responses to the environment are discussed in a recent review [55]. Within this group, a number of RLP containing the DOMAIN OF UNKNOWN FUNCTION 26 (DUF26), namely, the plasmodesmata-localized proteins (PDLPs) and cysteine-rich RLP kinases (CRKs), were identified. A mechanism involving the regulation of callose is triggered by modifications in PDLP expression in response to a wide range of stress challenges [6, 56, 57]. CYS-RICH RLK 2 (CRK2) was found to transiently localize to PD under osmotic stress and trigger callose induction [58]. This mechanism is dependent of phosphorylation of the RESPIRA TORY BURST OXIDASE HOMOLOG D (RBOHD), which is required for ROS (reactive oxygen species) production. PDLP1 and PDLP5 also require RBOHD-dependent ROS production to induce callose [59], supporting the links between PD receptor proteins, ROS, and the regulation of callose at PD [56, 57, 60]. Recent research highlights that PDLP5 is induced by auxin and localizes in cells overlying lateral root primordia to control branching, a process where PdBG1 and PdBG2 have been previously implicated [37, 61]. Thus, PDLPs are not only perceiving stress signals but have a significant role in plant development likely via the control of callose.

The importance of PD in hormone signaling is also being recognized. Abscisic acid (ABA), often considered the "stress" hormone, regulates PD and the expression of mobile factors involved in bud dormancy release [62]. Symplastic transport pathways are also important in auxin signaling [63]. Auxin distribution is generally attributed to the actions of families of auxin efflux carriers PIN-formed (PIN) and ATP-BINDING CASSETTE subfamily B (ABCB) and the influx carrier AUXIN1/LIKE-AUX1 (AUX/LAX) [64]. These factors create and modulate heterogeneous auxin distributions in cells by selective efflux and influx into/from the apoplast. Considering the contribution of PD in auxin transport in a computational model greatly improves predictions on auxin distribution as demonstrated using experimental reporters [63]. The significance of symplastic transport on auxin distribution is discussed in a recent review [65] where it is highlighted that such transport complements, and not supersede, the established mechanisms of auxin transport. Auxin signaling is intertwined with other signaling pathways involving sugars, brassinosteroids (BR), ABA, cytokinins (CK), and ethylene affecting organ development [1, 66, 67]. It is also integrated with other signaling mechanisms at PD domains. It is proposed, for example, that auxin-dependent expression of PDLP5 restricts symplastic movement of auxin to negatively regulate lateral root emergence [61].

Sucrose is also involved in symplasmic signaling. Sucrose/H+ symporters in phloem companion cells (CC) actively load sucrose into the phloem in source organs such as leaves from where it can be transported and unloaded into sink organs, such as the root meristem, mainly via PD-connecting CC and sieve elements (SE) [68]. The protein kinase TARGET OF RAPAMYCIN (TOR) has emerged as a major signaling hub for the integration of developmental and environmental signals, detecting glucose levels in the root meristem and regulating sucrose unloading by triggering modifications at PD [69].

Besides hormones, sugars, and PD receptors, symplasmic signaling domains can be formed via changes in PD lipid composition. Chemicals that modify the balance in PD lipid composition modify the targeting of glycosylphosphatidylinositol (GPI)-anchored proteins that regulate PD including PdBGs and PD CALLOSE-BINDING PROTEINS (PDCBs) involved in the regulation of callose [22]. By separate evidence, osmotic stress-induced lipid modification by a phospholipase D was found required for the transient PD localization of CRK2 [58]. A regulator of sphingolipid composition, named PHLOEM UNLOADING MODULATOR (PLM), was identified in a suppressor screen to revert transport impairment driven by a hyperactive CALS3 (*cals3m*) [70]. Mutants in PLM (*plm*) display enhanced symplasmic unloading and a presumed reduction in ER-PM tethering affecting PD transport capacity. Recent work has demonstrated that aberrations in PM-DT tethering in a triple *syt* mutant (*syt1,5,7*) lead to changes in PD structure and suppressed the cell-cell trafficking of Youcai mosaic virus movement protein [30]. Together, these examples demonstrate the importance of membrane composition on PD ultrastructure and signaling function. It remains to be seen whether plants can modulate PM-ER tethering dynamically to regulate transport through the CS.

The formation or disruption of symplasmic domains affects the transport of proteins and RNAs and thus their function in signaling and plant development. Proteomic analysis of heterografts between different Cucurbitaceae species and of the tissues involved in parasitic symbiosis between *A. thaliana* and *Cuscuta australis* demonstrates that a considerable portion of a plant's proteome is mobile [71–73]. The archetypal example of symplastic signaling via cell-to-cell mobile (non-cell autonomous) proteins in root development is SHORTROOT (SHR). SHR is a GRAS-family TF that is translated in the stele and moves to the root apical meristem quiescent center (QC) and endodermis [74]. SHR interacts with SCARE CROW (SCR) to promote the expression of factors (TFs and miRNAs) that control asymmetric cell divisions and the patterning of the vascular, endodermal, and cortical root cell layers [75–77]. The movement of both SHR and its negative regulator, miR165/6, is mediated by PD and affected by callose deposition [38]. Heterologous expression of SHR disrupts miRNA behavior leading to aberrant tissue patterning [77]. There are several other examples that demonstrate that PD movement of TFs is essential for proper root tissue patterning such as WUSCHEL RELATED HOMEOBOX 5 (WOX5) and TARGET OF MONOPTEROS 7 (TMO7) [78, 79]. Intriguingly, movement of TFs through PD is not always a function of their size, and specific domains of non-cell-autonomous proteins when fused to cell-autonomous proteins can bestow cell-to-cell mobility. This is exemplified by the analysis of the fusion of the homeodomain of KNOTTED1 (KN1HD), a mobile protein, to the cell-autonomous protein GLABROUS1 (GL1) [80]. GL1-KN1HD fusion movement from subepidermal to epidermal cell layer is assisted by a special class of chaperonins expressed in the receiving tissue [81].

Grafting experiments combined with transcriptomics have demonstrated that a high proportion of RNAs, both coding and noncoding, are transported long distance [71, 82–84]. Long-distance movement of RNAs from shoot to root is presumably via the phloem [82], but there is also mobility reported from root to shoot. In grafts between ecotypes of *A. thaliana*, approximately 20% of the mRNA population was found to be graft-transmissible [83]. The mobile transcriptome was shown to be significantly modulated under nitrogen- or phosphate-limiting conditions, which suggests that mRNAs are acting as long-range signals or effectors of responses under these conditions. There are also examples in which mRNA mobility was linked to organ growth such as the case of the CONTROLLED TUMOR PROTEIN 1 (TCTP1). The mobility of TCTP1 is important in the promotion of root growth, and it is dependent on the presence of a modified base 5-methylcytosine feature, essential for mRNA systemic movement [85]. There are also examples in which mobile miRNAs are linked to the promotion of flowering in *Solanum tuberosum* [86]. tRNAs are also phloem-mobile and have been associated with developmental mechanisms such as the regulation of CK [87].

To summarize, proteins, RNAs, hormones, and sugars are important developmental signaling factors moving via PD. This fact highlights the importance of searching for strategies to modify symplastic communication as potential target to modify plant development. One of these strategies is through modification of the PD proteome. A new tool to assist the screening of the proteome in several plant species is discussed in the next subsection.

5 Screening for PD Modifiers Using Proteomic Analysis

The design of strategies to modify PD as a tool to improve plant development is impeded by the lack of knowledge on the specific composition of the PD proteome. Over the last decade, PD proteomes for several plant model species, including *Arabidopsis thaliana*, *Nicotiana benthamiana*, and *Populus trichocarpa* [29, 52, 88, 89], have been determined. The methodology for the isolation of PD proteomes is summarized in Fig. 2. Briefly, cells grown in culture are purified to produce fractions enriched in PD which are then sequenced by mass spectrometry. Relative abundance in relation to the PM and the ER is then used to produce a list of likely candidates. Despite improvement in the strategies to purify PD, identification of a protein in the proteomes requires further confirmation as isolation of these domains often carries contamination with non-PD structures such as the bulk PM, the cell wall, and the endoplasmic reticulum (ER). So far, techniques are optimized for cell cultures; thus, the dynamics of PD proteomic composition associated with plant development and responses to the environment are often missed. Despite the technical difficulties, a number of verified PD proteins were initially identified in PD proteomics and localization confirmed using protein fusions with fluorescent reporters imaged by confocal microscopy and/or FRET-FLIM assays. The approach identified around 60 proteins in *A. thaliana* (Table 1) including callose metabolic enzymes, such as CALS3 and the PD-located β-(1,3)–glucanases PdBGs and AtBG_ppap [37, 38, 110] and signaling proteins such as the PDLPs and the LYSIN MOTIF DOMAIN- CONTAINING G LYCOSYLPHO SPHATIDYLINOSITOL -ANCHORED PROTEIN 2 (LYM2) [103, 106].

Most of the proteins identified to target PD are members of large families with multiple functions and no distinctive domains linked to PD localization. This poses challenges in predicting PD activities in plant species where the PD proteomes have not yet been experimentally determined. To overcome this limitation and to aid screening for PD proteins, we introduced an R-based tool named Plasmodesmata in silico Proteome 1 (PIP1) [111]. PIP1 integrates experimentally determined PD proteomes and enables the identification of new PD genes in a variety of plant species. This tool extracts candidates based on family or subfamily classification and groups them according to their occurrences in PD proteomes

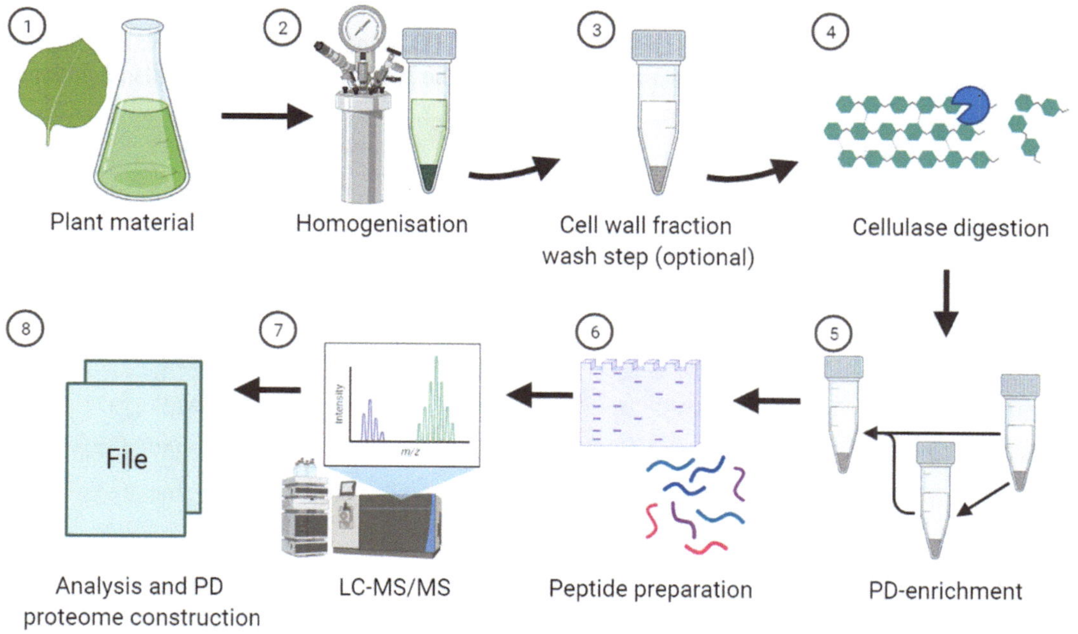

Fig. 2 Preparation of experimentally determined PD proteomes. (1) The starting material for PD proteomes is usually a batch of suspension-cultured cells. (2) Plant material is homogenized in a cell wall extraction buffer in the presence of protease inhibitors; high-pressure homogenization is the most common method employed (3). The pellet formed after centrifugation is considered the cell wall-enriched fraction (CEF). CEF may be washed in buffer several times (4). CEF is digested with cellulase in the presence of protease inhibitors (5). The digestion reaction is centrifuged and washed and the supernatants ultracentrifuged. The resulting pellet is considered the PD-enriched fraction (PEF) (6). Preparation of peptides differs from study to study but may involve solubilization of peptides with Laemmli buffer, cleanup in SDS-PAGE, gel extraction, and trypsin digestion (7). Peptide sequences are determined with liquid chromatography coupled to tandem mass spectrometry (LC-MS-MS) (8). (Data are processed to form a list of proteins that are present in the PEF. The diagram was created with BioRender.com)

and on the presence/absence of structural features, such as GPI anchor or transmembrane domains (TMD), that characterize proteins targeting PD. The pipeline uses PANTHER16 family and subfamily classifications and a variety of prediction tools for structural proteomics available online [112]. PIP1 generates lists of genes organized according to their probability to target PD facilitating the identification of unknown PD-localizing proteins. Co-expression analysis suggests that many PD proteins and candidates interact at the transcriptional level, which can be used in the identification of PD regulatory mechanisms.

PIP1, like experimental proteomes, has limitations but holds many advantages such as establishing an order of priorities to validate candidates identified in experimental proteomes. For species for which no experimental proteome exists, PIP1 can create a PD proteome in silico with far less resources than required to create one experimentally. PIP1 is publicly accessible (https://github.

Table 1
Genes encoding *Arabidopsis thaliana* proteins localized at PD. Gene ID, name, symbol, and Panther family (Hmmpanther) are included. References describing the protein localization are also mentioned

Gene ID	Gene name	Symbol	Hmmpanther	References
AT2G17390	ANKYRIN REPEAT-CONTAINING 2B	AKR2B	PTHR24203	[90]
AT4G35450	ANKYRIN REPEAT-CONTAINING PROTEIN 2	AKR2	PTHR24203	[90]
AT5G42100	BETA-1,3-GLUCANASE	Atbg_ppap	PTHR32227	[91]
AT3G13560	BETA-1,3-GLUCANASE 1	PDBG1	PTHR32227	[37]
AT2G01630	BETA-1,3-GLUCANASE 2	PDBG2	PTHR32227	[37]
AT1G66250	BETA-1,3-GLUCANASE 3	PDBG3	PTHR32227	[37]
AT1G22610	C2 CALCIUM/LIPID-BINDING PLANT P HOSPHORIBOSYLTRANSFERASE	MTCP6	PTHR45707	[92]
AT1G74720	C2 CALCIUM/LIPID-BINDING PLANT P HOSPHORIBOSYLTRANSFERASE	QUIRKY	PTHR45707	[93]
AT4G00700	C2 CALCIUM/LIPID-BINDING PLANT P HOSPHORIBOSYLTRANSFERASE	MTCP9	PTHR45707	[29]
AT5G06850	C2 CALCIUM/LIPID-BINDING PLANT P HOSPHORIBOSYLTRANSFERASE	MCTP1	PTHR45707	[94]
AT1G51570	CALCIUM-DEPENDENT LIPID-BINDING (CALB DOMAIN) PHOSPHORIBOSYL TRANSFERASE	C2_89	PTHR45707	[92]
AT3G57880	CALCIUM-DEPENDENT LIPID-BINDING (CALB DOMAIN) PLANT PHOSPHORI BOSYLTRANSFERASE	MCTP3	PTHR45707	[29]
AT3G16510	CALCIUM-DEPENDENT LIPID-BINDING (CALB DOMAIN) PROTEIN	Atsrc2.2	PTHR32246	[95]
AT3G50770	CALMODULIN-LIKE 41	CML41	PTHR23050	[96]
AT5G61790	CALNEXIN 1	CNX1	PTHR11073	[28]
AT5G07340	CALRETICULIN FAMILY PROTEIN	CNX2	PTHR11073	[28]
AT4G28540	CASEIN KINASE I-LIKE 6	PAPK1	PTHR11909	[97]
AT3G59420	CRINKLY4	ACR4	PTHR47460	[98]
AT1G70520	CYSTEINE-RICH RLK	CRK2	PTHR47973	[58]
AT3G25500	FORMIN HOMOLOGY 1	FH1	PTHR23213	[32]
AT2G43800	FORMIN HOMOLOGY 2	FH2	PTHR23213	[32]
AT1G02335	GERMIN-LIKE PROTEIN 2	PDGLP2	PTHR31238	[99]
AT1G09560	GERMIN-LIKE PROTEIN 5	PDGLP1	PTHR31238	[99]
AT5G13000	GLUCAN SYNTHASE-LIKE 12	GSL12/ CALS3	PTHR12741	[38]

(continued)

Table 1 (continued)

Gene ID	Gene name	Symbol	Hmmpanther	References
AT2G36850	GLUCAN SYNTHASE-LIKE 8	GSL8	PTHR12741	[39]
AT3G51740	INFLORESCENCE MERISTEM KINASE2	IMK2	PTHR48008	[100]
AT3G22790	KINASE INTERACTING (KIP1-LIKE) FAMILY PROTEIN	NET1A	PTHR32258	[101]
AT1G75820	CLAVATA 1	CLV1	PTHR48056	[98]
AT3G49670	BARELY ANY MERISTEM 2	BAM2	PTHR48053	[102]
AT5G65700	BARELY ANY MERISTEM 1	BAM1	PTHR48053	[102]
AT1G11130	LEUCINE-RICH REPEAT PROTEIN KINASE FAMILY PROTEIN	SUB	PTHR27001	[93]
AT1G56145	LEUCINE-RICH REPEAT TRANSMEM BRANE PROTEIN KINASE	LRR-RK	PTHR27006	[88]
AT2G17120	LYSM DOMAIN GPI-ANCHORED PROTEIN 2	LYM2	PTHR33734	[103]
AT2G23770	LYSM-CONTAINING RECEPTOR-LIKE KINASE 4	LYK4	PTHR45927	[104]
AT1G60870	MATERNAL EFFECT EMBRYO ARREST 9	MEE9	PTHR37707	[92]
AT5G61130	PLASMODESMATA CALLOSE-BINDING PROTEIN 1	PDCB1	PTHR31044	[105]
AT5G08000	PLASMODESMATA CALLOSE-BINDING PROTEIN 2	PDCB2	PTHR31044	[105]
AT1G18650	PLASMODESMATA CALLOSE-BINDING PROTEIN 3	PDCB3	PTHR31044	[105]
AT1G69295	PLASMODESMATA CALLOSE-BINDING PROTEIN 4	PDCB4	PTHR31044	[105]
AT3G58100	PLASMODESMATA CALLOSE-BINDING PROTEIN 5	PDCB5	PTHR31044	[105]
AT5G43980	PLASMODESMATA-LOCATED PROTEIN 1	PDLP1	PTHR32080	[106]
AT1G04520	PLASMODESMATA-LOCATED PROTEIN 2	PDLP2	PTHR32080	[106]
AT2G33330	PLASMODESMATA-LOCATED PROTEIN 3	PDLP3	PTHR32080	[106]
AT3G04370	PLASMODESMATA-LOCATED PROTEIN 4	PDLP4	PTHR32080	[106]
AT1G70690	PLASMODESMATA-LOCATED PROTEIN 5	PDLP5	PTHR32080	[106]
AT2G01660	PLASMODESMATA-LOCATED PROTEIN 6	PDLP6	PTHR32080	[106]
AT5G37660	PLASMODESMATA-LOCATED PROTEIN 7	PDLP7	PTHR32080	[106]
AT3G60720	PLASMODESMATA-LOCATED PROTEIN 8	PDLP8	PTHR32080	[106]

(continued)

Table 1
(continued)

Gene ID	Gene name	Symbol	Hmmpanther	References
AT5G24010	PROTEIN KINASE SUPERFAMILY PROTEIN	–	PTHR27003	[88]
AT3G15480	PROTEIN OF UNKNOWN FUNCTION (DUF1218)	CKL6	PTHR31769	[88]
AT3G02880	QIĀN SHŎU KINASE	QSK1	PTHR48010	[100]
AT4G21380	RECEPTOR KINASE 3	ARK3	PTHR27002	[88]
AT3G61260	REMORIN 2	REM1.2	PTHR31775	[25]
AT3G61560	RETICULON FAMILY PROTEIN	RTN6	PTHR10994	[27]
AT1G64090	RETICULON LIKE PROTEIN B3	RTNLB3	PTHR10994	[27]
AT5G15650	REVERSIBLY GLYCOSYLATED POLYPEP TIDE 2	RGP2	PTHR31682	[107]
AT2G20990	SYNAPTOTAGMIN A	SYTA	PTHR10774	[108]
AT3G45600	TETRASPANIN3	TET3	PTHR32191	[88]
AT3G60600	VESICLE ASSOCIATED PROTEIN	VAP27-1	PTHR10809	[109]

com/PhilPlantMan/PIP1) and can be modified by the user to add new sequenced proteomes and/or experimentally verified genes and thus will serve as a long-lasting resource for the plant community. The discovery of new PD proteins and experimental proteomes will improve the predictive power of PIP1, thus enhancing its ability to analyze future candidate PD proteomes of specific organs, tissues, and/or plant species.

6 Future Perspective

This review offers insights into the major findings in the last decade regarding the structural and regulatory features that characterize PD and PD domains and their role in determining cell-to-cell communication. Our focus was on the developmental role of these channels and on discussing the domains that regulate their function as well as on presenting new tools to enable their modification. There are still many gaps in knowledge including the contribution of PD in the formation of hormonal gradients and/or in other molecular signaling pathways in relation to more established apoplastic routes. There are also questions regarding the role of cell wall components modified around PD microdomains and on the significance and mechanism for function of type I vs type II PD. The identification of PD components using experimental or in silico proteomes holds promises in dissecting the mechanism underlying dynamic modifications in PD structure and transport capacity. Further work is required using these tools in combination with ultrastructural characterization to establish a mechanistic understanding of the dynamics in PD regulation and function in plant development.

Acknowledgments

PK was supported by a BBSRC DTP (BB/M011151/1). YBA is supported by a UKRI Future Leaders Fellowship (MR/T04263X/1).

References

1. Liu J, Moore S, Chen C, Lindsey K (2017) Crosstalk complexities between auxin, cytokinin, and ethylene in Arabidopsis root development: from experiments to systems modeling, and back again. Mol Plant 10: 1480–1496
2. Aldon D, Mbengue M, Mazars C, Galaud J-P (2018) Calcium signalling in plant biotic interactions. Int J Mol Sci 19:665
3. Chen K, Li GJ, Bressan RA, Song CP, Zhu JK, Zhao Y (2020) Abscisic acid dynamics, signaling, and functions in plants. J Integr Plant Biol 62(1):25–54
4. Zandalinas SI, Fichman Y, Devireddy AR, Sengupta S, Azad RK, Mittler R (2020) Systemic signaling during abiotic stress combination in plants. Proc Natl Acad Sci U S A 117(24):13810–13820
5. Amsbury S, Kirk P, Benitez-Alfonso Y (2018) Emerging models on the regulation of intercellular transport by plasmodesmata-associated callose. J Exp Bot 69(1):105–115. https://doi.org/10.1093/jxb/erx337
6. Li ZP, Paterlini A, Glavier M, Bayer EM (2021) Intercellular trafficking via plasmodesmata: molecular layers of complexity. Cell Mol Life Sci 78(3):799–816. https://doi.org/10.1007/s00018-020-03622-8
7. Goodwin P (1983) Molecular size limit for movement in the symplast of the Elodea leaf. Planta 157(2):124–130
8. Oparka KJ, Roberts AG, Boevink P, Santa Cruz S, Roberts I, Pradel KS, Imlau A, Kotlizky G, Sauer N, Epel B (1999) Simple, but not branched, plasmodesmata allow the nonspecific trafficking of proteins in developing tobacco leaves. Cell 97(6):743–754
9. Itaya A, Ma F, Qi Y, Matsuda Y, Zhu Y, Liang G, Ding B (2002) Plasmodesma-mediated selective protein traffic between symplasmically isolated cells probed by a viral movement protein. Plant Cell 14(9): 2071–2083
10. Benitez-Alfonso Y, Cilia M, Roman AS, Thomas C, Maule A, Hearn S, Jackson D (2009) Control of *Arabidopsis* meristem development by thioredoxin-dependent regulation of intercellular transport. Proc Natl Acad Sci U S A 106(9):3615–3620. https://doi.org/10.1073/pnas.0808717106
11. Burch-Smith TM, Zambryski PC (2010) Loss of INCREASED SIZE EXCLUSION LIMIT (ISE)1 or ISE2 increases the formation of secondary plasmodesmata. Curr Biol 20(11): 989–993. https://doi.org/10.1016/j.cub.2010.03.064
12. Gerlitz N, Gerum R, Sauer N, Stadler R (2018) Photoinducible DRONPA-s: a new tool for investigating cell-cell connectivity. Plant J 94(5):751–766. https://doi.org/10.1111/tpj.13918
13. Hawes C, Juniper B, Horne J (1981) Low and high voltage electron microscopy of mitosis and cytokinesis in maize roots. Planta 152(5):397–407
14. Hepler P (1982) Endoplasmic reticulum in the formation of the cell plate and plasmodesmata. Protoplasma 111(2):121–133
15. Kollmann R, Glockmann C (1991) Studies on graft unions. Protoplasma 165(1–3):71–85
16. Fischer K, Lachner LA-M, Olsen S, Mulisch M, Krause K (2021) The enigma of interspecific plasmodesmata: insight from parasitic plants. Front Plant Sci 12:641924

17. Ehlers K, Kollmann R (2001) Primary and secondary plasmodesmata: structure, origin, and functioning. Protoplasma 216(1):1–30
18. Burch-Smith TM, Stonebloom S, Xu M, Zambryski PC (2011) Plasmodesmata during development: re-examination of the importance of primary, secondary, and branched plasmodesmata structure versus function. Protoplasma 248(1):61–74
19. Faulkner C, Akman OE, Bell K, Jeffree C, Oparka K (2008) Peeking into pit fields: a multiple twinning model of secondary plasmodesmata formation in tobacco. Plant Cell 20(6):1504–1518
20. Ross-Elliott TJ, Jensen KH, Haaning KS, Wager BM, Knoblauch J, Howell AH, Mullendore DL, Monteith AG, Paultre D, Yan D, Otero S, Bourdon M, Sager R, Lee JY, Helariutta Y, Knoblauch M, Oparka KJ (2017) Phloem unloading in Arabidopsis roots is convective and regulated by the phloem-pole pericycle. eLife 6:e24125. https://doi.org/10.7554/elife.24125
21. Nicolas WJ, Grison MS, Trépout S, Gaston A, Fouché M, Cordelières FP, Oparka K, Tilsner J, Brocard L, Bayer EM (2017) Architecture and permeability of post-cytokinesis plasmodesmata lacking cytoplasmic sleeves. Nat Plants 3(7):1–11
22. Grison MS, Brocard L, Fouillen L, Nicolas W, Wewer V, Dormann P, Nacir H, Benitez-Alfonso Y, Claverol S, Germain V, Boutte Y, Mongrand S, Bayer EM (2015) Specific membrane lipid composition is important for plasmodesmata function in Arabidopsis. Plant Cell 27(4):1228–1250. https://doi.org/10.1105/tpc.114.135731
23. Harayama T, Riezman H (2018) Understanding the diversity of membrane lipid composition. Nat Rev Mol Cell Biol 19(5):281
24. Wang F, Muto A, Van de Velde J, Neyt P, Himanen K, Vandepoele K, Van Lijsebettens M (2015) Functional analysis of Arabidopsis TETRASPANIN gene family in plant growth and development. Plant Physiol 169(3):2200–2214
25. Huang D, Sun Y, Ma Z, Ke M, Cui Y, Chen Z, Chen C, Ji C, Tran TM, Yang L, Lam SM, Han Y, Shu G, Friml J, Miao Y, Jiang L, Chen X (2019) Salicylic acid-mediated plasmodesmal closure via Remorin-dependent lipid organization. Proc Natl Acad Sci U S A 116(42):21274–21284. https://doi.org/10.1073/pnas.1911892116
26. Wei Z, Tan S, Liu T, Wu Y, Lei J-G, Chen Z, Friml J, Xue H-W, Liao K (2020) Plasmodesmata-like intercellular connections by plant remorin in animal cells. bioRxiv:791137. https://doi.org/10.1101/791137
27. Knox K, Wang P, Kriechbaumer V, Tilsner J, Frigerio L, Sparkes I, Hawes C, Karl JO (2015) Putting the squeeze on PDs-a role for RETICULONS in primary plasmodesmata formation. Plant Physiol 168(4):1563–1572
28. Liu DY, Smith PM, Barton DA, Day DA, Overall RL (2017) Characterisation of Arabidopsis calnexin 1 and calnexin 2 in the endoplasmic reticulum and at plasmodesmata. Protoplasma 254(1):125–136
29. Brault ML, Petit JD, Immel F, Nicolas WJ, Glavier M, Brocard L, Gaston A, Fouché M, Hawkins TJ, Crowet JM, Grison MS, Germain V, Rocher M, Kraner M, Alva V, Claverol S, Paterlini A, Helariutta Y, Deleu M, Lins L, Tilsner J, Bayer EM (2019) Multiple C2 domains and transmembrane region proteins (MCTPs) tether membranes at plasmodesmata. EMBO Rep 20(8):e47182. https://doi.org/10.15252/embr.201847182
30. Ishikawa K, Tamura K, Fukao Y, Shimada T (2020) Structural and functional relationships between plasmodesmata and plant endoplasmic reticulum–plasma membrane contact sites consisting of three synaptotagmins. New Phytol 226(3):798–808. https://doi.org/10.1111/nph.16391
31. Tilsner J, Amari K, Torrance L (2011) Plasmodesmata viewed as specialised membrane adhesion sites. Protoplasma 248(1):39–60
32. Diao M, Ren S, Wang Q, Qian L, Shen J, Liu Y, Huang S (2018) Arabidopsis formin 2 regulates cell-to-cell trafficking by capping and stabilizing actin filaments at plasmodesmata. elife 7:e36316
33. Barton DA, Cole L, Collings DA, Liu DY, Smith PM, Day DA, Overall RL (2011) Cell-to-cell transport via the lumen of the endoplasmic reticulum. Plant J 66(5):806–817
34. Deinum EE, Mulder BM, Benitez-Alfonso Y (2019) From plasmodesma geometry to effective symplasmic permeability through biophysical modelling. eLife 8:e49000. https://doi.org/10.7554/elife.49000
35. Park K, Knoblauch J, Oparka K, Jensen KH (2019) Controlling intercellular flow through mechanosensitive plasmodesmata nanopores. Nat Commun 10(1):3564. https://doi.org/10.1038/s41467-019-11201-0
36. Knox JP, Benitez-Alfonso Y (2014) Roles and regulation of plant cell walls surrounding plasmodesmata. Curr Opin Plant Biol 22:93–100

37. Benitez-Alfonso Y, Faulkner C, Pendle A, Miyashima S, Helariutta Y, Maule A (2013) Symplastic intercellular connectivity regulates lateral root patterning. Dev Cell 26(2): 136–147. https://doi.org/10.1016/j.devcel.2013.06.010
38. Vatén A, Dettmer J, Wu S, Stierhof Y-D, Miyashima S, Yadav SR, Roberts CJ, Campilho A, Bulone V, Lichtenberger R (2011) Callose biosynthesis regulates symplastic trafficking during root development. Dev Cell 21(6):1144–1155
39. Saatian B, Austin RS, Tian G, Chen C, Nguyen V, Kohalmi SE, Geelen D, Cui Y (2018) Analysis of a novel mutant allele of GSL8 reveals its key roles in cytokinesis and symplastic trafficking in *Arabidopsis*. BMC Plant Biol 18(1):295. https://doi.org/10.1186/s12870-018-1515-y
40. Abou-Saleh RH, Hernandez-Gomez MC, Amsbury S, Paniagua C, Bourdon M, Miyashima S, Helariutta Y, Fuller M, Budtova T, Connell SD (2018) Interactions between callose and cellulose revealed through the analysis of biopolymer mixtures. Nat Commun 9(1):453841
41. Willats WG, McCartney L, Mackie W, Knox JP (2001) Pectin: cell biology and prospects for functional analysis. Plant Mol Biol 47(1): 9–27
42. Clausen MH, Willats WG, Knox JP (2003) Synthetic methyl hexagalacturonate hapten inhibitors of anti-homogalacturonan monoclonal antibodies LM7, JIM5 and JIM7. Carbohydr Res 338(17):1797–1800
43. Giovane A, Servillo L, Balestrieri C, Raiola A, D'avino R, Tamburrini M, Ciardiello M, Camardella L (2004) Pectin methylesterase inhibitor. Biochim Biophys Acta 1696(2): 245–252
44. Jarvis MC (1984) Structure and properties of pectin gels in plant cell walls. Plant Cell Environ 7(3):153–164
45. Casero P, Knox J (1995) The monoclonal antibody JIM5 indicates patterns of pectin deposition in relation to pit fields at the plasma-membrane-face of tomato pericarp cell walls. Protoplasma 188(1–2):133–137
46. Orfila C, Knox JP (2000) Spatial regulation of pectic polysaccharides in relation to pit fields in cell walls of tomato fruit pericarp. Plant Physiol 122(3):775–782. https://doi.org/10.1104/pp.122.3.775
47. Roy S, Watada AE, Wergin WP (1997) Characterization of the cell wall microdomain surrounding plasmodesmata in apple fruit. Plant Physiol 114(2):539–547. https://doi.org/10.1104/pp.114.2.539
48. Chen MH, Sheng J, Hind G, Handa AK, Citovsky V (2000) Interaction between the tobacco mosaic virus movement protein and host cell pectin methylesterases is required for viral cell-to-cell movement. EMBO J 19(5): 913–920
49. Wu H-C, Bulgakov VP, Jinn T-L (2018) Pectin methylesterases: cell wall remodeling proteins are required for plant response to heat stress. Front Plant Sci 9:1612
50. Carpita NC (2011) Update on mechanisms of plant cell wall biosynthesis: how plants make cellulose and other (1- 4)-β-D-glycans. Plant Physiol 155(1):171–184
51. Lopez-Sanchez P, Martinez-Sanz M, Bonilla MR, Wang D, Gilbert EP, Stokes JR, Gidley MJ (2017) Cellulose-pectin composite hydrogels: intermolecular interactions and material properties depend on order of assembly. Carbohydr Polym 162:71–81
52. Park S-H, Li F, Renaud J, Shen W, Li Y, Guo L, Cui H, Sumarah M, Wang A (2017) NbEXPA1, an α-expansin, is plasmodesmata-specific and a novel host factor for potyviral infection. Plant J 92(5):846–861. https://doi.org/10.1111/tpj.13723
53. Bayer E, Thomas C, Maule A (2008) Symplastic domains in the *Arabidopsis* shoot apical meristem correlate with PDLP1 expression patterns. Plant Signal Behav 3(10):853–855
54. Han X, Kim JY (2016) Integrating hormone- and micromolecule-mediated signaling with plasmodesmal communication. Mol Plant 9(1):46–56. https://doi.org/10.1016/j.molp.2015.08.015
55. Vu MH, Iswanto ABB, Lee J, Kim J-Y (2020) The role of plasmodesmata-associated receptor in plant development and environmental response. Plan Theory 9(2):216. https://doi.org/10.3390/plants9020216
56. Caillaud MC, Wirthmueller L, Sklenar J, Findlay K, Piquerez SJ, Jones AM, Robatzek S, Jones JD, Faulkner C (2014) The plasmodesmal protein PDLP1 localises to haustoria-associated membranes during downy mildew infection and regulates callose deposition. PLoS Pathog 10(10):e1004496. https://doi.org/10.1371/journal.ppat.1004496
57. Cui W, Lee J-Y (2016) Arabidopsis callose synthases CalS1/8 regulate plasmodesmal permeability during stress. Nat Plants 2: 16034
58. Hunter K, Kimura S, Rokka A, Tran HC, Toyota M, Kukkonen JP, Wrzaczek M

(2019) CRK2 enhances salt tolerance by regulating callose deposition in connection with PLDα1. Plant Physiol 180(4):2004–2021. https://doi.org/10.1104/pp.19.00560

59. Fichman Y, Myers RJ, Grant DG, Mittler R (2021) Plasmodesmata-localized proteins and ROS orchestrate light-induced rapid systemic signaling in Arabidopsis. Sci Signal 14(671): eabf0322
60. Wang X, Sager R, Cui W, Zhang C, Lu H, Lee JY (2013) Salicylic acid regulates plasmodesmata closure during innate immune responses in *Arabidopsis*. Plant Cell 25(6):2315–2329. https://doi.org/10.1105/tpc.113.110676
61. Sager R, Wang X, Hill K, Yoo B-C, Caplan J, Nedo A, Tran T, Bennett MJ, Lee J-Y (2020) Auxin-dependent control of a plasmodesmal regulator creates a negative feedback loop modulating lateral root emergence. Nat Commun 11(1):1–10
62. Tylewicz S, Petterle A, Marttila S, Miskolczi P, Azeez A, Singh R, Immanen J, Mähler N, Hvidsten T, Eklund D (2018) Photoperiodic control of seasonal growth is mediated by ABA acting on cell-cell communication. Science 360(6385):212–215
63. Mellor NL, Voß U, Janes G, Bennett MJ, Wells DM, Band LR (2020) Auxin fluxes through plasmodesmata modify root-tip auxin distribution. Development 147(6): dev181669. https://doi.org/10.1242/dev.181669
64. Kieffer M, Neve J, Kepinski S (2010) Defining auxin response contexts in plant development. Curr Opin Plant Biol 13(1):12–20
65. Paterlini A (2020) Uncharted routes: exploring the relevance of auxin movement via plasmodesmata. Biol Open 9(11):bio055541
66. Rock CD, Sun X (2005) Crosstalk between ABA and auxin signaling pathways in roots of *Arabidopsis thaliana* (L.) Heynh. Planta 222(1):98–106
67. Bishopp A, Lehesranta S, Vatén A, Help H, El-Showk S, Scheres B, Helariutta K, Mähönen AP, Sakakibara H, Helariutta Y (2011) Phloem-transported cytokinin regulates polar auxin transport and maintains vascular pattern in the root meristem. Curr Biol 21(11):927–932
68. Durand M, Mainson D, Porcheron B, Maurousset L, Lemoine R, Pourtau N (2018) Carbon source–sink relationship in *Arabidopsis thaliana*: the role of sucrose transporters. Planta 247(3):587–611
69. Brunkard JO, Xu M, Scarpin MR, Chatterjee S, Shemyakina EA, Goodman HM, Zambryski P (2020) TOR dynamically regulates plant cell-cell transport. Proc Natl Acad Sci U S A 117(9):5049–5058. https://doi.org/10.1073/pnas.1919196117
70. Yan D, Yadav SR, Paterlini A, Nicolas WJ, Petit JD, Brocard L, Belevich I, Grison MS, Vaten A, Karami L (2019) Sphingolipid biosynthesis modulates plasmodesmal ultrastructure and phloem unloading. Nat Plants 5(6): 604–615
71. Hu C, Ham B-K, El-Shabrawi HM, Alexander D, Zhang D, Ryals J, Lucas WJ (2016) Proteomics and metabolomics analyses reveal the cucurbit sieve tube system as a complex metabolic space. Plant J 87(5): 442–454. https://doi.org/10.1111/tpj.13209
72. Xu Y, Yuan Y, Du N, Wang Y, Shu S, Sun J, Guo S (2018) Proteomic analysis of heat stress resistance of cucumber leaves when grafted onto *Momordica* rootstock. Hortic Res 5(1):1–18
73. Liu N, Shen G, Xu Y, Liu H, Zhang J, Li S, Li J, Zhang C, Qi J, Wang L, Wu J (2020) Extensive inter-plant protein transfer between *Cuscuta* parasites and their host plants. Mol Plant 13(4):573–585. https://doi.org/10.1016/j.molp.2019.12.002
74. Koizumi K, Gallagher KL (2013) Identification of SHRUBBY, a SHORT-ROOT and SCARECROW interacting protein that controls root growth and radial patterning. Development 140(6):1292–1300. https://doi.org/10.1242/dev.090761
75. Carlsbecker A, Lee J-Y, Roberts CJ, Dettmer J, Lehesranta S, Zhou J, Lindgren O, Moreno-Risueno MA, Vatén A, Thitamadee S (2010) Cell signalling by microRNA165/6 directs gene dose-dependent root cell fate. Nature 465(7296): 316–321
76. Kim H, Zhou J, Kumar D, Jang G, Ryu KH, Sebastian J, Miyashima S, Helariutta Y, Lee J-Y (2020) SHORTROOT-mediated intercellular signals coordinate phloem development in arabidopsis roots. Plant Cell 32(5): 1519–1535
77. Wu S, Lee C-M, Hayashi T, Price S, Divol F, Henry S, Pauluzzi G, Perin C, Gallagher KL (2014) A plausible mechanism, based upon Short-root movement, for regulating the number of cortex cell layers in roots. Proc Natl Acad Sci 111(45):16184–16189
78. Pi L, Aichinger E, van der Graaff E, Llavata-Peris CI, Weijers D, Hennig L, Groot E, Laux T (2015) Organizer-derived WOX5 signal maintains root columella stem cells through chromatin-mediated repression of CDF4 expression. Dev Cell 33(5):576–588

79. Lu K-J, De Rybel B, Van Mourik H, Weijers D (2018) Regulation of intercellular TARGET OF MONOPTEROS 7 protein transport in the *Arabidopsis* root. Development 145(2): dev15289280
80. Kim J-Y, Rim Y, Wang J, Jackson D (2005) A novel cell-to-cell trafficking assay indicates that the KNOX homeodomain is necessary and sufficient for intercellular protein and mRNA trafficking. Genes Dev 19(7): 788–793
81. Xu XM, Wang J, Xuan Z, Goldshmidt A, Borrill PG, Hariharan N, Kim JY, Jackson D (2011) Chaperonins facilitate KNOTTED1 cell-to-cell trafficking and stem cell function. Science 333(6046):1141–1144
82. Buhtz A, Springer F, Chappell L, Baulcombe DC, Kehr J (2008) Identification and characterization of small RNAs from the phloem of *Brassica napus*. Plant J 53(5):739–749
83. Thieme CJ, Rojas-Triana M, Stecyk E, Schudoma C, Zhang W, Yang L, Minambres M, Walther D, Schulze WX, Paz-Ares J, Scheible WR, Kragler F (2015) Endogenous *Arabidopsis* messenger RNAs transported to distant tissues. Nat Plants 1(4):15025. https://doi.org/10.1038/nplants.2015.25
84. Ham B-K, Lucas WJ (2017) Phloem-mobile RNAs as systemic signaling agents. Annu Rev Plant Biol 68:173–195
85. Yang L, Perrera V, Saplaoura E, Apelt F, Bahin M, Kramdi A, Olas J, Mueller-Roeber-B, Sokolowska E, Zhang W, Li R, Pitzalis N, Heinlein M, Zhang S, Genovesio A, Colot V, Kragler F (2019) m5C methylation guides systemic transport of messenger RNA over graft junctions in plants. Curr Biol 29(15): 2465–2476. https://doi.org/10.1016/j.cub.2019.06.042
86. Martin A, Adam H, Díaz-Mendoza M, Żurczak M, González-Schain ND, Suárez-López P (2009) Graft-transmissible induction of potato tuberization by the microRNA miR172. Development 136(17):2873–2881
87. Mok DW, Mok MC (2001) Cytokinin metabolism and action. Annu Rev Plant Biol 52(1): 89–118
88. Fernandez-Calvino L, Faulkner C, Walshaw J, Saalbach G, Bayer E, Benitez-Alfonso Y, Maule A (2011) *Arabidopsis* plasmodesmal proteome. PLoS one 6(4):e18880
89. Leijon F, Melzer M, Zhou Q, Srivastava V, Bulone V (2018) Proteomic analysis of plasmodesmata from *Populus* cell suspension cultures in relation with callose biosynthesis. Front Plant Sci 9:1681
90. Fridborg I, Grainger J, Page A, Coleman M, Findlay K, Angell S (2003) TIP, a novel host factor linking callose degradation with the cell-to-cell movement of potato virus X. Mol Plant-Microbe Interact 16(2):132–140
91. Sagi G, Katz A, Guenoune-Gelbart D, Epel BL (2005) Class 1 reversibly glycosylated polypeptides are plasmodesmal-associated proteins delivered to plasmodesmata via the Golgi apparatus. Plant Cell 17(6):1788–1800
92. Kraner ME, Müller C, Sonnewald U (2017) Comparative proteomic profiling of the choline transporter-like1 (CHER 1) mutant provides insights into plasmodesmata composition of fully developed *Arabidopsis thaliana* leaves. Plant J 92(4):696–709
93. Vaddepalli P, Herrmann A, Fulton L, Oelschner M, Hillmer S, Stratil TF, Fastner A, Hammes UZ, Ott T, Robinson DG (2014) The C2-domain protein QUIRKY and the receptor-like kinase STRUBBELIG localize to plasmodesmata and mediate tissue morphogenesis in *Arabidopsis thaliana*. Development 141(21):4139-48 108
94. Liu L, Liu C, Hou X, Xi W, Shen L, Tao Z, Wang Y, Yu H (2012) FTIP1 is an essential regulator required for florigen transport. PLoS Biol 10(4):e1001313
95. Rodriguez A, Angel CA, Lutz L, Leisner SM, Nelson RS, Schoelz JE (2014) Association of the P6 protein of cauliflower mosaic virus with plasmodesmata and plasmodesmal proteins. Plant Physiol 166(3):1345–1358. https://doi.org/10.1104/pp.114.249250
96. Xu B, Cheval C, Laohavisit A, Hocking B, Chiasson D, Olsson TSG, Shirasu K, Faulkner C, Gilliham M (2017) A calmodulin-like protein regulates plasmodesmal closure during bacterial immune responses. New Phytol 215(1):77–84. https://doi.org/10.1111/nph.14599
97. Lee J-Y, Taoka K-I, Yoo B-C, Ben-Nissan G, Kim D-J, Lucas WJ (2005) Plasmodesmal-associated protein kinase in tobacco and Arabidopsis recognizes a subset of non-cell-autonomous proteins. Plant Cell 17(10): 2817–2831. https://doi.org/10.1105/tpc.105.034330
98. Stahl Y, Grabowski S, Bleckmann A, Kühnemuth R, Weidtkamp-Peters S, Pinto KG, Kirschner GK, Schmid JB, Wink RH, Hülsewede A (2013) Moderation of Arabidopsis root stemness by CLAVATA1 and ARABIDOPSIS CRINKLY4 receptor kinase complexes. Curr Biol 23(5):362–371
99. Ham B-K, Li G, Kang B-H, Zeng F, Lucas WJ (2012) Overexpression of *Arabidopsis*

plasmodesmata germin-like proteins disrupts root growth and development. Plant Cell 24(9):3630–3648. https://doi.org/10.1105/tpc.112.101063

100. Grison MS, Kirk P, Brault ML, Wu XN, Schulze WX, Benitez-Alfonso Y, Immel F, Bayer EM (2019) Plasma membrane-associated receptor-like kinases relocalize to plasmodesmata in response to osmotic stress. Plant Physiol 181(1):142–160

101. Deeks MJ, Calcutt JR, Ingle EK, Hawkins TJ, Chapman S, Richardson AC, Mentlak DA, Dixon MR, Cartwright F, Smertenko AP (2012) A superfamily of actin-binding proteins at the actin-membrane nexus of higher plants. Curr Biol 22(17):1595–1600

102. Rosas-Diaz T, Zhang D, Fan P, Wang L, Ding X, Jiang Y, Jimenez-Gongora T, Medina-Puche L, Zhao X, Feng Z (2018) A virus-targeted plant receptor-like kinase promotes cell-to-cell spread of RNAi. Proc Natl Acad Sci U S A 115(6):1388–1393

103. Faulkner C, Petutschnig E, Benitez-Alfonso-Y, Beck M, Robatzek S, Lipka V, Maule AJ (2013) LYM2-dependent chitin perception limits molecular flux via plasmodesmata. Proc Natl Acad Sci 110(22):9166–9170

104. Cheval C, Samwald S, Johnston MG, De Keijzer J, Breakspear A, Liu X, Bellandi A, Kadota Y, Zipfel C, Faulkner C (2020) Chitin perception in plasmodesmata characterizes submembrane immune-signaling specificity in plants. Proc Natl Acad Sci U S A 117(17):9621–9629

105. Simpson C, Thomas C, Findlay K, Bayer E, Maule AJ (2009) An *Arabidopsis* GPI-anchor plasmodesmal neck protein with callose binding activity and potential to regulate cell-to-cell trafficking. Plant Cell 21(2):581–594

106. Thomas CL, Bayer EM, Ritzenthaler C, Fernandez-Calvino L, Maule AJ (2008) Specific targeting of a plasmodesmal protein affecting cell-to-cell communication. PLoS Biol 6(1):e7

107. Epel BL, van Lent JW, Cohen L, Kotlizky G, Katz A, Yahalom A (1996) A 41 kDa protein isolated from maize mesocotyl cell walls immunolocalizes to plasmodesmata. Protoplasma 191(1–2):70–78

108. Levy A, Judy S (2015) Synaptotagmin SYTA forms ER-plasma membrane junctions that are recruited to Plasmodesmata for plant virus movement. Curr Biol 25(15):2018–2025. https://doi.org/10.1016/j.cub.2015.06.015

109. Wang P, Richardson C, Hawkins TJ, Sparkes I, Hawes C, Hussey PJ (2016) Plant VAP27 proteins: domain characterization, intracellular localization and role in plant development. New Phytol 210(4):1311–1326. https://doi.org/10.1111/nph.13857

110. Levy A, Erlanger M, Rosenthal M, Epel BL (2007) A plasmodesmata-associated β-1,3-glucanase in *Arabidopsis*. Plant J 49(4):669–682

111. Kirk P, Amsbury S, German L, Gaudioso-Pedraza R, Benitez-Alfonso Y (2021) Comparative meta-proteomic analysis for the identification of novel plasmodesmata proteins and regulatory cues. bioRxiv. https://doi.org/10.1101/2021.05.04.442592

112. Mi H, Ebert D, Muruganujan A, Mills C, Albou L-P, Mushayamaha T, Thomas PD (2021) PANTHER version 16: a revised family classification, tree-based classification tool, enhancer regions and extensive API. Nucleic Acids Res 49(D1):D394–D403

Chapter 2

Function of Plasmodesmata in the Interaction of Plants with Microbes and Viruses

Caiping Huang and Manfred Heinlein

Abstract

Plasmodesmata (PD) are gated plant cell wall channels that allow the trafficking of molecules between cells and play important roles during plant development and in the orchestration of cellular and systemic signaling responses during interactions of plants with the biotic and abiotic environment. To allow gating, PD are equipped with signaling platforms and enzymes that regulate the size exclusion limit (SEL) of the pore. Plant-interacting microbes and viruses target PD with specific effectors to enhance their virulence and are useful probes to study PD functions.

Key words Plasmodesmata, Immunity, Callose, Virus, Bacteria, Fungi, Flagellin, Chitin

1 Introduction

Plasmodesmata (PD) are intercellular nanochannels, often as small as 30 nm in diameter and 100 nm in length, that allow direct transport of molecules across the cell walls between cells (Fig. 1a). In providing cytoplasmic continuity between cells, the system of PD and connected phloem establishes a symplasmic short- and long-distance communication network that enables plants to transmit information between cells, tissues, and organs for developmental patterning and for orchestrating responses to changes in the biotic and abiotic environment [11–15]. Communication through PD involves the trafficking of various molecules that include small molecules like water, ions, metabolites, and hormones as well as macromolecules, such as proteins, long and short RNAs, and protein-RNA complexes. The molecular movement through the channel occurs either by passive diffusion or in a targeted manner employing specialized mechanisms [11, 16]. Molecular passage is generally restricted by structural features of PD that determine the SEL and, thus, the permeability of the channel. The permeability of PD is a highly dynamic property that depends on cell type as well as

Yoselin Benitez-Alfonso and Manfred Heinlein (eds.), *Plasmodesmata: Methods and Protocols*, Methods in Molecular Biology, vol. 2457, https://doi.org/10.1007/978-1-0716-2132-5_2,

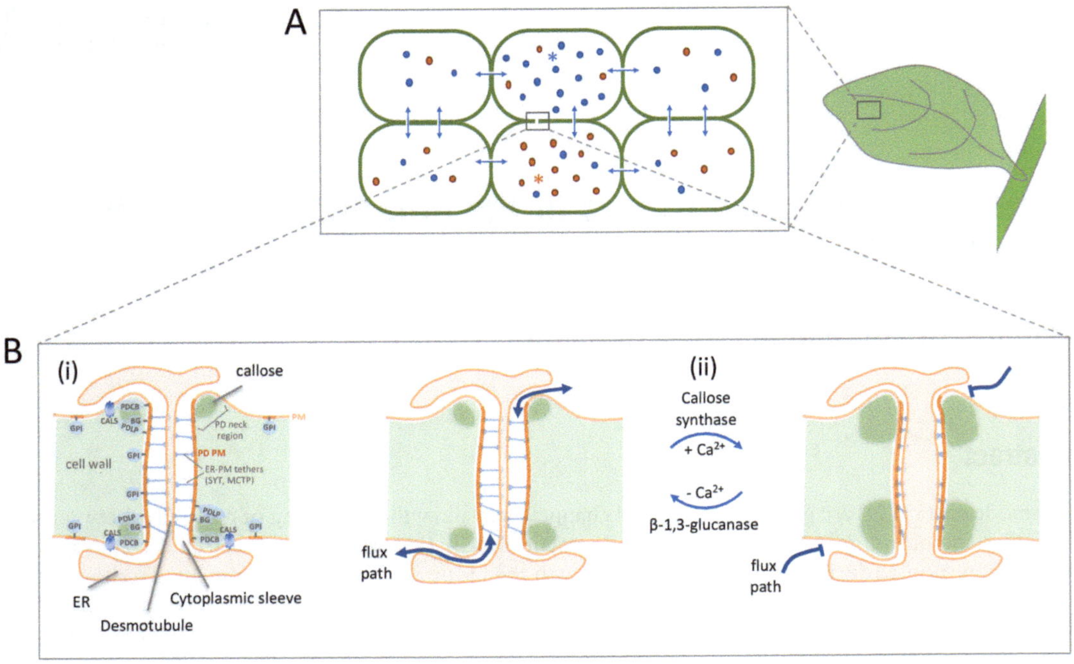

Fig. 1 Intercellular communication through PD. (**a**) Soluble molecules (blue dots and red dots) can move from the cells in which they are synthesized (blue and red asterisk, respectively) into surrounding cells via PD (arrows). (**b**) Basic composition (i) and regulation (ii) of PD. The cytoplasmic sleeve is delimited by the PM and the centrally located desmotubule [1]. The PM in PD (PD PM) and the PM in membrane rafts represent specific PM domains that differ in lipid and protein composition from the PM in other cellular regions [2] and are depicted in darker color. The PD PM is enriched with GPI-anchored cell wall proteins, and the cell wall regions surrounding the PD neck regions contain callose deposits. The size of these callose deposits is controlled by callose synthases (CALS), β-1,3-glucanases (BG), and other PD proteins, such as members of the GPI-anchored PD CALLOSE BINDING (PDCB) and PD-LOCATED PROTEIN (PDLP) protein families [3–5]. The desmotubule is connected to the PD PM by specific tether proteins (SYT, MCTP) [6]. Cells regulate the PD SEL and thus the intercellular trafficking of molecules by controlling the size of the cytoplasmic sleeve (ii). This occurs through callose synthase and β-1,3-glucanase, which increase and decrease the size of the callose deposits in the PD neck regions, respectively, and also through alterations in the levels of Ca^{2+} [7], which may alter the activity of callose synthases [8] or determine the length or membrane-binding activity of tether proteins that contain Ca^{2+}-dependent lipid-binding C2 domains and link the desmotubule to the PD PM [6, 9, 10]

on the physiological and developmental status of the cell. Although PD allow intercellular trafficking according to their SEL, they can also be permanently or transiently closed in order to establish symplasmic domains in which cells and tissues are isolated from surrounding cells to enable the determination of specific cell and tissue fates. PD are closed, or transiently closed, for example, in winter buds during dormancy [17], in epidermal cells determined to form and function as stomata [18] and matured root hairs [19], in cotton fiber cells during fiber elongation [20], between the developing embryo and the seed coat in *Arabidopsis* [21], during somatic *Arabidopsis* embryo induction [22], and for symplasmic isolation of phloem and companion cells from non-phloem cells in source tissues of apoplastic phloem loaders such as tobacco [21, 23]. PD are usually formed in new cell walls during cytokinesis (primary PD; [24, 25]) but can also be formed across already existing cell walls (secondary PD; [26–28]). During plant development, the PD may change in frequency (e.g., during floral induction; [29]) and also undergo alterations in structure and function. Cell wall extension and particularly the sink-to-source transition are paralleled by a "secondary modification" of existing PD from a simple structure with a single linear pore to a more complex (branched) structure with several pores leading into a shared wall cavity [28, 30–33]. Specific probes, such as fluorescent proteins fused to plasmodesmal proteins such as PD-LOCATED PROTEIN 1 (PDLP1), which generally localizes to PD upon expression, and the *Potato leaf roll virus* movement protein (MP17), which shows specific accumulation in PD cavities, can be used for the imaging of primary and secondary (or secondarily modified) PD [34].

In the following, we describe the features of PD in greater detail and then discuss the role of PD in controlling the interaction of plants with microbes and viruses.

2 Structure and Regulation of the PD SEL

2.1 General Features of PD

Due to the small size of the PD, only little is known about the internal structural organization of the channels. Transmission electron microscopy (TEM) studies have shown that PD are concentric pores that consist of the plasma membrane (PM) delimiting the channel against the cell wall and a strand of endoplasmic reticulum (ER), called "desmotubule," in the center (Fig. 1b). The PM and the ER as well as the cytoplasmic space between these membranes (the "cytoplasmic sleeve") are continuous through the channel between the connected cells. The PM of PD contains rafts and other types of microdomains characterized by the presence of specific lipids and which bind and concentrate proteins carrying lipophilic modifications (e.g., a glycosyl-phosphatidylinositol (GPI) "anchor") that are enriched in PD [2, 35]. The ER is highly fluid, and experiments using fluorescence redistribution after photobleaching (FRAP) and other approaches have shown that the desmotubule (but not the PM) can transport lipids and potentially also associated cargo molecules between cells [36, 37]. However, the major pathway for intercellular transport is likely provided by the cytoplasmic sleeve between the PM and ER components of the channel.

2.2 Potential Role of Membrane Tether Proteins in Regulating the SEL of PD

Early EM studies showed that the cytoplasmic sleeve is occluded by proteinaceous components which may represent transported cargo or internal spoke-like substructures that connect the desmotubule with the PM [1]. More recent evidence derived from cryo-EM and proteomic studies substantiated the view that PD contain tether proteins that link the two membranes [6, 38] (Fig. 1b). The tethering of the two membranes may facilitate exchange of lipids and contribute to the specific lipid composition of the PM within PD, which shows an enrichment for sphingolipids and sterols and is important for PD function [2]. In addition, these tether proteins may play an important role in determining the SEL by regulating the size of the cytoplasmic sleeve available for the transport of molecules. Several studies suggest the presence of the synaptotagmin SYT1/SYTA, a known plant ER-PM tethering factor [39, 40] as well as other members of the SYT protein family at PD [41–43]. These proteins are contractile and shorten their tether length in the presence of Ca^{2+}, as shown for SYT1/SYTA [9] and for members of the extended-synaptotagmin (E-Syt) family of animals [44, 45]. In regulating the SEL by reducing the size of the cytoplasmic space of PD upon contraction in the presence of Ca^{2+}, the presence of SYT1/SYTA and other members of the SYT family in PD may explain the early observation that PD are sensitive to Ca^{2+} and rapidly decrease their SEL in response to increased levels of this ion [7] (Fig. 1b). Recent studies in *Arabidopsis* revealed that members of the multiple C2 domains and transmembrane region proteins (MCTPs) localize to PD with tight ER-PM contact and exhibit the typical architecture of membrane tethers, with multiple Ca^{2+}-dependent lipid-binding C2 domains in the N-terminal and multiple transmembrane domains in the C-terminal region [6]. Among this class of proteins are AtMCTP1/FTIP1 and AtMCTP15/QKY, which have been already previously implicated as regulators of cell-to-cell signaling [46, 47]. Consistent with a role of tether proteins in regulating the size of the cytoplasmic sleeve, the desmotubule has been observed in EM studies to be constricted with no lumen or to be tubular with an open lumen. In some PD, the desmotubule was shown to be even partially or entirely dilated, thus leaving no room for intercellular transport via the cytoplasmic sleeve [38, 48–50]. In addition to a mechanism involving shortening of the ER-PM tether length, the size of the cytoplasmic sleeve may also be regulated in an opposite manner by plant reticulons (RTNs). These proteins induce curvature into membranes and contribute significantly to the tubularization of the ER [51–53]. RTN3 and RTN6 have been localized to PD [25, 54] and may play a role in constricting the diameter of a dilated desmotubule and, thus, increasing the size of the cytoplasmic sleeve. A proteomic assay showed that RTN3 and RTN6 interact with SYT1/SYTA [41], which may suggest that reticulons and

SYTs/MCTPs contribute to the regulation of the SEL in functional competition.

2.3 Role of Callose Deposition in Regulating the PD SEL

The major mechanism implicated in the regulation of the SEL involves the deposition of callose (β-1,3-glucan) in the cell wall surrounding the PD. Callose synthesis and accumulation by callose synthases in the cell wall leads to closure of the cytoplasmic compartment by forcing the PM against the desmotubule, whereas the degradation of callose by β-1,3-glucanase opens the cytoplasmic space for intercellular transport [55] (Fig. 1b). Large callose deposits were found around PD during wounding, plasmolysis, pathogen attack, and other stresses, during winter dormancy, as well as during organogenesis [3, 4, 56–58]. Mutations and conditions that affect the metabolism of callose at PD were shown to strongly affect the conductivity of PD for macromolecular transport [4, 5, 59–67]. Regulation of the SEL through callose deposition implies the function of hormone- and defense-related signaling cascades that involve PD-localized receptor-like kinases (RLKs) and receptor-like proteins (RLPs) and other PD-localized components such as the members of the PDLP family and callose-binding proteins [55] (Fig. 1b).

3 Movement of Macromolecules Through PD

3.1 Protein Movement

PD can be open or closed. However, dependent on developmental and physiological state of a given tissue, PD may occur in a predominantly open state and allow free symplastic coupling between cells. Fluorescent proteins, such as GFP, and large fluorescent dyes, such as dextrans labeled with fluorochromes, were observed to readily move from cell to cell [16, 30, 68], suggesting that PD provide ample opportunity for proteins and other molecules to move freely between cells. Indeed, numerous transcription factors were shown to move between cells. Classical examples are KNOTTED1 (KN1) of maize [69, 70], *Arabidopsis* LEAFY (LFY) [71] and SHORTROOT (SHR) [72], and the *Antirrhinum* proteins FLORICAULA (FLO) [73], GLOBOSA ((GLO) [74] and DEFICIENS (DEF) [73, 75]. Today, numerous transcription factors are known to move cell-to-cell in roots or meristems, or even between organs, like FLOWERING LOCUS T (FT) [12, 76]. Movement through PD with an open state may be prevented for proteins only (a) if they are targeted to other subcellular sites [77], (b) if they sequestered by interacting proteins [78–83], or (c) if they are simply too large for movement through PD with a given SEL [84, 85]. Movement of some proteins may depend on partial protein unfolding for movement and chaperone-assisted post-movement refolding in the recipient cells as has been indicated for the movement of KN1 [86].

3.2 RNA Movement

In addition to proteins and metabolites, also RNA molecules move between cells. Phloem sap contains numerous mRNAs which may have passed through PD to enter the enucleate phloem sieve elements. Phloem exudates also contain different classes of noncoding RNAs such as small interfering RNAs (siRNAs), microRNAs (miRNAs), tRNAs, and ribosomal RNAs (rRNAs) [87]. Evidence supporting that small RNA molecules move cell-to-cell by diffusion through PD is provided by the observation that siRNAs are unable to enter isolated stomatal guard cells and that the movement of a miRNA controlling radial root differentiation is inhibited in mutants with increase callose deposition at the PD [88]. The intercellular trafficking of the large mRNA molecules found in phloem, in contrast, may be limited to the trafficking through specialized PD connecting the sieve elements and the immediate sieve element-connected cells (e.g., companion cells), unless their cell-to-cell movement is assisted by a specialized PD targeting and PD-modifying cellular machinery that allows large mRNA molecules to move cell-to-cell. As it is further detailed below, RNA viruses encode PD-interacting movement proteins (MPs) that manipulate the PD SEL to facilitate the cell-to-cell transport of viral RNA [89, 90]. mRNA transport may depend on a similar mechanism for cell-to-cell transport as indicated by studies on KN1 mRNA, which was shown to move cell-to-cell in the presence of KN1 protein and to be dependent on the ability of this protein to increase the PD SEL [70, 91]. Interestingly, numerous mRNAs present in the phloem carry tRNA-like sequences [92] or polypyrimidine track protein binding sequences [93–95] or are m^5C methylated and require m^5C methylation for transport [96]. These structural features of mRNAs found in phloem suggest a role of specific RNA-binding proteins for phloem entry and mRNA transport to distant cells [96].

4 Role of PD During Plant-Microbe Interactions

4.1 PD Closure as an Immunity Response Against Bacteria and Fungi

Plant cells can perceive invading microorganisms through pathogen-recognition receptors (PRRs) that recognize and bind conserved, microorganism-specific molecules (e.g., microbe-associated molecular patterns, MAMPs) and activate intracellular defense signaling pathways [97] (Fig. 2). Sustained immunity responses against microbes depend on the synthesis and accumulation of salicylic acid (SA), which triggers the nuclear localization of NONEXPRESSOR OF PATHOGENESIS-RELATED GENES 1 (NPR1) to upregulate the transcription of defense-related genes [111]. Importantly, SA induces callose deposition and PD closure. The SA-induced pathway leading to PD closure involves both NPR1 and PLASMODESMATA-LOCATED PROTEIN 5 (PDLP5) [58, 109], thus placing PD in the center of plant innate

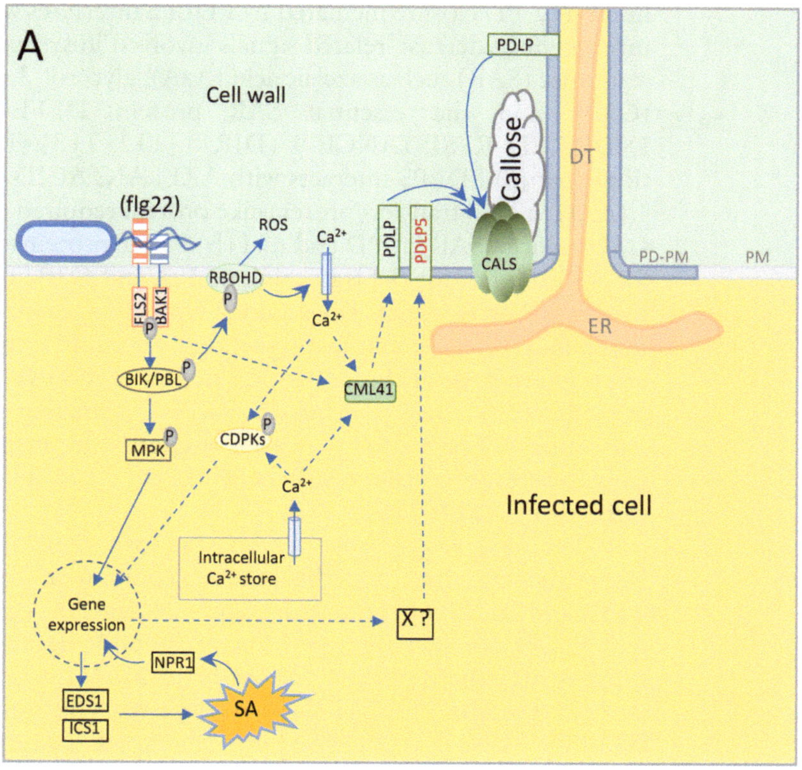

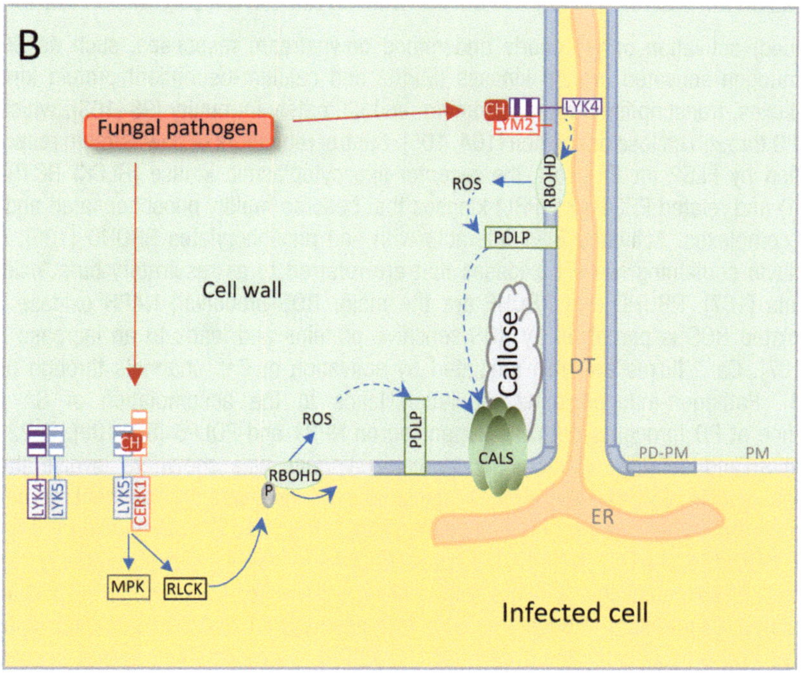

Fig. 2 Potential signaling pathways involved in the regulation of the PD SEL in the presence of microbial pathogens. (**a**) Cellular sensing of bacteria through bacterial flagellin or the flagellin-derived peptide flg22 by the PRR FLS2 induces the instant association of FLS2 with its coreceptor BRASSINOSTEROID INSENTIVE 1-ASSOCIATED RECEPTOR KINASE 1 (BAK1) and phosphorylation of both proteins. This in turn causes the

immunity. PD closure mediated by PDLP5 interferes with systemic movement of defense-related signals involved in systemic acquired resistance (SAR) such as azelaic acid (AzA), glycerol-3-phosphatase (G3P), and the essential SAR protein DEFECTIVE IN INDUCED RESISTANCE 1 (DIR1) [112, 113]. The observations that (a) PDLP5 interacts with AZELAIC ACID INDUCED 1 (AZI1, a lipid transfer- protein-like protein required for Aza- and G3P-induced SAR) at PD and that (b) PDLP proteins are required for SAR but not for the trafficking of AzA and G3P [113] indicate that PDLP proteins act as components of a PD-associated pathway that regulates movement and signaling through PD. Innate immunity signaling also includes the production of reactive oxygen species (ROS) that may also play a role as a signaling molecule in regulating PD permeability. This is suggested by studies showing close relationships between cellular redox status and trafficking through PD [114, 115].

The observation that chitin (a long-chain polymer of N-acetylglucosamine and the primary component of fungal cell walls) and flg22 (a peptide from bacterial flagellin) cause a reduction in PD conductivity for intercellular GFP trafficking further substantiated the view that PD closure is a MAMP-triggered innate defense response [116, 117]. Consistently, a GFP fusion to FLA GELLIN SENSING 2 (FLS2), the receptor kinase that perceives

Fig. 2 (continued) activation of yet poorly understood downstream responses, such as ROS production, activation of mitogen-activated protein kinases (MPKs) and calcium-dependent protein kinases (CPKs or CDPKs), Ca^{2+} fluxes, transcriptional reprogramming, and ultimately immunity [98–103], which also involves the closure of PD through callose deposition [104, 105]. Central regulators of downstream responses following PAMP perception by FLS2 (or EFR) are the receptor-like cytoplasmic kinase (RLCK) BOTRYTIS INDUCED KINASE 1 (BIK1) and related PBS1-like (PBL) kinases that become rapidly phosphorylated and then released from the PRR complexes. Activated BIK1 interacts with and phosphorylates RBOHD [106], one of several PM-localized flavin-containing NADPH oxidases that are referred to as respiratory burst oxidase homologs (RBOH) in plants [107]. RBOHD and RBOHF are the major ROS-producing NAPH oxidase in *Arabidopsis* [107]. Accumulated ROS is perceived by ROS-sensitive proteins and leads to an increase in cytoplasmic Ca^{2+} levels [107]. Ca^{2+} fluxes are also mediated by activation of Ca^{2+}channels through phosphorylation by BIK1 [108]. Pathogen-induced gene regulation leads to the accumulation of SA which causes callose deposition at PD through a pathway depending on NPR1 and PDLP5 [58, 109]. flg22-induced rapid callose deposition and defense against *Pseudomonas syringae* was shown to involve the PD-localized Ca^{2+} – binding protein CML41. The CML41-mediated response was independent of gene expression responses and occurred downstream or independent of flg22-triggered ROS production [104]. CALS, callose synthase; ICS1, ISOCHORISMATE SYNTHASE1, a crucial enzyme in the SA synthesis pathway. (**b**) LYM2 and the coreceptor kinase LYK4 establish a PD-localized sensing system for fungal chitin (CH) leading to callose deposition. Chitin is also sensed through the receptor kinase CERK1 and its coreceptor kinase LYK5. Both pathways involve ROS production by the NADPH oxidase RBOHD. However, PD localized signaling triggered by the LYM2-LYK4 receptor complex occurs through specific calcium-dependent kinases that activate RBOHD through a phosphorylation pattern which is different to that induced by CERK1. LYK4 and LYK5 form a complex in the PM and are recruited to their cognate LYM2 and CERK1 receptors in the presence of chitin [105, 110]

bacterial flagellin, localized to the PM of PD in addition to the cellular PM indicating that the PD themselves may have the capacity to perceive the presence of microbes. flg22-induced PD closure also depends on the PD-localized Ca^{2+}-binding protein CALMODULIN-LIKE PROTEIN 41 (CML41) [104] (Fig. 2a). Although CML41 expression is induced by flg22, the comparison of wild-type plants with mutants in which CML41 expression was either downregulated by an artificial miRNA (amiRNA) or overexpressed in transgenic plants showed that CML41 acts independently or downstream of other defense responses triggered by flg22 perception such as the production of ROS or the activation of specific flg22-induced defense-related genes. An increased deposition of callose that was observed in the CML41-overexpressing plants in the absence of flg22 could be reduced by the Ca^{2+}-chelator EGTA. Although the activity of several callose synthases is Ca^{2+}-dependent [8], this finding is consistent with the binding of Ca^{2+} by CML41 and is also in agreement with the known Ca^{2+} sensitivity of the PD [7]. The analysis of wild-type plants and CML41 overexpressing and CML41 downregulated plants also showed that CML41 does not interact with the pathway leading to PD closure triggered by chitin. Thus, CML41 may provide a Ca^{2+}-signaling specificity in PD closure that is specific for bacterial attack [104].

PD closure induced by chitin (Fig. 2b) is independent of the canonical chitin receptor CHITIN-ELICITOR RECEPTOR KINASE 1 (CERK1), which is present in the PM and dimerizes in the presence of chitin [118]. Instead, this response depends on the PD-located, RLP LYSIN MOTIF (LysM) DOMAIN-CONTAINING GLYCOSYLPHOSPHATIDYLINOSITOL-ANCHORED PROTEIN 2 (LYM2) [105] and two LysM receptor kinases, LysM-CONTAINING RECEPTOR-LIKE KINASE 4 (LYK4) and LYK5 [110]. The localization of a Citrine-LYM2 fusion to PD was increased in the presence of chitin, which is consistent with other reports indicating that stress-related receptors, such as QIAN SHOU KINASE (QSK1), INFLUORESCENCE MERISTEM KINASE 2 (IMK2), or CYSTEINE-RICH RECEPTOR KINASE 2 (CRK2), accumulate at PD in response to stress [119, 120]. Because LYK4, but not LYK5, was also detected at PD, the chitin-triggered PD signaling is proposed to be mediated by a LYM2-LYK4 complex. The chitin response at PD involves the activation of the ROS-producing NADPH oxidase RESPIRATORY BURST OXIDASE HOMOLOG PROTEIN D (RBOHD) and also depends on the calcium-dependent kinases CPK6 and CPK11, which activate RBOHD by site-specific phosphorylation that distinguishes the activation by CPK6 and CPK11 from the RBOHD phosphorylation pattern typically induced by CERK1 [110]. Consistently, the LYM2-mediated chitin perception and signaling at PD is independent of other chitin-elicited responses, such as mitogen-activated kinase

(MPK) activation, induced by CERK1 [105]. These findings establish the existence of a specific, PD-localized signaling platform that leads to PD closure upon perception of chitin.

4.2 Targeting of PD by Bacterial and Fungal Effectors

To overcome host immunity and enhance their virulence, microbes deliver effector proteins that inhibit specific steps in the host defense pathway or in the expression, synthesis, delivery, or subcellular localization of its components. There is evidence indicating that microbes use effectors to inhibit PD closure (Fig. 3). One example is the bacterium *Pseudomonas syringae*, which uses its type III secretion system to inject 36 effector proteins into host cells. Recently, one of the effectors, HopO1–1, was shown to target PD, to interact with PDLP5 and PDLP7, and to enhance the cell-to-cell mobility of co-expressed YFP. Interestingly, HopO1–1 causes the degradation of PDLP7 in a pathway that may involve a predicted ADP-ribosyltransferase activity of HopO1–1 [121]. PD are also targeted by the RxLR3 effector of the oomycete plant pathogen *Phytophthora brassicae*. The results of a recent study employing in vivo imaging and reciprocal co-immunoprecipitation suggests that this effector interacts with callose synthases at the PD and therefore induces a reduction in PD-associated callose levels and an increase in the cell-to-cell diffusion of GFP [122]. A recent analysis provided evidence indicating that at least 16 of the 36 - effectors of *P. syringae* pv. *tomato* DC3000 have the capacity to move cell-to-cell upon transient expression in *N. benthamiana* [123]. Movement was restricted upon inducing PD closure by expression of PDLP5 and PDLP7 which is consistent with movement through PD. Nevertheless, although the results are consistent with PD-mediated movement, it remains to be seen if these effectors moved by diffusion from the overexpressing cells into neighboring cells or whether they indeed move naturally under native expression conditions, either again by diffusion or using an active, PD targeting mechanism. It is indeed yet unclear why bacteria and oomycetes would depend on intercellular communication to enhance their virulence. This is more obvious for the rice blast fungus *Magnaporthe oryzae* that exploits the PD to grow its hyphae from cell to cell. *M. oryzae* penetrates the cell walls through pit fields in which PD are clustered together. During wall penetration, the fungal septins and F-actin are reorganized to locally constrict the diameter of the hyphae from 5.0 μm to 0.6 μm. To enhance the ability to invade neighboring cells, the fungus is able to overcome callose deposition at the pit fields. Invasive growth involves the fungal effector Bas4, which increases the PD SEL to enhance its spread between cells, as well as the fungal MPK Pmk1, which controls the constriction of invasive hyphae to traverse pit fields and also regulates the expression of fungal effectors to suppress plant immunity and therefore to suppress PD closure [124]. Another example is the root-infecting fungal pathogen

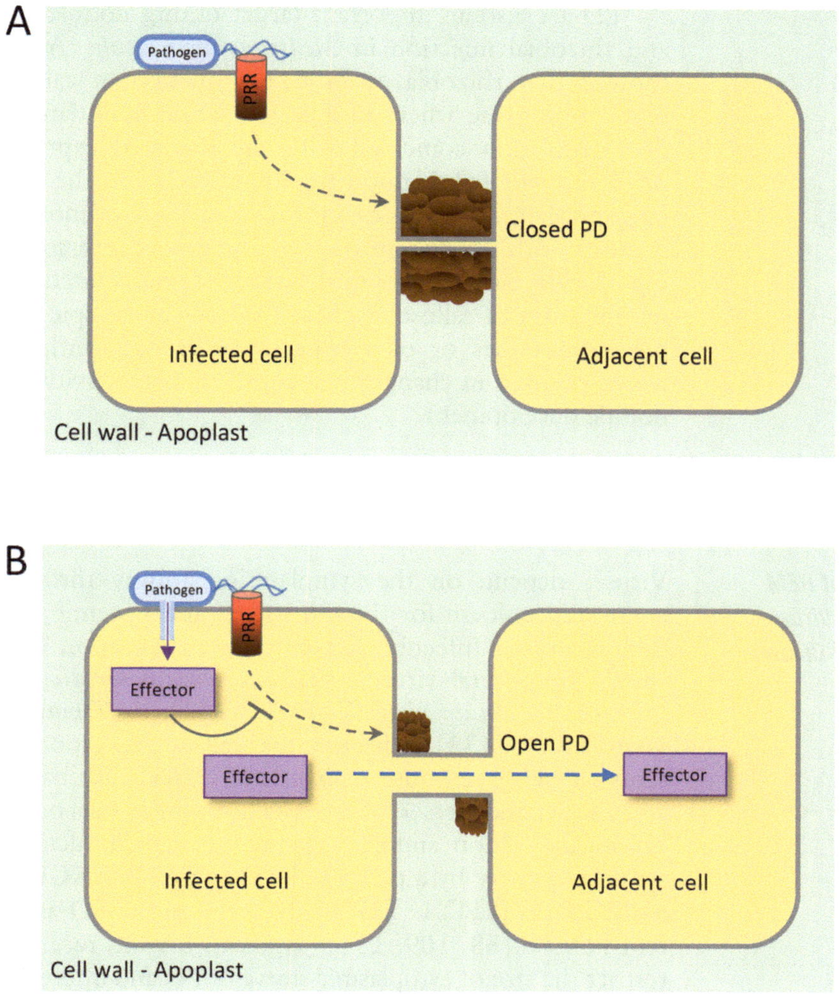

Fig. 3 Pathogens may encode effector proteins with the capacity to interfere with PD closure. (**a**) Sensing of a pathogen by a PRR induces host defense responses, including isolation of the infected cell by callose deposition (brown cumulus) and PD closure. (**b**) Some pathogens encode effector proteins that target PD [121–125] and may have the potential to inhibit the PD-localized host defense response to modify PD permeability. This could allow the effectors to move cell-to-cell and to favor infection by increasing the susceptibility of the surrounding cells

Fusarium oxysporum. This fungus secretes the effectors Avr2 and Six5 during infection in tomato [125], and it is assumed that these effectors are taken up by root endodermal and cortical cells, while the fungus colonizes the apoplastic spaces [126, 127]. Both effectors act together to alter the PD SEL and to allow Avr2 to move between cells [125]. As Avr2 is a potent suppressor of receptor-mediated immunity [127], its movement likely supports fungal virulence by suppressing plant defenses in cells surrounding the infection.

PD are also an important target during nodule development and rhizobial infection in *Medicago truncatula*. A recent study showed that rhizobia promote a reduction in callose levels in inner root tissue where nodule formation is initiated [66]. This downregulation coincided with the localized expression of the *M. truncatula* β-1,3-glucanase 2 (MtBG2) and the establishment of symplasmic domains as shown by analysis of intercellular connectivity with fluorescent tracers. The role of intercellular trafficking during nodulation was further demonstrated by selective modification of callose levels at PD sites by ectopic expression of β-1,3-glucanases or of a hyperactive callose synthase (cals3m), which resulted in changes in symplastic connectivity and aberrant nodule development.

5 Regulation of PD During Virus Infection

5.1 Role of REM Proteins in Virus-Induced PD Closure

Viruses depend on the symplasmic pathway through PD and connected phloem for the cell-to-cell and systemic movement of their genome. Infection causes callose deposition at PD as was shown for several viruses, such as *Tobacco mosaic virus* (TMV) [128–131], *Potato virus X* (PVX) [132], or *Tomato bushy stunt virus* (TBSV) [133], indicating that callose deposition at PD represents a defense response against virus attack. Like bacteria, viruses are known inducers of SA signaling which has been linked to callose deposition and PD closure through induction of callose synthase activity by a pathway involving ENHANCED DISEASE SUSCEPTIBILITY1 (EDS1), NPR1, and the PD-associated protein PDLP5 [58, 109, 134] (Fig. 2a). SA was recently shown to restrict the size of cytoplasmic sleeve of PD and intercellular mobility of a fluorescent tracer (carboxyfluorescein diacetate, CFDA) by promoting the assembly of remorin (REM)-dependent lipid nanodomains [135]. Such nanodomains are known to play a role as platforms for enrichment of membrane-resident proteins [136–138] and are also thought to harbor the GPI-anchored proteins playing a role in callose homeostasis at PD. Treatment of *Arabidopsis* plants with SA or with *Cucumber mosaic virus* (CMV), which primes the induction of SA biosynthesis [139], caused the assembly of REM-labeled nanodomains, and REM overexpression interfered with *Tobacco rattle virus* (TRV) cell-to-cell movement while also causing PD closure, thus providing strong evidence for the involvement of REM-dependent lipid raft organization in SA-triggered plant defense [135]. REM overexpression was shown to impair the movement of a number of other viruses, including PVX, TMV, and *Rice stripe virus* (RSV) [140–142]. REM was shown also to interact with the PM protein PCaP1, which impaired PCaP1-dependent movement of *Turnip mosaic virus* (TuMV) [143]. Altogether these findings emphasize the role of

REM proteins as central regulators of PD function and virus movement and may also indicate that viruses must target proteins in the PM to facilitate their movement. Interestingly, the transient expression of tobacco REM facilitated, rather than inhibited, the cell-to-cell movement of *Tomato mosaic virus* (ToMV) [144]. This may reflect a strong diversity in the activity of different REM proteins, which deserved further study.

5.2 Viruses Encode MPs to Facilitate Their Movement Through PD

To overcome the gating of PD, viruses encode specialized MPs that modify PD in different ways [145, 146]. Whereas the MP of TMV and the MPs of probably most other viruses cause a transient modification of the PD SEL, the MPs of "tubule-forming viruses," such as *Cowpea mosaic virus* (CPMV), *Grapevine fanleaf virus* (GFLV), or *Cauliflower mosaic virus* (CaMV), assemble a hollow tubule within PD and therefore drastically alter the internal structure and function of the PD channel (Fig. 4). The tubules formed within PD allow the cell-to-cell transport of whole virion particles, whereas viruses that modify the PD SEL move between cells rather in a non-virion form (ribonucleoprotein (RNP) complexes) and may even be independent of viral coat protein (CP), as in the case of TMV. The formation of tubules by the MPs of GFLV and CaMV depends on interaction with PDLP proteins at the PD [147]. Virion entry into tubules is mediated by direct or indirect interactions between the CP of the capsid and MP residues exposed in the inner tubule wall [148, 149]. Whereas the CPs and MPs of CPMV and GFLV interact directly, the binding of the CP of CaMV to the MP involves also other viral proteins [150]. Tubules formed within the PD have a given length suggesting that MP units and associated virions that have docked to the tubule end treadmill through the tubule length and are released at the distal end. However, whether tubules are dynamic or static remains to be investigated.

The cell-to-cell movement of non-tubule-forming viruses is linked to the ability of their MPs to alter the SEL of PD [90], which was first discovered for the MP of TMV [151]. Classical experiments using microinjection of fluorescent, membrane-impermeable size-specific dextran probes showed that MP-transgenic tobacco plants allowed the cell-to-cell diffusion of probes with an approximate size of 10 kD, whereas diffusion was restricted to a molecular size of 0.8 kD in control plants [151]. The protein itself has the capacity to move between cells upon injection [152] or upon transient expression from a plasmid introduced into tissue by microprojectile bombardment [16] or agroinfiltration [153]. The ability of MP to alter PD aperture is usually correlated with the accumulation of the protein at PD. However, the presence of accumulated MP in PD does not necessarily indicate that the SEL of PD is increased. Microinjection of cytoplasmic probes into cells within different radial zones of spreading TMV infection sites

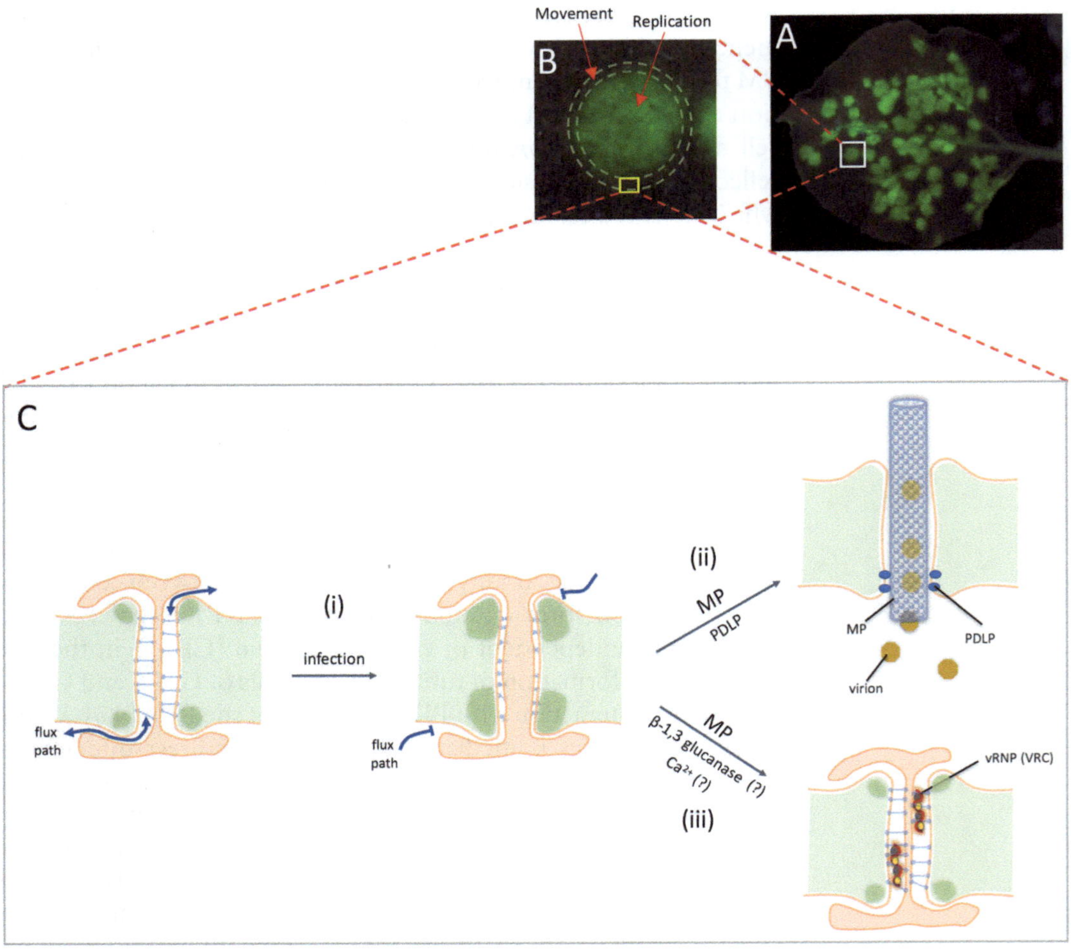

Fig. 4 Viruses exploit PD for their cell-to-cell movement by different mechanisms. (**a**) Local sites of viral infection caused by GFP-tagged TMV in a *N. benthamiana* leaf. Each of the infection sites originates from an initially infected cell and expanded radially by repetitive movement of the virus from infected cells to noninfected cells. Upon entry into the phloem of the vascular veins, the virus can spread systemically (not shown). (**b**) Individual infection site. The infection front where the virus moves from infected cell into noninfected cells via PD is indicated. (**c**) Infection causes callose deposition and PD closure (i). To overcome PD closure and allow movement, viruses encode one or more MPs. Some viruses encode MPs that bind to one or more PDLPs at PD and assemble into a hollow tube, allowing the cell-to-cell transport of whole virion particles (ii). Most viruses (e.g., TMV) encode MPs with the ability to increase the PD SEL by activating a callose-degrading enzyme or other pathways that interfere with callose deposition at PD (e.g., through yet unknown Ca^{2+}-dependent mechanisms [7]); these viruses can move cell-to-cell as a viral ribonucleoprotein complex (vRNP), which may represent an intact viral replication complex (VRC) that has all components to initiate infection in the newly infected cell

has demonstrated that MP alters PD only of cells at the infection front although the MP resides in PD throughout the infection site [154]. Accumulation of MP in PD and the modification of the SEL of PD may represent independent functions of MP, since TMV may spread through PD without MP being trapped in PD [155]. Plants that express the MP and are able to complement the movement function of movement-deficient TMV mutants do not show obvious growth defects, suggesting the existence of mechanisms that tightly control the ability of MP to modify PD aperture and thus prevent the continuous trafficking of signaling macromolecules between cells. The nature of these mechanisms still remains to be elucidated. However, the MP has several amino acids that are phosphorylated in vivo [156–160] and may play a role in regulating MP functions.

5.3 Potential Mechanisms Through Which MPs Alter the PD SEL

5.3.1 MPs May Increase the PD SEL Through Interaction with β-1-3-Glucanases

The mechanism through which MPs increase the PD SEL is still not understood. Electron micrographs suggested that the function of MP may be linked to the formation of fibrillary substructures within PD cavities [161–163], but it remains unknown if these structures are functional. Experimental evidence showing that the rate of TMV movement can be influenced by the expression of β-1,3-glucanase [131, 164, 165] supports the hypothesis that the MP may recruit a β-1,3-glucanase to interfere with the infection-induced callose deposition at PD. Indeed, whereas replication of a MP-deficient TMV replicon induced callose deposition at the PD, the levels of callose induced by this replicon were reduced in MP-transgenic plants [166]. A glucanase isoform that may play a role during virus infection in *Arabidopsis* is AtBG_pap. Similar to several recently isolated PD proteins, this protein is predicted to be a GPI-anchored protein. This protein was localized to the ER, the cell periphery, and PD, and mutation of the protein-encoding gene by a T-DNA insertion led to a reduction in GFP cell-to-cell movement and increased stress-induced callose deposition at PD [167]. Moreover, the cell-to-cell movement of GFP-tagged *Turnip vein clearing virus* (TVCV), a TMV-related tobamovirus, was reduced in *Arabidopsis atbg_pap* mutant plants. The same mutant also showed highly increased PD-associated callose levels in another study, thus confirming the role of AtBG_pap-mediated callose degradation during virus movement [65]. The MP of TMV as well as TGB2, one of the triple gene block (TGB) MPs of PVX, have the capacity to interact with members of the ankyrin repeat-containing proteins (ANK proteins) and to recruit these proteins from the cytoplasm to the cell periphery [168, 169]. The TGB-interacting proteins (TIP) of this family (termed TIP1, TIP2, TIP3) were shown to interact with a tobacco class I β-1,3-glucanase in the yeast two-hybrid assay [169] suggesting that MPs may modify callose levels by interacting with β-1,3-glucanase through ANK proteins. While a role of TIP proteins in PVX movement has not been reported, TMV cell-to-cell movement was promoted and restricted in tobacco lines overexpressing or suppressing a homologous ANK protein derived from another tobacco cultivar, respectively. However, while ANK overexpression enhanced the reduction in PD-associated callose levels in the presence of MP [168], it remains to be shown whether MP functions are indeed ANK-dependent and, if so, whether ANK-family proteins play a role also for viruses other than TMV and whether they are acting in plant virus movement also in other plant species than tobacco. In addition, also the role of the ANK-interacting class I β-1,3-glucanase is yet unclear. Unlike the PD-localized, GPI-anchored AtBG_pap in *Arabidopsis* [167], tobacco class I β-1,3-glucanases are localized in the vacuole. Although vacuolar glucanases might be directed to PD by interaction with MP [168], an interaction of MP with components already residing at PD might be more likely.

5.3.2 MPs May Target PD and Increase Their SEL Through Interaction with SYTA

Accumulating evidence suggests that the MP of TMV and those of several other viruses target the Ca^{2+} sensor SYT1/SYTA. However, while MP may target SYT1/SYTA to modify the PD SEL through altering the tether length of this protein, direct evidence for such activity is still lacking. SYTA is essential for regulating endocytosis and MP-directed cell-to-cell transport of diverse viruses, including TMV and another tobamovirus, TVCV [170, 171]. Virus movement as well as the cell-to-cell trafficking of tested viral MPs was inhibited in *syta-1* mutants, but it remains unclear if the mutant disables the MP function to increase the PD SEL or simply its targeting to PD. A more recent study [42] supports the role of SYTA in providing a structural environment required for PD targeting by MP and the virus. The analysis of *Arabidopsis syta*/SYTA-RFP rescue lines in this study confirmed the role of SYTA in tethering the ER to the PM and in stabilizing the normal reticulate ER network structure. SYTA-RFP localized to ER-PM junctions that corresponded to stationary nodes of the network. The same study also showed that SYTA is required for efficient targeting of MP to PD, whereas it is not required for PDLP1 that, unlike the MP of TMV, depends on the secretory pathway for PD targeting [172]. Moreover, TVCV infection promoted the SYTA-MP interaction and increased the localization of SYTA to PD together with MP. Based on the observation that SYTA colocalizes with viral replication sites and that the localization of SYTA to PD is promoted during infection, the SYTA-MP interaction may play a role in the formation of replication sites at the PD, thus coupling TVCV replication with movement through PD [42] (Fig. 5). A role of SYTA at ER-PM contact sites in altering PD permeability for virus cell-to-cell movement is not known but could be tested by determining the cell-to-cell movement of a fluorescent tracer such as GFP in virus-infected and noninfected *syta* versus wild-type plants.

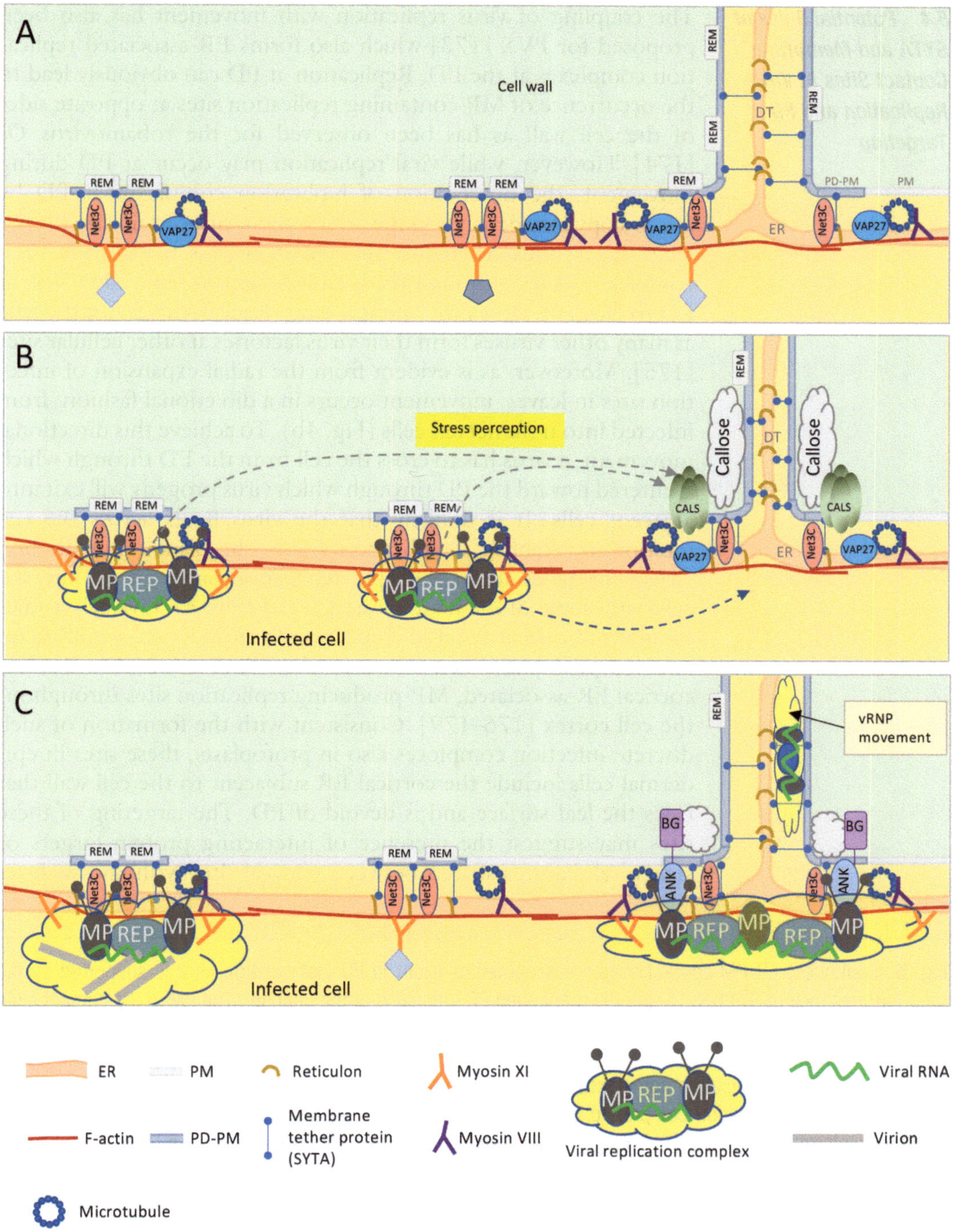

Fig. 5 Viruses may target PD by interacting with ER-PM contact sites. (**a**) Hypothetical model of the cell cortex at PD and adjacent to the PD as suggested by available evidence described in the text. ER-PM contact sites are marked by the presence of a complex with Net3C, which interacts with ER-actin, SYT1/SYTA, and ER reticulons, and the presence of a neighboring complex with VAP27, which interacts with ER-actin, microtubules, and also ER reticulons. The presence of the microtubules marks these ER-PM contact sites as potentially equivalent to cortical microtubule-associated ER sites (cMERs). The PM domains in these ER-PM contact sites and in PD have a specific composition and are highlighted in darker gray. The same or similar

5.4 Potential Roles of SYTA and Membrane Contact Sites in Virus Replication and PD Targeting

The coupling of virus replication with movement has also been proposed for PVX [173] which also forms ER-associated replication complexes at the PD. Replication at PD can obviously lead to the occurrence of MP-containing replication sites at opposite sides of the cell wall as has been observed for the tobamovirus *Ob* [174]. However, while viral replication may occur at PD during movement, the observation of replication complexes at PD by microscopy may depend on the achievement of a certain size during later stages, thus when the movement process has already been completed. Also, although the establishment of replication sites at PD to facilitate virus movement appears logical, this is not the rule as many other viruses form their virus factories at other cellular sites [175]. Moreover, as is evident from the radial expansion of infection sites in leaves, movement occurs in a directional fashion, from infected into noninfected cells (Fig. 4b). To achieve this directional movement, a virus has to cross the cell from the PD through which it entered toward the PD through which virus progeny will exit into adjacent cells. It is unlikely that the virus floats across the cell interior to establish a replication site at the distant PD. Rather, viruses likely establish multiple replication sites within infected cells, and some of them are located at PD (Fig. 5). Live cell studies on infected cells at the TMV infection front in leaves as well as on infected protoplasts demonstrated that TMV establishes discrete cortical ER-associated, MP-producing replication sites throughout the cell cortex [176–179]. Consistent with the formation of such discrete infection complexes also in protoplasts, these sites in epidermal cells include the cortical ER subjacent to the cell wall that faces the leaf surface and is devoid of PD. The targeting of these sites may suggest the presence of interacting protein targets or structural features shared between cortical sites with and without

Fig. 5 (continued) ER-PM contact sites are assumed to occur in PD (not shown), as it is supported by the reported presence of SYT1/SYTA and the mentioned ER reticulons at the channel. The PM in the contact sites may contain the membrane raft protein REM, as indicated. Myosin XI transports cargo along ER-actin, and myosin VIII holds the membrane raft/ER-PM contact site components in place. (**b**) Anchorage of the ER-associated TMV replication complexes to ER-PM contact sites (VAP27 not shown here). The MP of this virus anchors the replication complex to ER-PM contact sites by linking its ER association with its specific binding to microtubules and SYT1/SYTA. The anchored replication complex grows in size through replication and myosin XI-mediated recruitment of membranes and host factors. The perception of stress caused by infection results in callose deposition and PD closure. The anchored replication complex may detach from its position and transported along ER-actin to other ER-PM attachment sites. This may occur when myosin VIII no longer holds the complex in place or by active movement by this motor protein. (**c**) Anchorage of the replication complex at the PD. The MP interacts with β-1,3-glucanase (BG) via ANK and induces the degradation of callose to increase PD SEL and to allow subcomplexes (vRNP) to move through the channel into the adjacent cell to spread infection. Continued replication may allow more than one complex to move. However, as soon as the viral complex infects the next cell, the MP in the donor cell is phosphorylated and inactivated, and the CP is produced, thus allowing the virus to switch phases from movement to virion progeny production. The mechanism by which the vRNP is transported through the PD pore is unknown

PD. The protoplast is attached to the cell wall at PD but also at numerous other cortical sites, as can be seen upon plasmolysis [180, 181]. Although further studies are needed, it appears conceivable that these sites or other cortical sites targeted by TMV are ER-PM contact sites. The MP-interacting contact site marker SYT1/SYTA localizes to the cortical ER in a beads-on-a-string-like pattern [42, 43, 182], and this pattern appears to have similarity to the punctate pattern of distinct MP-expressing and ER-associated TMV replication sites observed in cells at the virus front or in infected protoplasts [176–179] and also to the pattern of distinct cortical ER-associated puncta observed in MP-expressing cells [179, 183, 184]. Remarkably, the cortical MP puncta contain PDLP1 [178], which is consistent with the viral targeting of membrane rafts similar to those present in PD. SYT1-associated ER-PM contact sites mark the location of immobile ER tubules [182]. These immobile ER tubules occur adjacent to tubule junctions that are capable of dynamic tubule formation and are associated with another type of ER-PM contact site that is characterized by the presence of membrane-tethering protein VESICLE-ASSOCIATED MEMBRANE PROTEIN-ASSO CIATED PROTEIN 27 (VAP27), the actin-binding protein NET WORKED SUPERFAMILY PROTEIN 3C (NET3C), and microtubules [39, 185]. The latter type of ER-PM contact site may be targeted by TMV because its MP has affinity to microtubules [186, 187]. Importantly, the distinct cortical sites targeted by TMV for replication and movement are microtubule-associated or occur in the vicinity of microtubules [179, 184]. Functional studies have shown that the ability of MP to localize to microtubules is tightly associated with its ability to form the cortical particles and to support virus movement [188–191]. The importance of microtubules for TMV is also supported by host mutations that affect the dynamic behavior of microtubules. In hosts with such mutations, virus movement is strongly affected. Moreover, these host mutations cause viral adaptation through amino acid mutations in the viral replicase, which underscores the importance of microtubule network functions for the virus [192, 193]. However, given that MP interacts with both SYT1/SYTA [42, 170] and microtubules [176, 186, 187, 194], the virus may target both types of ER-PM contact sites, i.e., the VAP27-NET3C-microtubule type and the SYT1 type. Both VAP27 and SYT1 have been localized to PD [42, 195] and were also shown to interact with the PD-localized reticulon proteins RTNLB3 and RTNLB8 [41]. Thus, these proteins and likely the two types of ER-PM contact site may occur adjacent to each other and represent common features of localized PM domains at PD and other cortical sites. When deprived of cell wall, the cortical MP-producing replication sites have the capacity to produce MP-containing membrane protrusions [176], suggesting that the cortical replication sites may have the capacity to initiate a dynamic motility event once the inhibitory cell wall is removed. It is tempting to speculate that

TMV targets ER complexes associated with microtubules and SYT1 at ER-PM contact sites and that PD represent such sites equipped with a channel into adjacent cells. Thus, in targeting many cortical ER-PM contact sites for replication, the virus may also find the PD for movement. Collective in vivo observations suggest that some of the viral complexes can detach from the cortical sites and undergo stop-and-go movements along the cortical ER-actin network (from microtubule to microtubule) [179, 184, 196]. Thus, if not initially already targeted to PD by chance, the viral complexes may find PD upon intracellular movement along the cortical network (Fig. 5).

5.5 Role of Actin and Myosin

The intracellular stop-and-go movements of the viral replication and movement complexes along the cortical ER-actin network and the interaction of these complexes with PD for movement is supported by myosin motor proteins. Subcellular trafficking of MP complexes and the intercellular spread of TMV infection were shown to depend on class VIII and class XI myosins [197, 198] and to be blocked by overexpression of actin-binding protein [199]. Whereas class XI myosins support the native structure and mobility of the ER network as a basis for the formation of replication and movement complexes, class VIII myosins provide a function required for MP to find and enter the PD [197]. Inhibition of myosin VIII proteins by myosin VIII tail expression resulted in the accumulation of MP within patches at the PM that were adjacent to similar-sized patches labeled with the PM raft marker REM, suggesting that myosin VIII may play a role in the intracellular transport of viral complexes from cortical sites to the PD. Myosin VIII tails were previously reported to localize to stationary puncta adjacent to three-way junctions of the ER [200] proposed to represent lipid rafts [201] and to PD [202], which is consistent for a role of myosin VIII in transporting viral complexes and likely also other cargo from cortical sites to other cortical sites with PD. This view may also be supported by other studies showing that myosin VIII interacts with microtubule ends and shows high, actin-dependent motility in the cortex [203].

The role of myosins in virus movement and in the targeting of MP to PD is supported by studies indicating the presence of actin at PD and that the PD SEL is directly or indirectly controlled by myosins [204]. Interestingly, the MPs of TMV and CMV were shown to sever actin filaments in vitro. Moreover, the stabilization of actin filaments with phalloidin inhibited the ability of MP to increase the PD SEL [205]. These observations suggest that MPs manipulate actin to control PD permeability. However, it remains to be clarified whether actin severing by MP also occurs in vivo and whether such activity of MP can be localized to PD or to other sites in the cell. Nevertheless, these findings may be consistent with the role of myosin VIII in the mechanism by which TMV MP targets and enters the PD channel [197].

5.6 A Model for PD Targeting by Viruses

The various pathways used by different viruses and their MPs to target PD have been summarized in previous reviews [145, 146, 150, 175, 206–208] and cannot be covered here. The above thoughts and current knowledge are combined in the hypothetical model for TMV replication and PD targeting shown in Fig. 5. In this model, the viral RNA that has entered a cell is translated on the cortical ER. The viral replication complex consisting of initially translated replicase, replicated viral RNA, and MP attaches to ER-PM contact sites identified by MP by the presence of a microtubule and SYTA to which it binds directly. The compactness, anchorage, and growth of the replication complex are supported by myosin XI. While anchored to the microtubule and the ER-actin surface by MP, the complex recruits additional ER membrane and host factors with the help of myosin XI, thereby expanding its size. Further growths of the replication site and the production of the viral CP initiate the packaging of viral progeny RNA into virus particles. Before the virus switches to CP production and virion formation, other replication complexes interact with PD for infecting the next cell. This process may involve the detachment of the complex and the potentially myosin VIII-mediated trafficking along the cortical ER-actin network from ER-PM contact site to ER-PM contact site, thus from microtubule to microtubule. SYTA is always targeted by MP together with microtubules and may also be recruited to enhance the binding to ER-PM contact sites at PD [42]. Once the viral complex is attached to a site near PD, a VRC-derived vRNP enters the channel and reaches the adjacent cell. Following movement of the replication complex, the donor cell switches phases from movement to the sustained production of progeny virions.

5.7 Role of PD in Antiviral Defense Signaling

Beyond allowing the movement of viruses themselves, PD are also essential for regulating the plant responses to virus infection. Apart from replication-triggered PD closure that is overcome by the viral MP at the virus front (*see* Subheading 5.3), the sustained virus replication in cells behind the virus front triggers antiviral RNA silencing in which primary and secondary small interfering (si)RNAs derived from the replicating virus genome (vsiRNAs) program AGO slicer proteins to target viral RNA for sequence-specific cleavage. To prevent cleavage of their RNA, viruses have evolved suppressor proteins (viral suppressor of RNA silencing, VSR) that can counteract against silencing by siRNA sequestration or by inhibition of other steps of the RNA silencing pathway. If the VSR is either weak [209] or expressed at low levels, as can be expected for cells at the virus front, the vsiRNAs and also host small RNAs can be active and target host genes as well as the viral RNA genome [210–214]. Moreover, vsiRNAs can move between cells to act as a systemic signal, thereby regulating the systemic spread of the virus as well as the level of resistance and disease tolerance in the systemic tissues [209, 215–218]. Consistent with its ability to increase the SEL of PD, the MP of TMV facilitates the cell-nonautonomous spread of RNA silencing [153]. Spreading RNA silencing ahead of the virus front may play an important role in host-virus compatibility as it may allow the virus to avoid its own overaccumulation by establishing a basal antiviral state and, at the same time, to increase virulence by targeting host defense genes. Conversely, the C4 protein of *Tomato yellow leaf curl virus* (TYLCV) inhibits the spread of siRNAs and promotes systemic infection though binding of the *Arabidopsis* RLKs BARELY ANY MERISTEM 1 (AtBAM1), a positive regulator of the cell-to-cell spread of RNA silencing, at PD [219]. Binding to AtBAM1 was also observed for the C4 protein of *Mungbean yellow mosaic virus* (MYMV) [220], and the C4 protein of TYLCV also interacts with the homologue AtBAM2 as well as with BAM1 orthologs from tomato and *N. benthamiana* [219]. Interestingly, BAM1 and BAM2 are targeted at PD also by the tombusvirus RNA silencing suppressor protein P19 [221]. A sRNA binding-compromised version of P19 maintained the ability to bind to BAM1 and BAM2 as well as to interfere with the cell-to-cell movement of RNA silencing. This suggests that p19 targets BAM1 and BAM2 similar to C4 in order to interfere with the transport of the systemic silencing signal through PD [221]. Thus, while PD are gateways for viruses to spread infection, they are associated with RLKs that function to control systemic plant defenses and therefore play a key role in determining the outcome of infection.

6 Conclusions

While PD play an important role in the orchestration of plant life by allowing movement of small and large signaling molecules between cells, they are able to downregulate their conductivity and to function as barriers against intrusion by microorganisms and viruses that target these intercellular communication channels for their own cell-to-cell transport (e.g., viruses) and for supporting their virulence with effector proteins. Cells downregulate the SEL of PD by deposition of callose in the surrounding cell wall and, presumably, also by regulating the length of tether proteins that bridge the space between the desmotubule and the PM. An increasing number of studies implies the roles of specific PD proteins in the signal transduction leading to PD closure. The localization of PRRs to PD [105] and other ligand-sensing receptor protein kinases [117, 222] suggest that PD are associated with signaling hubs that can sense pathogens as well as developmental and stress signals. However, given that PD respond to a variety of signals, including changes in temperature (e.g., cold stress), wounding, defense hormones (e.g., SA), ROS, and Ca^{2+}, it will be a challenge to understand how these different signals, and particularly also the signals triggered by microbes and viruses, are decoded at the PD. Although proteomic approaches lead to continued refinements in the identification of bona fide PD proteins [6], further studies are needed to integrate the PD proteins into a coherent picture of PD structure and PD regulation. Viruses are known to target PD to move from cell to cell. However, despite studies on viral MPs since decades, we still do not know the exact mechanism through which these proteins manipulate the PD. We even do not yet know how viruses trigger callose deposition and PD closure in the first place. Nevertheless, the recent finding that viruses elicit receptor-mediated immunity [223, 224] in addition to RNA silencing may offer a new starting point to elucidate virus-induced signaling mechanisms through which PD are regulated during infection (Huang and Heinlein, in preparation). Important insights may come also from the further identification and analysis of host proteins involved in the functional linkage between virus replication and movement. These proteins may be localized at PD as well as at other peripheral sites targeted by the virus. The viral interactions with ER-PM contact site component SYT1/SYTA at PD and within replication sites [42, 170, 225, 226], the interaction of PD reticulons with ER-PM contact site components SYT1/SYTA, VAP27, and NET3C [41], the presence of VAP27 and SYT1/SYTA at PD [42, 195] as well as at other cortical ER sites [42, 195], and also the presence of reticulons within replication sites [227] are just few of several hints suggesting that TMV and potentially also other viruses have affinity to components that are shared between PD and other cortical ER-PM contact sites in the cell. Distinct virus-targeted sites occurring in epidermal cells near the PM facing the plant leaf surface are easily accessible by confocal microscopy. Studying these sites and finding and characterizing protein markers that are also present at PD may advance our understanding of the mechanisms involved in PD targeting and manipulation by viruses.

Acknowledgments

C.H. was supported by a PhD fellowship from the Chinese Scholarship Council (CSC).

References

1. Ding B, Turgeon R, Parthasarathy MV (1992) Substructure of freeze-substituted plasmodesmata. Protoplasma 169:28–41
2. Grison MS, Brocard L, Fouillen L et al (2015) Specific membrane lipid composition is important for plasmodesmata function in Arabidopsis. Plant Cell 27:1228–1250
3. Zavaliev R, Ueki S, Epel BL et al (2011) Biology of callose (beta-1,3-glucan) turnover at plasmodesmata. Protoplasma 248:117–130
4. Vaten A, Dettmer J, Wu S et al (2011) Callose biosynthesis regulates symplastic trafficking during root development. Dev Cell 21: 1144–1155
5. De Storme N, Geelen D (2014) Callose homeostasis at plasmodesmata: molecular regulators and developmental relevance. Front Plant Sci 5:138
6. Brault ML, Petit JD, Immel F et al (2019) Multiple C2 domains and transmembrane

region proteins (MCTPs) tether membranes at plasmodesmata. EMBO Rep 20:e47182

7. Tucker EB, Boss WF (1996) Mastoparan induced intracellular Ca^{2+} fluxes may regulate cell-to-cell communication in plants. Plant Physiol 111:459–467
8. Leijon F, Melzer M, Zhou Q et al (2018) Proteomic analysis of plasmodesmata from populus cell suspension cultures in relation with callose biosynthesis. Front Plant Sci 9: 1681
9. Yamazaki T, Kawamura Y, Minami A et al (2008) Calcium-dependent freezing tolerance in Arabidopsis involves membrane resealing via synaptotagmin SYT1. Plant Cell 20: 3389–3404
10. Tilsner J, Nicolas W, Rosado A et al (2016) Staying tight: plasmodesmal membrane contact sites and the control of cell-to-cell connectivity in plants. Annu Rev Plant Biol 67: 337–364
11. Zambryski P, Crawford K (2000) Plasmodesmata: gatekeepers for cell-to-cell transport of developmental signals in plants. Annu Rev Cell Dev Biol 16:393–421
12. Sevilem I, Yadav SR, Helariutta Y (2015) Plasmodesmata: channels for intercellular signaling during plant growth and development. Methods Mol Biol 1217:3–24
13. Lu KJ, Danila FR, Cho Y et al (2018) Peeking at a plant through the holes in the wall - exploring the roles of plasmodesmata. New Phytol 218:1310–1314
14. Heinlein M (2002) Plasmodesmata: dynamic regulation and role in macromolecular cell-to-cell signaling. Curr Opin Plant Biol 5:543–552
15. Sager R, Lee JY (2014) Plasmodesmata in integrated cell signalling: insights from development and environmental signals and stresses. J Exp Bot 65:6337–6358
16. Crawford KM, Zambryski PC (2001) Non-targeted and targeted protein movement through plasmodesmata in leaves in different developmental and physiological states. Plant Physiol 125:1802–1812
17. Rinne PL, van der Schoot C (1998) Symplasmic fields in the tunica of the shoot apical meristem coordinate morphological events. Development 125:1477–1485
18. Palevitz BA, Hepler PK (1985) Changes in dye coupling of stomatal cells of *Allium* and *Commelina* demonstrated by microinjection of lucifer yellow. Planta 164:473–479
19. Duckett CM, Oparka KJ, Prior DAM et al (1994) Dye-coupling in the root epidermis of *Arabidopsis* is progressively reduced during development. Development 120:3247–3255
20. Ruan YL, Llewellyn DJ, Furbank RT (2001) The control of single-celled cotton fiber elongation by developmentally reversible gating of plasmodesmata and coordinated expression of sucrose and K^+ transporters and expansin. Plant Cell 13:47–60
21. Imlau A, Truernit E, Sauer N (1999) Cell-to-cell and long-distance trafficking of green fluorescent protein in the phloem and symplastic unloading of the protein into sink tissues. Plant Cell 11:309–322
22. Godel-Jedrychowska K, Kulinska-Lukaszek K, Horstman A et al (2020) Symplasmic isolation marks cell fate changes during somatic embryogenesis. J Exp Bot 71:2612–2628
23. Oparka KJ, Turgeon R (1999) Sieve elements and companion cells - traffic control centers of the phloem. Plant Cell 11:739–750
24. Hepler PK (1982) Endoplasmic reticulum in the formation of the cell plate and plasmodesmata. Protoplasma 111:121–133
25. Knox K, Wang P, Kriechbaumer V et al (2015) Putting the squeeze on plasmodesmata: a role for reticulons in primary plasmodesmata formation. Plant Physiol 168:1563–1572
26. Ding B, Haudenshield JS, Willmitzer L et al (1993) Correlation between arrested secondary plasmodesmal development and onset of accelerated leaf senescence in yeast acid invertase transgenic tobacco plants. Plant J 4:179–189
27. Evert RF, Russin WA, Bosabalidis AM (1996) Anatomical and ultrastructural changes associated with sink-to-source transition in developing maize leaves. Int J Plant Sci 157:247–261
28. Ehlers K, Kollmann R (2001) Primary and secondary plasmodesmata: structure, origin, and functioning. Protoplasma 216:1–30
29. Ormenese S, Havelange A, Bernier G et al (2002) The shoot apical meristem of Sinapis alba L. expands its central symplasmic field during the floral transition. Planta 215:67–78
30. Oparka KJ, Roberts AG, Boevink P et al (1999) Simple, but not branched, plasmodesmata allow the nonspecific trafficking of proteins in developing tobacco leaves. Cell 97: 743–754
31. Roberts AG, Oparka K (2003) Plasmodesmata and the control of symplastic transport. Plant Cell Environ 26:103–124
32. Faulkner C, Akman OE, Bell K et al (2008) Peeking into pit fields: a multiple twinning model of secondary plasmodesmata formation in tobacco. Plant Cell 20:1504–1518

33. Roberts IM, Boevink P, Roberts AG et al (2001) Dynamic changes in the frequency and architecture of plasmodesmata during the sink-source transition in tobacco leaves. Protoplasma 218:31–44
34. Fitzgibbon J, Beck M, Zhou J et al (2013) A developmental framework for complex plasmodesmata formation revealed by large-scale imaging of the Arabidopsis leaf epidermis. Plant Cell 25:57–70
35. Zavaliev R, Dong X, Epel BL (2016) Glycosylphosphatidylinositol (GPI) modification serves as a primary plasmodesmal sorting signal. Plant Physiol 172:1061–1073
36. Cantrill LC, Overall RL, Goodwin PB (1999) Cell-to-cell communication via plant endomembranes. Cell Biol Int 23:653–661
37. Grabski S, De Feijter AW, Schindler M (1993) Endoplasmic reticulum forms a dynamic continuum for lipid diffusion between contiguous soybean root cells. Plant Cell 5:25–38
38. Nicolas WJ, Grison MS, Trepout S et al (2017) Architecture and permeability of post-cytokinesis plasmodesmata lacking cytoplasmic sleeves. Nat Plants 3:17082
39. Siao W, Wang PW, Voigt B et al (2016) Arabidopsis SYT1 maintains stability of cortical endoplasmic reticulum networks and VAP27-1-enriched endoplasmic reticulum-plasma membrane contact sites. J Exp Bot 67:6161–6171
40. Schapire AL, Voigt B, Jasik J et al (2008) Arabidopsis synaptotagmin 1 is required for the maintenance of plasma membrane integrity and cell viability. Plant Cell 20:3374–3388
41. Kriechbaumer V, Botchway SW, Slade SE et al (2015) Reticulomics: protein-protein interaction studies with two plasmodesmata-localized reticulon family proteins identify binding partners enriched at plasmodesmata, endoplasmic reticulum, and the plasma membrane. Plant Physiol 169:1933–1945
42. Levy A, Zheng JY, Lazarowitz SG (2015) Synaptotagmin SYTA forms ER-plasma membrane junctions that are recruited to plasmodesmata for plant virus movement. Curr Biol 25:2018–2025
43. Ishikawa K, Tamura K, Fukao Y et al (2020) Structural and functional relationships between plasmodesmata and plant endoplasmic reticulum-plasma membrane contact sites consisting of three synaptotagmins. New Phytol 226:798–808
44. Fernandez-Busnadiego R, Saheki Y, De Camilli P (2015) Three-dimensional architecture of extended synaptotagmin-mediated endoplasmic reticulum-plasma membrane contact sites. Proc Natl Acad Sci U S A 112: E2004–E2013
45. Lin CC, Seikowski J, Perez-Lara A et al (2014) Control of membrane gaps by synaptotagmin-Ca2+ measured with a novel membrane distance ruler. Nat Commun 5: 5859
46. Liu L, Li C, Liang Z et al (2018) Characterization of multiple C2 domain and transmembrane region proteins in Arabidopsis. Plant Physiol 176:2119–2132
47. Vaddepalli P, Herrmann A, Fulton L et al (2014) The C2-domain protein QUIRKY and the receptor-like kinase STRUBBELIG localize to plasmodesmata and mediate tissue morphogenesis in *Arabidopsis thaliana*. Development 141:4139–4148
48. Glockmann C, Kollmann R (1996) Structure and development of cell connections in the phloem of *Metasequoia glyptostroboides* needles. I. Ultrastructural aspects of modified primary plasmodesmata in Strasburger cells. Protoplasma 193:191–203
49. Robinson-Beers K, Evert RF (1991) Fine structure of plasmodesmata in mature leaves of sugar cane. Planta 184:307–318
50. Waigmann E, Turner A, Peart J et al (1997) Ultrastructural analysis of leaf trichome plasmodesmata reveals major differences from mesophyll plasmodesmata. Planta 203:75–84
51. Chen S, Novick P, Ferro-Novick S (2012) ER network formation requires a balance of the dynamin-like GTPase Sey1p and the Lunapark family member Lnp1p. Nat Cell Biol 14:707–716
52. Tolley N, Sparkes IA, Hunter PR et al (2008) Overexpression of a plant reticulon remodels the lumen of the cortical endoplasmic reticulum but does not perturb protein transport. Traffic 9:94–102
53. Tolley N, Sparkes I, Craddock CP et al (2010) Transmembrane domain length is responsible for the ability of a plant reticulon to shape endoplasmic reticulum tubules in vivo. Plant J 64:411–418
54. Fernandez-Calvino L, Faulkner C, Walshaw J et al (2011) Arabidopsis plasmodesmal proteome. PLoS One 6:e18880
55. Wu SW, Kumar R, Iswanto ABB et al (2018) Callose balancing at plasmodesmata. J Exp Bot 69:5325–5339
56. Rinne PLH, Kaikuranta PM, van der Schoot C (2001) The shoot apical meristem restores its symplasmic organization during chilling-induced release from dormancy. Plant J 26: 249–264

57. Hughes JE, Gunning BES (1980) Glutaraldehyde-induced deposition of callose. Can J Bot 58:250–258
58. Wang X, Sager R, Cui W et al (2013) Salicylic acid regulates plasmodesmata closure during innate immune responses in Arabidopsis. Plant Cell 25:2315–2329
59. Benitez-Alfonso Y, Faulkner C, Pendle A et al (2013) Symplastic intercellular connectivity regulates lateral root patterning. Dev Cell 26:136–147
60. Guseman JM, Lee JS, Bogenschutz NL et al (2010) Dysregulation of cell-to-cell connectivity and stomatal patterning by loss-of-function mutation in Arabidopsis chorus (glucan synthase-like 8). Development 137:1731–1741
61. Han X, Hyun TK, Zhang M et al (2014) Auxin-callose-mediated plasmodesmal gating is essential for tropic auxin gradient formation and signaling. Dev Cell 28:132–146
62. Simpson C, Thomas C, Findlay K et al (2009) An Arabidopsis GPI-anchor plasmodesmal neck protein with callose binding activity and potential to regulate cell-to-cell trafficking. Plant Cell 21:581–594
63. Slewinski TL, Baker RF, Stubert A et al (2012) Tie-dyed2 encodes a callose synthase that functions in vein development and affects symplastic trafficking within the phloem of maize leaves. Plant Physiol 160:1540–1550
64. Zavaliev R, Sagi G, Gera A et al (2010) The constitutive expression of Arabidopsis plasmodesmal-associated class 1 reversibly glycosylated polypeptide impairs plant development and virus spread. J Exp Bot 61:131–142
65. Zavaliev R, Levy A, Gera A et al (2013) Subcellular dynamics and role of Arabidopsis beta-1,3-glucanases in cell-to-cell movement of tobamoviruses. Mol Plant-Microbe Interact 26:1016–1030
66. Gaudioso-Pedraza R, Beck M, Frances L et al (2018) Callose-regulated symplastic communication coordinates symbiotic root nodule development. Curr Biol 28:3562–3577.e6
67. Liu NJ, Zhang T, Liu ZH et al (2020) Phytosphinganine affects plasmodesmata permeability via facilitating PDLP5-stimulated callose accumulation in Arabidopsis. Mol Plant 13:128–143
68. Roberts AG, Santa Cruz S, Roberts IM et al (1997) Phloem unloading in sink leaves of *Nicotiana benthamiana*: comparison of a fluorescent solute with a fluorescent virus. Plant Cell 9:1381–1396
69. Jackson D, Veit B, Hake S (1994) Expression of the maize *KNOTTED1* related homeobox genes in the shoot apical meristem predicts patterns of morphogenesis in the vegetative shoot. Development 120:405–413
70. Lucas WJ, Bouche-Pillon S, Jackson DP et al (1995) Selective trafficking of KNOTTED1 homeodomain protein and its RNA through plasmodesmata. Science 270:1980–1983
71. Sessions A, Yanofsky MF, Weigel D (2000) Cell-cell signaling and movement by the floral transcription factors LEAFY and APETALA1. Science 289:779–782
72. Helariutta Y, Fukaki H, Wysocka-Diller J et al (2000) The SHORT-ROOT gene controls radial patterning of the *Arabidopsis* root through radial signaling. Cell 101:555–567
73. Mezitt LA, Lucas WJ (1996) Plasmodesmal cell-to-cell transport of proteins and nucleic acids. Plant Mol Biol 32:251–273
74. Kragler F, Monzer J, Shash K et al (1998) Cell-to-cell transport of proteins: requirement for unfolding and characterization of binding to a putative plasmodesmal receptor. Plant J 15:367–381
75. Perbal M-C, Haughn G, Saedler H et al (1996) Non-autonomous function of *Antirrhinum* floral homeotic proteins DEFICIENS and GLOBOSA is exerted by their polar cell-to-cell trafficking. Development 122:3433–3441
76. Yoo SC, Chen C, Rojas M et al (2013) Phloem long-distance delivery of FLOWERING LOCUS T (FT) to the apex. Plant J 75: 456–468
77. Crawford KM, Zambryski PC (2000) Subcellular localization determines the availability of non-targeted proteins to plasmodesmatal transport. Curr Biol 10:1032–1040
78. Cui H, Levesque MP, Vernoux T et al (2007) An evolutionarily conserved mechanism delimiting SHR movement defines a single layer of endodermis in plants. Science 316:421–425
79. Schiefelbein J, Huang L, Zheng X (2014) Regulation of epidermal cell fate in Arabidopsis roots: the importance of multiple feedback loops. Front Plant Sci 5:47
80. Bouyer D, Geier F, Kragler F et al (2008) Two-dimensional patterning by a trapping/depletion mechanism: the role of TTG1 and GL3 in Arabidopsis trichome formation. PLoS Biol 6:e141
81. Balkunde R, Bouyer D, Hulskamp M (2011) Nuclear trapping by GL3 controls intercellular transport and redistribution of TTG1

protein in Arabidopsis. Development 138: 5039–5048
82. Daum G, Medzihradszky A, Suzaki T et al (2014) A mechanistic framework for noncell autonomous stem cell induction in Arabidopsis. Proc Natl Acad Sci U S A 111:14619–14624
83. Yadav RK, Perales M, Gruel J et al (2011) WUSCHEL protein movement mediates stem cell homeostasis in the Arabidopsis shoot apex. Genes Dev 25:2025–2030
84. Kim I, Zambryski PC (2005) Cell-to-cell communication via plasmodesmata during Arabidopsis embryogenesis. Curr Opin Plant Biol 8:593–599
85. Kim I, Cho E, Crawford K et al (2005) Cell-to-cell movement of GFP during embryogenesis and early seedling development in Arabidopsis. Proc Natl Acad Sci U S A 102:2227–2231
86. Xu XM, Wang J, Xuan Z et al (2011) Chaperonins facilitate KNOTTED1 cell-to-cell trafficking and stem cell function. Science 333: 1141–1144
87. Kehr J, Kragler F (2018) Long distance RNA movement. New Phytol 218:29–40
88. Carlsbecker A, Lee JY, Roberts CJ et al (2010) Cell signalling by microRNA165/6 directs gene dose-dependent root cell fate. Nature 465:316–321
89. Deom MC, Lapidot M, Beachy RN (1992) Plant virus movement proteins. Cell 69:221–224
90. Lucas WJ (2006) Plant viral movement proteins: agents for cell-to-cell trafficking of viral genomes. Virology 344:169–184
91. Kragler F, Monzer J, Xoconostle-Cazares B et al (2000) Peptide antagonists of the plasmodesmal macromolecular trafficking pathway. EMBO J 19:2856–2868
92. Zhang W, Thieme CJ, Kollwig G et al (2016) tRNA-related sequences trigger systemic mRNA transport in plants. Plant Cell 28: 1237–1249
93. Ham BK, Brandom JL, Xoconostle-Cazares B et al (2009) A polypyrimidine tract binding protein, pumpkin RBP50, forms the basis of a phloem-mobile ribonucleoprotein complex. Plant Cell 21:197–215
94. Huang NC, Yu TS (2009) The sequences of Arabidopsis GA-INSENSITIVE RNA constitute the motifs that are necessary and sufficient for RNA long-distance trafficking. Plant J 59:921–929
95. Cho SK, Sharma P, Butler NM et al (2015) Polypyrimidine tract-binding proteins of potato mediate tuberization through an interaction with StBEL5 RNA. J Exp Bot 66:6835–6847
96. Yang L, Perrera V, Saplaoura E et al (2019) m (5)C methylation guides systemic transport of messenger rna over graft junctions in plants. Curr Biol 29:2465–2476.e5
97. Zipfel C, Oldroyd GE (2017) Plant signalling in symbiosis and immunity. Nature 543:328–336
98. Chinchilla D, Zipfel C, Robatzek S et al (2007) A flagellin-induced complex of the receptor FLS2 and BAK1 initiates plant defence. Nature 448:497–500
99. Heese A, Hann DR, Gimenez-Ibanez S et al (2007) The receptor-like kinase SERK3/BAK1 is a central regulator of innate immunity in plants. Proc Natl Acad Sci U S A 104: 12217–12222
100. Schulze B, Mentzel T, Jehle AK et al (2010) Rapid heteromerization and phosphorylation of ligand-activated plant transmembrane receptors and their associated kinase BAK1. J Biol Chem 285:9444–9451
101. Roux M, Schwessinger B, Albrecht C et al (2011) The Arabidopsis leucine-rich repeat receptor-like kinases BAK1/SERK3 and BKK1/SERK4 are required for innate immunity to hemibiotrophic and biotrophic pathogens. Plant Cell 23:2440–2455
102. Schwessinger B, Roux M, Kadota Y et al (2011) Phosphorylation-dependent differential regulation of plant growth, cell death, and innate immunity by the regulatory receptor-like kinase BAK1. PLoS Genet 7:e1002046
103. Sun Y, Li L, Macho AP et al (2013) Structural basis for flg22-induced activation of the Arabidopsis FLS2-BAK1 immune complex. Science 342:624–628
104. Xu B, Cheval C, Laohavisit A et al (2017) A calmodulin-like protein regulates plasmodesmal closure during bacterial immune responses. New Phytol 215:77–84
105. Faulkner C, Petutschnig E, Benitez-Alfonso Y et al (2013) LYM2-dependent chitin perception limits molecular flux via plasmodesmata. Proc Natl Acad Sci U S A 110:9166–9170
106. Kadota Y, Sklenar J, Derbyshire P et al (2014) Direct regulation of the NADPH oxidase RBOHD by the PRR-associated kinase BIK1 during plant immunity. Mol Cell 54:43–55
107. Castro B, Citterico M, Kimura S et al (2021) Stress-induced reactive oxygen species compartmentalization, perception and signalling. Nat Plants 7:403–412
108. Thor K, Jiang S, Michard E et al (2020) The calcium-permeable channel OSCA1.3

regulates plant stomatal immunity. Nature 585:569–573
109. Lee JY, Wang X, Cui W et al (2011) A plasmodesmata-localized protein mediates crosstalk between cell-to-cell communication and innate immunity in Arabidopsis. Plant Cell 23:3353–3373
110. Cheval C, Samwald S, Johnston MG et al (2020) Chitin perception in plasmodesmata characterizes submembrane immune-signaling specificity in plants. Proc Natl Acad Sci U S A 117:9621–9629
111. Mou Z, Fan W, Dong X (2003) Inducers of plant systemic acquired resistance regulate NPR1 function through redox changes. Cell 113:935–944
112. Carella P, Isaacs M, Cameron RK (2015) Plasmodesmata-located protein overexpression negatively impacts the manifestation of systemic acquired resistance and the long-distance movement of defective in induced Resistance1 in Arabidopsis. Plant Biol (Stuttg) 17:395–401
113. Lim GH, Shine MB, de Lorenzo L et al (2016) Plasmodesmata localizing proteins regulate transport and signaling during systemic acquired immunity in plants. Cell Host Microbe 19:541–549
114. Benitez-Alfonso Y, Jackson D, Maule A (2011) Redox regulation of intercellular transport. Protoplasma 248:131–140
115. Stonebloom S, Brunkard JO, Cheung AC et al (2012) Redox states of plastids and mitochondria differentially regulate intercellular transport via plasmodesmata. Plant Physiol 158:190–199
116. Faulkner C (2013) Receptor-mediated signaling at plasmodesmata. Front Plant Sci 4:521
117. Stahl Y, Faulkner C (2016) Receptor complex mediated regulation of symplastic traffic. Trends Plant Sci 21:450–459
118. Liu T, Liu Z, Song C et al (2012) Chitin-induced dimerization activates a plant immune receptor. Science 336:1160–1164
119. Grison MS, Kirk P, Brault ML et al (2019) Plasma membrane-associated receptor-like kinases relocalize to plasmodesmata in response to osmotic stress. Plant Physiol 181:142–160
120. Hunter K, Kimura S, Rokka A et al (2019) CRK2 enhances salt tolerance by regulating callose deposition in connection with PLDalpha1. Plant Physiol 180:2004–2021
121. Aung K, Kim P, Li Z et al (2020) Pathogenic bacteria target plant plasmodesmata to colonize and invade surrounding tissues. Plant Cell 32:595–611
122. Tomczynska I, Stumpe M, Doan TG et al (2020) A Phytophthora effector protein promotes symplastic cell-to-cell trafficking by physical interaction with plasmodesmata-localised callose synthases. New Phytol 227:1467–1478
123. Li Z, Variz H, Chen Y et al (2021) Plasmodesmata-dependent intercellular movement of bacterial effectors. Front Plant Sci 12:640277
124. Sakulkoo W, Oses-Ruiz M, Oliveira Garcia E et al (2018) A single fungal MAP kinase controls plant cell-to-cell invasion by the rice blast fungus. Science 359:1399–1403
125. Cao L, Blekemolen MC, Tintor N et al (2018) The *Fusarium oxysporum* Avr2-Six5 effector pair alters plasmodesmatal exclusion selectivity to facilitate cell-to-cell movement of Avr2. Mol Plant 11:691–705
126. Di X, Gomila J, Ma L et al (2016) Uptake of the fusarium effector Avr2 by tomato is not a cell autonomous event. Front Plant Sci 7:1915
127. Di X, Cao L, Hughes RK et al (2017) Structure-function analysis of the Fusarium oxysporum Avr2 effector allows uncoupling of its immune-suppressing activity from recognition. New Phytol 216:897–914
128. Wu JH, Dimitman JE (1970) Leaf structure and callose formation as determinants of TMV movement in bean leaves as revealed by UV irradiation studies. Virology 40:820–827
129. Moore AE, Stone BA (1972) Effect of infection with TMV and other viruses on the level of a −1,3-glucan hydrolase in leaves of *Nicotiana glutinosa*. Virology 50:791–798
130. Leisner SM, Turgeon R (1993) Movement of virus and photoassimilate in the phloem: a comparative analysis. BioEssays 15:741–748
131. Beffa RS, Hofer RM, Thomas M et al (1996) Decreased susceptibility to viral disease of [beta]-1,3-glucanase-deficient plants generated by antisense transformation. Plant Cell 8:1001–1011
132. Allison AV, Shalla TA (1973) The ultrastructure of local lesions induced by *Potato virus X*: a sequence of cytological events in the course of infection. Phytopathology 64:784–793
133. Pennazio S, D'Agostino G, Appiano A et al (1978) Ultrastructure and histochemistry of the resistant tissue surrounding lesions of *Tomato bushy stunt virus* in *Gomphrena globosa* leaves. Physiol Plant Pathol 13:165–168
134. Dong X, Hong Z, Chatterjee J et al (2008) Expression of callose synthase genes and its

connection with Npr1 signaling pathway during pathogen infection. Planta 229:87–98
135. Huang D, Sun Y, Ma Z et al (2019) Salicylic acid-mediated plasmodesmal closure via Remorin-dependent lipid organization. Proc Natl Acad Sci U S A 116:21274–21284
136. Jarsch IK, Konrad SS, Stratil TF et al (2014) Plasma membranes are subcompartmentalized into a plethora of coexisting and diverse microdomains in Arabidopsis and *Nicotiana benthamiana*. Plant Cell 26:1698–1711
137. Konrad SS, Popp C, Stratil TF et al (2014) S-acylation anchors remorin proteins to the plasma membrane but does not primarily determine their localization in membrane microdomains. New Phytol 203:758–769
138. Martinez D, Legrand A, Gronnier J et al (2019) Coiled-coil oligomerization controls localization of the plasma membrane REMORINs. J Struct Biol 206:12–19
139. Zhou T, Murphy AM, Lewsey MG et al (2014) Domains of the *Cucumber mosaic virus* 2b silencing suppressor protein affecting inhibition of salicylic acid-induced resistance and priming of salicylic acid accumulation during infection. J Gen Virol 95:1408–1413
140. Raffaele S, Bayer E, Lafarge D et al (2009) Remorin, a solanaceae protein resident in membrane rafts and plasmodesmata, impairs *Potato virus X* movement. Plant Cell 21: 1541–1555
141. Perraki A, Gronnier J, Gouguet P et al (2018) REM1.3's phospho-status defines its plasma membrane nanodomain organization and activity in restricting PVX cell-to-cell movement. PLoS Pathog 14:e1007378
142. Fu S, Xu Y, Li C et al (2018) *Rice stripe virus* interferes with S-acylation of remorin and induces its autophagic degradation to facilitate virus infection. Mol Plant 11:269–287
143. Cheng G, Yang Z, Zhang H et al (2020) Remorin interacting with PCaP1 impairs *Turnip mosaic virus* intercellular movement but is antagonised by VPg. New Phytol 225:2122–2139
144. Sasaki N, Takashima E, Nyunoya H (2018) Altered subcellular localization of a tobacco membrane raft-associated remorin protein by tobamovirus infection and transient expression of viral replication and movement proteins. Front Plant Sci 9:619
145. Niehl A, Heinlein M (2011) Cellular pathways for viral transport through plasmodesmata. Protoplasma 248:75–99
146. Heinlein M (2015) Plasmodesmata: channels for viruses on the move. Methods Mol Biol 1217:25–52
147. Amari K, Boutant E, Hofmann C et al (2010) A family of plasmodesmal proteins with receptor-like properties for plant viral movement proteins. PLoS Pathog 6:e1001119
148. Lekkerkerker A, Wellink J, Yuan P et al (1996) Distinct functional domains in the *Cowpea mosaic virus* movement protein. J Virol 70: 5658–5661
149. Belin C, Schmitt C, Gaire F et al (1999) The nine C-terminal residues of the grapevine fanleaf nepovirus movement protein are critical for systemic virus spread. J Gen Virol 80 (Pt 6):1347–1356
150. Schoelz JE, Angel CA, Nelson RS et al (2016) A model for intracellular movement of *Cauliflower mosaic virus*: the concept of the mobile virion factory. J Exp Bot 67:2039–2048
151. Wolf S, Deom CM, Beachy RN et al (1989) Movement protein of *Tobacco mosaic virus* modifies plasmodesmatal size exclusion limit. Science 246:377–379
152. Waigmann E, Lucas W, Citovsky V et al (1994) Direct functional assay for *Tobacco mosaic virus* cell-to-cell movement protein and identification of a domain involved in increasing plasmodesmal permeability. Proc Natl Acad Sci U S A 91:1433–1437
153. Vogler H, Kwon MO, Dang V et al (2008) *Tobacco mosaic virus* movement protein enhances the spread of RNA silencing. PLoS Pathog 4:e1000038
154. Oparka KJ, Prior DAM, Santa Cruz S et al (1997) Gating of epidermal plasmodesmata is restricted to the leading edge of expanding infection sites of *Tobacco mosaic virus*. Plant J 12:781–789
155. Kim I, Kobayashi K, Cho E et al (2005) Subdomains for transport via plasmodesmata corresponding to the apical-basal axis are established during Arabidopsis embryogenesis. Proc Natl Acad Sci U S A 102:11945–11950
156. Haley A, Hunter T, Kiberstis P et al (1995) Multiple serine phosphorylation sites on the 30 kDa TMV cell-to-cell movement protein synthesized in tobacco protoplasts. Plant J 8: 715–724
157. Tyulkina LG, Karger EM, Sheveleva AA et al (2010) Binding of monoclonal antibodies to the movement protein (MP) of *Tobacco mosaic virus*: influence of subcellular MP localization and phosphorylation. J Gen Virol 91:1621–1628
158. Kawakami S, Padgett HS, Hosokawa D et al (1999) Phosphorylation and/or presence of serine 37 in the movement protein of tomato mosaic tobamovirus is essential for

intracellular localization and stability *in vivo*. J Virol 73:6831–6840

159. Watanabe Y, Meshi T, Okada Y (1992) *In vivo* phosphorylation of the 30-kDa protein of *Tobacco mosaic virus*. FEBS Lett 313:181–184
160. Citovsky V, McLean BG, Zupan JR et al (1993) Phosphorylation of *Tobacco mosaic virus* cell-to-cell movement protein by a developmentally regulated plant cell wall-associated protein kinase. Genes Dev 7:904–910
161. Ding B, Haudenshield JS, Hull RJ et al (1992) Secondary plasmodesmata are specific sites of localization of the *tobacco mosaic virus* movement protein in transgenic tobacco plants. Plant Cell 4:915–928
162. Moore PJ, Fenczik CA, Deom CM et al (1992) Developmental changes in plasmodesmata in transgenic tobacco expressing the movement protein of *Tobacco mosaic virus*. Protoplasma 170:115–127
163. Heinlein M, Wood MR, Thiel T et al (1998) Targeting and modification of prokaryotic cell-cell junctions by *Tobacco mosaic virus* cell-to-cell movement protein. Plant J 14: 345–351
164. Bucher GL, Tarina C, Heinlein M et al (2001) Local expression of enzymatically active class 1 beta-1,3-glucanase enhances symptoms of TMV infection in tobacco. Plant J 28:361–369
165. Iglesias VA, Meins F Jr (2000) Movement of plant viruses is delayed in a β-1,3-glucanase-deficient mutant showing a reduced plasmodesmatal size exclusion limit and enhanced callose deposition. Plant J 21:157–166
166. Guenoune-Gelbart D, Elbaum M, Sagi G et al (2008) *Tobacco mosaic virus* (TMV) replicase and movement protein function synergistically in facilitating TMV spread by lateral diffusion in the plasmodesmal desmotubule of *Nicotiana benthamiana*. Mol Plant-Microbe Interact 21:335–345
167. Levy A, Erlanger M, Rosenthal M et al (2007) A plasmodesmata-associated beta-1,3-glucanase in Arabidopsis. Plant J 49:669–682
168. Ueki S, Spektor R, Natale DM et al (2010) ANK, a host cytoplasmic receptor for the *Tobacco mosaic virus* cell-to-cell movement protein, facilitates intercellular transport through plasmodesmata. PLoS Pathog 6: e1001201
169. Fridborg I, Grainger J, Page A et al (2003) TIP, a novel host factor linking callose degradation with the cell-to-cell movement of *Potato virus X*. Mol Plant-Microbe Interact 16:132–140
170. Lewis JD, Lazarowitz SG (2010) Arabidopsis synaptotagmin SYTA regulates endocytosis and virus movement protein cell-to-cell transport. Proc Natl Acad Sci U S A 107:2491–2496
171. Uchiyama A, Shimada-Beltran H, Levy A et al (2014) The Arabidopsis synaptotagmin SYTA regulates the cell-to-cell movement of diverse plant viruses. Front Plant Sci 5:584
172. Thomas CL, Bayer EM, Ritzenthaler C et al (2008) Specific targeting of a plasmodesmal protein affecting cell-to-cell communication. PLoS Biol 6:e7
173. Tilsner J, Linnik O, Louveaux M et al (2013) Replication and trafficking of a plant virus are coupled at the entrances of plasmodesmata. J Cell Biol 201:981–995
174. Padgett HS, Epel BL, Kahn TW et al (1996) Distribution of tobamovirus movement protein in infected cells and implications for cell-to-cell spread of infection. Plant J 10:1079–1088
175. Heinlein M (2015) Plant virus replication and movement. Virology 479-480:657–671
176. Heinlein M, Padgett HS, Gens JS et al (1998) Changing patterns of localization of the *Tobacco mosaic virus* movement protein and replicase to the endoplasmic reticulum and microtubules during infection. Plant Cell 10: 1107–1120
177. Heinlein M (2002) The spread of *Tobacco mosaic virus* infection: insights into the cellular mechanism of RNA transport. Cell Mol Life Sci 59:58–82
178. Niehl A, Pasquier A, Ferriol I et al (2014) Comparison of the *Oilseed rape mosaic virus* and *Tobacco mosaic virus* movement proteins (MP) reveals common and dissimilar MP functions for tobamovirus spread. Virology 456-457:43–54
179. Niehl A, Pena EJ, Amari K et al (2013) Microtubules in viral replication and transport. Plant J 75:290–308
180. Pont-Lezica RF, McNally JG, Pickard BG (1993) Wall-to-membrane linkers in onion epidermis: some hypotheses. Plant Cell Environ 16:111–123
181. Oparka KJ (1994) Plasmolysis: new insights into an old process. New Phytol 126:571–591
182. Ishikawa K, Tamura K, Ueda H et al (2018) Synaptotagmin-associated endoplasmic reticulum-plasma membrane contact sites are localized to immobile er tubules. Plant Physiol 178:641–653

183. Brandner K, Sambade A, Boutant E et al (2008) *Tobacco mosaic virus* movement protein interacts with green fluorescent protein-tagged microtubule end-binding protein 1. Plant Physiol 147:611–623
184. Sambade A, Brandner K, Hofmann C et al (2008) Transport of TMV movement protein particles associated with the targeting of RNA to plasmodesmata. Traffic 9:2073–2088
185. Wang P, Hawkins TJ, Richardson C et al (2014) The plant cytoskeleton, NET3C, and VAP27 mediate the link between the plasma membrane and endoplasmic reticulum. Curr Biol 24:1397–1405
186. Heinlein M, Epel BL, Padgett HS et al (1995) Interaction of tobamovirus movement proteins with the plant cytoskeleton. Science 270:1983–1985
187. Ashby J, Boutant E, Seemanpillai M et al (2006) *Tobacco mosaic virus* movement protein functions as a structural microtubule-associated protein. J Virol 80:8329–8344
188. Boyko V, Ashby JA, Suslova E et al (2002) Intramolecular complementing mutations in tobacco mosaic virus movement protein confirm a role for microtubule association in viral RNA transport. J Virol 76:3974–3980
189. Boyko V, Ferralli J, Heinlein M (2000) Cell-to-cell movement of TMV RNA is temperature-dependent and corresponds to the association of movement protein with microtubules. Plant J 22:315–325
190. Boyko V, Hu Q, Seemanpillai M et al (2007) Validation of microtubule-associated *Tobacco mosaic virus* RNA movement and involvement of microtubule-aligned particle trafficking. Plant J 51:589–603
191. Boyko V, van der Laak J, Ferralli J et al (2000) Cellular targets of functional and dysfunctional mutants of *Tobacco mosaic virus* movement protein fused to green fluorescent protein. J Virol 74:11339–11346
192. Ouko MO, Sambade A, Brandner K et al (2010) Tobacco mutants with reduced microtubule dynamics are less susceptible to TMV. Plant J 62:829–839
193. Peña EJ, Ferriol I, Sambade A et al (2014) Experimental virus evolution reveals a role of plant microtubule dynamics and TORTIFOLIA1/SPIRAL2 in RNA trafficking. PLoS One 9:e105364
194. Boyko V, Ferralli J, Ashby J et al (2000) Function of microtubules in intercellular transport of plant virus RNA. Nat Cell Biol 2:826–832
195. Wang P, Richardson C, Hawkins TJ et al (2016) Plant VAP27 proteins: domain characterization, intracellular localization and role in plant development. New Phytol 210:1311–1326
196. Pena EJ, Robles Luna G, Heinlein M (2020) In vivo imaging of tagged mRNA in plant tissues using the bacterial transcriptional antiterminator BglG. Plant J 105:271–282
197. Amari K, Di Donato M, Dolja VV et al (2014) Myosins VIII and XI play distinct roles in reproduction and transport of *Tobacco mosaic virus*. PLoS Pathog 10:e1004448
198. Harries PA, Park JW, Sasaki N et al (2009) Differing requirements for actin and myosin by plant viruses for sustained intercellular movement. Proc Natl Acad Sci U S A 106: 17594–17599
199. Hofmann C, Niehl A, Sambade A et al (2009) Inhibition of *Tobacco mosaic virus* movement by expression of an actin-binding protein. Plant Physiol 149:1810–1823
200. Golomb L, Abu-Abied M, Belausov E et al (2008) Different subcellular localizations and functions of Arabidopsis myosin VIII. BMC Plant Biol 8:3
201. Tilsner J, Amari K, Torrance L (2011) Plasmodesmata viewed as specialised membrane adhesion sites. Protoplasma 248:39–60
202. Reichelt S, Knight AE, Hodge TP et al (1999) Characterization of the unconventional myosin VIII in plant cells and its localization at the post-cytokinetic cell wall. Plant J 19:555–567
203. Wu SZ, Bezanilla M (2014) Myosin VIII associates with microtubule ends and together with actin plays a role in guiding plant cell division. elife 3:e03498
204. Radford JE, White RG (2011) Inhibitors of myosin, but not actin, alter transport through Tradescantia plasmodesmata. Protoplasma 248:205–216
205. Su S, Liu Z, Chen C et al (2010) *Cucumber mosaic virus* movement protein severs actin filaments to increase the plasmodesmal size exclusion limit in tobacco. Plant Cell 22: 1373–1387
206. Pitzalis N, Heinlein M (2017) The roles of membranes and associated cytoskeleton in plant virus replication and cell-to-cell movement. J Exp Bot 69:117–132
207. Schoelz JE, Harries PA, Nelson RS (2011) Intracellular transport of plant viruses: finding the door out of the cell. Mol Plant 4: 813–831
208. Wang A (2015) Dissecting the molecular network of virus-plant interactions: the complex roles of host factors. Annu Rev Phytopathol 53:45–66

209. Martin-Hernandez AM, Baulcombe DC (2008) *Tobacco rattle virus* 16-kilodalton protein encodes a suppressor of RNA silencing that allows transient viral entry in meristems. J Virol 82:4064–4071
210. Pitzalis N, Amari K, Graindorge S et al (2020) *Turnip mosaic virus* in oilseed rape activates networks of sRNA-mediated interactions between viral and host genomes. Commun Biol 3:702
211. Shimura H, Pantaleo V, Ishihara T et al (2011) A viral satellite RNA induces yellow symptoms on tobacco by targeting a gene involved in chlorophyll biosynthesis using the RNA silencing machinery. PLoS Pathog 7:e1002021
212. Smith NA, Eamens AL, Wang MB (2011) Viral small interfering RNAs target host genes to mediate disease symptoms in plants. PLoS Pathog 7:e1002022
213. Navarro B, Gisel A, Rodio ME et al (2012) Small RNAs containing the pathogenic determinant of a chloroplast-replicating viroid guide the degradation of a host mRNA as predicted by RNA silencing. Plant J 70:991–1003
214. Miozzi L, Gambino G, Burgyan J et al (2012) Genome-wide identification of viral and host transcripts targeted by viral siRNAs in *Vitis vinifera*. Mol Plant Pathol 14:30–43
215. Schwach F, Vaistij FE, Jones L et al (2005) An RNA-dependent RNA polymerase prevents meristem invasion by *Potato virus X* and is required for the activity but not the production of a systemic silencing signal. Plant Physiol 138:1842–1852
216. Kørner CJ, Pitzalis N, Peña EJ et al (2018) Crosstalk between PTGS and TGS pathways in natural antiviral immunity and disease recovery. Nat Plants 4:157–164
217. Ghoshal B, Sanfacon H (2015) Symptom recovery in virus-infected plants: revisiting the role of RNA silencing mechanisms. Virology 479-480:167–179
218. Paudel DB, Sanfacon H (2018) Exploring the diversity of mechanisms associated with plant tolerance to virus infection. Front Plant Sci 9:1575
219. Rosas-Diaz T, Zhang D, Fan P et al (2018) A virus-targeted plant receptor-like kinase promotes cell-to-cell spread of RNAi. Proc Natl Acad Sci U S A 115:1388–1393
220. Carluccio AV, Prigigallo MI, Rosas-Diaz T et al (2018) S-acylation mediates *Mungbean yellow mosaic virus* AC4 localization to the plasma membrane and in turns gene silencing suppression. PLoS Pathog 14:e1007207
221. Garnelo Gomez B, Rosas-Diaz T, Shi C et al (2021) The viral silencing suppressor P19 interacts with the receptor-like kinases BAM1 and BAM2 and suppresses the cell-to-cell movement of RNA silencing independently of its ability to bind sRNA. New Phytol 229:1840–1843
222. Vu MH, Iswanto ABB, Lee J et al (2020) The role of plasmodesmata-associated receptor in plant development and environmental response. Plants (Basel) 9:216
223. Niehl A, Wyrsch I, Boller T et al (2016) Double-stranded RNAs induce a pattern-triggered immune signaling pathway in plants. New Phytol 211:1008–1019
224. Kørner CJ, Klauser D, Niehl A et al (2013) The immunity regulator BAK1 contributes to resistance against diverse RNA viruses. Mol Plant-Microbe Interact 26:1271–1280
225. Yu H, Liu Y, Gulbranson DR et al (2016) Extended synaptotagmins are $Ca2^{+}$-dependent lipid transfer proteins at membrane contact sites. Proc Natl Acad Sci U S A 113:4362–4367
226. Yuan C, Lazarowitz SG, Citovsky V (2018) The plasmodesmal localization signal of tmv mp is recognized by plant synaptotagmin SYTA. mBio 9:e01314–e01318
227. Diaz A, Wang X, Ahlquist P (2010) Membrane-shaping host reticulon proteins play crucial roles in viral RNA replication compartment formation and function. Proc Natl Acad Sci U S A 107:16291–16296

Part II

Structure and Distribution of Plasmodesmata

Chapter 3

Plasmodesmata Ultrastructure Determination Using Electron Tomography

Jules D. Petit, Marie Glavier, Lysiane Brocard, and Emmanuelle M. F. Bayer

Abstract

Plant plasmodesmata (PD) are complex intercellular channels consisting of a thin endoplasmic reticulum (ER) tubule enveloped by the plasma membrane (PM). PD were first observed by electron microscopy about 50 years ago and, since, numerous studies in transmission and scanning electron microscopy have provided important information regarding their overall organization, revealing at the same time their diversity in terms of structure and morphology. However, and despite the fact that PD cell-cell communication is of critical importance for plant growth, development, cellular patterning, and response to biotic and abiotic stresses, linking their structural organization to their functional state has been proven difficult. This is in part due to their small size (20–50 nm in diameter) and the difficulty to resolve these structures in three dimensions at nanometer resolution to provide details of their internal organization.

In this protocol, we provide in detail a complete process to produce high-resolution transmission electron tomograms of PD. We describe the preparation of the plant sample using high-pressure cryofixation and cryo-substitution. We also describe how to prepare filmed grids and how to cut and collect the sections using an ultramicrotome. We explain how to acquire a tilt series and how to reconstruct a tomogram from it using the IMOD software. We also give a few guidelines on segmentation of the reconstructed tomogram.

Key words Plasmodesmata, Ultrastructure, Electron tomography, Cryofixation, Segmentation

1 Introduction

In order to obtain information about plasmodesmata (PD) distribution and architecture in different plant tissues, several protocols were established using electron microscopy-based techniques [1–5]. In the case of transmission electron microscopy, protocols involve chemical fixation [1] or cryofixation and cryo-substitution [2]. However, most of these protocols are not suitable to access close-to-native inner structures of PD. Chemical fixations can be very handy for researchers as it allows the preparation of

Yoselin Benitez-Alfonso and Manfred Heinlein (eds.), *Plasmodesmata: Methods and Protocols*,
Methods in Molecular Biology, vol. 2457, https://doi.org/10.1007/978-1-0716-2132-5_3,

many samples in a relatively short time, but it is also impacting the cell physiology. High-pressure freezing and freeze-substitution are much more adequate for the cell preservation, but the choice of resin is very important to optimize the observation of the PD architecture embedded in the cell wall.

On the other hand, the development of cryogenic electron microscopy where biological samples are cryofixed by vitrification (embedded in vitreous water) and imaged by cryo-electron tomography is very promising to uncover the extremely fine structural elements of PD and macromolecular organization, in a near-native state condition. However, despite their attractiveness, these approaches have, so far, mainly been applied to single particles and biomacromolecules [6, 7] and remain challenging for complex plant tissues. In addition, performing cryo-electron microscopy requires expensive equipment that will not be available to all laboratories.

In this chapter, we propose a methodology that relies on high-pressure freezing, freeze-substitution, and low-temperature staining, to study the inner architecture of PD using transmission electron tomography. This protocol is accessible to most researchers and can be used as a routine tool. We made the choice to describe the entire process, from sample preparation to tomogram analysis and the construction of three-dimensional models, to give a complete guideline of what is necessary material-wise as well as time involvement-wise. It is important to note that adaptations can be made, depending on the available equipment, especially for the tilt series acquisition and the use of the tomogram reconstruction and segmentation software. IMOD software [8] information and tutorials are easily accessible (https://bio3d.colorado.edu/imod/), making it possible to optimize usage at best depending on the dataset and the computational equipment.

This protocol was first described in Nicolas et al., Nature Plant (2017) [9], and Nicolas et al., Bio-protocols (2018) [10], and we acknowledge its complementarity with the current version, mostly by means of the visual support.

2 Materials

2.1 Equipment

1. Liquid nitrogen container and protective gear.
2. Air compressor.
3. Leica EMPACT1 machine.
4. Regular biomolecular pipets and tips from 2 μL to 1000 μL.
5. Binocular with transmitted light with a high working distance (more than 15 cm).

6. Binocular with lighting from above with a high working distance (more than 15 cm).
7. Heating surface/bench.
8. UV light adapted for AFS2.
9. Leica Automatic Freeze-Substitution 2 (EM AFS2) machine.
10. Ventilated chemical hood.
11. Bloc holder.
12. Diamond knives (Trim20 dry, Ultra 45° wet, HISTO 45° wet).
13. Leica Ultracut UC7.
14. Tabletop microscope.
15. Rubber bellow.
16. Computer with sufficient CPU (and GPU, optional) power.
17. Graphic tablet (optional).

2.2 High-Pressure Freezing

1. Liquid nitrogen.
2. Insulated tweezers.
3. Leica loading system with associated platelet holder.
4. 100 μm deep 1.5 mm diameter copper membrane carriers.
5. Sample transfer metal containers.
6. Pod holder.
7. Pods.
8. Torque screwdriver.
9. Glass slides.
10. Wooden toothpicks.
11. Razor blades.
12. 1/2 Murashige-Skoog (MS) liquid media (the same used for plant material growth).
13. 20% bovine serum albumin (BSA) solution in liquid MS medium (the same used for plant material growth).

2.3 Freeze-Substitution

1. Screw top 2 mL tubes.
2. Disposable plastic pipets of 1.5, 2, and 3 mL and thin tip pipets of 1 mL.
3. Glass sample vials with snap-cap.
4. Cartridge half mask.
5. Plastic pills Leica molds and plastic mold containers.
6. Plastic solvent containers with screw tops for mix with OsO4.
7. Glass containers (for dehydration solutions).
8. Aluminum foil.

9. Uranyl acetate solution 20% in pure methanol (store in dark at −20 °C).
10. Glutaraldehyde 10% EM-grade anhydrous solution in acetone.
11. Ultrapure 100% ethanol.
12. Ultrapure 100% acetone.
13. Osmium tetroxide 0.1 g vials.
14. Nail polish.
15. AFS tube holders.
16. Metal cups for holding cryomix tubes.
17. Metal socket for mold containers for polymerization.
18. Screwdriver for tube/containers transfer.
19. EMS microneedle and Ted Pella Inc. ultramicroneedle tools.
20. Liquid nitrogen.

2.4 Resin Embedding

1. HM20 resin EMS.

2.5 Ultramicrotomy

1. Glass crystallization dish.
2. Glass wand.
3. Fast-absorbent Whatman paper filters.
4. Copper grids (200 and/or 300 mesh thin bars).
5. Glass Pasteur pipet.
6. 0.2 μm syringe filters.
7. Syringe.
8. 0.5 mL Eppendorf tubes.
9. Solid Parlodion.
10. Isoamyl-acetate.
11. 0.05% toluidine blue solution in water.
12. MilliQ water.
13. Petri dishes of various sizes.
14. Grid containers.
15. EM-grade precision tweezers of various styles (2, 5, 5X, 7).
16. Razor blades.
17. Dog hair mounted on a long wooden stick.

2.6 Electron Microscopy

1. Electron microscope.
2. Tomo grid holder.
3. Screwdriver for the grid holder clip screws.
4. EM-grade tweezers.
5. Parafilm.

6. 5 nm colloidal gold solution.
7. BSA 0.5% in filtered MilliQ water.

2.7 Software

1. Tecnai Imaging and Analysis (TIA) software.
2. Xplore3D (FEI).
3. IMOD (Etomo and 3Dmod).

3 Methods

3.1 High-Pressure Freezing

1. Start the Leica EMPACT1 and prepare it for usage according to the user manual (*see* **Note 1**).
2. Once the system is ready (working pressure 4.8–5 bars), shoot an empty pod to remove potential air from the machine circuit.
3. Place a membrane carrier on the Leica loading device and fill it with 20% BSA in liquid MS medium using a 0.5–2 μL pipet (*see* **Note 2**) while avoiding air bubbles.
4. In the shortest time lapse possible (around 1–2 min): under a binocular with transmitted light, prepare a glass slide with a drop of MS 1/2 solution in order to place the *Arabidopsis* seedling root. Cut the root tip (~1–1.5 mm) with a sharp razor blade and carefully slide it up using a wooden toothpick. Drop the root by slightly moving the toothpick in the BSA contained in the membrane carrier (*see* **Note 3**).
5. Push the membrane carrier in the pod, which is positioned in its allocated socket of the Leica loading system. Screw the sapphire tight on the carrier with the TORX screwdriver (set on 2.5 cN.m-1) (*see* **Note 4**). Screw the white pod holder.
6. Load the ensemble in the Leica EMPACT loading stage and insert the piston. On the screen, tap "prepare," "lock," and then "start." The pod and the sample fall in the liquid nitrogen bath.
7. Verify the curves by checking that the maximum pressure (>1900 bars) is reached ~20 ms before the sample is at final temperature (*see* **Note 5**).
8. From this step, the sample has to be maintained in liquid nitrogen to avoid ice crystallization (occurs between −90 and −80 °C at standard atmosphere pressure). Carefully place the pod holder in the special socket present in the liquid nitrogen bath and unscrew it. Unscrew the clamping from the sapphire to release it and gently tap out the membrane carrier in the bath. Use pre-cooled insulated tweezers to manipulate the membrane carrier and place it in the metal sample container already present in the bath (*see* **Note 6**).

9. To avoid the blocking of the screw parts due to the freezing of residual water, use a heating surface or a hair dryer to remove all water on the pod and pod holder between each sample. Repeat the process for each sample.
10. When all the samples are ready to be transferred, place a metal lid on the sample container and quickly transfer the container to the already prepared, pre-cooled (−90 °C), and running AFS2 machine (*see* **Note 7**).

3.2 Freeze-Substitution

During all steps, solutions, containers, and tools need to be pre-cooled before entering in contact with samples. Exchange solutions slowly to avoid the loss of samples.

1. In a chemical hood, prepare the cryomix consisting of glutaraldehyde 0.5%, uranyl acetate 0.1%, and osmium tetroxide 2% in pure acetone. To avoid osmium tetroxide reaction, immediately dispatch 200–500 μL of the mix in screw top tubes (labeled with a permanent market) and dip them vertically in liquid nitrogen (*see* **Note 8**).
2. Place the cryomix tubes in the AFS well, in the metal cups, and dispatch one to three membrane carriers containing the frozen samples in every tube, as the mix is still in a solid state. Close the tubes with the screw top. The samples will sink as the mix thaw while reaching −90 °C.
3. Set the AFS program at −90 °C for 48 h, followed by a controlled rising of +3 °C/h for 13 h and a standby at −50 °C. To avoid uranyl acetate precipitation, ensure that no light reaches the samples by using an opaque lid or aluminum foil to cover the lid of the AFS2 (*see* **Note 9**).
4. When the AFS has reached −50 °C and the samples were infiltrated by the cryomix for around 10 h at this temperature, wash thoroughly every tube three times with ultrapure acetone and then three times with ultrapure ethanol (*see* **Note 10**). For the last wash, transfer the samples into a plastic container (or more if several conditions) containing ultrapure ethanol. Label the containers beforehand with several points of nail polish according to the different conditions (condition 1 with no point, condition 2 with 1 points, etc.)
5. Using the microtools, carefully unmold the frozen sample in BSA from the membrane carrier. To do so, gently poke the BSA at the carrier-sample interface, all around, several times (*see* **Note 11**).
6. Fill the plastic embedding molds with cold ultrapure ethanol and let them cool down to −50 °C, and then gently transfer the detached sample in the mold with a 3 mL plastic pipet. Sample might need to be repositioned in the center of the embedding

mold. Label the molds beforehand with nail polish according to the different conditions, as before.

3.3 Resin Embedding

During all steps, solutions, containers, and tools need to be pre-cooled before entering in contact with the samples. Exchange solutions slowly to avoid the loss of samples.

1. After all samples have been transferred, pipet out the excess ethanol and replace with a pre-cooled 25% HM20 in pure ethanol for ~2 h.
2. Always at a working temperature of −50 °C, change the solution for a 50% HM20 in pure ethanol, and let it impregnate ~2 h. Continue with 75% HM20 in pure ethanol overnight.
3. Replace the solution with fresh pre-cooled 100% HM20 and incubate at least 2 h. Replace again the resin by fresh pre-cooled 100% HM20 solution for 2 h. Replace the resin with fresh pre-cooled 100% HM20 one last time. Place the mold containers in the pre-cooled metal piece and adjust the position of samples in the center of the embedding mold if necessary.
4. Mount the UV lamp on top of the AFS well and set the following program: 8 h at −50 °C without UV (to continue resin embedding), 24 h at −50 °C with UV on, and 24 h at +20 °C with UV on.
5. The resin blocks can be removed from the molds by cutting the plastic with a razor blade and/or punched out from the bottom using a flat tip hexagonal screwdriver.

3.4 Grid Preparation

1. Under the hood, place a crystallization dish in a large glass Petri dish and fill it with distilled water until it overflows a little.
2. Roll a glass rod on the top of the crystallization dish to make the water surface even.
3. With a glass Pasteur pipet, let one drop of 2% parlodion solution (*see* **Note 12**) fall vertically on the water surface and let it spread on the water surface.
4. Using tweezers, remove the parlodion film to clean the water surface from eventual dust.
5. Ensure the hood flux is low and repeat **step 3**.
6. Gently place the grids on the parlodion film (*see* **Note 13**).
7. Carefully lay a filter paper on the floating film and grids. Let it slowly soak water.
8. Fold the excess parlodion film from the sides over the filter paper using tweezers.
9. Pinch, at two opposite points, the filter paper covered by the parlodion film using two auto-clamping tweezers and flip it

(without opening the tweezers) before transferring it into a Petri dish.

10. Place the Petri dish cover over the grids to prevent dust from falling onto them but keeping an opening on one side to allow airflow. Let it under the ventilated hood for 24 h to dry.

3.5 Ultramicrotomy

Important: Be careful to avoid deformations of the grid.

1. Select the resin block (*see* **Note 14**) and prepare its face for cutting. First, remove resin excess on each side of the sample with a razor blade. Adjust the cutting surface by using a diamond Trim 20 dry knife for precise trimming. The face must be as small as possible around the biological material and be trapezoidal (parallel sides up and down) with clear and clean edges, especially if serial sectioning is required.
2. When the block face is very close to the sample, switch to a diamond Histo 45° knife (*see* **Note 15**). Place and secure the knife on the machine, away from the block.
3. Fill the knife boat with filtered distilled water until the water surface is above the knife edges (dome shape). Move the knife manually closer to the block. Do not to touch the block with the knife edge, and keep a safe distance of a couple of micrometers.
4. Use a wooden stick mounted with a dog hair to gently guide the water to and along all edges and corners. At the diamond knife edge, slide the dog hair tip very gently. Be extra careful not to touch it with the stick.
5. Suck the excess water with a thin syringe needle. Stop removing water when, from the binocular, the water surface reflection is changing, passing through a white reflection to a darker one at the edges and along the diamond knife (*see* **Note 16**).
6. Approach the knife to the block using the device control panel and the binocular (*see* **Note 17**). Stop the approach when a very thin ray of light is still visible between the block face and the knife edge.
7. Cut 250–500 nm sections from the block, setting the ultramicrotome speed during the cutting window to 1 mm/s.
8. Once you have entered the sample at desired depth, stop the cutting. Collect the sections using a perfect loop and place them on a glass side.
9. On a heating bench, dry the glass slide, cover the section with a drop of 0.05% toluidine blue solution, and let it dry. Next, wash the excess by dipping the slide and gently moving it successively into three distilled water baths. Let it dry again. The section can be observed under a table microscope.

10. To make the final sections, either continue with the Histo 45° diamond knife for sections between 100 and 150 nm (or more if using a high-voltage microscope) or switch to Ultra 45° diamond knife for sections between 60 and 90 nm. To cut with the Ultra 45° diamond knife, prepare the knife as in **steps 4–7**. Cut the sections as required.
11. Collect the sections with the parlodion film-coated grids. This is done using auto-clamping tweezers to clamp the grid on one side and plunging it in the water perpendicularly. Then, gently push the sections toward the grid with the dog hair and slowly bring the grid up toward the surface with a small angle (*see* **Note 18**).
12. Once the sections are on the grid, let them dry still clamped in the tweezers before transferring it into a grid storage box.

3.6 Coating Sections with Fiducial Markers

Important: Be careful to avoid deformations of the grid.

1. Prepare the gold particle solution by mixing 1:1 (v:v) the colloidal gold solution with filtered 0.5% BSA in MilliQ water (*see* **Note 19**).
2. Clamp the grid with auto-clamping tweezers and place it on the lab bench.
3. Use a pipet to gently lay a drop of 20 μL of the gold solution on the grid without touching it. Wait 30 s.
4. Carefully blot the excess solution onto a filter paper from the side of the grid (*see* **Note 20**).
5. Repeat the **steps 3** and **4** on the other face of the grid and let it dry (*see* **Note 21**).
6. Place the grid back in the grid storage box.

3.7 Tilt Series Acquisition

Important: Be careful to avoid deformations of the grid.

1. Put the grid into the microscope grid holder, depending on the fixing method, and the grid can be clipped with a metal ring or tightened by screwing two metal claws.
2. Insert the holder in the electron microscope and wait for the vacuum to go down.
3. Screen the grid to find the root section at low magnification.
4. Set the eucentric height.
5. Perform a coarse alignment of the electron beam.
6. Screen the sample for PD at high magnification. Don't forget to add the diaphragm at the low-Mag/Mag transition to avoid destruction of the section.

7. When a PD is selected, magnify as desired, center it in the image frame for acquisition, and ensure that you have 10 to 30 gold beads in your field of view (*see* **Note 22**).
8. Set the eucentric focus again.
9. Precisely refine alignment in the following sequence: gun tilt and shift, the focus, the pivot points, and the rotation center (*see* **Note 23**).
10. Open the Xplore3D (FEI) software. It is possible to perform either single or batch acquisitions (*see* **Note 24**).
11. If you chose to perform batch acquisition, prepare the first area in the software. Continue screening and add as many areas as wanted. Then, or if you decide to perform a single acquisition, follow the software steps: select the working directory and options, choose the acquisition mode and start/end angles (*see* **Note 25**), and center the image when requested by the software.
12. To obtain more resolution, it is possible to perform a dual-axis tomogram. Depending on the grid holder and microscope, an automated or manual rotation can be performed. To do the manual rotation, the grid is rotated at a 90° angle from the first tilt series position. Then, start another acquisition along the second axis.

3.8 Tomogram Reconstruction

IMOD is an open source software available for LINUX, Windows, and MAC OS environments. Software, documentation, and tutorials are available on https://bio3d.colorado.edu/imod/.

Four files are important for the reconstruction: name.mrc, name.rawtlt, name.txt, and name_shifts.txt. They must be in the same working directory. We are here going to assume a reconstruction from a single axis acquisition.

1. Open the Etomo software. Click on "Build Tomogram."
2. Browse and select the .mrc file (*see* **Note 26**). Select "single axis" and click on "Scan Header" and enter the size for the fiducial marker (e.g., 5 for a 5 nm diameter fiducial marker). If the .rawtlt file is available, click "Tilt angles in existing rawtlt file. Then, click on "Create Com Scripts". If you chose to start the tilt series acquisition at 0°, enter 0 for the selected parameter "series was bidirectional from:".
3. On the new window, click on "Pre-processing" (*see* **Note 27**).
4. To ponderate extreme intensity pixels created during the acquisition step, click on "Create Fixed Stack." The progress bar is visible at the top of the Etomo window. Then, click on "Use Fixed Stack." Finish by clicking on "Done."
5. For the Coarse Alignment, start with "Calculate Cross-Correlation." Click on the little "A" button of the "Newstack" and

deselect the "Convert to bytes" tick box. Continue with "Generate Coarse Aligned Stack." You can verify the alignment with "View Aligned Stack In 3dmod." The Z bar at the top of the 3dmod is used to browse through all tilt angles. Finish by clicking on "Done."

6. Seed tracking method is powerful to optimize the final alignment. First, on the "Seed Model" panel, define which and how many beads will be used for the fine alignment step. Etomo offers two ways to define the fiducial marker: an automatic ("Generate seed model automatically") and a manual ("Make seed model manually") one. Select the automatic method. Then, type the number of beads ("Seed Points to Select/ Total number," e.g., 20) for the fiducial tracking. To select the fiducials present on both sides of the sample, click "Select beads on two surfaces." Then click on "Generate Seed Model." To verify the quality of the fiducial selection, click on "Open Seed Model" and check that the software selected only gold beads that are not overlapping and well distributed in the central frame of the series overview (*see* **Note 28**). Save the "Seed Model." Keep all the 3dmod windows open.
7. Before the fine alignment, the Etomo software has to track each fiducial on each image of the tilt series. Go on the "Track Beads" panel and click on "Track Seed Model" and "Fix Fiducial Model." The log window gives the missing points. If the missing point is different from 0, click on "Track with Fiducial Model as Seed" to allow iterative tracking. Stop to click on "Track with Fiducial Model as Seed" when the number of missing points stops to decrease.
8. If the number has not yet reached 0, open the 3dmod model by clicking on "Fix Fiducial model," for saving bead positions, before going to the Bead Fixer window (3dmod panel/Special/Bead Fixer) and filling manually every gap. To proceed gap by gap, use the button "Go to Next Gap" (in the Bead Fixer window). The concerned bead is highlighted with a green or purple and yellow circle (color code corresponds to the beads on the two surfaces of the sample). Moreover, an arrow indicates if the gap is on the previous or following image, depending on whether it is pointing up or down, respectively (*see* **Note 29**). Once the point is manually added, click on "Reattach to Gap point." Repeat for every gap until the main 3dmod window highlights "No more gaps found." Save the model in the 3dmod software window. Verify there are no missing points by clicking on "Track with Fiducial Model as Seed" one last time in the Etomo window and check the log file. Click on "Done."
9. In the Fine Alignment step, in the "Global Variables" panel, if the alignment doesn't show visible shifting and distortion,

select "Full solution" in the distortion Solution Type sub-window. If the alignment is not good, select "Disabled." Then, click on "Compute Alignment." If the residual error is too high (*see* **Note 30**), click on "View/Edit Fiducial Model" to correct the fiducial positions.

10. In the Bead Fixer window, click on "Go to Next Big Residual." The 3dmod visualization panel shows the bead to correct (highlighted with a yellow circle), and the red arrow shows the suggested correction. Magnify the image to have a better view of different fiducial positions. Click on "Move Point by Residual" to move the residual according to the automatic correction (*see* **Note 31**). Repeat for every residual until "No more residual" shows in the main 3dmod window. Save the model.
11. Click again on "Compute Alignment" in the Etomo window. The residual error in the log should decrease. New residuals may be needing examination in the 3dmod window. If so, repeat **steps 10–12** until there are no more residuals to examine or when the residual error and sd are low enough. Then click on "Done."
12. In the Tomogram Positioning step, change the default "Positioning tomogram thickness" to 1500 in order to access the whole of z view of your section. Click "Create Sample Tomogram" and "Create Boundary Model." In the 3dmod visualization window, create two parallel lines delimiting the top and bottom of the sample section (include beads). Do this for all three sample sections, top_rec.mrc, mid_rec.mrc, and bot_rec.mrc of one Z position (available with the black right/left arrowheads at the top of the window). Save the model.
13. In the Etomo window, change the "Added border thickness (unbinned)" to your sample thickness in nm and click "Compute Z Shift & Pitch Angles" followed by "Create Final Alignment." Finish by clicking on "Done."
14. In the Final Alignment Stack step, activate "Use linear interpolation" (select "Aligned image stack binning: 4" to generate a 4 × binned tomogram useful for illustration and no binning for fully resolved tomograms for image analysis) before clicking on "Create Full Aligned Stack." "View Full Aligned Stack" allows to open the Final Aligned Stack. Click on "Done."
15. Choose Tomogram Generation options (*see* **Note 32**) and click on "Generate Tomogram." "View Tomogram In 3dmod" enables verification and allows to note the beginning and end sections for the trimming (*see* **Note 33**). Click on "Done."
16. In the Post-processing step, unnecessary sections present before and after the sample could be trimmed away by defining the Z min and Z max in "Trim vol" panel, section "Volume

Trimming/Volume Range." Deselect "Convert to bytes" in the Scaling section. Then Click on "Trim Volume." Click on "Done."

17. The final window enables the deletion of temporary files created during the reconstruction process. Click on "Done."
18. Close Etomo and find the final reconstructed tomogram in the working directory with the .rec file extension (under Linux) or rec.mrc file extension (under Windows).

3.9 Three-Dimensional Segmentation

In this section, we give limited guidance since every segmentation is dependent on the sample and the researcher objectives. We still share important features to use in the software as well as a few tips. To perform segmentation on a regular basis, the use of a graphic tablet can save time. Segmentation step is relatively interpretative but permits access to the 3D visualization of individual PD compartments.

1. Open 3dmod and select the .rec file in the Image file browser and click ok. Press "V" on the keyboard to open the Model View window (*see* **Note 34**).
2. Under "Edit," "Object," go to "Type." Select the object type: Closed or Open (in which case we advise selecting "Start New contour at new Z").
3. In the 3dmod main window, search for "Drawing Tools" under "Special" and choose the Drawing Mode (*see* **Note 35**).
4. Trace the elements every five to ten sections and then interpolate for in-between sections using the "Interpolator" tool under "Special" of the main 3dmod window (*see* **Note 36**).
5. Render the segmentation from the 3dmod Model View window, under "Edit," "Objects" and select "Meshing" on the bottom left of the new window. Then, select the mesh type and click on "Mesh One" or "Mesh All" (*see* **Note 37**).

4 Notes

1. Keep liquid nitrogen source at proximity in case of need during the process, especially during long sessions of cryofixation. Be aware that liquid nitrogen is dangerous, and it can cause burns and asphyxiation; we thus recommend wearing a lab coat, protective gloves, protective goggles, and working in a well-ventilated space.
2. The solution must form a clean dome in the socket, without any air bubbles nor spilling on the edges.
3. Once the sample is cut, the steps until the sample is cryofixed must be done very rapidly (~1 min) to avoid bias due to tissue

damage and stress responses in the cells. Moreover, the sample has to be kept in solution as much as possible to avoid air-drying. When the sample is placed in the carrier, it is best if it is completely submerged in BSA, centered in the socket and that no air bubble is present.

4. The tightening is the best when hearing two or three "clicks" from the screwdriver.
5. From the moment the sample is plunged in liquid nitrogen, it must not be taken out of it.
6. Be attentive to prepare everything you need in advance in the liquid nitrogen bath to avoid bringing room temperature objects in the bath when samples are present. Be sure to cool down the tweezers in a remote corner of the bath before handling the carriers.
7. To easily move containers without transferring heat, a special handle can be screwed to the container while in the bath and then unscrewed once in the AFS2. If the distance from the EMPACT to the AFS2 is more than a few meters, use an intermediate container filled with liquid nitrogen to put the sample container without risk.
8. Starting from the preparation of the cryosubstitution mix, the procedure involves the handling of many very dangerous chemicals such as acetone, osmium tetroxide, glutaraldehyde, uranyl acetate, and Lowicryl resin. We advise clear awareness of the risks and the related safety rules and a handling of the chemicals in appropriate conditions (vented hood, special gloves, cartridge mask for manipulation outside the hood and in the AFS, protective eyewear, etc.) as well as appropriate waste disposal and management.
9. We encourage checking the tubes the next day, to ensure all samples are in the solution and not covering each other.
10. Every solution and tool that will be used must be pre-cooled in the AFS well for a good 15–20 min. In order to remove and add new solution, pipet out the solution with a thin tip 1-mL pipet (avoid pipetting in and out or any other turbulence; always leave some solution in the bottom as a security against air-drying), and then add the fresh solution gently with a classic plastic pipet, by letting it drip on the side of the tube. This is true for the entire protocol, but it is very important here that everything is done with calm and patience because of both the sensitivity of the samples and the dangerousness of the chemicals.
11. This step is certainly the most tedious task of the protocol. Depending on the number of samples, experience, and preciseness of the person, it can take up to several hours. If the

unmolding step is difficult with a risk of breaking the sample, skipping this step can be considered. Nevertheless, the resin embedding will require an increase of the times in Subheading 3.3, **steps 1** and **2**. The carrier can later be removed from the polymerized block by doing repetitive cold/heat shocks.

12. 2% parlodion solution is made from solid parlodion (nitrocellulose) dissolved in isoamyl-acetate. This solution needs to be done at least 24 h in advance to allow proper dissolution. Note that this solution is very easily contaminated with water from the air, which makes bubbles on the film of the grids. To avoid this, it is best to prepare under the vented hood, in a container that won't have much air remaining on top of the solution, to avoid agitation procedures that would bring air into the solution, to limit as much as possible the opening time of the container, and to seal the lid with Parafilm between uses. It is also recommended that the container is made of stained glass and wrapped in tinfoil.
13. As the parlodion is transparent, it can be difficult to see the film. However, the edges of the film can be seen by the presence of tiny wrinkles. If the parlodion is not flat or presents several visible folds in the middle, remove it and repeat **step 3** until it is adequate. For the placement of the grids, the best is to select a homogeneous area on the film and arrange the grids as tight as possible. We advise you to choose one side of the grid (shiny or opaque) to face the film and always work that way. The choice of the grids is as desired, but 200 or 300 mesh/μm^2 thin bars is, to us, optimal for tomography.
14. Check every block under a binocular and select the ones containing a biological sample in a good condition (some roots might have disappeared or being damaged during the sample preparation process). It is possible to carefully smooth the block surface using a dry diamond knife, to see the sample more clearly.
15. Diamond knives are extremely fragile, and very expensive, and must be handled with extra care and calm. Any hit or bump of the knife or tool against the knife might cause irremediable damage to the knife edge and then cause unwanted stripes on the sections. It is crucial to keep that in mind during the whole process, from taking the knife out of the box to the cleaning at the end.
16. If the water is disrupted from the diamond knife edge during this process, maybe there was too much water removed. Fill again the knife boat with new filtered distilled water and level the water again.
17. For the approach phase, the reflection of the knife edge onto the block face can be used to monitor and optimize distances.

The light reflected on the block permits to adjust the lateral inclination and orientation of the block. The block needs to be parallel to the diamond knife blade, and every point of the block face must be at the same distance from the blade as the block is moved up and down.

18. This necessitates calm and patience as the process is very delicate. This step could be facilitated by hydrophilizing the grids using a glow discharge device. The process takes 2 min, and the grids will be hydrophile for about 30 min.
19. This solution can be stored at −20 °C between uses.
20. It is easier to use gravity in the blotting process by folding the paper at a 90° angle and placing it with the fold up on the bench; then, bring the grid on top of one of the paper faces (90° angle between the paper face and the grid).
21. Another method consists in putting a drop of the solution onto a Parafilm placed on a flat surface, and then carefully lay the grid on top of the drop. The grid should not fall in the drop nor touch the Parafilm. After 30 s, remove the grid, blot the excess solution, and let it dry. Repeat on the other side of the grid.
22. If there are not enough beads, you can repeat the coating section. We usually work between 30,000× and 42,000× for tilt series acquisition on PD.
23. We advise to refine the microscope alignment in an area close to the area of interest but not to do it directly above it.
24. The Batch/Single processes are fairly similar, but the choice needs to be made when opening the software. Note that single acquisition can be done using either continuous or discontinuous mode; batch acquisition can only be done using the discontinuous mode. The holder calibration file must be chosen accordingly to the selected mode.
25. Tilt angles can be verified manually before starting the acquisition to ensure that there are no grid bars passing on the area of interest at high angles.
26. Because the Etomo software modifies the original files, we advise to copy all original files as "archives," to be able to start again from them if necessary. The most important file for the reconstruction is the .mrc.
27. All reconstruction steps are visible on the left of the Etomo window; they have a color code and a mention of their advancement (red, step not processed; purple, step is being processed; green, step processed).
28. We typically use around 20–25 nicely segregated and well-spread fiducials, but this number can be adjusted depending on the number of beads present on the section (the minimum

is 3) and the precision of the reconstruction that is needed. Selecting more fiducials can improve the tomogram alignment. Problematic fiducials can be eliminated by suppressing corresponding "Contour" in the 3dmod tool window before saving the model.

29. In 3dmod, keyboard shortcuts come very handy for rapidly switching from one image to the other in the tilt series. Mouse settings for adding and moving objects are important to know.
30. The log window indicates the residual error mean and sd which represent the number of pixels of incertitude for the image alignment. Thus, the acceptable values for residual error and standard deviation depend on the pixel size of your image. In our case, the pixel size is usually 0.22 nm since we acquire the tilt series at 4096 × 4096 resolution. In this condition, correct fine alignment is obtained when the mean residual error is around or below 1 and the sd around or below 0.6 (non-binned images).
31. If the automatic correction isn't correct, it is possible to manually adjust the residual position or to keep the position as it is. Moves can be undone by clicking "Undo Move." When the bead is difficult to see, tracking its movement by going up and down in the tilt series can help.
32. We typically use the Back Projection Method, using the "SIRT-like filter equivalent to . . . iterations." The number of iterations to choose depends on the signal-to-noise ratio. We generally use from 10 to 50 iterations.
33. The quality of the general reconstruction is evaluated by looking at the shape of reconstructed beads. In the XZ view, beads have to have a symmetric X shape.
34. Under "Edit," "Options," it is possible to change the mouse and behavior preferences.
35. New straight lines are made with "Normal," whereas "Warp" and "Sculpt" can be used to adjust lines afterward. Pressing CTRL+Z undo the last action. Depending on the number and the continuity of the elements (typically the ER and PM membranes), it is possible to create as many objects as necessary. New objects can be created under "Edit," "Object," "New," and the object color can be modified in "Edit," "Object," "Color." Working with different colors for different organelles is usually more convenient. In the object window, each object can be off (no visible) or on (visible).
36. We generally use the linear interpolation with a Z-Bridge of 10 or 5. Note that the interpolation only works for one object at the time. When the object switches from a continuous to a discontinuous line, the interpolation can easily go wrong. This

usually happens for the PM, which is split in two, at the PD site. To avoid any problem, trace the object on two consecutive sections at the "continuous to discontinuous" line transition.

37. We usually work with "Tube" meshing (with a diameter of 3 or 5, for open objects only) or "Surface" meshing (for closed objects only). It is important to note that object rendering can be different depending on the object type (open or closed).

Acknowledgments

This work was supported by the National Agency for Research (Grant ANR-18-CE13-0016 STAYING-TIGHT to E.M.B), the European Research Council (ERC) under the European Union's Horizon 2020 research and innovation program (grant agreement No 772103-BRIDGING) to E.M.B, and the EMBO Young Investigator Program to E.M.B. J.D.P. is funded by the Belgian "Formation à la Recherche dans l'Industrie et l'Agriculture" (FRIA grant no. 1.E.096.18). The development of this protocol was performed at the Bordeaux Imaging Centre (http://www.bic.u-bordeaux.fr/). The Region Aquitaine supported the acquisition of the electron microscope (grant no. 2011 13 04 007 PFM).
The authors would also like to thank William J. Nicolas, who actively participated in the development of the original protocol, and Sylvain Trépout for his help.

References

1. Hepler PK (1982) Endoplasmic reticulum in the formation of the cell plate and plasmodesmata. Protoplasma 111:121–133. https://doi.org/10.1007/BF01282070
2. Ding B, Turgeon R, Parthasarathy MV (1992) Substructure of freeze-substituted plasmodesmata. Protoplasma 169:28–41. https://doi.org/10.1007/BF01343367
3. Bell K, Oparka K (2011) Imaging plasmodesmata. Protoplasma 248:9–25. https://doi.org/10.1007/s00709-010-0233-6
4. Brecknock S, Dibbayawan TP, Vesk M et al (2011) High resolution scanning electron microscopy. Planta 234:749–758. https://doi.org/10.1007/s00425-011-1440-x
5. Barton DA, Overall R (2015) Imaging plasmodesmata with high resolution scanning electron microscopy. In: Heinlein M (ed) Plasmodesmata: methods and protocols. Humana Press, New York
6. Frank J (2017) Advances in the field of single-particle cryo-electron microscopy over the last decade. Nat Protoc 12:209–212. https://doi.org/10.1038/nprot.2017.004
7. Luque D, Castón JR (2020) Cryo-electron microscopy for the study of virus assembly. Nat Chem Biol 16:231–239. https://doi.org/10.1038/s41589-020-0477-1
8. Kremer JR, Mastronarde DN, McIntosh JR (1996) Computer visualization of three-dimensional image data using IMOD. J Struct Biol 116:71–76. https://doi.org/10.1006/jsbi.1996.0013
9. Nicolas WJ, Grison MS, Trépout S et al (2017) Architecture and permeability of post-cytokinesis plasmodesmata lacking cytoplasmic sleeves. Nat Plants 3:17082. https://doi.org/10.1038/nplants.2017.82
10. Nicolas WJ, Bayer E, Brocard L (2018) Electron tomography to study the three-dimensional structure of plasmodesmata in plant tissues - from high pressure freezing preparation to ultrathin section collection. Bio Protoc 8(1):e2681. https://doi.org/10.21769/BioProtoc.2681

Chapter 4

Ultrastructural Analysis and Three-Dimensional Reconstruction of Plasmodesmata

Kamila Godel-Jędrychowska, Tilman Franke, and Ewa Kurczyńska

Abstract

Array tomography (AT) is a new high-throughput imaging method for high-resolution imaging of ultrastructure and for 3-D reconstruction of cells and organelles. Here, we describe the entire procedure for obtaining a spatial image of the distribution of plasmodesmata (PD). As example, the protocol is applied here to reconstruct the number and arrangement of PD between cells undergoing differentiation during *Arabidopsis* somatic embryogenesis.

Key words Array tomography, Plasmodesmata, 3-D reconstruction, Ultrastructure

1 Introduction

Plasmodesmata (PD) are cytoplasmic channels in the plant cell walls that enable the control of cell differentiation and cellular functions by regulating the movement of ions, some hormones, signaling molecules, and proteins including transcription factors as well as various types of RNA between cells. Understanding the mechanisms that regulate the function of PD in controlling intercellular transport is fundamental for explaining the role of PD in cell differentiation. Although the ultrastructure, molecular composition, and function of PD are intensely studied, we are also still far from even knowing the distribution of PD within the cell walls of different cell types. It is difficult to determine the arrangement of PD in the walls based on the analysis of single images obtained by transmission electron microscopy (TEM), which by itself is a laborious and time-consuming method. However, recent and older literature demonstrates the use of this technique [1–4]. The requirement of analyzing individual serial sections has been overcome by electron tomography which is an extension of TEM and has been initially developed for animal cells [5–7] and, so far,

Yoselin Benitez-Alfonso and Manfred Heinlein (eds.), *Plasmodesmata: Methods and Protocols*,
Methods in Molecular Biology, vol. 2457, https://doi.org/10.1007/978-1-0716-2132-5_4,

applied to only few plant species [8–12]. In contrast to traditional TEM analysis, electron tomography involves the imaging of the sample from different angles around the specific target structure [13]. The information obtained is collected and used to assemble a three-dimensional image of the target [9, 12, 13].

Array tomography (AT) allows to assemble three-dimensional images based on two-dimensional arrays of ultrathin sections. Ribbons of serial ultrathin sections representing substantial tissue volumes can be cut and collected on a single silicon (Si) wafer and simultaneously processed for 3-D reconstruction of the target.

With respect to the imaging of PD, AT allows to overcome original limitations of electron micrographs, because PD distribution can be traced through the serial high- resolution images taken from consecutive ultrathin sections with high precision [12]. An additional advantage of AT compared to other techniques of 3-D reconstruction is the possibility of returning to the scanned material and the possibility of pre-saving the ribbons. The following protocol can be used for obtaining 3-D distribution data for PD in various plant species. As an example, the AT method is here applied to reconstruct the number and arrangement of PD between cells undergoing differentiation during *Arabidopsis* somatic embryogenesis.

2 Materials

2.1 In Vitro Culture

1. *Arabidopsis thaliana* Col-0 seeds.
2. 70% Ethanol.
3. 20% ACE (commercial agent containing sodium hypochlorite) in demineralized water (dH_2O). Add one or two droplets of TWEEN-80 per 100 ml.
4. 2,4D stock solution: Dissolve 50 mg of 2,4D (2,4-dichlorophenoxyacetic acid) in 1 ml of 95% ethanol. Add 59 ml sterile water and stir until a homogenous solution is obtained (*see* **Note 1**).
5. Sterile water: Sterilize 1 l of dH_2O before use.
6. E5 solid medium: Measure 3.2 g of Gamborg's B-5 Basal Salt Mixture, 1 ml of 2,4D stock solution, 20 g saccharose, and 3.6 g of Phytagel® (plant cell culture tested). Complete to 1 l with dH_2O. Adjust to pH = 5.8 with 0.1 M HCl. Sterilize by autoclaving before use.

2.2 Solutions

1. 3% Glutaraldehyde (GA) fixative: Use 1,2 ml of 25% GA (Grade I) to prepare 10 ml 3% GA solution in cacodylate buffer (v/v).
2. Cacodylate buffer: Dissolve 21.4 g of sodium cacodylate trihydrate in 1 l dH_2O. Adjust to pH = 7.2 with 0.1 M NaOH.

3. 2 mM $CaCl_2$: Weigh 0.11 g and complete to 500 ml with cacodylate buffer (w/v). Store at 4 °C (*see* **Note 2**).
4. 3% Potassium ferricyanide solution: Weigh 0.3 g $K_3[Fe_3(CN)_6]$ and complete to 10 ml with cacodylate buffer (w/v).
5. 4% Osmium tetroxide solution: Carefully open the ampoule and drop the content (0,25 g OsO_4) into a dark flask with 6,25 ml of cacodylate buffer (w/v).
6. Pre-contrast solution: Mix 3% potassium ferricyanide solution with 4% osmium tetroxide solution in an equal volume ratio (1: 1, v/v).
7. Thiocarbohydrazide (TCH): Add 0.1 g of TCH to 10 ml of dH_2O and keep at 60 °C; gently mix every 10 min until completely dissolved. Filter through 0.22 μm Millipore before use.
8. 2% Osmium tetroxide solution: Carefully open the ampoule and drop the content (0,25 g OsO_4) into a dark flask with 12,5 ml of dH_2O (w/v).
9. Uranyl acetate solution: Dissolve 0.1 g uranyl acetate in 10 ml dH_2O.
10. L-Aspartic stock solution: Dissolve 0.998 g of L-aspartic acid in 250 ml of dH_2O. It will dissolve faster in $pH = 3.8$ (*see* **Note 3**).
11. *En bloc* Walton's fixative: Dissolve 0.66 g of lead nitrate in 10 ml of the L-aspartic acid stock solution and adjust the pH to 5.5 with 1 N KOH. Warm the solution at 60 °C for 30 min before use (no precipitate should form).
12. Ethanol series solutions (5%, 10%, 15%, 20%, 30%, 40%, 50%, 60, 70%, 80%, 90%, in dH_2O (v/v) and 100% ethanol.
13. 50% Propylene oxide solution: 50 ml of propylene oxide completed to 100 ml by 100% ethanol (v/v).
14. 100% Propylene oxide.
15. Epon 812 embedding medium mixture: Add 8.86 ml Poly Bed 812®, 6.23 ml dodecenylsuccinic anhydride (DDSA), and 3.0 ml nadic methyl anhydride (MNSA) into a falcon tube. Add 1.8% 2,4,6-tris(dimethylaminomethyl)phenol (DMP-30) (v/v).
16. Solution 3:1 (v/v) propylene oxide/Epon 812: 30 ml of propylene oxide mixed with 10 ml Epon 812 embedding medium mixture.
17. Solution 1:1 (v/v) propylene oxide/Epon 812: 20 ml of propylene oxide mixed with 20 ml Epon 812 embedding medium mixture.

18. Solution 1:3 (v/v) propylene oxide/Epon 812: 10 ml of propylene oxide mixed with 30 ml Epon 812 embedding medium mixture.

2.3 Other Accessories

1. Ice from freezer.
2. Peat pots.
3. 2 ml Eppendorf tubes.
4. Preparation needles.
5. Sterile culture dishes.
6. Silicone molds.
7. Silicon (Si) wafers.
8. TWEEN80.
9. Filter 0.22 μm Millipore.
10. Double-sided carbon sticky tape.
11. Aluminum half-inch SEM pin.
12. Electrically conductive glue.
13. Aluminum tape.
14. Razor blades.

2.4 Equipment

1. Scanning electron microscope (SEM; *see* **Note 4**).
2. Ultramicrotome.
3. Stereomicroscope.
4. Ultra 45° Jumbo diamond knife (Diatome, Switzerland).
5. Advanced substrate holder (ASH-100, RMC Boeckeler).
6. Desiccator.

2.5 Software

1. Fiji ImageJ Software.
2. Amira Software.
3. Microscopy Image Browser (MIB).
4. Thermo fisher maps array tomography software.

3 Methods

3.1 In Vitro Culture

1. Sow *Arabidopsis* Col-0 seeds in moist peat pots. Place one to three seeds into each moist peat pot.
2. Leave seeds to germinate in a greenhouse at a temperature of 21 °C and a 16 h photoperiod with a light intensity of 40 μmol m^{-2} s^{-1}.
3. After 6 weeks, collect the siliques (seed pods) with immature zygotic embryos (IZEs: *see* **Note 5**) and surface sterilize them

in 70% ethanol for 2 min and in 20% ACE with two droplets of a TWEEN80 for 15 min.

4. Wash several times in sterile water.
5. Select IZEs at the late cotyledonary stage of development from the siliques using sterile preparation needles. Place the IZEs on sterile culture dishes with E5 solid medium for 2 weeks to the somatic embryogenesis process.
6. Keep the culture in a greenhouse at a temperature of 24 °C and under a 16 h photoperiod with a light intensity of 40 μmol m^{-2} s^{-1} for 2 weeks.
7. Analyze the explants and developed somatic embryos under a stereomicroscope and select the appropriate explants and somatic embryos for the following procedures: the fixation, dehydration, and embedding of the samples (*see* **Note 6**).

3.2 Fixation

1. Prepare the 3% GA fixative (*see* **Note** 7).
2. Transfer the sample into a 2-ml Eppendorf tube with 1.5 ml of the 3% GA fixative (Fig. 1a: *see* **Note 8**).
3. Deaerate the samples several times inside the desiccator chamber (10–15 cycles, 2–3 min each; the safest pressure is 0.006 Pa) under the vacuum until the samples sink completely to the bottom of the Eppendorf tube (remember to make holes in the Eppendorf lid to permit the air to escape; Fig. 1b).
4. Leave the sample with 3% GA fixative for 24 h at 4 °C.

3.3 Postfixation Method: Pre-contrast Procedure

1. Carefully remove the 3% GA fixative from the Eppendorf tubes.
2. Wash the sample in cacodylate buffer, three times for 5 min.
3. Incubate the sample with pre-contrast solution for 1 h on ice.
4. Wash the sample with dH_2O, three times for 5 min at room temperature (RT).
5. Transfer the sample to freshly prepared TCH. Keep the sample in this solution for 30 min at RT.
6. Wash the sample with dH_2O, three times for 5 min at RT.
7. Transfer the sample to 2% osmium tetroxide. Keep the sample in this solution for 30 min at RT.
8. Wash the sample with dH_2O, three times for 5 min at RT.
9. Transfer the sample to 1% uranyl acetate. Keep the sample in this solution overnight at 4 °C.
10. Wash the sample with dH_2O, three times for 5 min at RT.
11. Transfer the sample to *en bloc* Walton's fixative. Keep the sample in this solution for 30 min at 40 °C.
12. Wash the sample with dH_2O three times for 5 min at RT. The sample becomes dark after this postfixation step (Fig. 1c).

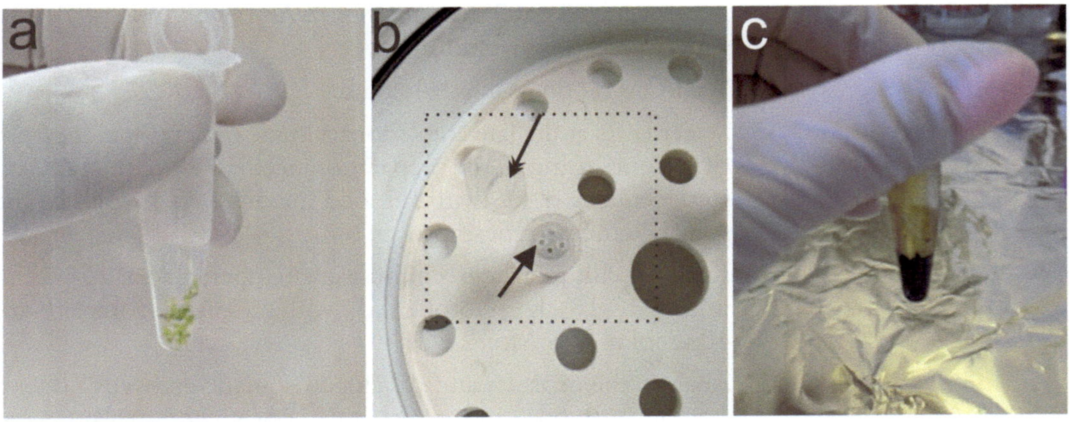

Fig. 1 Sample preparation for electron microscopy analyses. (**a**) Samples in the fixation solution before evacuation from air. (**b**) Eppendorf tube in the desiccator where the samples are evacuated from air; Eppendorf tube with holes (arrow) in its lid to permit the air to escape; the double arrow shows the original lid of the Eppendorf tube. (**c**) Fixed and contrasted sample (black) at the bottom of the Eppendorf tube

3.4 Dehydration

1. Dehydrate the sample in increasing ethanol from 5% to 100% series. Keep the samples in each concentration for 30 min at RT (*see* **Note 9**).
2. Replace the ethanol with a mixture of ethanol/propylene oxide (1:1 v/v). Keep for 15 min at RT.
3. Replace the ethanol/propylene oxide mixture with the pure propylene oxide. Keep for 15 min at RT.

3.5 Embedding

1. Prepare the Epon 812 resin (*see* **Note 10**).
2. Infiltrate each sample into Epon 812 resin by incubation in the following mixtures for 24 h each at 4 °C:

   Solution 3:1 (v/v) propylene oxide/Epon 812.

   Solution 1:1 (v/v) propylene oxide/Epon 812.

   Solution 1:3 (v/v) propylene oxide/Epon 812.

   100% Epon 812.
3. Embed the samples in silicone molds with freshly prepared Epon 812 resin and polymerize the resin for 24 h at 60 °C.

3.6 Sectioning

1. Prepare the samples embedded in Epon 812 for sectioning. Use razor blades to get smooth shape of the block face and the straightness of the edges of the trimmed sample. Parallel edges top and bottom are the best shape for the ultrathin sections (Fig. 2a: *see* **Note 11**).
2. Place the trimmed sample into the ultramicrotome holder (Fig. 2b).

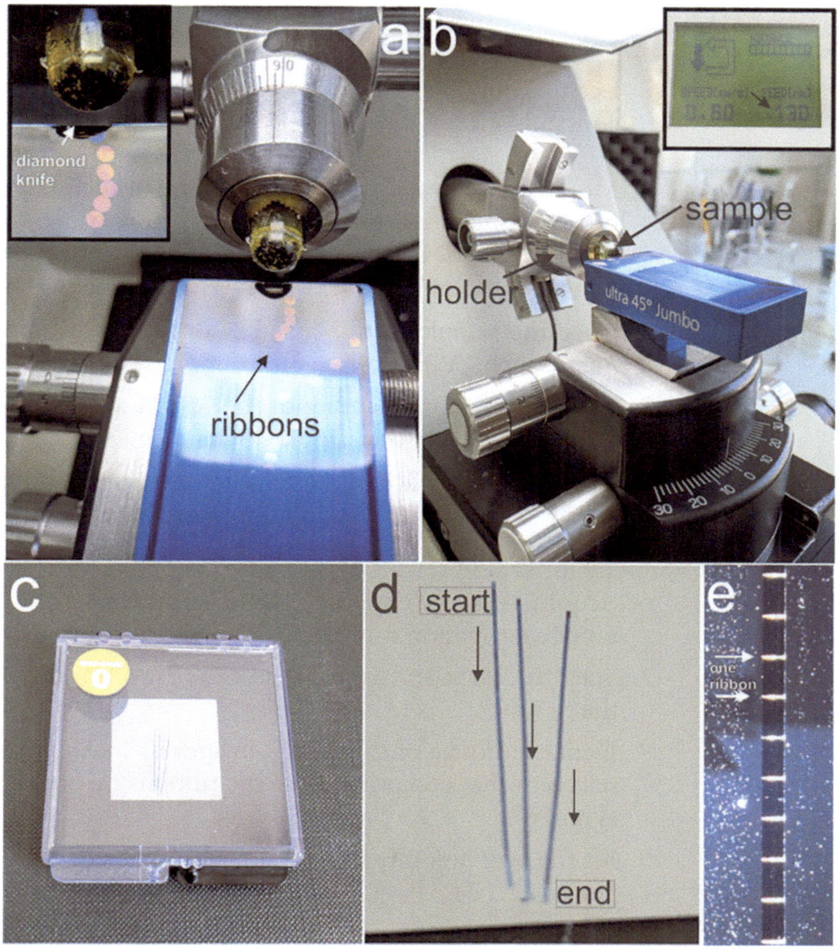

Fig. 2 Sample sectioning and section collection. (**a**) Ribbons on the surface of water in the knife "boat." (**a inset**) A sample at the front of diamond knife surface. (**b**) Part of the ultramicrotome with a sample in the holder and the sets for cutting with an ultra 45° Jumbo knife. (**b inset**) Parameter settings for cutting the sample and the thickness of the sections (indicated by an arrow). (**c**) Si wafer with the collected ribbons. (**d**) Ribbons on the Si wafer. The ribbons are oriented next to each other, and the "start" and the "end" of the sectioning is indicated. (**e**) The image shows a higher magnification of image 2d; the simple arrow marks the "start" of one ribbon, and the double arrow marks the "end" of the same ribbon

3. Cut the sample into 130-nm-thick sections using a 45° Jumbo diamond knife (Fig. 2b inset) and collect the ribbons in the section collecting space of the ultra 45° Jumbo knife (Fig. 2a, an inset).
4. Place a ribbon of ultrathin sections on an Si wafer. Although using an advanced substrate holder, it is also possible to use an Si wafer without a holder (in this case, manually submerge the wafer in a water bath and then lift it out) (Fig. 2c, d).
5. Mark the "start" and the "end" of the ultrathin sections ribbon on an Si wafer (Fig. 2d).

6. Make sure that each single ribbon is placed behind the previous one (Fig. 2e).
7. Let the ribbons dry at RT. Cover with Petri dishes to avoid contamination.

3.7 Apreo Scanning Electron Microscopy

1. Mount the Si wafer on an aluminum half-inch pin using electrically conductive glue. Figure 3a shows an Si wafer with ribbons glued onto a half-inch pin (Fig. 3a inset) using double-sided carbon sticky tape (*see* **Note 12**).
2. Place the sample in the chamber of Apreo scanning electron microscope (ASEM) (Fig. 3b).
3. Vent the vacuum chamber of the ASEM and place the aluminum half-inch on the sample holder (Fig. 3c).
4. Connect the surface of the Si wafer to the metal of the sample holder with a strip of aluminum tape (Fig. 3c).
5. Close the chamber door and pump a vacuum.
6. Acquire an overview image of the sample using the in-chamber sample navigation camera—NavCam (Fig. 3d).
7. Start the Maps software and use the NavCam image to locate the Si wafer.
8. Define a mosaic of the ASEM images by making a rectangular selection that comprises all of the ribbons on one of the wafers (Fig. 3e).
9. Set the overview mosaic with similar settings to those in Table 1 (*see* **Note 13**).
10. Determine the location of each section in the ASEM overview image mosaic (Fig. 3f: *see* **Note 14**).

3.8 Maps AT Workflow

All ASEM image series presented here were recorded using the Thermo Fisher Maps Array Tomography software, which automates this process to a high degree.

1. Open the AT workflow panel and go to the Section Finder step. Zoom in on any section that is visible in the ASEM overview mosaic and outline it with a rectangle in order to define the search template. Then, run the Section Finder step. After a moment, all of the sections that were found are outlined with brown rectangular frames.
2. Add/remove sections during the workflow steps, e.g., when sections are not identified properly. At the end, each section should be marked and outlined by a brown rectangle (*see* **Note 15**).
3. Get higher resolution images of all of the sections in an automated procedure, by using "Section Previews" (*see* **Note 16**).

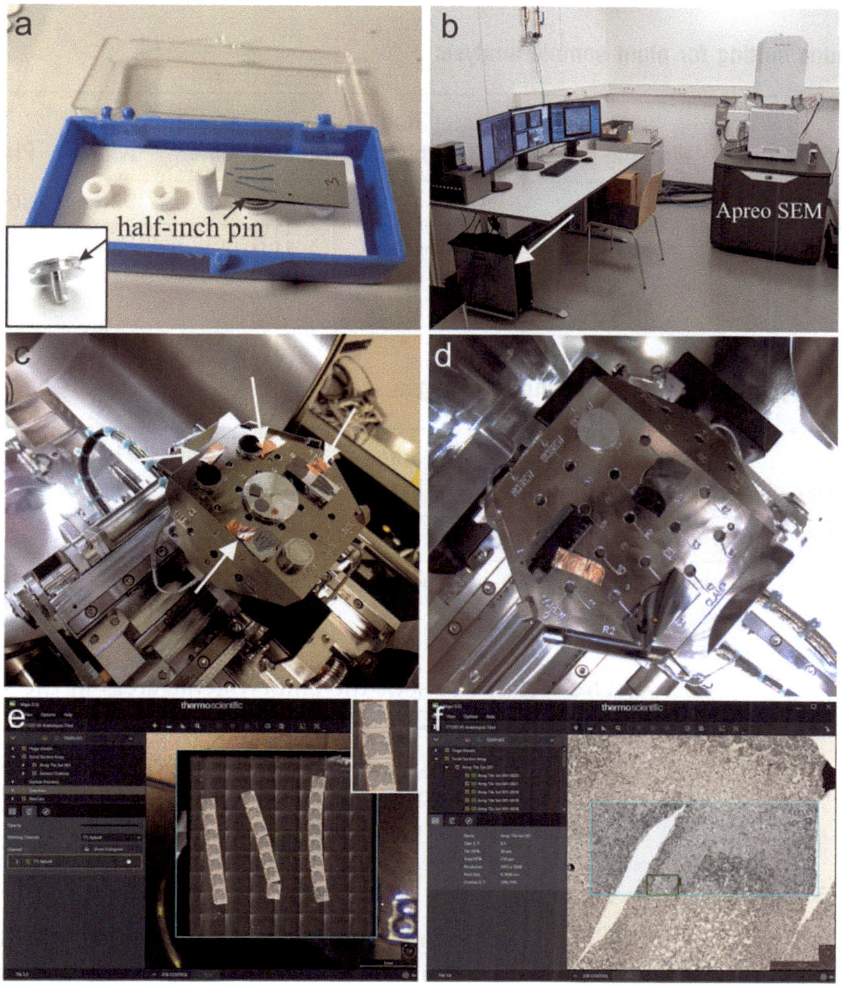

Fig. 3 Apreo SEM. (**a**) An Si wafer with ribbons of ultrathin sections glued onto a half-inch pin. (**a inset**) Half-inch pin without the Si wafer. (**b**) Room view with the equipment showing the ASEM, PC, and network-attached storage (arrow). (**c**) Multipurpose holder with half-inch pins with multiple samples. The samples on the Si wafers were connected to the metal surface of the sample holder using double-sided carbon sticky tape (arrows). (**d**) A holder with a sample as seen in the NavCam image. (**e**) Maps program window with a NavCam image in the background, an overview mosaic of ASEM images, and the sections that were found (brown frames) by the software; inset shows close-up of sections with frames). (**f**) Maps view of a part of a single section. The area in the blue rectangle was scan at a high resolution in all of the sections

Table 1
Recommended settings for mosaic of ASEM images overview

Mode	Detector	Working distance	HT	Stage bias	Current	Dwell time	Field width	Pixels
Optiplan	T1 (inlens)	4 mm	3 kV	–4 kV	100 pA	3 µs	1.5 mm	1024 x 884

mm, millimeter; kV, kilovolts; pA, pico Ampère; µs, microseconds

Table 2
Recommended setting for plant sample analysis

Mode	Detector	Working distance	HT	Stage bias	Current	Dwell time	Field width	Pixels
Optiplan	T1 (in lens)	4 mm	3 kV	−4 kV	100 pA	3 μs	950 μm	2048 × 1876

mm, millimeter; kV, kilovolts; pA, pico Ampère; μs, microseconds

Table 3
Recommended parameters for acquiring the images

Mode	Detector	Working distance	HT	Stage bias	Current	Dwell time	Field width	Pixels
Optiplan	T1 (in lens)	4 mm	3 kV	−4 kV	100 pA	3 μs	30 μm	3076 × 2048

mm, millimeter; kV, kilovolts; pA, pico Ampère; μs, microseconds

4. Use settings to spot the embryogenic areas in the *Arabidopsis* explants (Table 2: *see* **Note 17**).
5. Use the section preview images to improve the precision of the local coordinate systems that were mentioned above. Use "refinement step," which outlines the regions of interest (ROIs), which is then traced through the section set (*see* **Note 18**).
6. Create imaging regions in any of the sections, which will automatically be cloned to the same relative position in all of the other sections (*see* **Note 19**).
7. If size and pixel resolution of an imaging area exceed certain limits, Maps will automatically segment the area into so-called "tile sets", which are rectangular arrangements of images which can be stitched into larger images later. Use following settings for imaging "tile sets" (Table 3).
8. Stitch the image mosaics using the "Stitch All" function (*see* **Note 20**).

3.9 Manual Segmentation of the Cells and PD

*3.9.1 Stack Alignment and Cropping in Fiji Image J (See **Note 21**)*

1. Import a stack of images into Fiji ImageJ as a virtual stack.
2. Normalize the contrast/brightness by selecting "enhanced contrast" from the "process" menu" and set "saturated pixels" up to 0.1 and check "process all slices."
3. From the plugins menu, select "registration – linear stack alignment" with the "Shift" button.

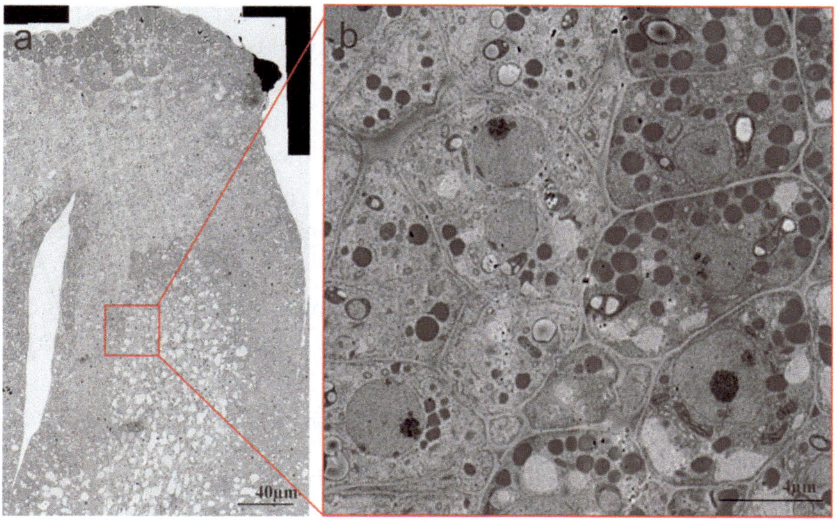

Fig. 4 Selection of the area for analysis. (**a**) A large overview scan of a part of one ultrathin section at a low magnification. The image was acquired as a mosaic of image tiles in the ASEM. (**b**) The magnified view shows that despite the large imaging area, the resolution was sufficient to area used for the 3-D reconstruction

4. Select "rigid" from the "expected transformation" drop-down menu. Otherwise, keep the default settings. Start the alignment by clicking "OK" (*see* **Note 22**).
5. Crop the image stack so that it contains only the ROI by making a rectangular selection and selecting "crop" from the "image" menu (Fig. 4a, b).
6. Save the stack as a series of .tiff images.

3.9.2 Stack Alignment and Cropping in Amira (See ***Notes 23*** *and* ***24****)*

1. Open a data stack using the "drag&drop" function in the program window.
2. Enter the pixel size when prompted.
3. Select "in memory" if Amira asks whether to load data into memory (*see* **Note 25**).
4. Click the right mouse button to the "green data" object in the project pool (top left) and select "align slices" from the context menu.
5. Click on the red "align slices" module in the project pool and select "edit."
6. Locate the slider at the top of the Alignment Editor. Move it to the left (first image). If only a part of the image is shown, zoom out.
7. Click the "align current slice pair." The algorithm used here is based on a least-squares algorithm that works well for undistorted, clear images (*see* **Note 26**).

8. Click "drag&drop" in the Viewer to move the second image until the difference between images becomes as small as possible.
9. Rotate the image if needed by holding down the control key while click-dragging with the left mouse button (*see* **Note 27**).
10. Once the stack is aligned, click the "resample" button in the properties area of the Align Slices module. This will create an aligned output stack.
11. Stack alignment leaves black areas where images had to be moved. At the same time, images get partially cropped where they stick out of the canvas (*see* **Note 28**).
12. Save the result using "export data" in the file menu.

3.10 Reconstructing the Structure in Microscope Image Browser (MIB)

1. Open MIB software (GNU General Public License v2) [14].
2. Open your file with all of the scans (segmentation) to manage individual sections.
3. In the window called "path" at the top of the screen, find and open the stack of the series of .tiff images from the book list (visible on the left side of the window), and select your files.
4. Define the voxel size in the "data_set" parameters (*see* **Note 29**).
5. Mark the structure of interest (Fig. 5: *see* **Notes 30** and **31**).
6. Use brush tools for mark the nucleus and nucleolus area on every third section. Use: I—"Interpolation" to fast marks areas on all sections (Fig. 5 nucleus, green, and nucleolus, blue).
7. Mark each PD on a single section by brush in high-resolution files (Fig. 5, PD—red color: *see* **Note 32**).
8. Mark the cell wall using the brush to visualize the number and arrangement of the PD in cell wall (Fig. 5, cell wall-light blue).
9. Add each of the marked structures to the "mask"—each structure should have its own "mask." On the left side on the window view, select "Mask" (Fig. 5 items highlighted in pink and blue) to add painting to a mask, and press Shift+A.
10. Save each created mask (nuclei, nucleoli, PD, cell wall) as a mask in an Amira (.am) file.

3.11 Three-Dimensional (3-D) Model

1. Open several files (saved as a mask .am in MIB) in Amira Software by "drag&drop."
2. Click the right mouse button on "green data object" and from the menu select "generate surface" (*see* **Note 33**).
3. A red "generate surface" object appears. Select it and click "apply" in the viewer options at the bottom left of the Amira

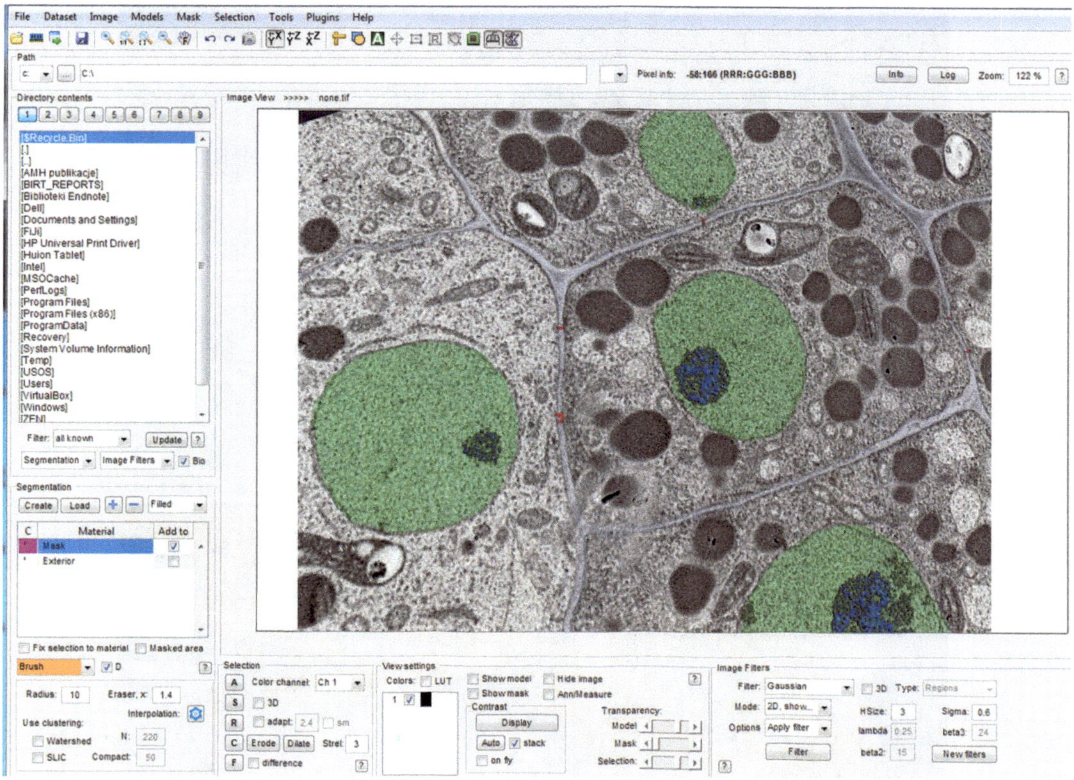

Fig. 5 Screen image during the reconstruction of the PDs at the 3-D level. In MIB, the PD (marked in red) between cells are shown. Nucleus, green; nucleolus, blue; cells walls, blue

program window. A new "green data object" appears (suffix: surf).

4. Click right button on the mouse and select "surface view" from the context menu.
5. Select the colors and transparency for the different parts of the structure using the tools in the viewer options of the yellow surface view objects.
6. Create final surface by changing the smoothing and render mode (*see* **Note 34**).
7. Rotate and change position of the 3-D model to get the right position images.
8. Create a snapshot using the "camera" in the button bar.
9. In the snapshot tool, use the "off-screen" function to produce snapshots with the desired resolution (pixel count).
10. Save snapshots as .png. Select "RGB alpha" to create an image with a transparent background (Fig. 6: *see* **Note 35**).

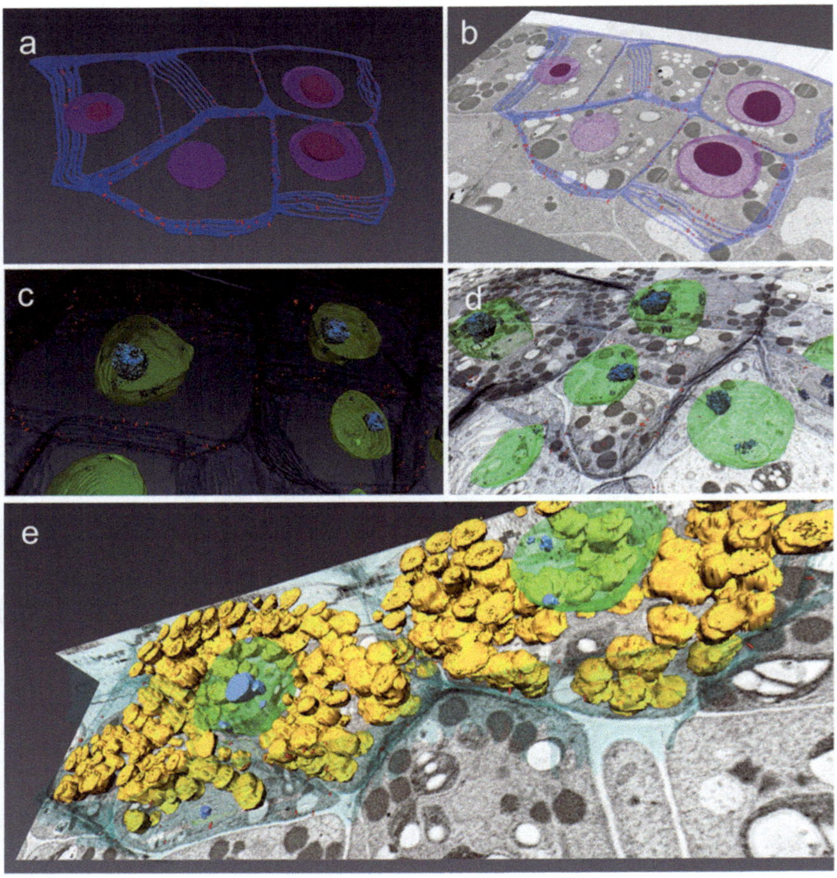

Fig. 6 Three-dimensional (3-D) model of a few cells from an *Arabidopsis* explant. (**a**) Model of the embryogenic area; reconstruction of the embryogenic cells with a large nucleus (pink) and nucleolus (dark pink) and with PD (red) and walls (blue). (**b**) The model obtained during the 3-D reconstruction shown against the background of one section of the area for which the reconstruction was performed for the distribution of the PD between the embryogenic cells (reconstructed—about 1 μm; cell walls, blue; nucleus, pink; and nucleolus, violet). (**c**) 3-D model of pluripotent cells; reconstruction of nucleus (green) and nucleoli (blue); PD (red) are less numerous than in the embryogenic area. The 3-D models in 6a and 6c are without the background of one of the cross sections to better visualize the distribution of the PD. (**d**) Reconstructed cells and distribution of the PD in the cell walls between pluripotent cells shown against the background of one of the cross sections showing the ultrastructure of pluripotent cells (reconstructed—about 2.9 μm; cell walls, dark gray; nucleus, green; nucleolus, blue). (**e**) 3-D reconstruction of the protodermal cells of the somatic embryos that had developed from the explant during the culture; reconstruction of the nucleus (green), nucleolus (blue), PD (red), and the lipid bodies (yellow) that characterized this somatic embryo tissue. PD are present in the anticlinal and periclinal cell walls but were not as numerous as they were in the embryogenic or pluripotent cells. Additionally, it is a very convenient method for the statistical calculations that are required to show the different number of PD between the cells that realize different development programs [12]

4 Notes

1. Solution can be warmed up to 40 °C during stirring.
2. If you know that the analyzed sample contains many lipids, add 2 mM calcium chloride ($CaCl_2$) to the fixative solution.
3. The stock can be frozen (−20 °C) and stored in a freezer for 1–2 months.
4. In our studies, Apreo scanning electron microscope (ASEM) with dedicated software was used and described precisely for AT procedure. It is possible to use another SEM, but the basic idea is using large solid substrates to deposit ribbons of sections and to image serial sections at ultrastructural resolution using the SEM instead of a TEM. This availability of SEM should be possible to acquire image data of ultrathin sections in SEM in high quality. The most important feature of the software is scanning two-dimensional image tiles that can be reconstructed into three-dimensional volume images for visualization. The steps in this procedure are very accurate for Apreo SEM and Maps Array Tomography Workflow, but the main goals in the following steps are equally important for other microscopes and software.
5. IZEs are embryos used as an explant. Once an IZE is placed on the medium, it is an "explant."
6. Appropriate explant means that you choose for analysis explant with morphological features important for studies, or you can use explant after appropriate time of culture.
7. Glutaraldehyde (GA glutaraldehyde; glutaraldehyde solution, Grade I, 25% in H_2O, especially purified for use as an electron microscopy fixative) is the most commonly used fixative. The fixation cross-links the molecules in the tissue and preserves cellular structure. In preserving structure fixation prevents the formation of histological artifacts. GA/PFA (paraformaldehyde) mixtures may be used alternative fixation medium. In this case, it is recommended to use solutions that contain 2.5% GA and 2% PFA (pH = 7.2). Generally, it is important to keep the time between collecting the sample and placing it into the fixative solution as short as possible. Moreover, all fixation, pre-contrast, and embedded procedures should be performed under a well-working fume hood.
8. The volume of fixative solution should be twice that of the sample volume.
9. The term "dehydration" is used to describe the removal of water from a sample. Moreover, some samples may contain nonaqueous liquids such as organic fluids, oils, and polymers. Thus, it is recommended that all of the fixating solution and

tissue fluids should be removed and replaced with organic solvents that are miscible with the embedding medium. In applying the ethanol series solutions, you may keep samples in 30% ethanol overnight at 4 °C. Samples in 70% ethanol can be kept overnight up to several weeks at 4 °C.

10. Epon 812 resin is miscible with propylene oxide but not with ethanol. Therefore, the ethanol must be replaced with propylene oxide.
11. The samples can be also embedded in the gelatin capsules. Upon sectioning, the sections will be circular. The preferred shape for ultrathin sections is squares which allow better adhesion between sections, which stabilizes the ribbons.
12. Instead of electrically conductive glue, you may also use a strip of aluminum tape (to achieve conductivity to electrical ground).
13. It is not necessary to acquire these images with a high resolution.
14. "Mosaic" means an array of images, for example, a field of 4 × 4 images that each overlap by 10%, so that a larger area is covered at a high resolution. In the Maps software, such an image set is called a "tile set," and it is created by simply making a rectangular selection around the section in the viewing area, which is very specific to Thermo Fisher Scientific software.
15. Later in the process, when the ROIs for high-resolution imaging need to be selected in multiple sections, the brown frames act as local coordinate systems. The ROIs are selected simultaneously in all of the sections simply by placing ROIs in the same position relative to the local coordinate system in each section.
16. The imaging settings must be suitable for detecting the ROIs.
17. The effective pixel size of just below 500 nm enables the tissue morphology to be recognized in sufficient detail. In order to illustrate, Fig. 4f shows the preview image of a section of an *Arabidopsis* explant. In of the event that many more sections need to be imaged, or if a higher resolution is needed, the acquisition time is noticeably longer. In such cases, reducing the pixel dwell time and compensating for this by increasing the beam current is advised.
18. In this process, the coordinate systems are shifted so that the selected area is in the same relative position in all of the sections. As a result, the images that are acquired in consecutive sections overlap much better. The remaining mismatch is typically decreased from >50 μm to 5–10 μm.
19. The imaging settings as well as the size and position of the imaging region can be adjusted, and any changes are always synchronized. There is no limit to the number of imaging

regions, and only a few clicks are required to define them. Therefore, while it is possible to scan entire sections at high resolution in principle, we recommend that several local imaging areas are defined rather than the entire section, which saves a lot of recording time and greatly simplifies handling the data. The total size of each of the mosaics of the ASEM images in our example was 219 μm × 92 μm at a pixel resolution of 9.7 nm. We recommend to use 10% tile overlap. The acquisition time per mosaic was 14 min. Autofocus was used in the first tile in each mosaic, i.e., once per section. The total acquisition time was 5 h 8 min. Below are some guidelines for selecting useful image acquisition strategies:

When setting up the image acquisition, it is important to consider the impact of the size and resolution of an image on the acquisition time. Use the following formula to calculate the acquisition time (Formula 1):

Formula 1 Principles for Calculating the Acquisition Time

$$\text{Scan time [hours]} = \frac{\text{Image size X}}{\text{Pixel resolution}} \times \frac{\text{Image size Y}}{\text{Pixel resolution}} \times \frac{\text{Pixel dwell time [μs]}}{3600\text{s/h} \times 1{,}000{,}000}$$

Scanning an entire section of 1 mm × 1 mm at a resolution of 10 nm with a 1-μs dwell time takes 2.78 h, which was calculated from the formula below:

$$\begin{aligned}\text{Scan time [hours]} &= \frac{1\text{mm}}{10\text{nm}} \times \frac{1\text{mm}}{10\text{nm}} \times \frac{1\,\mu\text{s}}{3600 \times 1{,}000{,}000}\\ &= 100{,}000 \times 100{,}000 \times \frac{1\,\mu\text{s}}{3{,}600{,}000{,}000}\\ &= 2.78\text{h}\end{aligned}$$

Please note that this is the time per section. Recording 22 sections at these settings would take more than 2 days.

The acquisition time is much shorter when ten areas of 80 μm × 80 μm are scanned in each section. Such a setup is in many cases sufficient to capture all of the important areas that are of interest (Formula 2).

Formula 2 Principles for Acquiring the Time Calculation for a Smaller Area

$$\text{Scan time [min]} = \frac{80\mu m}{10nm} \times \frac{80\mu m}{10nm} \times \frac{1\mu s}{60s/\min \times 1{,}000{,}000}$$
$$= 1.067\,\min$$

At these settings, recording ten images in a section takes roughly 11 min. With all 22 sections, the acquisition takes 4 h.

20. In Amira, there is no data export because Maps stores all of the images as standard .tiff images with their metadata in a folder in the project save location. The arrangement of the image files corresponds to the data tree in Maps and is generally as follows:

    Project directory

    "LayersData" folder that contains all of the images.

    (a) Several folders for Overviews, Snapshots, the NavCam images, etc., arranged as in the Maps data tree.

    (b) A "Serial Section Array" folder that contains all of the images that were recorded in multiple sections.

       - Array image series 1.
       - Array image series 2.
       - Array images of the subsequent series until the end of the scan.

21. Usually, AT data are not perfectly aligned. There are several automatic and semiautomatic methods in numerous software environments that can be used to align them, for example, in Fiji ImageJ [15] IMOD [16, 17] or Amira (Thermo Fisher Scientific™). In order to account for the different software equipment and user preferences, we will describe the process of aligning and cropping AT data for Fiji ImageJ and Amira.
22. Loading the data as a Virtual Stack enables Fiji ImageJ to handle stacks of any size. The output of the alignment is created in Random Access Memory (RAM). However, stack registration and thus the maximum size of the stacks that can be processed is limited in this format. Alternatively, use the "register virtual stack slices," which is a folder-to-folder implementation of the same registration algorithm. Once registration is complete, load the output data as a virtual stack.
23. The workflow described in the following steps is available in any Amira version after 3.x version and enables the operator to align image stacks semiautomatically. The described workflow is especially useful in cases in which the stack cannot be aligned using the automatic functions in other software.

24. Amira 2020.3 and newer versions have a new feature called the "EM toolbox," which contains new alignment features for rigid, affine, and elastic stack alignment. Stacks that are larger than the amount of RAM in the processing PC can be batch-processed folder-to-folder.
25. You can convert these data to "large data access" files.
26. The differential image display very clearly shows whether the alignment was successful. If it was, hit the space bar to advance to the next image pair and repeat. Otherwise, proceed to the next step, and if the automatic alignment does not work satisfactory, try to automatically adjust the entire stack by clicking the "align all slices." You can follow the procedure and stop it if there are errors.
27. If the image pair moves out of the Viewer, the pair (and the rest of the stack) can be re-centered by holding down the shift key and click-dragging. Once the alignment is satisfactory, go to the next image pair by hitting the space bar. Continue from **step 10**. In some cases, only a part of an image pair can be aligned perfectly, while other parts are still not aligned well. This is usually caused by a variable compression or other distortions of the sections. Image stacks that are recorded from such section sets are very challenging for most fully automated stack alignment functions including the ones in Fiji ImageJ. Using the manual process described here makes it possible to salvage at least some of these stacks.
28. If you do not want to crop, check the "automatic size" function in the "align menu-options," output tab. Here, you can also select a different color for the canvas. Now, when resampling the stack, the canvas of the output data will be enlarged so that there is no cropping.
29. The voxel parameter settings are very important because they enable you to accurately create a 3-D model with all of the sizes in three dimensions: x, y, and z. For each of the dimensions (x, y, z), enter the values that were obtained during the scanning.
30. In MIB, the most useful tool is the "brush," which is used to mark and paint the analyzed structures; however, each structure requires an individual approach. It is the most precise and time-consuming part for marking the area. The "BW thresholder" tools can be used to mark the structures with a high contrast, for example, the lipid bodies (Fig. 5; however, it is best to select the appropriate tools for the analyzed sections by watching the official MIB website tutorial available on YouTube).
31. Left scrolling in MIB enables you to move between individual sections, which is especially important while coloring the selected structures.

32. In this software, it is possible scroll sections and to zoom in and zoom out on the images.
33. If "generate surface" is not in the list, open the "search window" at the top of the menu.
34. Amira's toolbox is equipped with a large number of tools for editing and deriving statistics from the surface objects. User manual and tutorials for more information are recommended.
35. Amira can also be used to filter, align, and reconstruct the structures in the microscope images. Amira offers manual, semiautomatic, and fully automatic methods for all of the steps in the data evaluation workflow.

References

1. Sowiński P, Rudzińska-Langwald A, Kobus P (2003) Changes in plasmodesmata frequency in vascular bundles of maize seedling leaf induced by growth at sub-optimal temperatures in relation to photosynthesis and assimilate export. Environ Exp Bot 50:183–196
2. Gunning BES, Hardham AR, Hughes JE (1978) Evidence for initiation of microtubules in discrete regions of the cell cortex in Azolla root-tip cells, and an hypothesis on the development of cortical arrays of microtubules. Planta 143:161–179
3. Zhu T, Rost TL (2000) Directional cell-to-cell communication in the Arabidopsis root apical meristem III. Plasmodesmata turnover and apoptosis in meristem and root cap cells during four weeks after germination. Protoplasma 213:99–107
4. Schubert M, Koteyeva NK, Zdyb A et al (2013) Lignification of cell walls of infected cells in Casuarina glauca nodules that depend on symplastic sugar supply is accompanied by reduction of plasmodesmata number and narrowing of plasmodesmata. Physiol Plantarum 147:524–540
5. Briggman KL, Bock DD (2012) Volume electron microscopy for neuronal circuit reconstruction. Curr Opin Neurobiol 22:154–161
6. Wacker I, Chockley P, Bartels C et al (2015) Array tomography: characterizing FAC-sorted populations of zebrafish immune cells by their 3D ultrastructure. J Microsc 259:105–113
7. Russell MR, Lerner TR, Burden JJ et al (2017) 3D correlative light and electron microscopy of cultured cells using serial blockface scanning electron microscopy. J Cell Sci 130:278–291
8. Furuta KM, Yadav SR, Lehesranta S et al (2014) Arabidopsis NAC45/86 direct sieve element morphogenesis culminating in enucleation. Science 345:933–937
9. Płachno BJ, Świątek P, Jobson RW et al (2017) Serial block face SEM visualization of unusual plant nuclear tubular extensions in a carnivorous plant (Utricularia, Lentibulariaceae). Ann Bot 120:673–680
10. Zechmann B, Zellnig G (2017) 3D reconstruction of zucchini-and tobacco yellow mosaic virus induced ultrastructural changes in plants. Microsc Microanal 23:1220–1221
11. Reagan BC, Kim PJY, Perry PD et al (2018) Spatial distribution of organelles in leaf cells and soybean root nodules revealed by focused ion beam-scanning electron microscopy. Funct Plant Biol 45:180–191
12. Godel-Jędrychowska K, Kulinska-Lukaszek K, Horstman A et al (2020) Symplasmic isolation marks cell fate changes during somatic embryogenesis. J Exp Bot 71:2612–2628
13. Paterlini A, Belevich I, Jokitalo E et al (2020) Computational tools for serial block electron microscopy reveal plasmodesmata distributions and wall environments. Plant Physiol 184:53–64
14. Belevich I, Joensuu M, Kumar D et al (2016) Microscopy image browser: a platform for segmentation and analysis of multidimensional datasets. PLoS Biol 14:e1002340
15. Schindelin J, Arganda-Carreras I, Frise E et al (2012) Fiji: an open-source platform for biological-image analysis. Nat Methods 9:676–682
16. Mastronarde DN (2006) Tomographic reconstruction with the IMOD software package. Microsc Microanal 12:178–179
17. O'Toole E, van der Heide P, McIntosh JR et al (2018) Large-scale electron tomography of cells using SerialEM and IMOD. In: Cellular imaging. Springer, pp 95–116

Chapter 5

Serial Block Electron Microscopy to Study Plasmodesmata in the Vasculature of *Arabidopsis thaliana* Roots

Andrea Paterlini and Ilya Belevich

Abstract

Serial block electron microscopy (SB-EM) is a technique that enables acquisition and reconstruction of 3D cellular volumes. The approach is valuable for the study of plasmodesmata (PD) as the relative positions of these structures are contained in the datasets. In this chapter, we describe how to prepare plant roots for SB-EM via fixation, embedding, and trimming steps. We also provide details and recommendations for later image acquisition and processing. The procedure is suitable to work on root vascular tissues.

Key words Plasmodesmata, Phloem, *Arabidopsis thaliana*, Root, Serial block electron microscopy, Imaging

1 Introduction

In the study of plasmodesmata (PD), the small channels connecting plant cells, an experimental tension exists between capturing the structural details of individual PD [1—as an example] and describing PD populations at a cellular scale [2—as an example]. The two goals can be achieved using microscopy techniques differing in their resolution and volume capturing powers. Electron microscopy approaches, when applied to PD, offer tools to tackle both scales (see this and other chapters in the book). Here, we provide a protocol to generate PD-containing datasets using serial block electron microscopy (SB-EM) [3]. Densities and spatial arrangements of PD over entire interfaces can be extracted, along with information on the cell wall the PD span, even for cells deep in plant tissues [4–6].

The defining feature of SB-EM is the presence of a remotely controlled microtome in the chamber of the microscope. Imaging the uppermost surface of the sample block and trimming it away with a controllable z-step, to expose a new surface, are alternating

Yoselin Benitez-Alfonso and Manfred Heinlein (eds.), *Plasmodesmata: Methods and Protocols*,
Methods in Molecular Biology, vol. 2457, https://doi.org/10.1007/978-1-0716-2132-5_5,

phases of data acquisition [3]. In the final dataset, the obtained images are stitched together with each being spaced from the other by the predefined thickness of the cut slice. This distance can be very small (40 nm or even smaller), resulting in good preservation of information. A central advantage of SB-EM over separate EM images is that the relative positions of multiple cellular objects (PD in our case) are retained within the full volumes of the datasets. The objects and their coordinates can then be rendered in 3D [reviewed in 7] or extracted for analyses [6].

We display such properties in the context of cell types associated with the phloem, one of the long-distance transport systems in plants [8—and associated contributions to that volume]. The protocol is especially useful to investigate PD in the area of the root tip where phloem unloading occurs [4]; however, it can be broadly applied to all PD within the root stele.

2 Materials

2.1 Sample Preparation

1. *Plant growth medium*: 0.22% Murashige and Skoog Basal Salts, 0.05% 2-(N-morpholino)ethanesulfonic acid (MES), 1% sucrose, 1% Agar, pH 5.7.
2. *Fixative solution*: 2.5% glutaraldehyde and 2% formaldehyde in 0.1 M sodium-cacodylate buffer, pH 7.4 supplemented with 2 mM calcium chloride ($CaCl_2$).
3. *Storage solution*: 0.1 M sodium-cacodylate buffer, pH 7.4.
4. *Washing solution*: 0.1 M sodium-cacodylate buffer, pH 7.4 supplemented with 2 mM $CaCl_2$.
5. *Osmium ferrocyanide solution*: 2% osmium tetroxide (OsO_4), 1.5% potassium ferrocyanide ($K_4[Fe(CN)_6]$), 2 mM $CaCl_2$ in 0.1 M ice-cold sodium-cacodylate buffer.
6. *Osmium solution*: 2% OsO_4.
7. *Uranyl solution*: 1% uranyl acetate.
8. *Thiocarbohydrazide (TCH) solution*: add 100 mg of TCH to 10 ml of distilled water and place the mixture in a 60 °C oven for 1 h. Swirl every 15 min to facilitate dissolution of the powder. Filter the cooled solution through a 0.22-μm syringe filter before use. Prepare fresh.
9. *Walton's lead aspartate solution*: prepare an initial aspartic acid stock solution by dissolving 1.0 g of L-aspartic acid in 250 ml of distilled water. To dissolve aspartic acid, raise pH to 3.8. The stock solution is stable for 1–2 months, if refrigerated. To make the staining solution dissolve 66 mg of lead nitrate in 10 ml of aspartic acid stock, adjust pH to 5.5. Place the obtained

solution in a 60 °C oven for 30 min before using it in the protocol (ensure that no precipitate forms).

10. *Ethanol solutions*: 100% ethanol and 30%, 50%, 70%, 90%, and 96% solutions in distilled water.
11. *Durcupan ACM resin*: Durcupan is prepared using four components named A (epoxy resin), B (hardener), C (accelerator), and D (dibutyl phthalate). Weight the following amounts for the various components: A 11.4 g; B 10 g; C 0.3 g (or 300 μl equivalent); D 0.05–0.1 g (50–100 μl equivalent). During the first part of the procedure, only use a mixture of A + B + D components. Mix the components with extreme caution to avoid formation of bubbles. If bubbles still develop, remove them using vacuum suction. Dilute the A + B + D mixture with acetone in a 1:3 ratio (embedding solution 1), 1:1 ratio (embedding solution 2), and 3:1 ratio (embedding solution 3). Use a pure A + B + D mixture as embedding solution 4. Use the full A + B + C + D pure mixture (embedding solution 5) only in the last steps, as the accelerator component C makes the resin very viscous, affecting the penetration of the resin into the sample.
12. *Silicone holders.*
13. *Cutting equipment*: an EM TRIM2 trimming machine (Leica Microsystems) and an EM Ultracut UC6i ultramicrotome (Leica Microsystems) equipped with a diamond knife. Equivalent machines might be compatible with the protocol. Parafilm tape and a razor blade are also required.
14. *Light and confocal microscopes.*
15. *Toluene blue stain.*
16. *Sample pin*: 3View pin (EM resolutions).
17. *Silver paint.*
18. *Platinum.*
19. *Coating equipment*: Q150T Plus coater (Quorum Technologies, UK). Equivalent machines might be compatible with the protocol.

2.2 Sample Imaging

1. *Imaging equipment.* A Quanta 250 FEG Scanning Electron Microscope (SEM) (Thermo-Fisher, Waltham, MA) with a backscattered electron detector (Gatan, Pleasanton, CA). The microscope needs to be equipped with a 3View system (Gatan, Pleasanton, CA). Alternative SEM setups supplied with SB-EM equipment are also compatible with this protocol.
2. *Imaging software.* Gatan Microscopy Suite 3 Software (GMS 3) supplied with the 3View system.

2.3 Image Processing

1. *Image processing software*. Use the *Microscopy Image Browser* (MIB) [9], a freely available image processing and segmentation software designed for SB-EM workflows. Download and install it from http://mib.helsinki.fi. Find extensive support, including videos and tutorials, at http://mib.helsinki.fi/tutorials.html and http://mib.helsinki.fi/features_all.html or from the *Help* section within the MIB software.

2.4 Data Visualization

1. *3D rendering software*. Amira (Thermo Fisher Scientific) and Imaris (Oxford Instruments) commercial options are recommended, but open-source alternatives are also possible.

3 Methods

3.1 Sample Preparation

1. An overview of the sample preparation steps is provided in Fig. 1a.
2. *Plant material*. Grow *Arabidopsis thaliana* seedlings for 5 days in plastic plates containing plant medium, under controlled conditions (16 h light, 23 °C) (*see* **Note 1**).
3. *Fixation*. Cut the seedlings below the root-hypocotyl junction and immediately fix the root containing segment (about 0.5–1 cm) for 5 min under vacuum. After the vacuum step, fix the roots for 2–3 h at normal pressure (*see* **Note 2**). Post-fixation, store the roots in the storage solution until further processing.
4. *Staining*. Wash roots five times (3 min each) with ice-cold washing solution; incubate them for 1 h in osmium ferrocyanide solution; wash them five times (3 min each) with distilled water; incubate them for 20 min in the 1% TCH solution; wash them five times (3 min each) with distilled water; incubate them for 30 min in osmium solution; wash them five times (3 min each) with distilled water; incubate them overnight in uranyl solution at 4 °C; wash them five times (3 min each) with distilled water; incubate them for 30 min in Walton's lead aspartate solution at 60 °C; wash them five times (3 min each) with distilled water (*see* **Note 3**).
5. *Dehydration*. Expose samples to an ethanol gradient: 5 min in 30%, 50%, 70%, 90%, 96%, and 100% ethanol solutions. Afterward, treat samples for 10 min in ice-cold acetone.
6. *Embedding*. During this step, samples might need to be transferred between multiple containers. Manipulate them using a wooden stick or a brush as roots easily adhered to these tools without experiencing damage. Embed the dehydrated specimens: 2 h in embedding solution 1; 2 h in embedding solution 2; overnight in embedding solution 3; 1–2 h in fresh

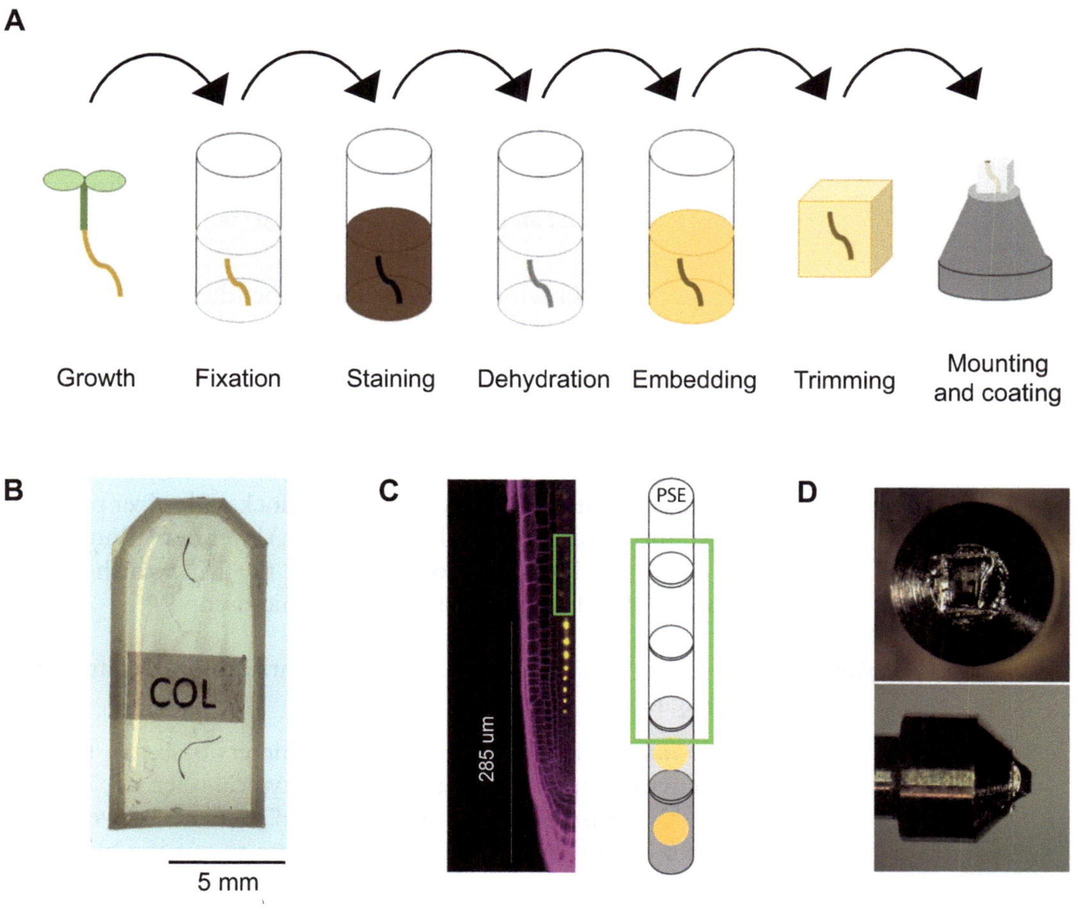

Fig. 1 Preparation of *Arabidopsis* roots for SB-EM. (**a**) Cartoon depicting the sample processing steps described in the protocol. (**b**) Col-0 roots in polymerized Durcupan resin at the end of the embedding procedure. (**c**) *CALS7:H2B-YFP* expression in Col-0 roots, imaged using confocal microscopy. Signal is restricted to nucleated proto sieve element cells (PSE), and it disappears after cell enucleation (green box). (**d**) Stained block mounted on a 3View pin (frontal and side views)

embedding solution 4 at 50 °C; 1–2 h in embedding solution 4 at 50 °C and 2 h in embedding solution 5 at 50 °C. Ultimately, transfer the roots to silicone holders containing embedding solution 5 and bake them for 48 h at 60 °C. An example of final block is visible in Fig. 1b.

7. *Shootward trimming*. Remove excess of plastic next to the root tip using a trimming machine. Cut the area around the tip in the shape of a flat-top pyramid, for ease of manipulation in following steps. Mount the block in an ultramicrotome equipped with a diamond knife. Trim the exposed root toward the mounted back of the root. This trimming enables to reach the area of interest in a controlled way, monitoring the reached distance from the root tip. In the case of enucleated sieve elements (SE)—those performing phloem unloading in the

root—remove around 240 μm. Determine the specific length studying the *CALS7:H2B-YFP* marker (nuclear fluorescence in the proto SE) under confocal microscopy (Fig. 1c) (*see* **Note 4**). Cut a histological section, stain it with toluene blue, and observe it under a light microscope to ensure that enucleation has not yet occurred, in order to fully acquire the first cell of interest. Enucleated cells would appear white upon toluene staining.

8. *Rootward trimming*. Shorten the embedded roots to a final approximate length of 500–600 μm (from the exposed, trimmed root tip) by placing them on a piece of a Parafilm tape and cutting them, at the rootward end, using a razor blade (*see* **Note 5**). Mount the trimmed blocks by the ground side to a 3View pin (Fig. 1d).
9. *Coating*. Brush the sides of the final block with silver paint and coat the whole block with a thin (5 nm) layer of platinum using a coater machine (*see* **Note 6**). Insert the pin with the coated root in the chamber of the electron microscope.

3.2 Sample Imaging

The cartoon in Fig. 2a depicts the sequential process of trimming and imaging.

1. *Parameters*. Define the imaging parameters based on the conductivity properties of the specimen; typical reference ones are 2.5-kV beam voltage, spot size 3, and pressure between 0.15 and 0.3 Torr. Set the 3View microtome to cut the block with 40-nm increments (a finer step of 30 nm improves Z resolution but at the expense of imaging time), moving from the root tip stump to the back of the root. Use a SEM imaging XY resolution of 7–10 nm. For calibration purposes, note the Z position when the microtome inside the SEM chamber starts to cut. This information will be later used to recalculate coordinates of each collected EM volume to physical root coordinates from the tip.
2. *Run*. Although image acquisition by the software is automated, perform an initial visual evaluation (and later occasional checks during the run) to ensure that the block is devoid of cracks and well stained throughout. Image the roots using the *region of interest (ROI) mode* of GMS 3, which includes a survey image (Fig. 2b) (*see* **Note 7**). If the area of interest (Fig. 2c) drifts during acquisition, stop the run whenever the area approaches the edge of the ROI and perform manual drift adjustment. Shift the ROI and launch a new run (label each of these sub-runs “Run1a,” “Run1b,” and so forth). To place the high-resolution ROI images into the context of the whole root, take low-resolution overviews of the whole root cross section at 1 μm intervals (every 25th slice with the settings

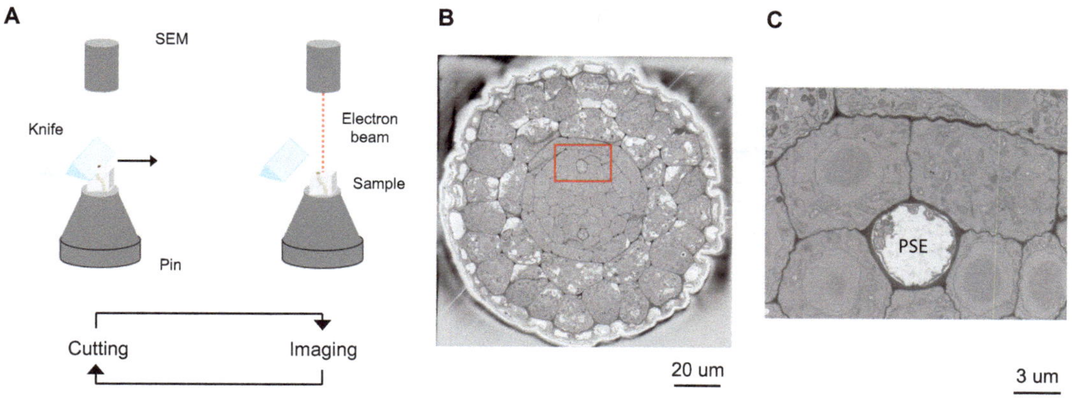

Fig. 2 SB-EM image acquisition. (**a**) Cartoon depicting the alternating phases of image acquisition within the EM chamber. (**b**) Overview of the whole root section, at the start of a run, with a ROI highlighted in red. Cells of the stele, but not those in more external cell layers, displays good preservation. (**c**) ROI displaying an enucleated PSE

described here). Store all collected images in Gatan DM4 format as 16-bit unsigned integers.

3.3 Image Processing

As it would be impossible to fully describe the processing of the images here, only the essential steps users should consider and the appropriate tools/functions to perform them are highlighted here:

1. *Format conversion and extraction of metadata.* Convert the obtained images from a proprietary *.dm4* format to a 16-bit . *tif* format, to process them. Generate text header files (*.hdr*) with metadata (regarding the parameters of image acquisition) for each image as well (Fig. 3a). Batch perform these two steps for multiple images using a custom script, compatible with the GMS 3 software, available at https://www.helsinki.fi/en/infrastructures/bioimaging/3d-electron-microscopy-imaging#section-47617. As individual runs may be small (about 100 sections), move images from multiple sub-runs into a single directory containing a total of about 1000 files in each (*see* **Note 8**).
2. *Assembling images into stacks and calibrating them.* Load the . *tif* files in each directory in MIB and assemble them using the *Combine selected dataset* function (Fig. 3b). If a frame made of empty pixels is generated in the process (*see* **Note 9**), define the color of the frame but note that the default value of 65535, corresponding to while color, is recommended. Calibrate the generated stack (in order to embed spatial coordinates in the images) using the *Bounding box* function and the metadata in the text header files. Convert the Z coordinate of the stage from the one present in the first metadata file to the physical

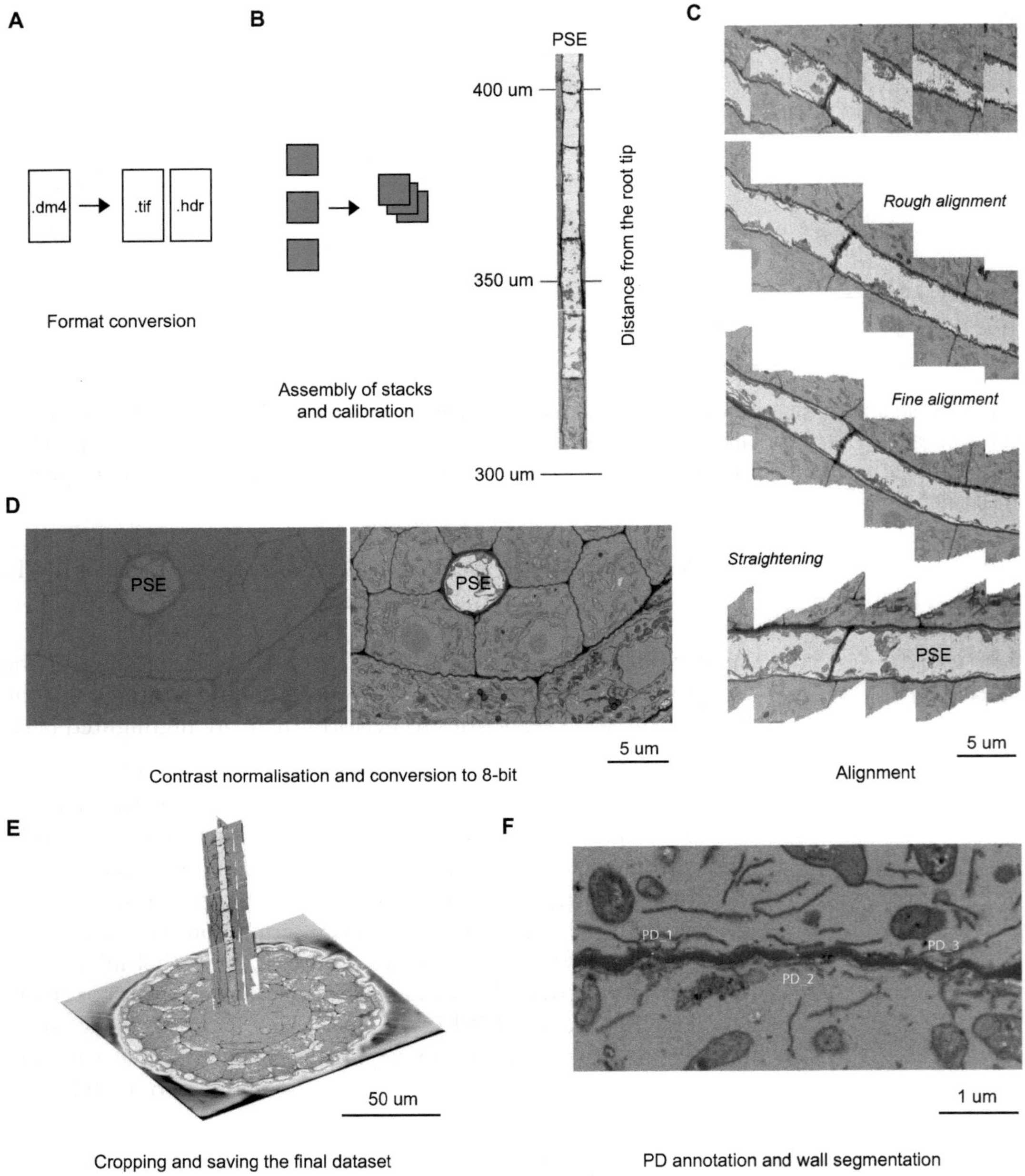

Fig. 3 Relevant image processing steps for SB-EM datasets. (**a**) Cartoon depicting file format conversion. (**b**) Cartoon displaying stack assembly. An example of a final dataset embedded with coordinates relative to the root tip is shown to the right. The dataset focuses on a cell file of PSEs. (**c**) Stack alignment steps. (**d**) Contrast normalization and conversion to 8-bit file format. (**e**) Final dataset loaded in Amira. (**f**) Section with a segmented wall and annotated PD

distance from the tip, as noted at the start of the run (Fig. 3b showing a dataset with embedded root spatial coordinates).

3. *Aligning stacks.* Align the stacks using the *Alignment tool* of MIB (Fig. 3c). In most cases, perform the alignment in three

stages: a rough alignment for sub-runs, a fine one to compensate for local drift between images, and a final one to straighten the root. Initially match individual sub-runs to each other using the *Single landmark point algorithm*: use the *brush tool* to mark corresponding areas at junctions between sub-runs and launch the procedure to align those areas. For the fine alignment, use the *brush tool* to draw an area around the cell of interest on the first and last slice of the dataset. Interpolate the area (*interpolate as a shape*) across all slices of the loaded dataset. Perform the actual alignment using the *Drift correction algorithm* (Subarea: selection), which detects drifts using cross-correlation in Fourier space and matches the images using translation transformation across the entire stack. Lastly, compensate for the issue of natural sample drift (*see* **Note 10**) post-acquisition using the *Single landmark point alignment* algorithm. In practice, draw a circle in the middle of the ROI on the first and last slice of a run and interpolate the positions of the spots on the intermediate slices. The images are then aligned to straighten the resulting points in a vertical line, achieving the vertical views visible in Fig. 3c. Store the translation coefficients to redeploy them if needed.

4. *Contrast normalization and conversion to 8-bit.* To ensure organelles throughout the stack have homogeneous intensity signals, perform a contrast normalization procedure for the dataset (*see* **Note 11**). Select areas of the dataset that have similar distribution of intensities (based on the histograms), with the *brush* and *interpolation tools*. Launch the *Normalize layers based on masked areas operation* function, which will then adjust the contrast across them. Convert the normalized datasets to 8-bit unsigned integer classes, which occupy half the memory space and are loaded and processed twice as fast (Fig. 3d).
5. *Cropping and saving.* Crop away the frame generated during assembly (and any other area not of interest) using the *Crop dataset* tool of MIB and save the resulting dataset. Use Amira mesh binary (*.am*) format for saving images as it can be quickly loaded in MIB and is compatible with embedded coordinates for bounding boxes (Fig. 3e as an example of a dataset loaded in Amira) (*see* **Note 12**).
6. *Annotation of PD.* Mark PD positions (X, Y, and Z coordinates) on the image stack using the *Annotation tool* of MIB (Fig. 3f). Place the annotation near the center of the PD, or as close to that as possible, if studies on PD distributions are of interest. In order to avoid duplicate counts and achieve conservative yet reliable total estimates, do not place any new PD within a set Z-window above and below an existing annotation. With the acquisition settings employed, 160 nm (+2, −2 slices)

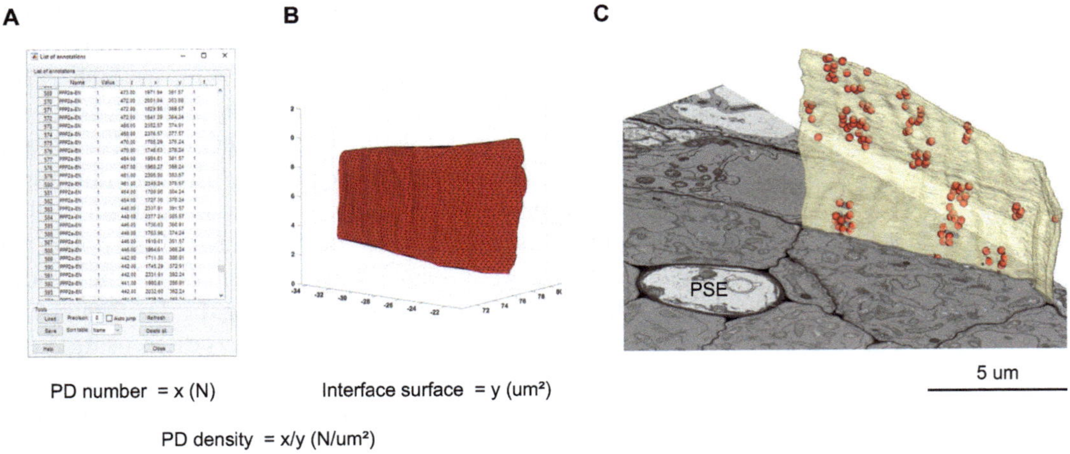

Fig. 4 PD data analysis and 3D visualization. (**a**) *Annotation list* panel enumerating the PD annotated at a cellular interface. (**b**) Visual output of the *SurfaceArea3D* plug-in, when applied to a segmented wall. (**c**) 3D rendering in Amira software of a segmented wall (pale yellow) and its PD (red dots). The segmented interface is that between a phloem pole pericycle cell and an endodermis one. The PSE is indicated on the image

seemed effective. Save the annotations in MIB *.ann* format, which can be reloaded at later stages (from the *Annotation list* interface) (Fig. 4a). Annotations can also be exported to other formats (such as Microsoft Excel).

7. *Segmentation of cell walls.* Prior to segmentation, filter the images to minimize noise using the *Image filter* function. Use edge-preserving denoising algorithms, as it is important and MIB has a large collection of those (*see* **Note 13**). To segment cell walls, use black-and-white (BW) thresholding (Fig. 3f) applying the *BW algorithm* on masked areas spanning the entire z-volume containing the wall. To do so, define the mask manually on some slices using the *brush tool* and derive it with the *interpolation function* on intermediate slices. Determine the precise values for the BW thresholding visually, inspecting the preview across the volume being segmented (*see* **Note 14**). After segmentation, perform manual removal of spurious segmented areas with the *brush tool*. After cleaning, temporarily convert the resulting models back to selections and smooth them to uniform edges (*Smooth selection* function). Empirically define values for smoothing (X, Y, Z kernel sizes). Filter the resulting materials (using *Model statistics*) to remove unreliable small fragments (a cutoff area of 5 pixels with 4/6 connectivity is advisable). Save segmentations in MIB *.model* format.

3.4 Data Analysis

1. Once careful PD annotations and high-quality (wall) segmentations have been achieved, perform quantitative analyses. The simplest of them is a PD density assessment, calculated dividing the number of PD annotations at a particular interface by the

surface of the same (in um^2). Read off the former from the *Annotation list* (Fig. 4a), while calculate the latter using the *SurfaceArea3D* plug-in available from the *Plugin* menu. The tool produces a 3D rendered surface within the MIB environment and a value for its area (Fig. 4b) (*see* **Note 15**).

3.5 Data Visualization

1. Export object features (such as PD and wall surfaces) directly from the MIB environment to compatible software using the *Export function*. Render them with color and aspect features of preference (Fig. 4c).

4 Notes

1. Seedlings of different ages, grown in alternative, specific conditions, can also be employed with minimal to no adjustments. Complications are more likely to arise when imaging a different area of the root, other than the root tips described here. Secondary metabolites or highly vacuolated cells in more mature parts of the root could impact fixation and staining. More substantial protocol adjustments would be required in those cases.
2. Cells within the *Arabidopsis* root vascular cylinder, the area of focus of this chapter, are well preserved upon chemical fixation. However, more external cell layers display clear damage (Fig. 2b). Should those tissues be of interest to the reader, different fixation protocols might be required. In principle, cryo-fixation (with potential better retention of cellular features and overall root preservation) is now compatible with SB-EM [10].
3. The staining protocol is based on the NCMIR method [11].
4. The *CALS7:H2B-YFP* fluorescent signal fades around enucleation [12], at 285 μm from the root tip in 5-day-old seedlings. This is because the nucleus (and its histone H2B proteins) is being degraded (Fig. 1c). The seedlings for confocal microscopy should experience the same growth conditions as the roots employed for fixation. Different markers can provide positional information on different cell types/developmental stages. The choice is heavily dependent on the biological process of interest to the reader. Overall, markers can inform the trimming step and reduce the time later spent imaging areas that are not of interest.
5. It is important to trim the root as much as possible to minimize charging and maximize conductivity of the specimen.
6. The coating treatment improves electron conductivity and reduces charging.

7. The *ROI mode* of GMS 3 is highly useful as it allows the user to place multiple ROIs on the image (typically closely fitting the areas of interest in order to speed up the imaging) and quickly adjust the same during the run should it be necessary (to compensate for drifts due to the tilt of the root within the block).
8. Combining sub-runs in a single directory is not compulsory, but we recommend doing this as it allows to simultaneously speed up the conversion process and to later create stacks of 1000 images, which are ideal for further processing.
9. As MIB can combine images with different width and height dimensions, a frame made of empty pixels might be introduced at some edges of the assembled dataset (to match image dimensions across the stack).
10. While the initial alignment steps are especially relevant when multiple sub-runs need to be stitched together, the final straightening steps are useful to decrease memory consumption, speed up workflows, and generate better visual renderings. As mentioned in the protocol, the area of interest is likely to drift across the field of view during acquisition. This is a normal effect caused by the angle of the root within the block. Even after correctly stitching sub-runs and performing the fine alignment, the effect would remain. The straightening step resolves this aspect.
11. The *Contrast* submenu of MIB provides multiple methods depending on the complexity of contrast disparities.
12. Alternative formats (*.tif*, *.nrrd*, *.mrc*) may also be used, but it is important to double-check that the bounding box information is preserved.
13. The choice of denoising algorithms largely depends on the amount of noise present. Deep neural network algorithms [13] or a BM3D filter [14] is useful for the type of images described here.
14. MIB enables multiple segmentation approaches, in addition to BW thresholding. Built-in procedures for training convolutional neural networks for image segmentation have recently been introduced [15]. The approach is in principle applicable for the cell walls described in this protocol.
15. For more elaborate applications, such as assessments of PD distributions and wall thicknesses, additional plug-ins and analysis pipelines are required. Refer to the detailed examples provided in [6]. Confident users can also develop their own plug-ins for data extraction (in MATLAB programming language), fit for their specific questions. MIB is an open-source software and developers welcome contributions. Computational approaches to interrogate the data are also manifold.

Acknowledgments

We would like to thank Yrjö Helariutta (Sainsbury Laboratory—University of Cambridge) and Eija Jokitalo (University of Helsinki) for their academic and professional support. We want to acknowledge Mervi Lindman and Antti Salminen for their insights into sample preparation.

References

1. Nicolas WJ, Grison MS, Trépout S, Gaston A, Fouché M, Cordelières FP, Oparka K, Tilsner J, Brocard L, Bayer EM (2017) Architecture and permeability of post-cytokinesis plasmodesmata lacking cytoplasmic sleeves. Nat Plants 3:17082
2. Danila FR, Quick WP, White RG, Furbank RT, von Caemmerer S (2016) The metabolite pathway between bundle sheath and mesophyll: quantification of plasmodesmata in leaves of C3 and C4 monocots. Plant Cell 28:1461–1471
3. Denk W, Horstmann H (2004) Serial block-face scanning electron microscopy to reconstruct three-dimensional tissue nanostructure. PLoS Biol 2:e329
4. Ross-Elliott TJ, Jensen KH, Haaning KS, Wager BM, Knoblauch J, Howell AH, Mullendore DL, Monteith AG, Paultre D, Yan D, Otero S, Bourdons M, Sager R, Lee J, Helariutta Y, Knoblauch M, Oparka KJ (2017) Phloem unloading in Arabidopsis roots is convective and regulated by the phloem-pole pericycle. eLife 6:e24125
5. Yan D, Yadav SR, Paterlini A, Nicolas WJ, Petit JD, Brocard L, Belevich I, Grison MS, Vaten A, Karami L, El-Showk S, Lee J, Murawska GM, Mortimer J, Knoblauch M, Jokitalo E, Markham JE, Bayer EM, Helariutta Y (2019) Sphingolipid biosynthesis modulates plasmodesmal ultrastructure and phloem unloading. Nat Plants 5:604–615
6. Paterlini A, Belevich I, Jokitalo E, Helariutta Y (2020) Computational tools for serial block electron microscopy reveal plasmodesmata distributions and wall environments. Plant Physiol 184:53–64
7. Kittelmann M, Hawes C, Hughes L (2016) Serial block face scanning electron microscopy and the reconstruction of plant cell membrane systems. J Microsc 263:200–211
8. Holbrook NM, Knoblauch M (2018) Phloem: a supracellular highway for the transport of sugars, signals, and pathogens. Curr Opin Plant Biol 43:3–7
9. Belevich I, Joensuu M, Kumar D, Vihinen H, Jokitalo E (2016) Microscopy image browser: a platform for segmentation and analysis of multidimensional datasets. PLoS Biol 14:e1002340
10. Tsang TK, Bushong EA, Boassa D, Hu J, Romoli B, Phan S, Dulcis D, Su C, Ellisman MH (2018) High-quality ultrastructural preservation using cryofixation for 3D electron microscopy of genetically labeled tissues. eLife 7:e35524
11. Deerinck TJ, Bushong EA, Thor A, Ellisman MH (2010) NCMIR methods for 3D EM: a new protocol for preparation of biological specimens for serial block face scanning electron microscopy. National Center for Microscopy and Imaging Research (NCMIR) (v7_01_10). https://ncmir.ucsd.edu/sbem-protocol
12. Furuta KM, Yadav SR, Lehesranta S, Belevich I, Miyashima S, Heo JO, Vaten A, Lindgren O, De Rybel B, Van Isterdael G, Somervuo P, Lichtenberger R, Rocha R, Thitamadee S, Tahtiharju S, Auvinen P, Beeckman T, Jokitalo E, Helariutta Y (2014) Arabidopsis NAC45/86 direct sieve element morphogenesis culminating in enucleation. Science 345:933–937
13. Zhang K, Zuo W, Chen Y, Meng D, Zhang L (2017) Beyond a gaussian denoiser: residual learning of deep CNN for image denoising. IEEE Trans Image Process 26:3142–3155
14. Dabov K, Foi A, Katkovnik V, Egiazarian K (2007) Image denoising by sparse 3-D transform-domain collaborative filtering. IEEE Trans Image Process 16:2080–2095
15. Belevich I, Jokitalo E (2021) DeepMIB: user-friendly and open-source software for training of deep learning network for biological image segmentation. PLoS Comput Biol 17:e1008374

Chapter 6

Focused Ion Beam-Scanning Electron Microscopy for Investigating Plasmodesmal Densities

Brandon C. Reagan, John R. Dunlap, and Tessa M. Burch-Smith

Abstract

Plasmodesmata (PD) facilitate the exchange of nutrients and signaling molecules between neighboring plant cells, and they are therefore essential for proper growth and development. PD have been studied extensively in efforts to elucidate the ultrastructure of individual PD nanopores and the distribution of PD in a variety of cell walls. These studies often involved the use of serial ultrathin sections and manual quantification of PD by transmission electron microscopy (TEM). In recent years, a variety of techniques that offer more amenable approaches for quantifying PD distribution have been reported. Here, we describe the quantification of PD densities using the serial scanning electron microscopy technique called focused ion beam-scanning electron microscopy (FIB-SEM). For this, resin-embedded samples prepared by standard TEM methods undergo successive rounds of imaging by SEM interspersed with milling of the sample surface by a focused beam of gallium ions to reveal a new surface. In this way, the details of the sample are sequentially revealed and imaged. Over the course of a few hours, repetitive milling and imaging facilitates the automated collection of nanometer-resolution data of several μm of sample depth. FIB-SEM can be targeted to interrogate specific cell walls and cell wall junctions, and the subsequent three-dimensional renderings of the data can be used to visualize the ultrastructural details of the sample. PD densities can then be rapidly quantified by calculating the number of PD per μm^2 of cell wall observed in the renderings.

Key words Focused ion beam-scanning electron microscopy, Plasmodesmata, Plant cell wall, Milling, Electron microscopy, Tomography, 3D reconstruction

1 Introduction

Intercellular communication is critical for the development of multicellular organisms. Plasmodesmata (PD) are membrane-lined pores that perforate plant cell walls and form direct cytoplasmic connections between the neighboring cells, facilitating the movement of metabolites and signaling molecules between cells. PD are delimited by the plasma membranes of the connected cells, and they contain a second membranous continuity called the desmotubule, a highly derived tightly furled tubule of endoplasmic

Yoselin Benitez-Alfonso and Manfred Heinlein (eds.), *Plasmodesmata: Methods and Protocols*,
Methods in Molecular Biology, vol. 2457, https://doi.org/10.1007/978-1-0716-2132-5_6,

reticulum (ER) that connects the ER networks of adjacent cells. The current view is that most, if not all, PD in higher plants contain a desmotubule, and some macromolecules use its surface for intercellular spread [1, 2]. These two membranes surround the cells' cytoplasmic continuum, the cytoplasmic sleeve, through which the bulk of diffusive trafficking occurs [3–6].

The typical textbook image or illustration of a plasmodesma presents a single cylindrical pore with a diameter between 20 and 50 nm. These representations of PD structure come from years of imaging PD by TEM [7]. This stereotypical structure is called a simple plasmodesma, and while such PD are indeed common in many plant tissues, PD can also occur in more elaborate structures. Twinned PD are two pores found within 100 nm of each other, and they are likely sites of de novo PD formation [8]. PD can also adopt branched structures with multiple pores that converge at the middle lamella of the cell wall, often forming a dilated region called the central cavity [9]. These different structural classes of PD likely have distinct trafficking capacities [10, 11]. In some instances, simple PD have higher trafficking capacity than do branched PD [10], but an association between twinned and branched PD and increased intercellular trafficking has also been reported [12].

PD can be randomly distributed in a cell wall but often cluster together into groups called pit fields [8, 13, 14]. In recent years, it has become apparent that the number of PD in a cell wall (PD density) and PD trafficking capacity are inherently linked [15–17]. Several studies have investigated the distribution of PD in different tissues and across plant development [18–22]. These studies have used transmission electron microscopy (TEM) to image thin sections and often serial thin sections, because PD, with their 20–50-nm diameters, are too small to visualize with light microscopy. While this is a powerful method for quantifying PD densities, it is technically difficult to undertake and requires extensive training or expertise as well as significant hours of data collection and analysis. Recently, alternative methods for quantifying PD densities have been introduced that are less time-consuming and that can image larger samples. This flexibility in sample size allows for the collection of longer lengths of cell wall with higher z-resolution than other approaches.

While individual PD are too small to image with light microscopy, pit fields can be imaged with confocal microscopy using fluorescently labeled PD marker proteins like the PLASMODES MATA LOCATED PROTEINS (PDLPs) or viral movement proteins [17, 23]. Fitzgibbon et al. first demonstrated this approach for studying pit field densities by developing a high-throughput quantification method in transgenic *Arabidopsis* plants [23]. More recently, we developed a transient expression assay to measure pit field distributions in *Nicotiana benthamiana* leaves [17]. Combining our assay with the reverse genetics approach of virus-induced gene silencing, we generated some of the strongest evidence to date supporting a positive correlation between PD density and intercellular trafficking. While fluorescence microscopy-based techniques allow for the evaluation of pit field distributions in living tissues, they cannot reveal details of the structures of the individual PD in the pit fields.

Two forms of serial scanning electron microscopy (SEM) have recently emerged as powerful tools for measuring PD densities at EM resolutions [24–26]. Both methods use SEM to image the newly exposed surface of a sample that has been prepared for TEM. The first method, serial block face scanning electron microscopy (SBF-SEM), uses a specialized setup with a built-in microtome to do serial sectioning within the microscope [27–31]. This approach has been used in a few studies to quantify PD distributions in *Arabidopsis* roots [24, 26]. Additionally, recent papers outline a toolkit for the analysis of PD densities with SBF-SEM [26, 32]. The second method, focused ion beam-scanning electron microscopy (FIB-SEM), uses a narrow beam of metal ions to cut away, or mill, the surface of the sample to expose a new surface for SEM imaging [25]. In the FIB-SEM instrument, the electron beam is oriented at 54° to the ion beam, and secondary or backscattered electrons produced by the electron beam-sample interactions are used to generate an image of the newly exposed surface of the sample [33]. By repeating the milling and imaging process over a few hours, image data representing several μm of sample thickness can be collected from a single FIB-SEM run [34].

Here, we describe a method for measuring PD densities with FIB-SEM. This technique has several advantages over traditional serial-TEM approaches. It can image relatively large samples, in the range of μm^3. It also allows for fine control of the step size between images with the theoretical limit of 5 nm, resulting in increased resolution in the z-axis and limiting the number of PD that are missed during image collection using larger step sizes or thicker sections. Using these images, 3D reconstructions of the cellular ultrastructure can be generated with segmentation software such as IMOD (University of Colorado) [35]. PD densities can then be determined from the 3D models of the cell walls under investigation.

2 Materials

1. Samples fixed for TEM and embedded in epoxy resin.
2. Razor blades.
3. Ultramicrotome.
4. Glass knives.

5. Compound microscope.
6. Glass slides.
7. Toluidine blue stain: 1 g of Toluidine blue, 2 g of sodium borate, and 100 mL ddH_2O.
8. Jeweler's saw.
9. SEM pin stub mounts.
10. Super glue.
11. Carbon paint.
12. Paint brush or wooden applicator.
13. Plasma sputtering coater for gold deposition.
14. SPI gold target (for gold deposition on block).
15. A scanning electron microscope with a dual (FIB) beam system.
16. Zeiss platinum target (for depositing platinum on block).
17. ImageJ (http://rsbweb.nih.gov/ij/).
18. A computer with enough processing power to handle 3D reconstructions in IMOD. Download and install IMOD from https://bio3d.colorado.edu/imod/ (*see* **Note 1**).

3 Methods

3.1 Biological Sample Preparation

1. Prepare samples as for TEM (*see* **Note 2**).

3.2 Mounting of Samples for FIB-SEM

1. Block trimming: Trim blocks for standard ultrathin sectioning using a glass knife and an ultramicrotome.
   (a) Using a razorblade, trim away excess resin around the sample. Shape the tip of the block into a trapezoid to help with orientation within the microscope.
   (b) Load the block into the ultramicrotome chuck and begin coarse trimming with a glass knife. Set the microtome to cut 1-μm-thick sections and trim away excess resin on the tip of the block.
   (c) Periodically, collect a few 1-μm-thick sections and mount on a glass slide.
   (d) Stain the sections with Toluidine blue and image with a compound microscope (*see* **Note 3**).
   (e) Fine trim and cut ultrathin sections with a diamond knife when the 1-μm sections contain the tissue or structures of interest.

(f) Carefully cut a few 70–100-nm sections with the diamond knife and collect sections and mount on a TEM grid. Image with TEM to determine sample quality (*see* **Note 4**).

2. Block tip removal:

(a) Immobilize the trimmed block using a chuck from the ultramicrotome.

(b) Using the Jeweler's saw, carefully saw through the block, a few millimeters below the sample. Be sure to make a clean, straight cut through the block to prevent the base of the block from being slanted (*see* **Note 5**).

3. Mounting of block tip on SEM stub:

(a) Apply a small drop of super glue to a pin stub for SEM.

(b) Place the block tip on the glue so that the base of the block is flush with the surface of the stub.

(c) Let the glue dry for about 15–30 min or until completely dry.

4. Carbon painting:

(a) Use a paintbrush or wooden applicator to apply a thin layer of conductive carbon paint to the base of the block to prevent charging of the sample. Do not apply paint to the cleaned and sectioned surface (Fig. 1a).

(b) Allow the paint to dry completely before moving on to the next steps (*see* **Note 6**).

5. Sputter coating samples:

(a) Use the sputter coater to apply a thin layer of gold to the sample surface in order to further limit charging of the sample (*see* **Note** 7).

3.3 Microscopy

1. Sample loading and positioning:

(a) Load the sample into the microscope as per the user manual for your system.

(b) Position the sample so that the block face is perpendicular to the ion beam and at a 54-degree angle to the electron beam (Fig. 1b).

2. Identification of regions of interest:

(a) Locate the cleaned block face while imaging in SEM mode (Fig. 1c).

(b) Identify regions of interest by increasing the magnification to view the block face (*see* **Note 8**).

3. Deposition of platinum coat:

Insert the deposition nozzle of the microscope and apply a thin layer of platinum to the block surface at the region of

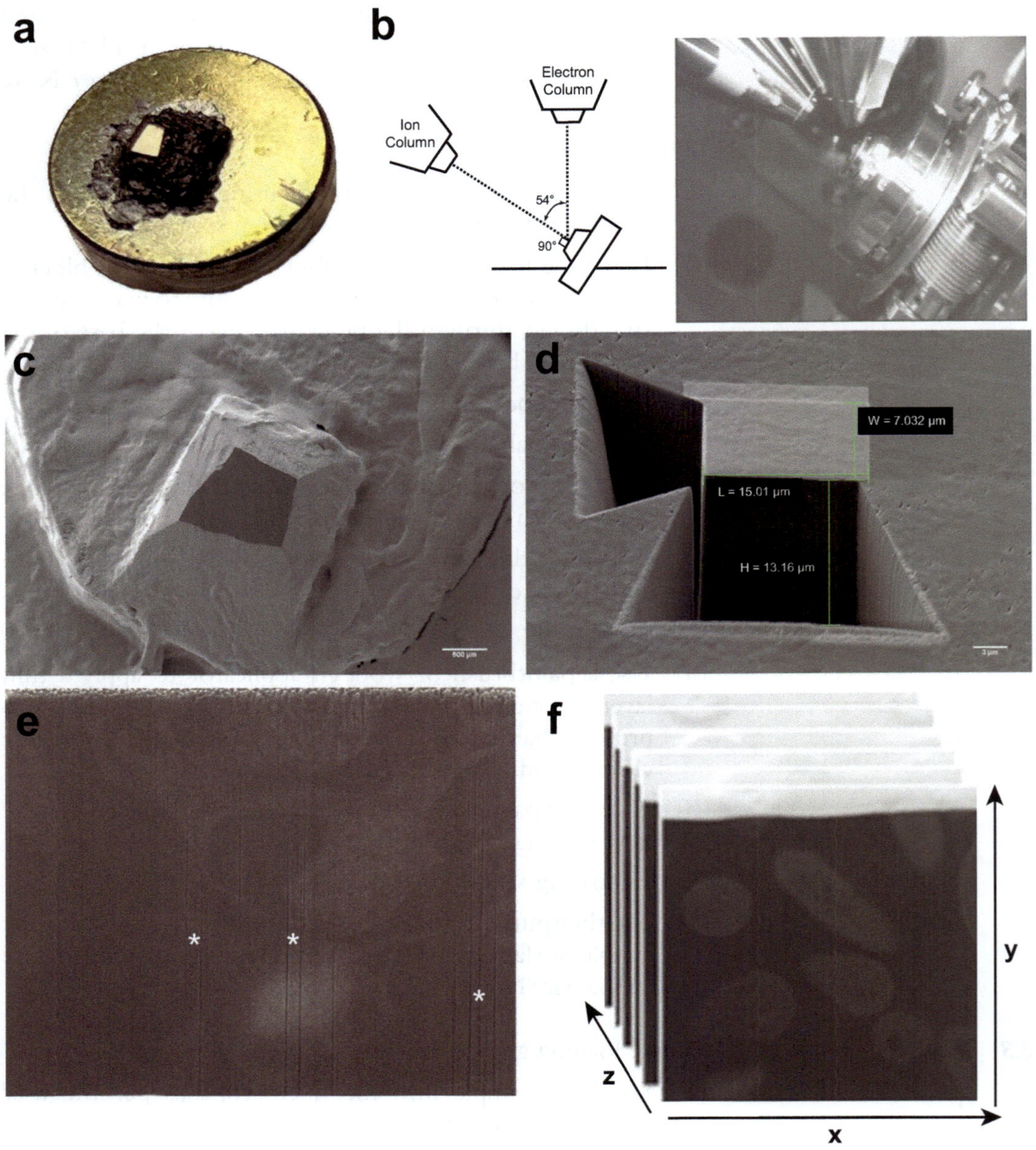

Fig. 1 Representative images of mounted blocks, milled trenches, and the exposed block faces. (**a**) Resin block tip mounted on an SEM stub, coated with carbon paint and coated with gold. (**b**) Diagram (left) and image (right) of the FIB-SEM configuration. The stage and sample are positioned in the microscope so that the ion beam is oriented at 90° to the block face and the electron beam is positioned at 54° to the block face. (**c**) Low-magnification SEM image of the prepared block face (scale bar = 500 μm). (**d**) SEM image of milled trenches in the block face with recommended dimensions for length (L), width (W), and height (H). The light gray portion of the block face (width) corresponds to the layer of platinum deposited prior to milling. (**e**) Representative image of an uncleaned block face image highlighting examples of curtaining (asterisks), a surface artifact that can occur during FIB milling. (**f**) Depiction of how sequentially collected images are combined into a 3D stack of images (x, y, and z)

interest in order to provide an extra layer of protection for your sample and a smooth surface for the ion beam to impact. Imperfections in the surface can result in curtaining of the surface during milling (Fig. 1e) (*see* **Note 9**).

4. Trenching:
   (a) Using the setting for a large probe to generate a larger ion beam size (e.g., 30 kV 2 nA) and the "mill for depth" feature, mill a trapezoid into the block face (Fig. 1d) with the small side of the trapezoid (at least 10 μm across) positioned just below the region of interest. The height of the trapezoid should be no less than 20 μm and a depth of 15–20 μm. The length of the short side of the trapezoid will determine how much tissue will be imaged in each section.
   (b) Mill two additional trapezoids on either side of the initial trapezoid (Fig. 1d) (*see* **Note 10**).
5. Cleaning of the block face:
   (a) Select the setting for a smaller probe to generate a smaller ion beam size (30 kV 50–100 pA) to clean up the milled block face.
   (b) Mill another trapezoid into the sample that overlaps with the top of the first trapezoid (*see* **Note 11**).
6. Milling and image acquisition:
   (a) Select a FIB probe of the desired size for the appropriate step size for imaging. The step size for a FIB probe is equal to half the radius of the probe (*see* **Note 12**).
   (b) Define a rectangular milling area that overlaps with the deposited platinum (*see* **Note 13**).
   (c) Collect an image in SEM mode with the SEM operating at 3 kV after each step.
   (d) Save Images as Tagged Image File Format (TIFF) during acquisition.
   (e) Use the "Slice and View" program from Zeiss Microscopy for serial imaging (*see* **Note 14**).

3.4 Image Processing

1. Generation of image stacks: Once the images have been collected, the raw images need to be converted to a stack of images (Fig. 1f).
   (a) Import the image sequence into ImageJ (http://rsbweb.nih.gov/ij/) to generate a stack (File>Import>Image sequence).
   (b) Set the scale using the scale bar from the first image collected (*see* **Note 14**).

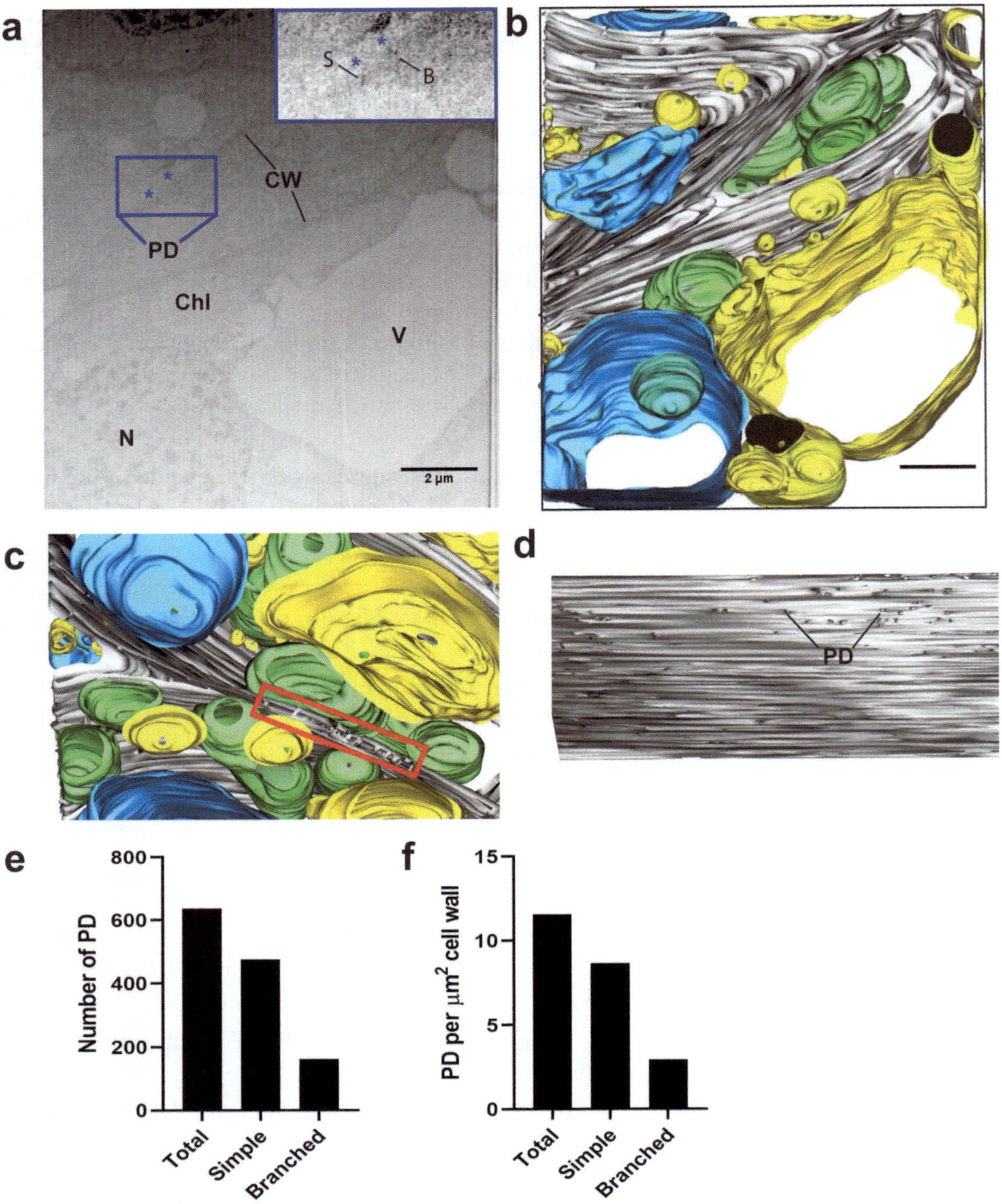

Fig. 2 3D reconstruction of FIB-SEM data and quantification of PD density in a representative FIB-SEM dataset. (**a**) A single SEM image of cells of a *N. benthamiana* leaf taken with FIB-SEM. Portions of four cells are visible in the image. Cell wall (CW), PD, chloroplasts (Chl), vacuole (V), and nuclei are visible. The structural class of PD can be determined. The PD labeled with asterisks are simple (left) and one branched (right). Inset is the region of cell wall with PD denoted by a blue box. The brightness and contrast of the inset image were adjusted to improve visibility of the simple (S) and branched (B) PD. (**b**) 3D reconstruction of the FIB-SEM dataset in (**a**). The cell wall (gray), chloroplast (green), vacuole (yellow), and nuclei (blue) are shown. (**c**) The model shown in (b) is rotated 180° along the vertical axis and approximately 90° along the horizontal axis and finally zoomed into to give a view of a cluster of PD (red box). This demonstrates that individual PD can be discerned from the 3D model. The cluster in the red box extends for approximately 4 μm and could probably be

(c) Once the scale is set, scroll through the image stack. Samples tend to drift during image acquisition, and therefore the images need to be aligned.

(d) Run the TurboReg plugin for ImageJ and select the Rigid Body option in the menu (http://bigwww.epfl.ch/thevenaz/turboreg/) to align the images.

(e) Using the rectangle tool in ImageJ, draw a rectangle that will capture as much of the region of interest as possible.

(f) Make a copy of the region of interest by using the duplicate function (Image>Duplicate). Make sure to duplicate the entire stack by selecting the "Duplicate stack" option.

2. Image processing: Because the images were collected with the secondary electron detector, the images will look like a negative of a TEM image.

   (a) Invert the image in order to generate images that have dark stained membranes and lightly stained cytoplasm that look more like standard TEM images (Fig. 2a).

   (b) Save the stack as a stack of TIFFs for further analysis with IMOD.

3. Adjusting for step size: In order for the 3D renderings to accurately depict the cellular ultrastructure, the z-scale for the tomogram needs to be set in 3Dmod (*see* **Note 15**).

4. Image segmentation:

   (a) Open the stack with IMOD (*see* **Note 16**). This will open the stack in the ZAP window.

   (b) Trace the structures of interest in the ZAP window using the drawing tools (Special>Drawing Tools). The progress of contouring can be monitored with the Model View window.

   Segmentation in IMOD can be performed manually by hand tracing individual contours on each slice of the stack or by semiautomated segmentation using the interpolator function (Special>Interpolator) in IMOD (*see* **Note 17**).

   (c) Apply a skin to the model by generating a mesh of the model for visualization of cellular structures in 3D (Image>Model View>Edit>Object>Meshing) (Fig. 2b–d).

Fig. 2 (continued) observed as a single large spot or a cluster of a few spots in confocal fluorescence microscopy if PD are labeled with a fluorescently tagged protein that localizes to PD. (**d**) Image of 3D rendering of the cell wall showing the distribution of PD in the wall. (**e**) The image stack was used to quantify total numbers of PD in the dataset. (**f**) Using the image stack, the density of simple and branched PD were measured (PD/μm^2 cell wall)

   (d) Repeat the segmentation steps to include additional objects in the 3D model (Fig. 2b–d).

5. Quantification of PD densities: PD density can be quantified as the number of PD in 1 μm^2 of cell wall (Fig. 2e, f).
   (a) Measure the length of all cell wall in the stack using the measure function in ImageJ (*see* **Note 18**).
   (b) Multiply the sum of the cell wall measurements by the step size used during milling and image acquisition to calculate the cell wall area.
   (c) Count the number of PD present in each image (Fig. 2e).
   (d) Divide the total number of PD by the cell wall area to calculate the PD density (Fig. 2f) (*see* **Note 19**).

4 Notes

1. The computer used for segmentation should meet the following standards:
   (a) Processor: Any up-to-date Intel or AMD Processor, but we recommend a computer with at least four processors for speedy segmentation and model reconstruction.
   (b) Video Cards: Any modern Nvidia-based card; however, 256 MB or higher cards are highly recommended. AMD cards will work for Windows or Mac systems.
   (c) A minimum of 2 GB of memory are required for Linux or Mac systems and 4 GB for Windows systems. Computers with more memory are highly recommended.
   (d) Operating system:
      - Linux: Red hat Workstation 6 or higher (equivalent levels for Fedora, Ubuntu, or Suse will also work).
      - Windows: Windows 7, 8, or 10 is recommended; however, Vista or XP will also run IMOD.
      - Mac: OS X 10.8 or higher is recommended.
   (e) For Windows: Cygwin (a Unix package) or several Python packages must be installed on the computer. Links for these downloads and the different versions of IMOD can be found at https://bio3d.colorado.edu/imod/download.html.
2. Samples presented in this method were fixed by high-pressure freezing and freeze substitution (HPF/FS) following the protocol used for TEM sample preparation [25, 36]. However, samples prepared by traditional chemical fixation following standard protocols can also be used. Biological samples used for FIB-SEM tend to be soft and can be easily damaged by the

ion beam so they are typically embedded in resin in order to provide a solid surface for milling. Embed sample in hard resin following the same protocol as used for sample preparation for TEM.

3. The semi-thin 1-μm section stained with Toluidine blue is very useful as a low-resolution map of the sample that can be used to help in orientation after the sample is viewed in the microscope.
4. Surface imperfections such as knife marks may make it difficult to identify regions of interest in your sample and will result in an uneven surface that can cause curtaining of your sample during milling. Additionally, we recommend that the block should be thin sectioned and imaged with TEM to assess the fixation quality of the sample, but this step will not be discussed in this method; instead, we refer the reader to publications [36–38].
5. It is very important to cut straight through the block. If the block containing the sample is mounted at an angle, the electron and ion beam angles will be off in the microscope and may negatively influence the milling and image acquisition.
6. The carbon paint dries rapidly so you will want to work very quickly but carefully. Alternately, a carbon marker can be used to cover the base of the block with a thin layer of carbon.
7. The time required for coating may vary between the models of sputter coaters, but we typically coat for about 45 s.
8. The outlines of cell walls are typically visible at this stage. You will want to identify a region where you think there is a lot of cell wall to maximize the length of wall collected in the final image stack. At this stage, it is very helpful to have a map of your block face such as the light microscopy image of a semi-thin section stained with Toluidine blue. You can also attempt to identify specific cell-cell interfaces such as epidermis-epidermis, epidermis-mesophyll, or vascular cell contacts for specialized imaging.
9. Curtaining is an artifact that occurs during FIB milling when imperfections in the block face interfere with milling and results in an unsmooth surface. Curtaining typically looks like lines running top to bottom of the sample surface.
10. These trapezoids should be perpendicular to the initial trapezoid and line up with the sides of the deposited platinum. The addition of these trapezoids clears the line of sight between the image surface and the electron detector. These trapezoids can have shorter heights, but the length of the short side should extend further than you intend to mill and image.

11. The height of this trapezoid should be substantially smaller (~5 μm) because the milling will take much longer with the smaller probe.
12. A step size of 25–50 nm should be used to minimize PD loss during image acquisition and maximize the amount of tissue collected over the course of several hours, but this can be adjusted to suit the user's experimental goals.
13. The milling area should start in front of the block face, just ahead of the actual exposed surface. This will ensure that the milling area does not start behind the surface of the sample. If the beam starts milling slightly behind the surface, this will result in a barrier between the newly exposed surface and the electron beam and will generate multiple images of the initial surface and not the underlying material.
14. An image captured before milling begins can be used to set the scale. We typically run our samples for 6 h (measured in seconds). With the 25-nm step size and collecting an image after every step, we can typically expect to collect images covering 9–10 μm in 6 h.
15. The Z-scale is equal to the ratio of the section thickness to the pixel size. The section thickness is determined during image acquisition when the probe used for milling is selected. The pixel size is dependent on the image quality and the detector used for image acquisition. The pixel size can be calculated by setting the scale for the tomogram in ImageJ.
16. A very useful tutorial for 3D segmentation using IMOD can be found at (https://bio3d.colorado.edu/imod/doc/3dmodguide.html). Careful review of this tutorial should be completed prior to segmentation to ensure that all parameters are set properly during segmentation.
17. The interpolator function will use information from a few contours spaced throughout the tomogram (approximately one contour every 20 images) to predict the position of each contour on each section between the reference contours. These predictions work pretty well for simple objects that have high contrast but do not work as well when then the contrast is low. Each interpolated contour must be manually adjusted using the drawing tools. Once an interpolated contour has been adjusted, the predicted contour (dashed line) can be converted into a true contour by pressing ENTER on the keyboard. This will rerun the interpolation and adjust the remaining predicted contours based on the modified position of the adjusted contour.
18. The cell wall length is determined by measuring the length of cell wall in each image. This can be achieved by drawing a line

along the length of the cell wall and using the measure function in ImageJ after the scale has been set.

19. High-quality FIB-SEM images allow for the distinction between simple and branched PD. Ratios of simple to branched PD can also be calculated by quantifying distinct population densities (Fig. 2e, f). PD distribution in the cell wall can be visualized in the model, but segmentation is not required for quantification of PD density (Fig. 2d).

Acknowledgments

This work was supported by the National Science Foundation through MCB 1846245.

References

1. Guenoune-Gelbart D, Elbaum M, Sagi G, Levy A, Epel BL (2008) Tobacco mosaic virus (TMV) replicase and movement protein function synergistically in facilitating TMV spread by lateral diffusion in the plasmodesmal desmotubule of *Nicotiana benthamiana*. Mol Plant-Microbe Interact 21:335–345. https://doi.org/10.1094/MPMI-21-3-0335
2. Epel BL (2009) Plant viruses spread by diffusion on ER-associated movement-protein-rafts through plasmodesmata gated by viral induced host beta-1,3-glucanases. Semin Cell Dev Biol 20:1074–1081. https://doi.org/10.1016/j.semcdb.2009.05.010
3. Crawford KM, Zambryski PC (2001) Non-targeted and targeted protein movement through plasmodesmata in leaves in different developmental and physiological states. Plant Physiol 125:1802–1812
4. Schonknecht G, Brown JE, Verchot-Lubicz J (2008) Plasmodesmata transport of GFP alone or fused to potato virus X TGBp1 is diffusion driven. Protoplasma 232:143–152
5. Brunkard JO, Burch-Smith TM, Runkel AM, Zambryski P (2015) Investigating plasmodesmata genetics with virus-induced gene silencing and an agrobacterium-mediated GFP movement assay. Methods Mol Biol 1217: 185–198. https://doi.org/10.1007/978-1-4939-1523-1_13
6. Nicolas WJ, Grison MS, Trepout S, Gaston A, Fouche M, Cordelieres FP, Oparka K, Tilsner J, Brocard L, Bayer EM (2017) Architecture and permeability of post-cytokinesis plasmodesmata lacking cytoplasmic sleeves. Nat Plants 3:17082. https://doi.org/10.1038/nplants.2017.82
7. Bell K, Oparka K (2011) Imaging plasmodesmata. Protoplasma 248:9–25. https://doi.org/10.1007/s00709-010-0233-6
8. Faulkner C, Akman OE, Bell K, Jeffree C, Oparka K (2008) Peeking into pit fields: a multiple twinning model of secondary plasmodesmata formation in tobacco. Plant Cell 20: 1504–1518. https://doi.org/10.1105/tpc.107.056903
9. Ding B, Turgeon R, Parthasarathy MV (1992) Substructure of freeze-substituted plasmodesmata. Protoplasma 169:28–41
10. Oparka KJ, Roberts AG, Boevink P, Santa Cruz S, Roberts I, Pradel KS, Imlau A, Kotlizky G, Sauer N, Epel B (1999) Simple, but not branched, plasmodesmata allow the nonspecific trafficking of proteins in developing tobacco leaves. Cell 97:743–754
11. Roberts IM, Boevink P, Roberts AG, Sauer N, Reichel C, Oparka KJ (2001) Dynamic changes in the frequency and architecture of plasmodesmata during the sink-source transition in tobacco leaves. Protoplasma 218:31–44
12. Burch-Smith TM, Zambryski PC (2010) Loss of INCREASED SIZE EXCLUSION LIMIT (ISE)1 or ISE2 increases the formation of secondary plasmodesmata. Curr Biol 20: 989–993. https://doi.org/10.1016/j.cub.2010.03.064
13. Seagull RW (1983) Differences in the frequency and disposition of plasmodesmata resulting from root cell elongation. Planta 159:497–504. https://doi.org/10.1007/BF00409138
14. Danila FR, Quick WP, White RG, Furbank RT, von Caemmerer S (2016) The metabolite

pathway between bundle sheath and mesophyll: quantification of plasmodesmata in leaves of C3 and C4 monocots. Plant Cell 28: 1461–1471. https://doi.org/10.1105/tpc.16.00155

15. Zambryski P (1995) Plasmodesmata: plant channels for molecules on the move. Science 270:1943–1944
16. Dmitrieva VA, Ivanova AN, Tyutereva EV, Evkaikina AI, Klimova EA, Voitsekhovskaja OV (2017) Chlorophyllide-a-oxygenase (CAO) deficiency affects the levels of singlet oxygen and formation of plasmodesmata in leaves and shoot apical meristems of barley. Plant Signal Behav 12:e1300732. https://doi.org/10.1080/15592324.2017.1300732
17. Ganusova EE, Reagan BC, Fernandez JC, Azim MF, Sankoh AF, Freeman KM, McCray TN, Patterson K, Kim C, Burch-Smith TM (2020) Chloroplast-to-nucleus retrograde signalling controls intercellular trafficking via plasmodesmata formation. Philos Trans R Soc Lond Ser B Biol Sci 375:20190408. https://doi.org/10.1098/rstb.2019.0408
18. Livingston LG (1935) The nature and distribution of plasmodesmata in the tobacco plant. Am J Bot 22:75–87. https://doi.org/10.1002/j.1537-2197.1935.tb05009.x
19. Juniper BE, Barlow PW (1969) The distribution of plasmodesmata in the root tip of maize. Planta 89:352–360. https://doi.org/10.1007/BF00387235
20. Evert RF, Eschrich W, Heyser W (1977) Distribution and structure of the plasmodesmata in mesophyll and bundle-sheath cells of *Zea mays* L. Planta 136:77–89. https://doi.org/10.1007/BF00387929
21. Tilney LG, Cooke TJ, Connelly PS, Tilney MS (1991) The structure of plasmodesmata as revealed by plasmolysis, detergent extraction, and protease digestion. J Cell Biol 112: 739–747
22. Schubert M, Koteyeva NK, Wabnitz PW, Santos P, Buttner M, Sauer N, Demchenko K, Pawlowski K (2011) Plasmodesmata distribution and sugar partitioning in nitrogen-fixing root nodules of Datisca glomerata. Planta 233: 139–152. https://doi.org/10.1007/s00425-010-1285-8
23. Fitzgibbon J, Beck M, Zhou J, Faulkner C, Robatzek S, Oparka K (2013) A developmental framework for complex plasmodesmata formation revealed by large-scale imaging of the Arabidopsis leaf epidermis. Plant Cell 25:57–70. https://doi.org/10.1105/tpc.112.105890
24. Ross-Elliott TJ, Jensen KH, Haaning KS, Wager BM, Knoblauch J, Howell AH, Mullendore DL, Monteith AG, Paultre D, Yan D, Otero S, Bourdon M, Sager R, Lee JY, Helariutta Y, Knoblauch M, Oparka KJ (2017) Phloem unloading in Arabidopsis roots is convective and regulated by the phloem-pole pericycle. eLife 6:e24125. https://doi.org/10.7554/eLife.24125
25. Reagan BC, Kim PJ, Perry PD, Dunlap JR, Burch-Smith TM (2018) Spatial distribution of organelles in leaf cells and soybean root nodules revealed by focused ion beam-scanning electron microscopy. Funct Plant Biol 45:180–191. https://doi.org/10.1071/FP16347
26. Yan D, Yadav SR, Paterlini A, Nicolas WJ, Petit JD, Brocard L, Belevich I, Grison MS, Vaten A, Karami L, El-Showk S, Lee JY, Murawska GM, Mortimer J, Knoblauch M, Jokitalo E, Markham JE, Bayer EM, Helariutta Y (2019) Sphingolipid biosynthesis modulates plasmodesmal ultrastructure and phloem unloading. Nat Plants 5:604–615. https://doi.org/10.1038/s41477-019-0429-5
27. Denk W, Horstmann H (2004) Serial block-face scanning electron microscopy to reconstruct three-dimensional tissue nanostructure. PLoS Biol 2:e329. https://doi.org/10.1371/journal.pbio.0020329
28. Deerinck TJ, Bushong EA, Lev-Ram V, Shu X, Tsien RY, Ellisman MH (2010) Enhancing serial block-face scanning electron microscopy to enable high resolution 3-D nanohistology of cells and tissues. Microsc Microanal 16: 1138–1139. https://doi.org/10.1017/S1431927610055170
29. Hughes L, Hawes C, Monteith S, Vaughan S (2014) Serial block face scanning electron microscopy—the future of cell ultrastructure imaging. Protoplasma 251:395–401. https://doi.org/10.1007/s00709-013-0580-1
30. Kremer A, Lippens S, Bartunkova S, Asselbergh B, Blanpain C, Fendrych M, Goossens A, Holt M, Janssens S, Krols M, Larsimont JC, Mc Guire C, Nowack MK, Saelens X, Schertel A, Schepens B, Slezak M, Timmerman V, Theunis C, VAN Brempt R, Visser Y, Guerin CJ (2015) Developing 3D SEM in a broad biological context. J Microsc 259:80–96. https://doi.org/10.1111/jmi.12211
31. Kittelmann M, Hawes C, Hughes L (2016) Serial block face scanning electron microscopy and the reconstruction of plant cell membrane systems. J Microsc 263:200–211. https://doi.org/10.1111/jmi.12424
32. Paterlini A, Belevich I, Jokitalo E, Helariutta Y (2020) Computational tools for serial block electron microscopy reveal plasmodesmata

distributions and wall environments. Plant Physiol 184:53–64. https://doi.org/10.1104/pp.20.00396

33. Milani M, Drobne D, Tatti F, Batani D, Poletti G, Orsini F, Zullini A, Zrimec A (2005) Read-out of soft X-ray contact microscopy microradiographs by focused ion beam/scanning electron microscope. Scanning 27:249–253
34. Drobne D, Milani M, Zrimec A, Zrimec MB, Tatti F, Draslar K (2005) Focused ion beam/scanning electron microscopy studies of *Porcellio scaber* (Isopoda, Crustacea) digestive gland epithelium cells. Scanning 27:30–34
35. Kremer JR, Mastronarde DN, McIntosh JR (1996) Computer visualization of three-dimensional image data using IMOD. J Struct Biol 116:71–76. https://doi.org/10.1006/jsbi.1996.0013
36. Bobik K, Dunlap JR, Burch-Smith TM (2014) Tandem high-pressure freezing and quick freeze substitution of plant tissues for transmission electron microscopy. J Vis Exp 92:e51844. https://doi.org/10.3791/51844
37. Lonsdale JE, McDonald KL, Jones RL (1999) High pressure freezing and freeze substitution reveal new aspects of fine structure and maintain protein antigenicity in barley aleurone cells. Plant J 17:221–229. https://doi.org/10.1046/j.1365-313X.1999.00362.x
38. McDonald KL (2014) Out with the old and in with the new: rapid specimen preparation procedures for electron microscopy of sectioned biological material. Protoplasma 251:429–448. https://doi.org/10.1007/s00709-013-0575-y

Chapter 7

Measuring Plasmodesmata Density on Cell Interfaces of Monocot Leaves Using 3D Immunolocalization and Scanning Electron Microscopy

Florence R. Danila

Abstract

Quantification of plasmodesmata density on cell interfaces of plant tissues, particularly of leaves, has been a long-standing challenge. Using electron microscopy alone to quantify plasmodesmata is difficult because of the limited surface area coverage per image and hence the need to examine large numbers of sections for robust quantification. Fluorescence microscopy provides the larger surface area coverage per image but can only visualize pit fields and not individual plasmodesma. Moreover, in pigmented tissue like leaves, imaging cell interfaces beyond the epidermal layer would also require accurate sectioning. The advent of tissue clearing techniques such as PEA-CLARITY provided the opportunity to capture all pit fields within the leaf without resorting to sectioning. This paved the way toward the development of a more robust and precise plasmodesmata density quantification method by combining the three-dimensional immunolocalization fluorescence microscopy with scanning electron microscopy (SEM). Here, I describe a protocol to quantify plasmodesmata density on cell interfaces between mesophyll and bundle sheath in C_3 and C_4 monocot leaves.

Key words Three-dimensional fluorescence imaging, Scanning electron microscopy, Plasmodesmata density, Pit field area, Photosynthesis

1 Introduction

Plasmodesmata are nanochannels between cells known to facilitate symplastic transport. In plants, particularly in leaves, plasmodesmata play a role in short-distance transport of assimilates from the mesophyll to the phloem, an important aspect of photosynthesis [1]. To enable the CO_2 concentrating mechanism, C_4 photosynthesis requires the rapid movement of metabolites between the mesophyll cells and bundle sheath cells of the leaf [2]. It has been predicted that to support this flux, C_4 leaves have a greater number of intercellular connections, i.e., plasmodesmata, than C_3 leaves [2].

Yoselin Benitez-Alfonso and Manfred Heinlein (eds.), *Plasmodesmata: Methods and Protocols* ,
Methods in Molecular Biology, vol. 2457, https://doi.org/10.1007/978-1-0716-2132-5_7,

Transmission electron microscopy or TEM has commonly been used for plasmodesmata studies due to its high intrinsic resolution and therefore its ability to capture and resolve the substructure of individual plasmodesma at the nanometer scale. It has proven very useful in elucidating the origin, formation, and substructure of plasmodesmata (reviewed in refs. 3–10). However, the two-dimensional nature of the electron micrographs and very small area of capture per image impose a significant limitation in quantifying the plasmodesmatal connection at a tissue or organ level. To accurately quantify plasmodesmatal frequency and dimensions (i.e., cross-sectional area) would require serial sectioning, preferably with the cell wall sectioned en face, and meticulous data analysis including three-dimensional reconstruction. Serial sectioning and three-dimensional reconstruction are both labor-intensive and time-consuming activities, although this can now be achieved through serial block face scanning electron microscopy (SEM) [11].

The first attempt at plasmodesmata quantification used *Azolla pinnata* roots [12]. Gunning [12] argued that when randomly distributed plasmodesmata are cut longitudinally, the number of cylindrical profiles seen per unit length of wall, multiplied by a proportionality constant, is equal to their frequency per unit area. Consequently, from electron micrographs of *Azolla* root plasmodesmata, one can estimate plasmodesmata density by implementing the Gunning constant $1/(t + 1.5R)$, where t is the section thickness and R is the average radius of plasmodesmata [12]. Although this proportionality constant provided good estimates of *Azolla* root plasmodesmata density, it is not appropriate for quantifying plasmodesmata density in higher plant tissues [12]. In contrast to the randomly distributed plasmodesmata in *Azolla* roots, pit fields or clusters of plasmodesmata are common in higher plant root [13] and leaf tissues [14–16]. These pit fields add another layer of complexity to quantification because their sizes and shapes are variable [17]. Moreover, the pattern of distribution of pit fields in leaves, whether the pit fields are in clusters or evenly distributed on the cell wall interfaces, is still unknown for almost all leaf tissues. Despite this, a number of studies have attempted to quantify plasmodesmata density between leaf cells using TEM either with [18, 19] or without [13–15, 20] incorporating the Gunning constant in their calculations. Most of these studies analyzed monocots, comparing a C_3 photosynthetic leaf to a C_4 counterpart with minimal replication. Of particular interest is the work of Botha [14] which showed that leaves of C_4 grasses have more plasmodesmatal connection than C_3 grasses. While this result was consistent with a C_4 mechanism, Botha's measurements, although empirical, are limited because the measurements are based on cell wall length or wall interface length rather than on the three-dimensional wall interface area. Consequently, numerical values from these

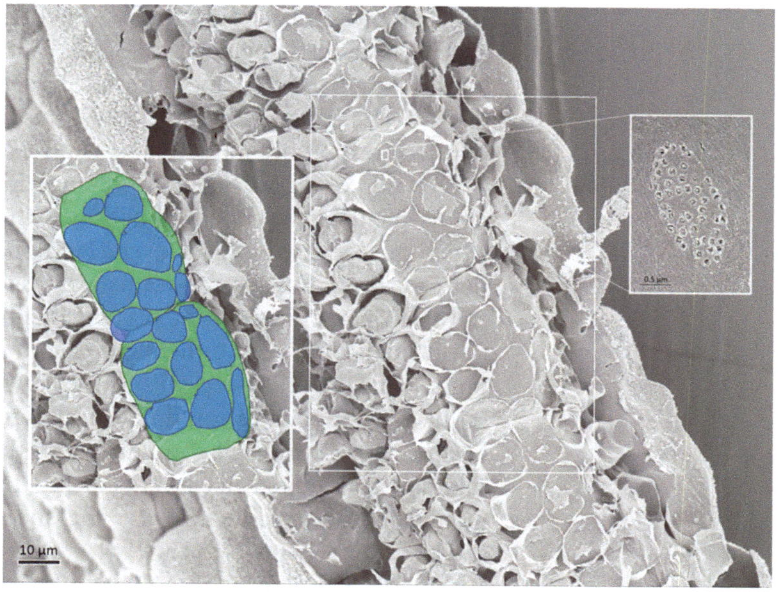

Fig. 1 Scanning electron micrograph showing the interface between mesophyll and bundle sheath cells in a monocot leaf. Left inset is the boxed area with underlying bundle sheath cell areas in green and mesophyll cell attachment sites in blue. When magnified, pit fields (right inset) can be found within the mesophyll cell attachment sites

measurements [14] would be very difficult to convert to an estimate of plasmodesmata per wall or interface area. Similarly, using these two-dimensional estimates per wall length for semi-quantification of plasmodesmata density in plasmodesmograms [21] cannot reflect the plasmodesmata density throughout a whole leaf.

One way to circumvent the problem of localized patches of plasmodesmata is to capture pit fields in leaves using SEM. Especially with monocot leaves where cells are arranged in parallel, random tearing of the leaf after critical point drying or high-pressure freezing can expose cell interfaces and allow capture of whole pit fields [16] (Fig. 1) and, thus, provide a measure of plasmodesmata frequency within pit fields. An earlier study combined TEM and SEM to quantify plasmodesmata density between the mesophyll and bundle sheath cells of a C_4 leaf [17]. However, similar to TEM, SEM can only capture a very small area representative of the whole tissue and would still require hundreds of images to obtain sufficient data for accurate quantification of plasmodesmata density for each cell interface in a tissue or organ.

The genes involved in plasmodesmata development are not yet known, but the close association of callose with plasmodesmata allows indirect labeling, quantification, and characterization of plasmodesmata in plant tissues [22]. Aniline blue staining [23] and β-1,3-glucan antibody hybridization [24] permit plasmodesmata-associated callose localization using fluorescence microscopy. However, because fluorescence microscopes can only resolve objects as small as pit fields, this imaging technique cannot be used to quantify individual plasmodesmata within cell walls. As noted above, SEM can capture individual plasmodesma within whole pit fields, so in combination with fluorescence microscopy, which can capture pit fields over a large surface area, we can obtain a better estimate of plasmodesmata density in leaves. Although we can tear open critical-point-dried leaves to observe plasmodesmata in pit fields, it is very difficult to obtain fluorescence images from bundle sheath cells deep within leaves. The main limitation is the reduced light penetration due to leaf pigments and light scattering by the cell wall and cell contents. With the exception of the almost pigment-free epidermal layer, cutting the leaf into thin sections would appear necessary to see the successive tissue layers inside the leaf.

In 2015, a new technique for three-dimensional micro-visualization of plant organs, PEA-CLARITY, was published [25]. Clear Lipid-exchanged Acrylamide-hybridized Rigid Imaging/Immunostaining/In situ hybridization-compatible Tissue-hYdrogel or CLARITY is a whole tissue clearing method followed by fluorescence immunolocalization that was first developed to image and map connections within an intact mouse brain [26]. It involves fixing and binding tissues within an acrylamide gel prior to extraction of optically interfering lipids with detergent. This results in an optically transparent tissue suitable for deep imaging using confocal microscopy. Adapting the CLARITY method with an additional enzyme digestion step to permeabilize the cell wall sufficiently to allow entry of antibodies and clearing agents led to a plant-based technique called Plant Enzyme-Assisted-CLARITY or PEA-CLARITY [27]. Implementing the PEA-CLARITY method by clearing pieces of leaf tissue followed by hybridization using an antibody against callose led to the visualization of pit fields within the whole leaf thickness under a laser scanning confocal microscope (Fig. 2) providing a measure of pit field area on cell interfaces deep within the leaf tissue [28]. As a result, a new plasmodesmata quantification method that combines three-dimensional immunolocalization fluorescence microscopy and SEM was developed [28] (Fig. 3). This method has allowed comparison of plasmodesmata density on cell interfaces between mesophyll and bundle sheath among 25 grass species including representatives from the C_3 and C_4 photosynthetic types providing further understanding on the important role leaf plasmodesmata play in C_4 photosynthesis [29]. While the details presented here are optimized for leaf, this method could also be used in other research fields to characterize the symplastic domain in a range of tissues.

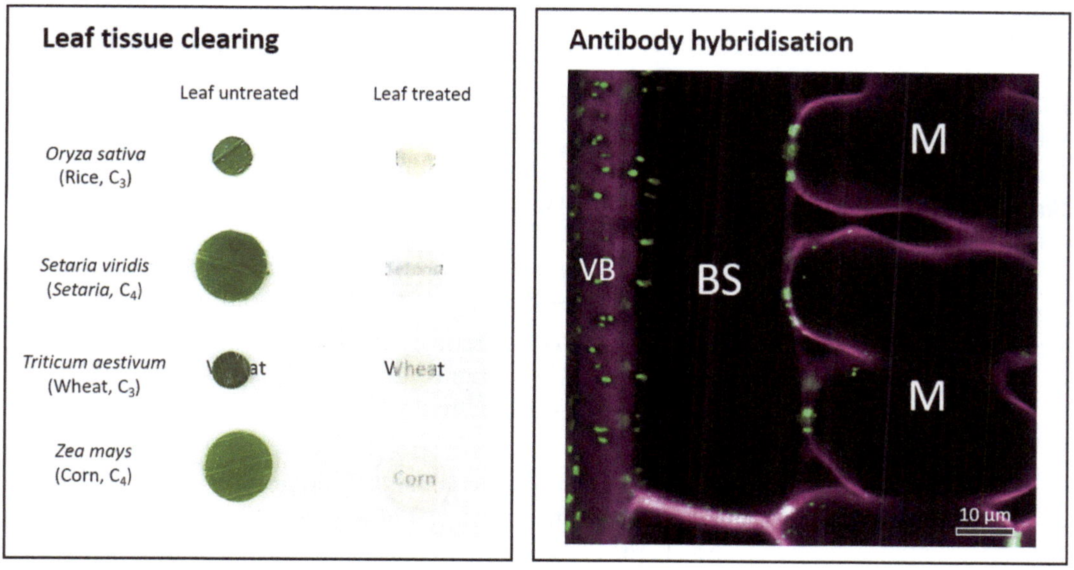

Fig. 2 Leaf tissue clearing and antibody hybridization for 3D immunolocalization. Clearing removes leaf pigments that interfere with light penetration resulting to a semitransparent leaf tissue, while additional enzyme treatment makes the cell wall permeable. The use of an antibody against callose, a plasmodesmata-associated polysaccharide, allows the localization of pit fields (green fluorescence) throughout the leaf tissue. Staining the cell wall with Calcofluor white (magenta) helps in navigating throughout the tissue. *VB* vascular bundle, *BS* bundle sheath, *M* mesophyll. Image on the right is obtained from a wheat leaf z-stack

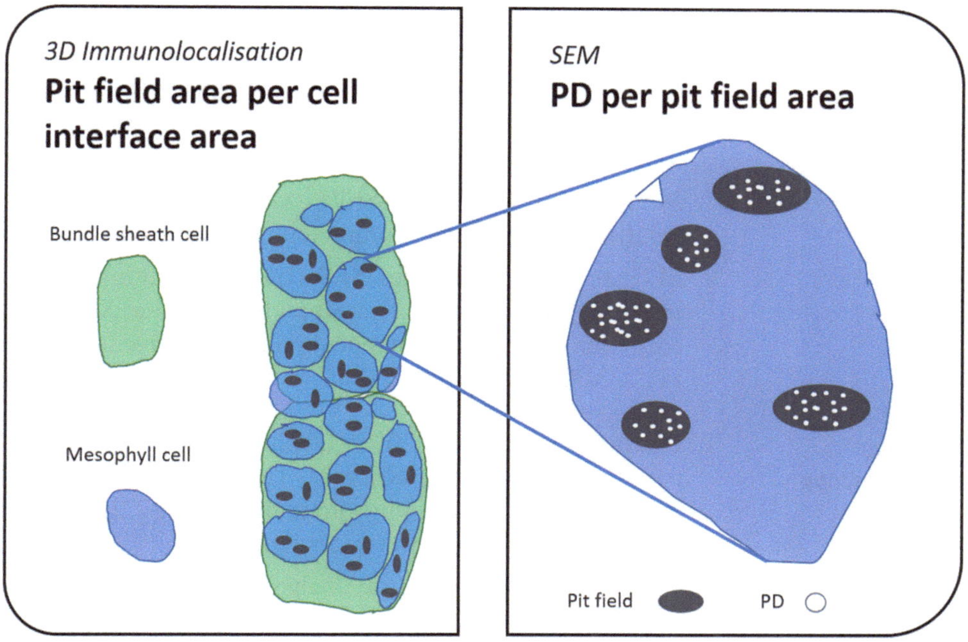

Fig. 3 3D immunolocalization and SEM provide a way to quantify plasmodesmata density in leaves. Visual explanation of how combining measurements from 3D immunolocalization fluorescence microscopy, which gives pit field area per cell interface area value, and SEM, which gives the plasmodesmata (PD) per pit field area value, provide a measure of plasmodesmata density per bundle sheath-mesophyll cell interface area

2 Materials

2.1 3D Immunolocalization Fluorescence Microscopy

1. Hydrogel fixative: Make this solution in the cold room (*see* **Note 1**). For 200 mL fixative, combine 105 mL distilled water, 20 mL 10× phosphate-buffered saline, 50 mL 16% paraformaldehyde pH 6.9 (*see* **Note 2**), 20 mL 40% acrylamide solution, 5 mL 2% bis-acrylamide solution, and 0.5 g VA-044 polymerization initiator. Stir until the VA-044 initiator is completely dissolved. Store the fixative solution at 4 °C.
2. Clearing solution: Dissolve 12.37 g of boric acid and 40 g of sodium dodecyl sulfate in 1 L distilled water. Adjust pH to 8.5 using 5 N sodium hydroxide solution. Store at room temperature.
3. Digestion enzyme cocktail: Prepare this fresh. In 1× phosphate-buffered saline pH 7.4 with 0.005% sodium azide and 1 mM calcium chloride (*see* **Note 3**), add 1 μL/mL β-xylanase M6, 1 μL/mL α-L-arabinofuranosidase, 1 μL/mL pectate lyase, 1 μL/mL α-amylase, 1 μL/mL pullulanase, and 0.5 μL/mL of 2 mg/mL cellulase. Vortex to mix.
4. Primary antibody: Prepare this fresh. Add 1 μL of β-1,3-glucan antibody per 500 μL TBST buffer pH 7.4 (*see* **Note 4**). Vortex to mix.
5. Secondary antibody: Prepare this fresh. Add 1 μL of goat anti-mouse IgG H&L Alexa Fluor® 488 per 500 μL TBST buffer pH 7.4 (*see* **Note 4**). Vortex to mix. Protect from light by wrapping with foil.
6. Cell wall stain: Add distilled water to make 0.05% calcofluor white solution. Mix and protect from light by wrapping with foil. Store at room temperature.
7. Mounting media: Add distilled water to make 80% glycerol solution. Vortex to mix completely. Store at room temperature.
8. Microscopy: Use a Leica SP8 confocal/multiphoton microscope or equivalent, Leica Application Suite X software or equivalent, and long-distance dipping objective lens preferably HCX APO L U-V-I 40x/0.80 water or equivalent.

2.2 Scanning Electron Microscopy

1. Fixative solution: For 200 mL solution, combine 50 mL 16% paraformaldehyde (*see* **Note 2**), 1.6 mL 25% glutaraldehyde, 50 mL 0.1 M sodium phosphate buffer pH 7.2 (*see* **Note 5**), and 98.4 mL distilled water. Store at 4 °C.
2. Tissue dehydration: Dilute analytical grade ethanol with distilled water to make 15%, 30%, 50%, 70%, 80%, and 90% ethanol solutions.
3. Tissue drying: Use analytical grade ethanol, Balzers critical point dryer CPD 030, and liquid carbon dioxide.

4. Mounting tools: Prepare forceps, copper holders, copper holder rack with lid or empty 200 μL pipette tip box, and nail polish.
5. Metal coating: Use an Emitech EMS550X sputter or equivalent and platinum sputter coater target.

Microscopy: Use a Zeiss Ultra Plus field emission scanning electron microscope or equivalent.

3 Methods

An overview summary of the sample preparation involved for 3D immunolocalization and SEM is illustrated in Fig. 4.

3.1 3D Immunolocalization Fluorescence Microscopy

3.1.1 Tissue Collection and Fixation

1. Collect leaf tissue from a 9-day-old seedling into ice-cold distilled water.
2. While submerged in distilled water, cut leaf tissue into 2 mm × 3 mm leaf pieces using a razor blade.
3. Collect leaf pieces into a screw cap 5 mL tube containing 16% paraformaldehyde with 0.01% Tween 20 and vacuum infiltrate until tissues sink (*see* **Note 6**).
4. Discard 16% paraformaldehyde with Tween 20 and replace with hydrogel fixative solution (*see* **Note 1**).
5. After overnight incubation at 4 °C, replace the solution with fresh hydrogel fixative filling up to the brim to minimize air bubbles.
6. Seal with parafilm and incubate at 37 °C overnight to facilitate hydrogel polymerization.

3.1.2 Tissue Clearing

1. While working under the hood, carefully separate the leaf tissue from the hydrogel using dry lint-free paper wipes. Make sure to remove as much gel sticking to the tissue as possible.
2. Immediately submerge the tissue into 50 mL clearing solution to avoid drying out.
3. Incubate the polymerized tissue in the clearing solution at room temperature, 60 rpm until tissue becomes clear or transparent. This usually takes at least 6–8 weeks (*see* **Note 7**).
4. For the first week, change the clearing solution every other day and then once a week for the succeeding weeks.

3.1.3 Cell Wall Permeabilization

1. Gently transfer cleared gel-embedded tissue into a 2 mL microcentrifuge tube containing 1× phosphate-buffered saline pH 7.4 and let sit for 1 min to rinse out excess clearing solution.

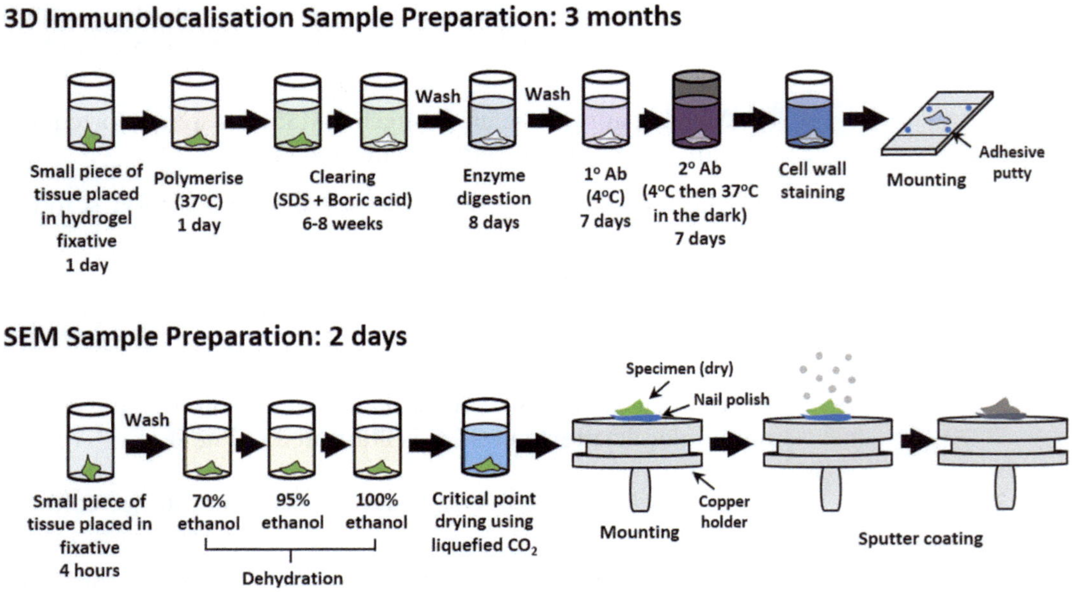

Fig. 4 Overview of the sample preparation involved in 3D immunolocalization fluorescence microscopy and SEM for plasmodesmata density quantification in leaves. Sample preparation for 3D immunolocalization can have shorter or longer clearing period depending on the age and composition of tissue used (*see* **Note 7**)

2. Discard the buffer solution and replace with digestion enzyme cocktail just enough to completely submerge the tissue.
3. Vacuum infiltrate for 15 min, gently releasing the vacuum every 2 min.
4. Incubate at 37 °C, 30 rpm for at least 8 days, replenishing the solution with fresh digestion enzyme cocktail every 2 days (*see* **Note 8**) and performing vacuum infiltration for 15 min each day.

3.1.4 Antibody Hybridization and Cell Wall Staining

1. Discard the digestion enzyme solution and rinse the digested gel-embedded tissue 3 × 20 min in 1× phosphate-buffered saline pH 7.4 (*see* **Note 9**).
2. Carefully submerge the tissue into the primary antibody and vacuum infiltrate 3 × 5 min.
3. Incubate at 4 °C for 7 days and perform 3 × 5 min vacuum infiltration each day.
4. Carefully wash with TBST buffer pH 7.4 5 × 30 min and leave at 4 °C overnight.
5. Discard the TBST buffer and incubate in secondary antibody at 4 °C for 7 days (*see* **Note 10**). Perform 3 × 5 min vacuum infiltration each day.
6. On the last day, prior to washing, incubate the tissue at 37 °C for 2 h.

7. Carefully wash with TBST buffer pH 7.4 5 × 30 min and leave at 4 °C overnight.
8. Discard the TBST buffer and post-stain with 0.05% calcofluor white for 30 min at room temperature with 3 × 5 min vacuum infiltration.
9. Carefully wash with distilled water 6 × 10 min and incubate in 80% glycerol at 4 °C overnight.

3.1.5 Mounting

1. Carefully mount the hybridized gel-embedded tissue onto a glass slide containing 80% glycerol.
2. Use a coverslip with adhesive putty placed downside onto the four corners to facilitate elevation and prevent tissue from being smashed.
3. Store mounted samples at 4 °C in the dark for up to 1 month.

3.1.6 Imaging

1. Examine the tissue using a Leica SP8 confocal/multiphoton microscope under a long-distance dipping objective lens.
2. Turn on the excitation wavelength at 488 nm using the Argon laser and visualize the pit fields at emission wavelength 522–550 nm. To simultaneously visualize the cell wall, turn on the 405-diode laser and collect emission at 434–445 nm. Use either PMT or HyD detectors, whichever is available.
3. Set the format to 512 × 512, speed to 400 Hz, zoom factor to 0.75, and line average to 16. Rotate scan field so the veins are vertically oriented.
4. Start a Z-stack imaging by setting the begin position and end position to capture the whole thickness of the leaf tissue. Set z-step size to 0.8 μm.
5. Hit Start to begin imaging. One complete leaf z-stack usually takes 2 h to finish.
6. Save z-stack file.

3.2 Scanning Electron Microscopy

3.2.1 Tissue Collection and Fixation

1. Collect leaf tissue from a 9-day-old seedling into ice-cold distilled water.
2. While submerged in distilled water, cut leaf tissue into 3 mm × 10 mm leaf strips using a razor blade. The longer side must be parallel to the veins.
3. Collect leaf strips into 2 mL tubes containing fixative solution with 0.01% Tween 20 and vacuum infiltrate until tissues sink (*see* **Note 6**).
4. Discard fixative solution with 0.01% Tween 20 and replace with fresh fixation solution.

5. Incubate at room temperature for 4 h and rinse fixed tissue 4 × 15 min using 25 mM sodium phosphate buffer pH 7.2 (*see* **Note 11**).

3.2.2 Tissue Dehydration and Critical Point Drying

1. Incubate fixed tissue into graded ethanol series (15%, 30%, 50%, 70%, 80%, 90%, 3 × 100%) at room temperature, 100 rpm for 10 min each.
2. Pour 100% ethanol into the Balzers critical point dryer CPD 030 specimen pressure chamber.
3. Transfer dehydrated tissue into the sample holder making sure the tissue is fully submerged in ethanol.
4. Proceed with drying according to CPD 030 critical point dryer operating instructions using CO_2 as a transition liquid.

3.2.3 Mounting and Metal Coating

1. Mount three to five leaf tissues, longer side down, on the copper holder using just enough nail polish to keep the tissue in an upright position.
2. Store copper holders in a rack with lid and allow the nail polish to completely dry overnight.
3. Using forceps, rip the dry leaf tissue parallel to the veins to expose cell interfaces (*see* **Note 12**).
4. Following Emitech EMS550X sputter operating instructions, coat mounted tissue with platinum for 4 min.

3.2.4 Imaging

1. Load the specimen onto the stage following the operating instructions of Zeiss Ultra Plus field emission scanning electron microscope.
2. Set EHT target or voltage to 3.00 kV and aperture to 20.00 μm.
3. Find the area of interest using Fig. 1 as a reference image. This is usually at 4–6 mm working distance.
4. Save image as tiff file.

3.3 Image Processing and Measurements

3.3.1 Image Processing Workflow for Pit Field Area Measurement Using 3D Immunolocalization Fluorescence Micrographs (See Note 13)

1. To generate 2D images from Z-stack (Fig. 5), initiate the Leica Application Suite X or LAS X software (*see* **Note 14**) and, under Process, click Open projects tab, and select a z-stack file to open.
2. Click Process Tools tab and, under Edit, select Crop.
3. Specify Z Stack Selection by indicating the start (From) and end (To) position of the focal planes to be cropped and click Apply. This step will automatically create a new Crop file.
4. Select the Crop file and, on the top right side, click the Max icon to generate a maximum intensity projection image. Click Ch1 or Ch2 icon to turn on and off the channels (*see* **Note 15**).

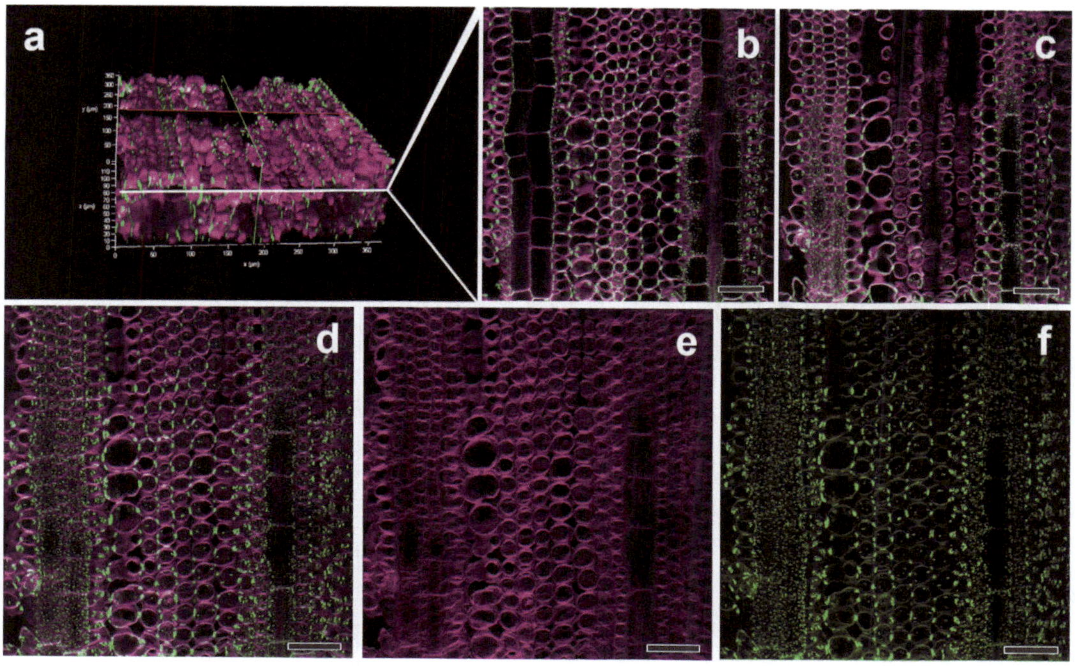

Fig. 5 Confocal image processing using Leica Application Suite X software. (**a**) 3D reconstructed image (from 154 single focal planes) of *Z. mays* leaf hybridized with primary antibody to β-1,3-glucan and secondary antibody tagged with Alexa Fluor® 488 (green) and stained with calcofluor white (magenta) to show cell walls. (**b**) A single optical section derived from (**a**) showing pit fields (green) between cells. This section is focused at the plane below which no further M-BS pit fields are detected and marks the start of the z-stack to be extracted. (**c**) A single optical section derived from (**a**), focused at the top surface of the two files of BS cells highlighted in (**b**), above which no additional M-BS pit fields are detected, and marks the end of the z-stack to be extracted. (**d**) Maximum intensity projection image of the 19 consecutive single focal planes from (**b**) to (**c**) comprising that region of BS cell surface in contact with the neighboring M cells. (**e**) Calcofluor white channel from (**d**). (**f**) Alexa Fluor® 488 channel from (**d**). Bars = 50 μm. Reproduced from [28], www.plantcell.org. Copyright American Society of Plant Biologists

5. Right click on the image and select Export Image to save it either as tiff or as jpeg.
6. For pit field area measurement (Fig. 6), initiate the ImageJ software and, under File tab, click Open and select an image file (*see* **Note 16**).
7. Under Image tab, click Adjust, and select Color Threshold. This will open a parameter box, where the following should be specified (*see* **Note 17**):

   Hue: Default.

   Saturation: Default.

   Brightness: 26 (*see* **Note 18**).

   Thresholding method: Triangle.

   Threshold color: B&W.

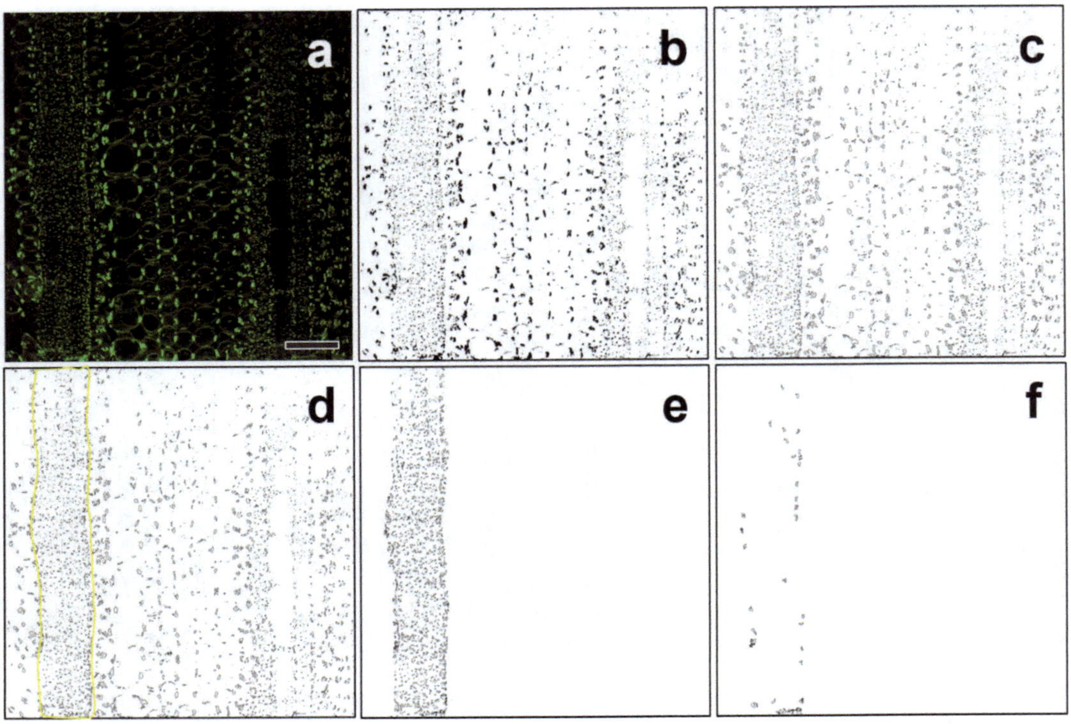

Fig. 6 Confocal image processing and pit field area measurement using ImageJ software. (**a**) Alexa Fluor® 488 channel from Fig. 5d. (**b**) Converted image of (**a**) after Color Threshold adjustment. (**c**) 8-bit image of (**b**) after Binary processing. (**d**) Image from (**c**) showing the selected area of interest outlined in yellow. (**e**) Image showing the area of interest analyzed. (**f**) Image showing the outliers from (**e**) (*see* **Note 21**). Bar = 50 μm. Reproduced from [28], www.plantcell.org. Copyright American Society of Plant Biologists

Color space: HSB.

Dark background ticked.

8. Under Image tab, click Type, and select 8-bit (*see* **Note 19**).
9. Under Process tab, click Binary, and select Outline (*see* **Note 20**).
10. Select the area of interest using Freehand selections tool (*see* **Note 21**).
11. Under Analyze tab, select Analyze Particles, and specify the following parameters (*see* **Note 22**):

    Size ($inch^2$): 0 to Infinity.

    Circularity: 0.00–1.00.

    Show: Bare Outlines.

    Summarize ticked.

3.3.2 Plasmodesmata per Pit Field Area Measurement from SEM Images Using ImageJ Software

1. Initiate the ImageJ software and, under File tab, click Open and select an SEM image file.
2. Under Plugins tab, select Cell Counter, and click Initialize.
3. In the Cell Counter window, under Counters, select Type 1 and count the number of plasmodesmata within the first pit field by clicking on the image (*see* **Note 23**).
4. Select Type 2 and so on to count plasmodesmata from the succeeding pit fields in the image (*see* **Note 24**). *See* example in Fig. 7a.
5. To measure individual pit field area in the active SEM image file, set the scale by clicking the line icon tool and trace the length of the scale bar embedded with the image.
6. Under Analyze tab, select Set Scale, type the value of the scale bar in the box beside Known distance, and click OK.
7. Select the Freehand selections icon tool and trace the periphery of a pit field.
8. Hit M from the keyboard to open the Results table (*see* **Note 25**). *See* example in Fig. 7b.
9. To generate the plasmodesmata per pit field area value, tabulate the measured values from **step 4** and **step 8** in Microsoft Excel.
10. Plot the number of plasmodesmata per pit field values against the corresponding individual pit field area values.
11. Generate a linear trendline equation, $y = mx + b$, where m is the slope, b is the intercept, and y is the plasmodesmata per pit field area at $x = 1\ \mu m^2$. *See* Fig. 7c as example.

3.3.3 Plasmodesmata Density per Cell Interface Area Calculation

1. Multiply the pit field area per cell interface area value obtained from Subheading 3.3.1 with the plasmodesmata per pit field area value obtained from Subheading 3.3.2 to get the plasmodesmata density per cell interface area value.

4 Notes

1. Hydrogel fixative can polymerize at room temperature, and therefore it is required to make sure all the components are cold (4 °C) before mixing them together to avoid premature polymerization. Making up the solution inside a cold room will ensure that everything will be kept cold.
2. To make 16% paraformaldehyde solution from powder, add 16 g paraformaldehyde in 100 mL 1× phosphate-buffered saline. Stir at 60 °C and gradually add sodium hydroxide pellet until clear. Adjust pH to 6.9 using hydrochloric acid. Store at 4 °C.

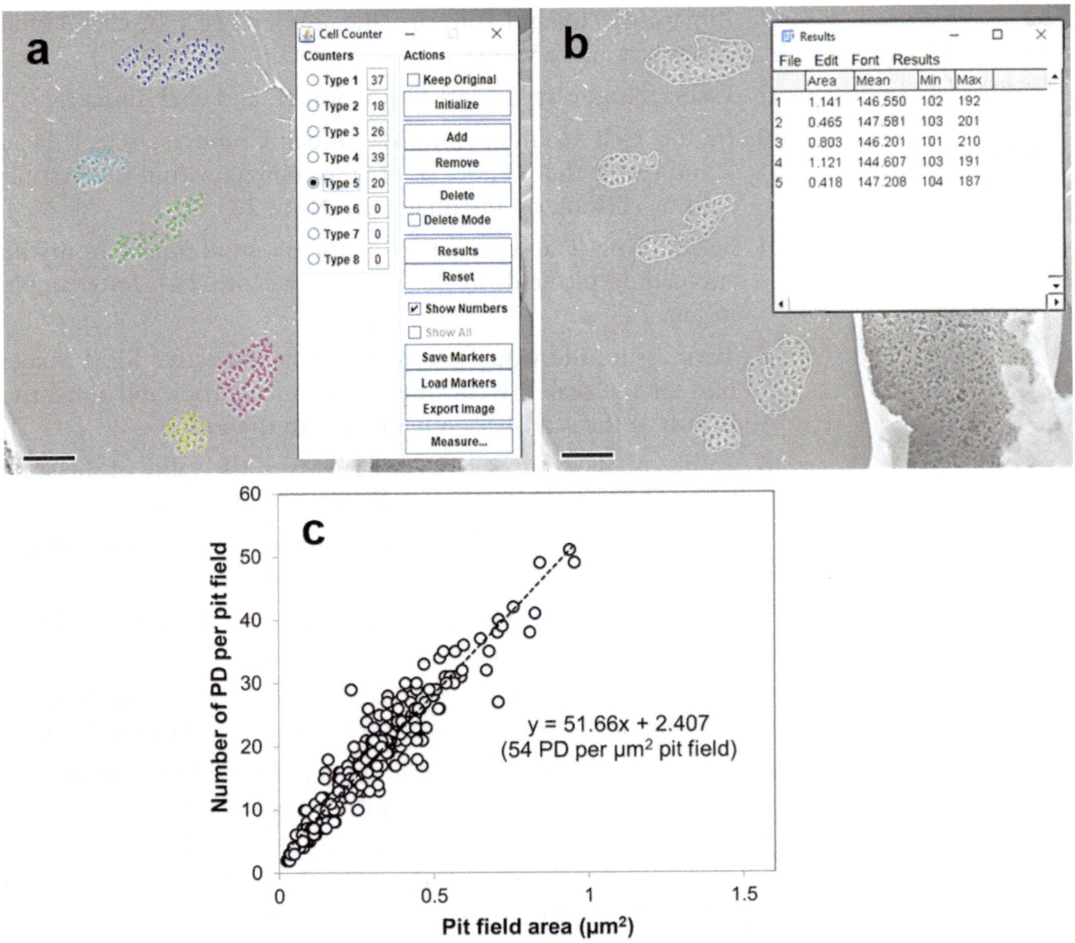

Fig. 7 Quantification of the number of plasmodesmata per pit field area in scanning electron micrographs using ImageJ software. (**a**) Plasmodesmata per pit field counting. (**b**) Individual pit field area measurement. (**c**) Scatter plot of plasmodesmata (PD) per pit field against pit field area generated in Microsoft Excel to determine PD per pit field area value. Bars = 1 μm

3. To make 1× phosphate-buffered saline, pH 7.4 with 0.005% sodium azide and 1 mM calcium chloride, dissolve 2.5 mg of sodium azide and 7.35 mg of calcium chloride dihydrate in 50 mL 1× phosphate-buffered saline, pH 7.4. Addition of sodium azide will prevent bacterial growth, while calcium chloride will provide the Ca^{2+} requirement of pectate lyase.
4. To make 200 mL TBST buffer, pH 7.4, dissolve 0.484 g Tris base, 1.798 g sodium chloride, and 2 mL 10% Tween 20 in distilled water. Adjust pH to 7.4 using hydrochloric acid.
5. Make 0.1 M sodium phosphate buffer pH 7.2 by mixing 23 mL of 0.2 M Solution A with 77 mL of 0.2 M Solution B into 100 mL distilled water. To make 0.2 M Solution A, dissolve 2.4 g of sodium phosphate monobasic (MW = 119.98) into

100 mL of distilled water. To make 0.2 M Solution B, dissolve 2.84 g of sodium phosphate dibasic (MW = 141.96) into 100 mL of distilled water. Store at room temperature.

6. Leaf tissue floats in the solution because of the presence of trapped air usually in the form of CO_2 and O_2. Vacuum infiltration removes the air inside the tissue making the tissue sinks and, therefore, facilitating the complete penetration of the fixative agent into the tissue.
7. The length of time it takes for clearing depends on the tissue used. Young tissues are often faster to clear, while older tissues would take longer clearing. The composition of tissue used also matters. For instance, young rice leaves could still take up to 10 months of clearing because of the presence of silica. Putting the tissue back into clearing solution after enzyme digestion step also helps to improve the clearing.
8. Enzyme activity degrades overtime so it is important to replenish the solution with freshly made digestion enzyme cocktail in between the whole incubation period.
9. After digestion, the gel-embedded leaf tissue is very fragile so extra care with handling is required from this point onward. To reduce the risk of damaging the tissue during transfer or mounting, it is advised to gently pour the gel-embedded leaf tissue from the tube into a small petri dish first before using a spatula to gently pick the tissue from the solution. If not proceeding with hybridization immediately, gel-embedded tissue can be kept in 1× phosphate-buffered saline pH 7.4 with 0.005% sodium azide (*see* **Note 3**).
10. Addition of the fluorophore makes the sample light-sensitive; therefore, from here onward, sample must be kept in the dark. This can be achieved by wrapping the tube with aluminum foil.
11. Make 25 mM sodium phosphate buffer pH 7.2 by diluting 25 mL of 0.1 M sodium phosphate buffer pH 7.2 (*see* **Note 5**) with 75 mL of distilled water.
12. It is very important to make sure that the tissue orientation is correct by checking under a dissecting microscope. The ripped side should be directly in the field of view. Otherwise, carefully and without touching the ripped side, adjust the angle of the leaf tissue using the forceps until the correct orientation is achieved.
13. One 3D stack (*see* Fig. 5a) can contain between 150 and 300 single focal planes depending on the sample leaf thickness. Since the interfaces between cell files are not flat, extracting several focal planes is necessary to obtain in focus images that cover the entire mesophyll-bundle sheath or mesophyll-mesophyll interface to be analyzed. This workflow shows how

all the pit fields contained in the mesophyll-bundle sheath cell interface of the leftmost vein in Fig. 5 were captured in a single image for subsequent analysis and quantification.

14. While this section is specific for the Leica SP8 confocal microscope software, LAS X, similar processing software is available with most major brands of confocal microscopes.
15. In Fig. 5, the maximum intensity projection image (*see* Fig. 5d) was obtained by combining 19 consecutive single focal planes derived as a subset from the whole 3D stack in Fig. 5a, where Fig. 5b is the first focal plane and Fig. 5c is the 19th focal plane. The maximum intensity projection image with only the calcofluor white channel (cell wall) is shown in Fig. 5e, while the maximum intensity projection image with only the Alexa Fluor® 488 channel (pit fields) is shown in Fig. 5f.
16. For pit field area measurement, use the maximum intensity projection image showing the Alexa Fluor®488 channel (*see* Fig. 6a).
17. This step will convert image Fig. 6a to image Fig. 6b.
18. This value can vary depending on the signal intensity of the pit fields. The rule is to make sure the converted image reflects the original image.
19. This step is necessary to allow further processing of the image.
20. This step will convert image Fig. 6b to image Fig. 6c.
21. In Fig. 6d, the selected area is derived from the bundle sheath cell outlines detected in Fig. 5b, leftmost vein, representing the subset of bundle sheath cells in direct contact with the overlying mesophyll cells seen in Fig. 5c.
22. This step will quantify the total pit field area as shown in Fig. 6e. For outlier pit field area quantification or pit fields that are larger than the maximum area for this mesophyll-bundle sheath interface, as seen in SEM images (Fig. 6f), under Analyze tab, select Analyze Particles, and specify the following parameters:

    Size (inch^2): 0.004 to Infinity (value depends on the maximum size pit fields at the specific interface to be analyzed).

    Circularity: 0.00–1.00.

    Show: Bare Outlines.

    Summarize ticked.

    Then subtract the outlier pit field area value from the total pit field area value to get the pit field area value reflecting only the area of interest.
23. Each click will be marked by a dot to help keep track of the counting. The total number of clicks, which corresponds to the

total number of plasmodesmata for that particular pit field, will be reflected in the box next to the selected Type.

24. Each Type will have a unique color assignment to differentiate one pit field from another (*see* Fig. 7a).
25. This step will display the measured value under Area column (*see* Fig. 7b).

Acknowledgments

This work was supported by the Australian Research Council Centre of Excellence for Translational Photosynthesis (CE1401000015). Electron microscopy facility was provided by Microscopy Australia at ANU Centre for Advanced Microscopy. Confocal microscopy facility was provided by CSIRO Black Mountain MicroImaging Centre, Canberra, Australia.

References

1. Gunning BES (1976) The role of plasmodesmata in short distance transport to and from the phloem. In: Gunning BES, Robards AW (eds) Intercellular communication in plants: studies on plasmodesmata. Springer, Berlin, pp 203–227. https://doi.org/10.1007/978-3-642-66294-2_11
2. Hatch MD, Osmond CB (1976) Compartmentation and transport in C_4 photosynthesis. In: Stocking CR, Heber U (eds) Transport in plants III, Encyclopedia of plant physiology, vol 3. Springer, Berlin, pp 144–184. https://doi.org/10.1007/978-3-642-66417-5_5
3. Beebe DU, Turgeon R (1991) Current perspectives on plasmodesmata: structure and function. Physiol Plant 83:194–199
4. Cilia ML, Jackson D (2004) Plasmodesmata form and function. Curr Opin Plant Biol 16(5):500–506
5. Crawford KM, Zambryski PC (1999) Plasmodesmata signaling: many roles, sophisticated statutes. Curr Opin Plant Biol 2:382–387
6. Ehlers K, Kollmann R (2001) Primary and secondary plasmodesmata: structure, origin, and functioning. Protoplasma 216:1–30. https://doi.org/10.1007/BF02680127
7. Faulkner CR, Oparka KJ (2016) Plasmodesmata. eLS. https://doi.org/10.1002/9780470015902.a0001681.pub3
8. Maule AJ (2008) Plasmodesmata: structure, function and biogenesis. Curr Opin Plant Biol 11:680–686
9. Maule AJ, Benitez-Alfonso Y, Faulkner C (2011) Plasmodesmata—membrane tunnels with attitude. Curr Opin Plant Biol 14:683–690. https://doi.org/10.1016/j.pbi.2011.07.007
10. Oparka KJ, Roberts AG (2001) Plasmodesmata. A not so open-and-shut case. Plant Physiol 125:123–126. https://doi.org/10.1104/pp.125.1.123
11. Hughes L, Hawes C, Monteith S, Vaughan S (2014) Serial block face scanning electron microscopy—the future of cell ultrastructure imaging. Protoplasma 251:395–401. https://doi.org/10.1007/s00709-013-0580-1
12. Gunning BES (1978) Age-related and origin-related control of the numbers of plasmodesmata in cell walls of developing Azolla roots. Planta 143:181–190. https://doi.org/10.1007/bf00387789
13. Seagull RW (1983) Differences in the frequency and disposition of plasmodesmata resulting from root cell elongation. Planta 159:497–504. https://doi.org/10.1007/bf00409138
14. Botha CEJ (1992) Plasmodesmatal distribution, structure and frequency in relation to assimilation in C_3 and C_4 grasses in southern Africa. Planta 187:348–358. https://doi.org/10.1007/bf00195658
15. Botha CEJ, Evert RF (1988) Plasmodesmatal distribution and frequency in vascular bundles and contiguous tissues of the leaf of *Themeda triandra*. Planta 173:433–441. https://doi.org/10.1007/bf00958954

16. Sage TL, Sage RF (2009) The Functional anatomy of rice leaves: implications for refixation of photorespiratory CO_2 and efforts to engineer C_4 photosynthesis into rice. Plant Cell Physiol 50:756–772. https://doi.org/10.1093/pcp/pcp033
17. Olesen P (1975) Plasmodesmata between mesophyll and bundle sheath cells in relation to the exchange of C_4-acids. Planta 123:199–202. https://doi.org/10.1007/bf00383870
18. Sowiński P, Bilska A, Barańska K, Fronk J, Kobus P (2007) Plasmodesmata density in vascular bundles in leaves of C_4 grasses grown at different light conditions in respect to photosynthesis and photosynthate export efficiency. Env Exp Bot 61:74–84. https://doi.org/10.1016/j.envexpbot.2007.03.002
19. Sowiński P, Rudzińska-Langwald A, Kobus P (2003) Changes in plasmodesmata frequency in vascular bundles of maize seedling leaf induced by growth at sub-optimal temperatures in relation to photosynthesis and assimilate export. Env Exp Bot 50:183–196. https://doi.org/10.1016/S0098-8472(03)00021-2
20. Botha C, Cross R (1997) Plasmodesmatal frequency in relation to short-distance transport and phloem loading in leaves of barley (Hordeum vulgare). Phloem is not loaded directly from the symplast. Physiol Plant 99:355–362
21. Botha CEJ, van Bel AJE (1992) Quantification of symplastic continuity as visualised by plasmodesmograms: diagnostic value for phloem-loading pathways. Planta 187:359–366. https://doi.org/10.1007/bf00195659
22. Lee J-Y (2014) New and old roles of plasmodesmata in immunity and parallels to tunneling nanotubes. Plant Sci 221-222:13–20. https://doi.org/10.1016/j.plantsci.2014.01.006
23. Zavaliev R, Epel BL (2015) Imaging callose at plasmodesmata using aniline blue: quantitative confocal microscopy. Methods Mol Biol 1217: 105–119. https://doi.org/10.1007/978-1-4939-1523-1_7
24. Faulkner CR, Blackman LM, Collings DA, Cordwell SJ, Overall RL (2009) Anti-tropomyosin antibodies co-localise with actin microfilaments and label plasmodesmata. Eur J Cell Biol 88:357–369
25. Palmer WM, Martin AP, Flynn JR, Reed SL, White RG, Furbank RT, Grof CPL (2015) PEA-CLARITY: 3D molecular imaging of whole plant organs. Sci Rep 5:13492. https://doi.org/10.1038/srep13492
26. Chung K, Wallace J, Kim S-Y, Kalyanasundaram S, Andalman AS, Davidson TJ, Mirzabekov JJ, Zalocusky KA, Mattis J, Denisin AK (2013) Structural and molecular interrogation of intact biological systems. Nature 497:332–337
27. Palmer WM, Flynn JR, Martin AP, Reed SL, Grof CPL, White RG, Furbank RT (2018) 3D clearing and molecular labeling in plant tissues. In: Covshoff S (ed) Photosynthesis: methods and protocols. Springer, New York, NY, pp 285–304. https://doi.org/10.1007/978-1-4939-7786-4_17
28. Danila FR, Quick WP, White RG, Furbank RT, von Caemmerer S (2016) The metabolite pathway between bundle sheath and mesophyll: quantification of plasmodesmata in leaves of C_3 and C_4 monocots. Plant Cell 28:1461–1471. https://doi.org/10.1105/tpc.16.00155
29. Danila FR, Quick WP, White RG, Kelly S, von Caemmerer S, Furbank RT (2018) Multiple mechanisms for enhanced plasmodesmata density in disparate subtypes of C_4 grasses. J Exp Bot 69:1135–1145. https://doi.org/10.1093/jxb/erx456

Chapter 8

Super-Resolution Imaging of Plasmodesmata Using 3D Structured Illumination Microscopy

Kirsten Knox

Abstract

Plasmodesmata (PD) have a diameter of around 30–50 nm which is well below the 200 nm limit of optical resolution, making analysis by light microscopy difficult and resolving internal structures of the PD such as the desmotubule impossible. Modern super-resolution methods such as 3D structured illumination microscopy (3D-SIM) can increase the lateral and axial resolution and work well on fixed, sectioned material. However, imaging in live plant cells requires careful optimization. Here we present a method to image PD using 3D-SIM in live BY2 cells.

Key words Plasmodesmata, Desmotubule, 3D-SIM, Super-resolution, BY2, Fluorescence

1 Introduction

The diffraction limit of light means that the achievable lateral resolution using optical microscopy is 200 nm and this has remained relatively unchallenged for over 130 years [1]. During the past decade, a number of different super-resolution techniques have come to the fore, and these circumvent the resolution limit using various approaches [2]. Many of the super-resolution techniques require lengthy or light-intensive imaging protocols with significant computational post-processing, and 3D structured illumination (3D-SIM) is no exception [3]. This technique uses sinusoidal striped patterned excitation light to take multiple overlapping images of the same z-plane in rapid succession. The phase of the light pattern is changed five times, before being rotated through 60°, usually three times. This results in 15 images for each z-plane [4]. Shifting the phase of the stripes results in moiré interference fringes because the fine patterns are superimposed on one another [3]. Such structured illumination gives back specific information about the sample via the emitted fluorescence [4, 5]. The information is then processed computationally to build up a 3D

Yoselin Benitez-Alfonso and Manfred Heinlein (eds.), *Plasmodesmata: Methods and Protocols*,
Methods in Molecular Biology, vol. 2457, https://doi.org/10.1007/978-1-0716-2132-5_8,

reconstruction of the field of view. The theoretical resolution of 100 nm in the lateral plane and 200 nm in the axial is also affected by sample-specific issues [5, 6]. A primary requirement is a stable fluorophore labeling of the required structures in either fixed or live samples [4, 5]. Imaging live cells introduces its own challenges, and the 3D-SIM system must be optimized to reduce the photon budget and imaging time required in order to avoid either photobleaching of the fluorophores or phototoxicity of the cell [7]. The OMX system offers rapid shuttering and fast stable focusing resulting in delays of just 1–2 ms, making it as fast and precise as possible meaning live imaging is feasible [4, 7].

With a diameter of around 30–50 nm, plasmodesmata (PD), the pores symplastically linking plant cells, are well below the resolution limit. In their simplest form, PD comprise of a plasma membrane-lined channel with a central rod called a desmotubule [8–10]. At just 15 nm, the desmotubule is one of the most tightly constricted membrane structures in nature [8, 11]. Most of the detailed information about PD structure has come from using EM techniques [8–10]. These require fixation processes which can potentially damage sensitive cellular components such as the endoplasmic reticulum (ER) which forms the desmotubule. Fixed sections have been analyzed using 3D-SIM which provided detailed images of PD labeled with a fluorescent Tobacco Mosaic Virus Movement Protein (TMV-MP), but of course this provided no dynamic data [6]. Here we present a method which we developed to image the desmotubule connecting the ER through the PD of adjacent live BY2 cells using RETICULON 6—GREEN FLUO RESCENT PROTEIN (RTN6-GFP) (Fig. 1) [12–14]. This method is also applicable to other PD localized fluorescent tags. The BY2 cell line is derived from tobacco and provides a useful model for imaging PD in live cells as they routinely divide forming connected chains in a single plane and are easily transformable both transiently and stably [15]. Choice of fluorophore and labeled structure is also important for imaging live cells. The labeled structure—here the ER and thus desmotubule, must not move during the time required to capture an image. We have optimized this protocol to minimize bleaching of GFP and minimize acquisition time although this reduces the achievable resolution. This method is adaptable to other super-resolution systems and would allow for dynamic experiments.

2 Materials

2.1 Culture and Preparation of BY2 Cells

1. BY2 culture media: 3% sucrose, 0.43% Murashige and Skoog (MS) Basal Salts (*see* **Note 1**); add 2,4-dichlorophenoxyacetic acid (2,4-D) stock solution to a final concentration of 2 μg/ml and make up to volume with ddH_2O. Autoclave. Store at 4 °C for up to 4 weeks.

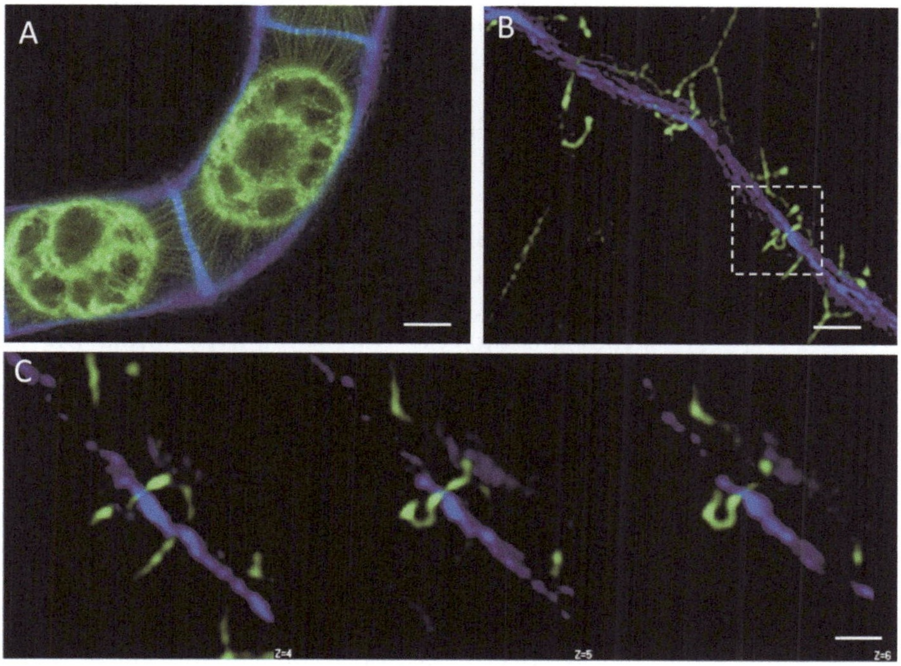

Fig. 1 Plasmolyzed BY2 cells showing ER labeled with RTN6-GFP. (**a**) Confocal image where RTN6-GFP is shown in green and cell walls labeled with Calcofluor White (blue). Scale = 10 μm. (**b**) 3D-SIM image focused on the cell walls between two connected BY2 cells where PD are located. Scale = 2 μm. (**c**) Detailed view, across subsequent z-sections, of the boxed area in **b**, showing the narrowing of the ER as it forms the desmotubule passing through the PD. Scale = 0.5 μm

2. 2,4-D stock solution: Dissolve 200 mg of 2,4-dichlorophenoxy acetic acid sodium salt monohydrate in 3 ml of ethanol. Once fully dissolved, add sterile ddH_2O to 10 ml final volume. Filter sterilize through a 0.45 μm filter, aliquot, and store at −20 °C.
3. Mannitol: Add D-mannitol to ddH_2O to a final concentration of 1 M. Mannitol will fully dissolve during autoclaving. Store at room temperature.
4. DiOC6 (3,3′-dihexyloxacarbocyanine iodide) stock solution: Dissolve in DMSO to a final concentration of 100 mM. Store at −20 °C and protect from light.
5. Calcofluor White stock solution: Dissolve Calcofluor White in ethanol to a concentration of 170 μg/ml. Store at −20 °C and protect from light.
6. Microscope slides and 0.17 mm thickness coverslips.

2.2 3D-Sim Imaging

1. PersonalDV DeltaVision Epi-fluorescence Inverted Microscope.
2. DeltaVision OMX Blaze 3D-SIM microscope fitted with an Olympus PlanApo N 100× 1.42 NA oil objective and an Edge sCMOS camera (PCO AG).

3 Methods

3.1 Preparation of BY2 Cells for Live Imaging

1. Maintain BY2 cells in 40 ml of BY2 culture media in 250 ml wide necked Erlenmeyer flasks sealed with a sponge bung and foil. Culture at 28 °C with rotary shaking at 140 rpm (*see* **Note 2**).
2. Aliquot 1 ml of cell suspension to a sterile Eppendorf tube and add Calcofluor White to a final concentration of 3.5 μg/ml to label cell walls. Incubate for 5 min at room temperature before gently rinsing cells twice with fresh media (*see* **Note 3**). *Optional: If imaging wild type cells, a stain such as DiOC6 can be used to label the ER and thus the desmotubule at a final concentration of 50 μM (see* **Note 4***). Incubate for 10 min and then rinse twice with fresh media.*
3. Induce plasmolysis by removing 450 μl of media and replacing with 450 μl 1 M mannitol. Incubate for 10 min at room temperature before gently inverting the tube to resuspend the cells (*see* **Note 5**).
4. Transfer 40 μl of cell suspension onto the center of a clean slide. Gently lower the coverslip avoiding air bubble formation. Apply gentle even pressure using a folded paper towel to remove excess liquid from the slide (*see* **Note 6**). Carefully wipe all edges, ensuring the coverslip does not become smeared. Seal the edges of the coverslip with nail varnish.

3.2 3D-SIM Imaging

This protocol was designed for use with an OMX DeltaVision Blaze, but the crucial steps are transferrable to other 3D-SIM setups.

1. Working quickly, initially identify candidate cells using the lowest intensity of bright-field light possible (*see* **Note 7**) on the auxiliary microscope (PersonalDV DeltaVision), which has stage coordinates synchronized with the OMX. Mark their position on the slide with the point visiting tool. Expose all candidate cells briefly to their excitation wavelength and check that both the calcofluor stain and the GFP reporter (or DIOC6) are sufficiently bright (*see* **Note 7**).
2. Continuing to work quickly, transfer the slide to the OMX microscope stage and carefully apply immersion oil (*see* **Note 8**). Find the previously marked cells using the Point List and then center the cell in the image using the Spiral Mosaic function.
3. Carry out initial optimization of the 3D-SIM settings on a sacrificial set of cells. Establish the minimal level of laser power required to produce optimal image intensities (between 1000 and 3000 counts in the raw image acquired by a 15-bit

dynamic range Edge sCMOS camera). Determine the shortest exposure times for each channel, typically between 100 and 200 ms (*see* **Note 9**).

4. Following optimization acquire image stacks of appropriate cells. Again avoid excess light exposure by setting the start and end positions for a minimal z-stack rapidly.
5. Adjust the images from each channel using the SoftWorx 6.0 alignment tool. Reconstruct 3D image stacks using SoftWorx 6.0 with channel specific OTFs and Wiener filter settings of 0.002 (*see* **Note 10**).

4 Notes

1. There are many variations of MS; here, we used MS-basal salts from Sigma.
2. Subculture cells weekly using a 1:40 dilution in fresh media. In cell lines stably transformed with a fluorescent marker such as RTN6-GFP, ensure that the signal is strong and relatively homogenous in the line of choice; we found it best to continue using the selective agent (in this case Kanamycin 40 μg/ml) during culture maintenance. Best results are usually obtained with cells which were subbed 3–4 days prior to imaging.
3. Treat the cells as gently as possible. Use pipette tips with the ends trimmed off to provide a wider bore and reduce shearing forces. To rinse, allow cells to settle to the bottom of the Eppendorf tube, carefully remove as much media as possible before adding fresh, and mix by gentle inversion.
4. DiOC6 will label the ER (including the desmotubule) as well as mitochondria and vesicle membranes. However, it is highly phototoxic so care must be taken to keep cells in the dark once DiOC6 is added and illumination minimized to reduce ER damage.
5. Depending on the marker used, plasmolysis is not essential for imaging PD with 3D-SIM, but it does allow much clearer visualization of the desmotubules when the ER of the Hechtian strands is labeled with a suitable marker or stain, e.g., RTN6-GFP or DiOC6, and the membrane is retracted away from the cell wall.
6. For image optimization with a 3D-SIM inverted scope, the cells need to be as close to the coverslip and as still as possible.
7. Photobleaching of fluorescent markers and/or phototoxicity from DiOC6 must be minimized by keeping all light exposure as brief and low intensity as possible before SIM occurs.

8. The Refractive Index of the sample should closely match that of the immersion oil in order to reduce spherical aberrations. Theoretically, the cells should be mounted in a high concentration of glycerol/antifade mountant rather than media, but we found this affected the ER structure in live cells. However, this may remain an option for other PD imaging where the desmotubule is less crucial.
9. It is advisable to carry out the optimization on a sacrificial cell—by the time the process is complete, the GFP is likely to have bleached sufficiently to prohibit good image capture during the 3D-SIM process.
10. It is vital to ensure that the microscope is regularly calibrated. This can be achieved by checking the optical transfer functions (OTFs) for each channel and adjusting for optimal resolution in both lateral and axial images. The alignment tools also require regular calibration, and we used 100-nm-diameter TetraSpeck beads to calibrate measurements.

References

1. Abbe E (1873) Beiträge zur Theorie des Mikroskops und der mikroskopischen Wahrnehmung. Arkiv Mikroskop Anat 9:413–468
2. Komis G, Šamajová O, Ovečka M, Šamaj J (2015) Super-resolution microscopy in plant cell imaging. Trends Plant Sci 20:834–843
3. Gustafson MGL (2000) Surpassing the lateral resolution limit by a factor of two using structured illumination microscopy. J Microsc 198:82–87
4. Dobbie IM, King E, Parton RM, Carlton PM, Sedat JW, Swedlow JR, Davis I (2011) OMX: a new platform for multimodal, multi-channel wide-field imaging. Cold Spring Harb Protoc 8:899–909
5. Huang B, Bates M, Zhuang XW (2009) Super-resolution fluorescence microscopy. Annu Rev Biochem 78:993–1016
6. Fitzgibbon J, Bell K, King E, Oparka K (2010) Super-resolution imaging of plasmodesmata using 3-dimensional structured illumination microscopy. Plant Physiol 153:1453–1463
7. Komis G, Mistrik M, Šamajová O, Ovečka M, Bartek J, Šamaj J (2015) Superresolution live imaging of plant cells using structured illumination microscopy. Nat Protoc 10:1248–1263
8. Overall RL, Blackman LM (1996) A model of the macro-molecular structure of plasmodesmata. Trends Plant Sci 1:307–311
9. Roberts AG (2005) Plasmodesmal structure and development. In: Oparka KJ (ed) Plasmodesmata: annual plant review, vol 18. Blackwell Publishing, Oxford, pp 1–32
10. Nicolas WJ, Grison S, Trépout A, Gaston M, Fouché M, Cordelières FP, Oparka K, Tilsner J, Brocard L, Bayer EM (2017) Architecture and permeability of post-cytokinesis plasmodesmata lacking cytoplasmic sleeves. Nat Plants 3: 17082
11. Tilsner J, Amari K, Torrance L (2010) Plasmodesmata viewed as specialised membrane adhesion sites. Protoplasma 248:39–36
12. Knox K, Wang P, Kriechbaumer V, Tilsner J, Frigerio L, Sparkes I, Hawes C, Oparka K (2015) Putting the squeeze on plasmodesmata: a role for reticulons in primary plasmodesmata formation. Plant Physiol 168:1561572
13. Bell K, Oparka K, Knox K (2016) Super-resolution imaging of live BY2 cells using 3D-structured illumination microscopy. Bio-protocol 6(1):e1697
14. Bell K, Oparka K, Knox K (2018) Preparation and imaging of specialised ER using super-resolution and TEM techniques. Methods Mol Biol 1691:33–34
15. Nagata T, Nemoto Y, Hasezawa S (1992) Tobacco BY-2 cell line as the "HeLa" cell in the cell biology of higher plants. Int Rev Cytol 132:1–30

Part III

Compositional Analysis of Plasmodesmata

Chapter 9

In Vivo Aniline Blue Staining and Semiautomated Quantification of Callose Deposition at Plasmodesmata

Caiping Huang, Jerôme Mutterer, and Manfred Heinlein

Abstract

The deposition and turnover of callose (beta-1,3 glucan polymer) in the cell wall surrounding the neck regions of plasmodesmata (PD) controls the cell-to-cell diffusion rate of molecules and, therefore, plays an important role in the regulation of intercellular communication in plants.

Here we describe a simple and fast in vivo staining procedure for the imaging and quantification of callose at PD. We also introduce *calloseQuant*, a plug-in for semiautomated image analysis and non-biased quantification of callose levels at PD using ImageJ.

Key words Callose, Plasmodesmata, Aniline blue, Confocal microscopy, ImageJ, *flg22*

1 Introduction

Plasmodesmata (PD) are nanopores in plant cell walls that function as cytoplasmic bridges between adjacent cells. They play a central role in intercellular communication by allowing the cell-to-cell transport of molecules. Plants regulate the cell-to-cell movement of a wide range of molecules by changing PD frequency and alterations in PD structure or, more rapidly, by modifying the size exclusion limit (SEL) of the pore. The SEL of the PD channel, i.e., the maximal size of the molecules transported between cells, is predominantly regulated by the degree of callose deposition within the cell wall surrounding the PD. This deposition is controlled by several PD-localized proteins [1, 2] such as callose synthases [3, 4], PD-callose binding proteins (PDCBs) [5], and β-1,3-glucanases (BGs) [6–8]. The molecular composition of PD is still unknown. However, proteomic and functional studies have shown that PD contain receptor kinases and other receptor proteins implicated in regulating the gatekeeping function of PD, and the PD-specific function of several PD-localized receptors has been experimentally confirmed [2, 9–11]. Numerous studies have shown that

Yoselin Benitez-Alfonso and Manfred Heinlein (eds.), *Plasmodesmata: Methods and Protocols*,
Methods in Molecular Biology, vol. 2457, https://doi.org/10.1007/978-1-0716-2132-5_9,

PD-localized proteins regulate callose levels and the cell-to-cell diffusion of molecules in plant tissues in response to stress or pathogens [12] and to intracellular signals linked to development and leaf age [13], light and the circadian clock [14], nutrition status [15], the redox status in chloroplasts and mitochondria [16], reactive oxygen species [17], cytosolic Ca^{2+} concentration [18], and hormones [1, 19, 20]. Callose deposition at PD plays an important role in isolating cells from communication with surrounding cells during dormancy [21, 22] and cell differentiation [23], whereas the degradation of callose at PD is associated with the cell-to-cell movement of viruses [24] or the establishment of nitrogen-fixing nodules upon rhizobial infection [25].

Because of the central role of PD-localized callose in regulating PD permeability [7], it is useful to image and quantify the level of callose at PD and to correlate the result with the permeability of PD under different physiological conditions, developmental stages, or in the presence and absence of biotic or abiotic factors. A simple way to image and quantify callose in plant tissues is by staining with aniline blue, a sulfonated triarylmethane dye. Aniline blue has long been employed as ultimate dye for callose staining in plant tissues [26] and has been widely used to detect callose associated with pollen tubes, microsporocytes, cell plate, phloem tissue, pathogen-associated molecular pattern (PAMP)-induced papillae, and PD. Commercial aniline blue is a heterogeneous mixture of various components, including a fluorochrome that has its excitation maximum at around 380 nm and a maximum emission at around 455–520 nm [27, 28]. Staining of callose with this dye gives rise to a bright yellow-green fluorescence when irradiated with ultraviolet light that can be easily observed by confocal fluorescence microscopy using a 405 nm laser for excitation and a suitable band-pass filter.

We describe here a simple procedure for comparative callose staining and quantification at PD in life plant tissues using aniline blue and confocal microscopy. Acquired images are analyzed with ImageJ using *calloseQuant*. This novel plug-in provides two different methods for semiautomated thresholding and detection as well as for quantification of the fluorescence levels associated with each individual PD within individual images, z-stacks of images, or images collected in a folder. For illustration, the two methods are applied to images of aniline blue-stained epidermis of *Nicotiana benthamiana* leaf tissues treated either with buffer or with the flagellin-derived defense elicitor peptide *flg22*. The results obtained with the two methods demonstrate an increase in PD-associated callose levels upon *flg22*-treatment, which is consistent with previous observations in *Arabidopsis* [18].

2 Materials

1. Plant material: The protocol described here has been successfully applied to leaf tissues of 4-week-old *N. benthamiana* and *Arabidopsis thaliana* plants. The plants should be grown under controlled conditions (light, humidity, temperature) and handled with care to avoid wounding as this will cause callose formation. For comparative experiments to determine PD-localized callose deposition in response to specific treatments, use plants with the same age and include appropriate controls.
2. Double-distilled water (DDW).
3. 0.2 M potassium phosphate buffer, pH 9.0. To make this buffer, mix 5.3 ml of 0.2 M KH_2PO_4 with 94.7 ml of 0.2 M K_2HPO_4.
4. Aniline blue solution: 0.01% (w/v) aniline blue in 50 mM potassium phosphate buffer (pH 9.0) (*see* **Note 1**). Clear the solution through a Millipore filter (0.45 μm) and store in the dark at room temperature. Aniline blue solution is blue/violet in color when freshly prepared. The solution must be kept for at least 48 h at room temperature until it assumes a bright yellow color (*see* **Note 2**).
5. Microscopy glass slides (26 × 76 mm) and coverslips (22 × 40 mm).
6. 6 mm cork borer.
7. Flat-ended filter forceps.
8. 12-well culture plates with cover.
9. Sticky tape (e.g., Scotch, 19 mm).
10. Vacuum desiccator with pump.
11. Confocal microscope equipped with a 405 nm diode laser for aniline blue excitation and a 475–525 nm band-pass filter for specific aniline blue emission imaging (*see* **Note 3**).
12. Image acquisition software: The protocol described here employs the publicly available image analysis program ImageJ. Download and install ImageJ from the NIH website (http://rsbweb.nih. gov/ij/). Launch ImageJ, go to *Help* > *Update ImageJ*, and update the program into its latest version.

3 Methods

3.1 Sample Preparation and Staining

1. Use the cork borer to isolate leaf disks. Use filter forceps and carefully place the leaf disks individually into the wells of 12-well culture plates containing 1 ml DDW and incubate the

covered plates overnight in the same plant growth room in which the plants were raised.

2. Wash the leaf disks in the 12-well plates three times with DDW.
3. Use filter forceps to place a leaf sample (leaf disk) onto a microscope slide, cover the sample with a coverslip, and fix the coverslip on two sides with tape to the slide.
4. Add 200 μl aniline blue solution into the space between glass slide and coverslip (*see* **Note 4**).
5. Place the glass slide with the sample(s) into a vacuum desiccator and evacuate for 1–2 min (<0.8 pa) followed by slow release of the pressure (*see* **Note 5**).
6. Remove the sample and use it for microscopy (*see* **Note 6**).

3.2 Confocal Microscope Settings and Image Acquisition

1. Place the slide with the stained leaf sample onto the microscope stage to observe the epidermis on the abaxial side of the leaf.
2. Set the confocal microscope to single-track mode and select the 405 nm diode laser for excitation and the 475–525 nm emission filter for aniline blue fluorescence (*see* **Note 3**). Choose white pseudo-color for the display of aniline blue emission. Keep the channel for transmitted light switched off.
3. Set the scanning parameters to frame-scanning mode, frame size 1024 × 1024 pixels, scan speed 9, pixel depth 8 bit (*see* **Note 7**); line scanning, scan average 4, zoom 1×.
4. Observe the samples (*see* **Note 8**) with a 20× or 40× objective and adjust the values of laser intensity, detector gain, amplifier offset, and amplifier gain to achieve strong, but not oversaturated, PD-associated fluorescent signal against a low background signal. View the histogram of pixel intensities along a drawn line connecting the most strongly fluorescent PD-associated spots. The peaks of intensities from brightest PD sites should not reach saturation (255 maximum gray scale level for 8-bit images and 65,536 maximum gray scale level for 16-bit images) but remain in the upper fifth section of the histogram. Disregard unrelated objects (e.g., stomata) that may show saturating intensities at these settings.
5. Acquire images of the samples (*see* **Note 9**). Choose areas where the leaf surface is flat and several PD can be focused in one view. The raw microscope image files (.lsm files) are single-channel 8-bit or 16-bit resolution RGB files, which can be directly used for subsequent image analysis with ImageJ (*see* **Note 10**).
6. Create a folder containing all the image files for the specific experiment. The set of images should contain images of control and test treatments acquired under the same conditions (dataset).

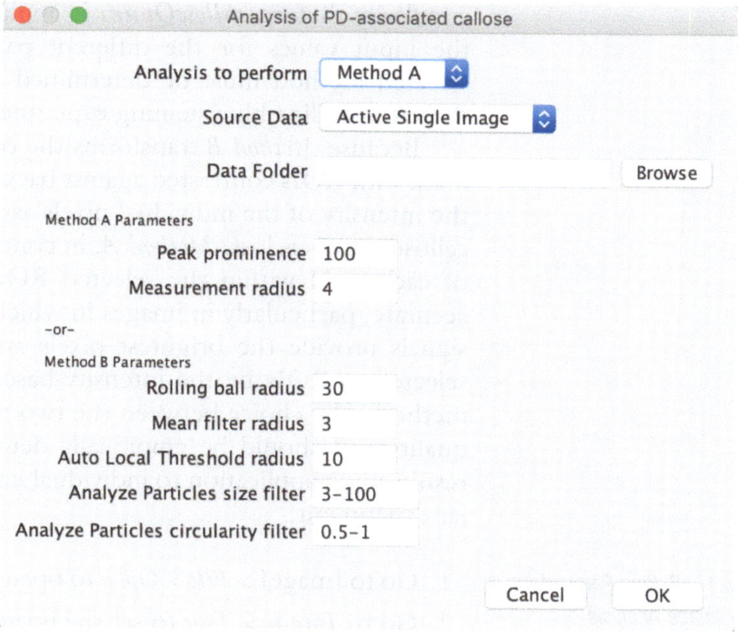

Fig. 1 Screenshot of the *calloseQuant* plug-in

3.3 Image Analysis and Data Acquisition with calloseQuant

calloseQuant is an ImageJ plug-in which localizes fluorescent callose spots and quantifies callose fluorescence intensity in each spot automatically.

Settings in this software (Fig. 1) allow to select for the automatic processing of a single specific image (setting "Active Single Image"), of an image z-stack (setting "Active Image Stack"), or of images collected in a specified folder of images (setting "Folder of Images"). Moreover, it is possible to choose between two semiautomated analysis methods. *Method A* finds spots of maximal fluorescence and measures fluorescence intensity in each spot according to a specified size of the region of interest ("Measurement radius") that will be applied around the peak fluorescence intensity of each spot. An input value for "Peak prominence" will determine the degree of selectivity for finding the maxima against the fluorescence background. *Method B* is based on a previously described protocol [29] which uses image thresholding, whereby all pixels with fluorescence intensity values above a certain threshold are assigned as a foreground signal representing regions of interests (ROIs) and the pixels below the threshold as a background signal. Before thresholding the image is processed for increased signal contrast by subtraction of the background fluorescence around brightest pixels within a user-defined "Rolling ball radius." Moreover, a noise reduction filter is applied by setting a "Mean filter radius." The signals identified as belonging to ROIs according to specific thresholding are then considered as particles which will be measured according to additional inputs such as the particle size range and circularity. Before *calloseQuant* is applied for automated analysis, the input values for the different parameters required for the selected method must be determined using sample images from the given aniline blue staining experiment.

Because *Method B* transforms the original image into a binary mask with ROIs contrasted against background, information about the intensity of the individual pixels associated with the individual callose signals is lost. *Method A*, in contrast, measures the intensity of each pixel within the selected ROIs and thus may be more accurate, particularly in images in which the PD-associated callose signals provide the brightest pixels and therefore are accurately selected as ROIs by the intensity-based approach applied in this method. The choice between the two methods depends on image quality and should be empirically determined by comparing the results upon application to individual image samples of the particular experiment.

3.3.1 Setting Parameters for Image Analysis and Data Acquisition with Method A

1. Go to ImageJ > *File* > *Open* to open an image file (*see* **Note 11**).
2. Go to *Image* > *Type* to set the image depth to 8-bit.
3. If the image is in color, go to *Image* > *Lookup Tables* and select "Grays" (*see* **Note 12**).
4. Determination of "Peak prominence value": Go to *Process* > *Find Maxima...*, enter a value for *Prominence* (e.g., 150; *see* **Note 13**) and under *Output style* select *Point Selection*. Check the box for *Preview point selection* and determine if the PD-localized callose spots at the plant cell walls are correctly selected in the image (Fig. 2a, b). If background spots are selected, increase the *Prominence* value until only the callose spots remain specifically highlighted (*see* **Notes 14** and **15**). Take note of this *Prominence* level.
5. Deselect *Preview point selection* and click *Cancel.*
6. Determination of "Measurement radius": Enlarge the image until individual pixels can be recognized (Fig. 2c). Count the number of pixels across the diameter of a typical fluorescent spot (e.g., 8 pixels) and divide this number by 2 (*see* **Note 16**). Take note of this number of pixels.

3.3.2 Setting Parameters for Image Analysis and Data Acquisition with Method B

1. Go to ImageJ > *File* > *Open* to open an image file (*see* **Notes 11** and **17**).
2. Go to *Image* > *Type* to set the image depth to 8-bit.
3. If the image is in color, go to *Image* > *Lookup Tables* and select "Grays" (*see* **Note 12**).
4. Determination of "Rolling ball radius": Go to *Process* > *Subtract Background* and enter a value for *Rolling ball radius* (e.g., 6; *see* **Note 18**). Check *Preview* to see the result. Modify the value for

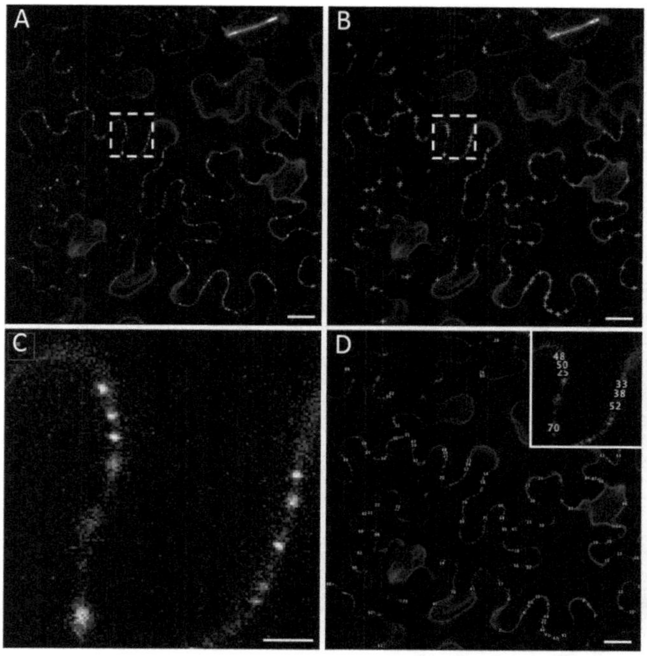

E

	Label	Area	Mean	StdDev	Min	Max	IntDen	Median	RawIntDen
1	PBS1.tiff	2.541	101.558	82.240	0	255	258.044	66	5281
2	PBS1.tiff	2.541	99.212	83.087	6	255	252.083	68	5159
3	PBS1.tiff	2.541	134.192	89.044	2	255	340.965	115	6978.000
4	PBS1.tiff	2.541	107.135	81.164	19	255	272.215	81	5571.000
5	PBS1.tiff	2.541	129.885	82.957	9	255	330.019	111	6754.000
6	PBS1.tiff	2.541	123.577	89.849	2	255	313.992	96	6426.000
7	PBS1.tiff	2.541	104.635	62.760	21	255	265.863	88	5441.000
8	PBS1.tiff	2.541	85.269	65.040	8	255	216.658	61	4434.000
9	PBS1.tiff	2.541	86.558	79.218	6	255	219.931	62	4501.000
10	PBS1.tiff	2.541	89.385	76.073	2	255	227.114	62	4648.000
11	PBS1.tiff	2.541	108.596	85.409	9	255	275.928	79	5647.000
12	PBS1.tiff	2.541	111.615	94.390	8	255	283.600	68	5804.000
13	PBS1.tiff	2.541	91.462	76.494	2	255	232.392	68	4756.000
14	PBS1.tiff	2.541	90.923	79.847	1	255	231.023	52	4728.000
15	PBS1.tiff	2.541	143.481	95.939	7	255	364.565	116	7461.000
16	PBS1.tiff	2.541	69.827	72.733	2	255	177.421	47	3631.000
17	PBS1.tiff	2.541	96.231	90.942	4	255	244.509	62	5004.000
18	PBS1.tiff	2.541	74.462	71.937	7	255	189.197	42	3872.000
19	PBS1.tiff	2.541	68.788	70.937	1	255	174.782	42	3577.000
20	PBS1.tiff	2.541	105.115	88.613	5	255	267.084	74	5466.000
21	PBS1.tiff	2.541	98.462	83.637	3	255	250.178	59	5120.000
22	PBS1.tiff	2.541	113.404	71.284	11	255	288.144	105	5897.000
23	PBS1.tiff	2.541	116.327	82.172	1	255	295.571	104	6049.000
24	PBS1.tiff	2.541	107.615	84.970	10	255	273.436	68	5596.000
25	PBS1.tiff	2.541	61.462	65.066	4	255	156.166	38	3196.000
26	PBS1.tiff	2.541	98.827	77.050	12	255	251.106	71	5139.000
27	PBS1.tiff	2.541	200.442	45.227	76	255	509.297	211	10423.000
28	PBS1.tiff	2.541	235.404	34.914	96	255	598.130	255	12241.000
29	PBS1.tiff	2.541	116.577	66.171	15	255	296.206	116	6062.000
30	PBS1.tiff	2.541	98.346	75.909	4	253	249.884	77	5114.000
31	PBS1.tiff	2.541	95.346	71.399	1	253	242.262	77	4958.000
32	PBS1.tiff	2.541	75.635	67.948	2	253	192.177	55	3933.000
33	PBS1.tiff	2.541	69.038	61.593	10	252	175.417	46	3590.000
34	PBS1.tiff	2.541	77.365	69.015	5	252	196.575	59	4023.000
35	PBS1.tiff	2.541	84.269	62.022	9	252	214.117	64	4382.000
36	PBS1.tiff	2.541	90.865	70.644	1	249	230.877	69	4725.000
37	PBS1.tiff	2.541	65.788	52.130	3	248	167.160	51	3421.000
38	PBS1.tiff	2.541	76.192	54.864	1	248	193.594	70	3962.000
39	PBS1.tiff	2.541	106.558	65.199	2	248	270.749	98	5541.000
40	PBS1.tiff	2.541	66.269	69.162	0	246	168.381	41	3446.000
41	PBS1.tiff	2.541	109.750	61.232	29	244	278.860	98	5707.000
42	PBS1.tiff	2.541	56.058	60.805	1	242	142.435	40	2915.000
43	PBS1.tiff	2.541	72.769	60.542	13	239	184.897	44	3784.000
44	PBS1.tiff	2.541	50.596	47.870	1	238	128.558	33	2631.000

Fig. 2 Key steps in image analysis applying parameters of *Method A*. (**a**) Original 8-bit single channel image (1024 × 1024 pixel) of aniline blue-stained *N. benthamiana* epidermal cells showing fluorescent callose signal. The area delimited with the dashed line is enlarged in (**c**) and (**d**). (**b**) The same image seen in the preview window of the *Find Maxima*. . .function with 150 as a value for *Prominence.* (**c**) Enlargement of the area highlighted in (**a**). Individual pixels can be seen to determine the *Measurement radius.* (**d**) Image after running *calloseQuant* using 150 as value for *Prominence* and 4 pixels for *Measurement radius.* Each peak prominence spot is indexed with a number that corresponds to an individual measurement. (**e**) Section of the *calloseQuant* data table showing the result of measuring each individual spot shown in (**d**). The running number corresponds to the individual peak prominence spots indexed in (**d**). Scale bars, (**a**), (**b**) and (**d**) = 20 μm, (**c**) = 5 μm

Rolling ball radius to optimize (*see* **Note 19**). Once the optimal radius is found, make sure that all the other options are unchecked and select *OK*. Take note of the selected radius value.

5. Determination of "Mean filter radius": Go to *Process* > *Filters* > *Mean*. Enter a radius value in the pop-up menu (e.g., 2) and check *Preview* to see the result. Optimize by entering different values. Take note of the selected radius value and click *OK*.

6. Determination of "Auto Local Threshold radius": Go to *Image* > *Adjust* > *Auto Local Threshold*. In the pop-up window, choose the "Bernsen" method. Enter a value for radius (e.g., 180), keep values for *Parameter 1* and *Parameter 2* as zero, check *White objects on black background*, and click *OK*. Use *Edit* > *Undo* to repeat this step to optimize the radius of thresholding until a value is found that provides a clear

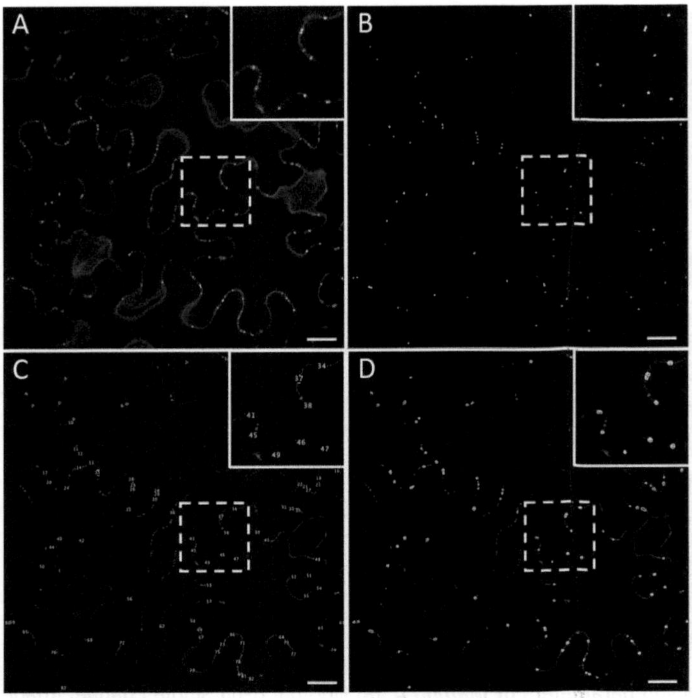

E

	Label	Area	Mean	Min	Max	IntDen	RawIntDen
1	PBS1.8bit.tif	0.635	98.692	37	191	62.691	1283
2	PBS1.8bit.tif	2.736	104.286	8	240	285.359	5840
3	PBS1.8bit.tif	0.733	47.600	23	116	34.888	714
4	PBS1.8bit.tif	1.319	48.444	0	151	63.913	1308
5	PBS1.8bit.tif	2.248	41.283	0	101	92.790	1899
6	PBS1.8bit.tif	1.466	48.667	0	112	71.340	1460
7	PBS1.8bit.tif	1.515	56.194	7	101	85.119	1742
8	PBS1.8bit.tif	1.906	55.205	2	140	105.202	2153
9	PBS1.8bit.tif	1.173	85.125	5	209	99.827	2043
10	PBS1.8bit.tif	1.417	87.000	23	160	123.281	2523
11	PBS1.8bit.tif	0.782	92.375	22	163	72.219	1478
12	PBS1.8bit.tif	1.661	91.206	0	240	151.524	3101
13	PBS1.8bit.tif	2.394	90.224	0	238	216.022	4421
14	PBS1.8bit.tif	1.906	74.282	25	154	141.556	2897
15	PBS1.8bit.tif	4.398	85.989	2	212	378.149	7739.000
16	PBS1.8bit.tif	2.932	75.267	0	240	220.664	4516.000
17	PBS1.8bit.tif	2.981	88.557	5	237	263.957	5402.000
18	PBS1.8bit.tif	1.515	87.000	15	216	131.783	2697.000
19	PBS1.8bit.tif	0.733	73.133	33	125	53.603	1097.000
20	PBS1.8bit.tif	1.564	95.312	31	203	149.032	3050.000
21	PBS1.8bit.tif	2.785	77.754	4	204	216.560	4432.000
22	PBS1.8bit.tif	1.368	79.750	3	213	109.111	2233.000
23	PBS1.8bit.tif	0.782	101.188	1	240	79.109	1619.000
24	PBS1.8bit.tif	2.590	86.226	0	239	223.303	4570.000
25	PBS1.8bit.tif	0.684	79.786	30	143	54.580	1117.000
26	PBS1.8bit.tif	0.489	91.400	56	121	44.661	914.000
27	PBS1.8bit.tif	2.590	68.302	0	150	176.883	3620.000
28	PBS1.8bit.tif	0.733	120.200	28	236	88.100	1803.000
29	PBS1.8bit.tif	1.319	96.000	13	232	126.652	2592.000
30	PBS1.8bit.tif	0.831	103.235	3	215	85.754	1755.000
31	PBS1.8bit.tif	0.586	64.917	0	176	38.064	779.000
32	PBS1.8bit.tif	3.274	85.209	0	191	278.958	5709.000
33	PBS1.8bit.tif	2.394	83.429	4	171	199.751	4088.000
34	PBS1.8bit.tif	2.101	108.558	0	241	228.092	4668.000
35	PBS1.8bit.tif	5.179	67.132	3	217	347.708	7116.000
36	PBS1.8bit.tif	0.684	96.429	17	211	65.965	1350.000
37	PBS1.8bit.tif	3.909	102.812	0	238	401.897	8225.000
38	PBS1.8bit.tif	1.368	118.964	0	233	162.762	3331.000
39	PBS1.8bit.tif	0.489	86.900	58	115	42.462	869.000
40	PBS1.8bit.tif	2.687	99.473	6	239	267.328	5471.000
41	PBS1.8bit.tif	1.759	103.167	10	194	181.476	3714.000
42	PBS1.8bit.tif	2.639	89.463	16	195	236.056	4831.000
43	PBS1.8bit.tif	0.831	99.294	33	176	82.480	1688.000
44	PBS1.8bit.tif	2.101	100.186	12	242	210.501	4308.000

Fig. 3 Key steps in image analysis applying parameters of *Method B*. This method is here applied to the same image as in Fig. 2. An enlargement of the image area delimited by the dashed line is shown in each panel. (**a**) Original 8-bit single channel image (1024 × 1024 pixel) of aniline blue-stained *N. benthamiana* epidermal cells showing fluorescent callose signal. (**b**) The same image after background subtraction using a *Rolling ball radius* of 6 pixels, filtering the image with a *Mean filter radius* of 2 pixels, and applying an *Auto Local Threshold radius* of 180. (**c**) Image after running *calloseQuant* with the following settings: *Rolling ball radius*, 6 pixels; *Mean filter radius*, 2 pixels; *Auto Local Threshold radius*, 180; *Analyze Particles size filter*, 25–240 pixels^2; *Analyze Particles circularity filter*, 0.5–1. Each callose spot is indexed with a number that corresponds to an individual measurement. (**d**) By clicking off "Labels" in the ROI manager window, which appears together with the image showing ROI index numbers (**c**), the underlying ROIs that were measured can be visualized. (**e**) Section of the *calloseQuant* data table showing the result of measuring each individual spot shown in (**b**) to (**d**). The running numbers in the table correspond to the individual ROI index numbers in (**c**). Scale bars = 20 μm

segmentation of the image by optimized selection of callose signals against the background (*see* **Note 20**). As compared to the original image (Fig. 3a), the thresholded, segmented image is reduced to binary information showing the ROI in white on a black background (Fig. 3b). The optimized radius value should in the best case produce an image in which the ROIs highlight only the fluorescent callose spots at the cell walls. Take note of the selected radius value and click *OK*.

7. Determination of "Analyze particle size filter": Enlarge the image until individual pixels can be recognized. Determine the size range of ROIs (in pixel area) that shall be measured (*see* **Note 21**) by multiplying the number of pixels in x and y direction for several ROIs that are in places of the original PD-associated callose signal. Take note of this size range (e.g., 25–240).

3.3.3 Running calloseQuant

1. Download the code for calloseQuant from https://raw.githubusercontent.com/mutterer/callose/main/calloseQuant_.ijm as a raw text file with name *calloseQuant_.ijm* and place this file into the ImageJ *plugins* folder on your computer.
2. Start or restart ImageJ, or alternatively apply the *ImageJ > Help > Refresh Menus* command.
3. Make sure that all images to be analyzed are in 8-bit.
4. Go to ImageJ > *File* > *Open* and open the file with the specific image or image stack to analyze. If a folder of images shall be analyzed, select the folder in **step 7**.
5. Go to *ImageJ* > *Plugins* > *calloseQuant* to start the user interface.
6. In the pop-up window (Fig. 1), choose the *Analysis to perform* ("Method A" or "Method B") and the type of *Source Data* ("Active single image," "Active Image Stack," or "Folder of images") (*see* **Note 22**).
7. If "Folder of Images" was selected in **step 6**, specify the source directory.
8. Fill in the appropriate parameters. If *Method A* was selected, enter the values for *Peak prominence* as determined in Subheading 3.3.1, **step 4**, and for *Measurement radius* as determined in Subheading 3.3.1, **step 6**. If *Method B* was selected, enter the values for *Rolling ball radius*, *Mean filter radius*, *Auto Local Threshold radius*, and *Analyze Particle size filter*, as determined in Subheading 3.3.2, **steps 4–7**. Also enter into the field *Analyze Particles circularity filter* a value for circularity of the particles to be measured (e.g., 0.5–1.0; *see* **Note 23**).
9. Press "OK" to perform the automated analysis (*see* **Note 24**). Following completion of the analysis, a results table will be shown on the screen (Figs. 2e and 3e) (*see* **Note 25**).
10. Choose *File* > *Save* to save the results table (comma separated values, .csv). Import the data into Excel or other software for data analysis.
11. A comparison of *Method A with Method B* for automated measurement of PD-associated callose levels is shown in Fig. 4. In this experiment, *N. benthamiana* leaf tissues were treated with the flagellin-derived defense elicitor peptide *flg22* or with buffer alone. Three representative images for each treatment were separately analyzed, and the results were then combined for data analysis. The results demonstrate that *flg22* treatment causes an increase in PD-associated callose levels in *N. benthamiana* as previously observed in *Arabidopsis* [18]. Moreover, both automated analysis methods show equal results, although the number of PD selected as ROIs for measurement was lower for *Method B* than for *Method A*.

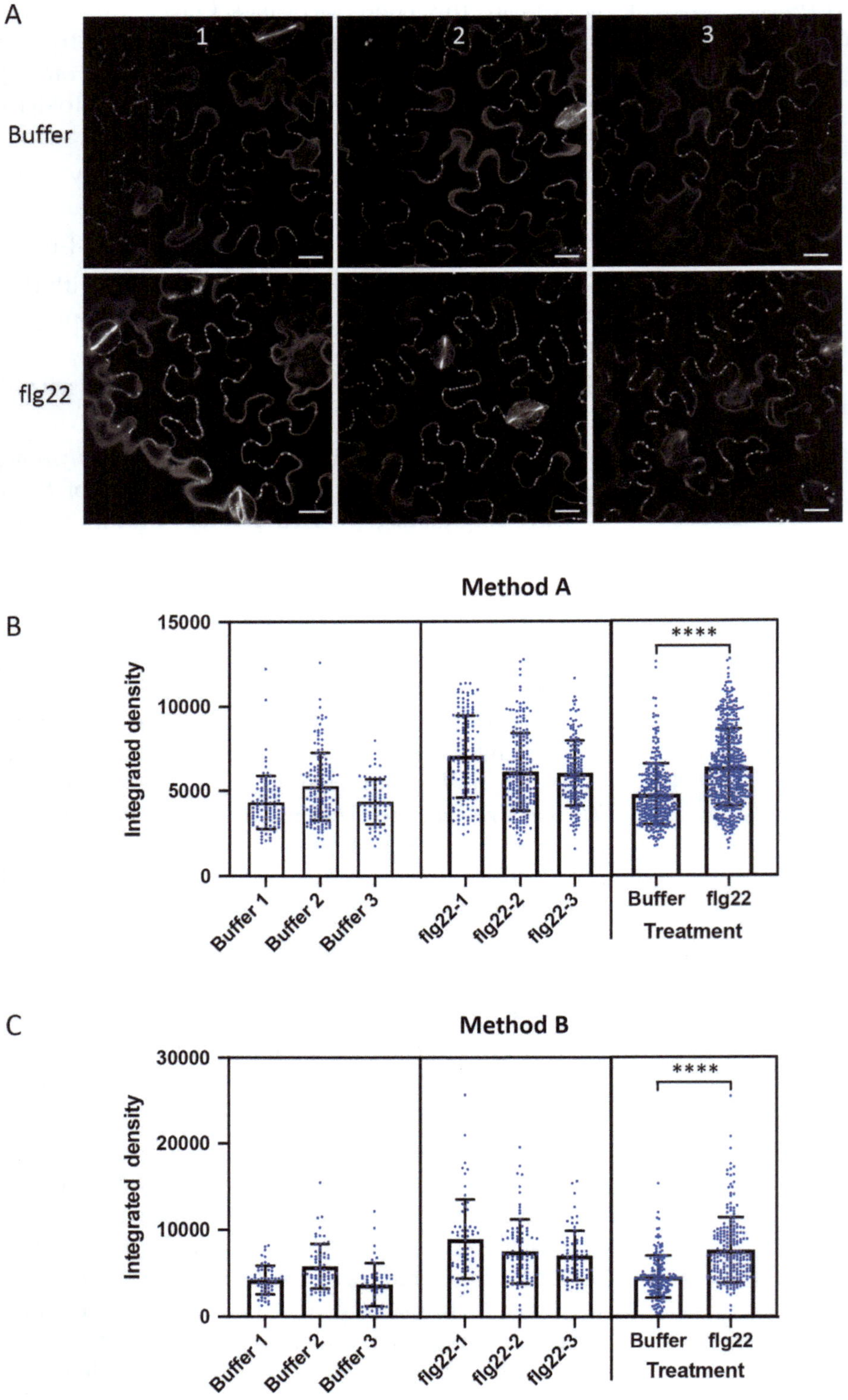

Fig. 4 Comparison of *Method A* and *Method B* applied to the analysis of images of aniline blue-stained leaf disks that were treated either with buffer or with buffer containing the flagellin elicitor peptide *flg22* (100 μM). (**a**) Three images from buffer-treated tissue and three images from *flg22*-treated tissue were used for analysis. Scale bars, 20 μm. (**b**) and (**c**) Analysis of PD-callose signal by *Method A* (**b**) and by *Method B* (**c**).

4 Notes

1. Aniline blue is also called "water blue" and commercially available from several manufacturers as a water-soluble sodium salt. Because the quality may vary between manufacturers, it is advisable to use the same batch of aniline blue in all your experiments.
2. Freshly prepared, blue-colored aniline blue solution must be decolorized for staining callose. Make sure that the solution has a pH slightly higher than pH 8.5. Solutions with lower pH will not decolorize and will not stain callose. Store the decolorized aniline blue solution in the dark and at room temperature. Avoid using solutions that are older than 1 year.
3. The range of fluorescence emission by aniline blue is 420–570 nm. Thus, also other emission filters can be used. The choice of filters may be important for samples in which also other fluorescent markers shall be observed. For example, in samples expressing green fluorescent protein (GFP, emission peak at 510 nm), you may want to choose a 420–480 nm emission filter.
4. The aniline solution may be mixed with other compounds, such as elicitors of plant defense (e.g., *flg22*) to test their effect on callose deposition. In this case, samples treated with the same added solution but without elicitor are required as control.
5. To avoid tissue stress, which may also lead to callose formation, the vacuum pressure should be kept at less than 0.8 pa. The leaf disk should assume a dark green color indicating penetration of the dye solution into the tissue.
6. The best observation time window is 30–120 min after staining. The staining result starts to fade after 60–120 min. Thus, make sure that all samples are imaged within the same time window.
7. Images can also be taken at higher depth but will be converted to 8-bit before image analysis.
8. If you compare samples of plants with different genetic backgrounds (wild type and mutant) or samples from treated and non-treated plants, always start observation and image

Fig. 4 (continued) Signal intensity associated with individual PD is shown by blue spots. Error bars represent the standard deviation. The data for individual and for the combined images per treatment are shown. The data show that *flg22* treatment causes a significant increase in callose levels at the PD (*********P*-value < 0.0001; two-tailed, unpaired, nonparametric Mann-Whitney test). Due to stringent selection, *Method B* selected less ROIs for measurement than *Method A*. However, both methods show the same trends

acquisition with the control samples (wild type, non-treated). Also, always choose cells in the center of the leaf disk, as cells at the rims of the leaf disks are often overstained.

9. Taking many images from the same sample is not necessary. Rather, have several samples and take few images from each sample. Observe rules for random image sampling.
10. Use the raw microscope images or image files in formats that keep the pixel data (256 gray levels in 8-bit images), such as .tif. Files in RGB format (three channels) must be split before analysis. In this case, go to *Image > Color > Split Channels*. Choose the channel with the relevant signal (gray scale) and save it for further analysis.
11. If you want to analyze callose signal in more images than one, set the parameters for the image with the strongest signal.
12. Although the color lookup table does not affect the measurements, we always use gray scale for better image contrast. This facilitates identification of appropriate parameter settings for measurement.
13. Dependent on the version of *ImageJ* you are using, this parameter may also appear as "*Noise Tolerance*."
14. It is helpful to switch off *Preview point selection* to see the original image before testing a new *Prominence* value. Also, verify the optimal value selection by switching *Preview point selection* on and off.
15. To facilitate the determination of the *Prominence* value, you may use a plug-in called *Prominence finder tool*. Download this plug-in from https://github.com/mutterer/callose/blob/main/Prominence_Finder_Tool.ijm and save it as a raw text file with the name *Prominence_Finder_Tool.ijm* into the ImageJ plug-ins folder. Activate this plug-in by applying *ImageJ > Help > Refresh Menus* followed by *ImageJ > Plugins > Prominence Finder tool*. Upon clicking on the open image window, the activated *Prominence Finder tool* will propose a value for *Prominence*. This value can be adjusted by dragging the mouse inside the image window. The fluorescent spots selected by the proposed *Prominence* value are shown in the open image window. Select the most appropriate *Prominence* value for use in the main dialog.
16. Alternatively, use ImageJ to measure the radius. Go to *Analyze > Set measurements*. Check "Area" in the pop-up menu. Next, draw a circle around a callose deposit signal using a *freehand* tool from the drawing control panel. Go to *Analyze > measure*, which creates a table in which the area (A) of the callose signal is shown (in pixel^2). The pixel radius can then be calculated by using the equation $A = \pi r^2$.

17. To compare outputs of *Methods* A and B, choose the same image file as for *Method A*.
18. The value for the *Rolling ball radius* should always be larger than the average fluorescence signal radius of the callose deposits. *See* Subheading 3.3.1, **step 6**, for a protocol to measure this radius.
19. This step is applied before thresholding. Thus, the rolling ball radius will be applied to all objects. Avoid effects of the chosen *Rolling ball radius* value on the PD-associated callose signal, which should be maintained.
20. The radius value should be higher than the radius of the PD-associated callose signal. *See* Subheading 3.3.1, **step 6**, for a protocol to measure this radius. It is important to note that a radius that is too small causes selection of background signals, whereas those that are too large will miss many relevant ROIs. Identification of the radius value that provides the best selection of relevant ROIs (selecting as much as possible only the callose signal) is critical as the sizes of all ROIs selected in this step will be measured.
21. Despite of background reduction, the threshold-based segmented image for measurement may contain ROIs of background fluorescence emanating from autofluorescent structures such as cell walls and stomata. Thus, selecting here a value equal to the size of spots of fluorescence signal emanating from PD-associated callose is critical to increase selectivity. In setting the size range, be careful to include the sizes of small ROIs that originate from PD with low levels of callose (compare the binary ROI mask image with the original image). Otherwise, the automated measurement using *Method B* will not be accurate.
22. Before running a batch analysis of stacks of images or of images collected in a folder, perform the analysis on single images first for testing the parameters.
23. The fluorescent signals emanating from PD-localized callose deposits are mostly circular as compared to other objects in the image. The circularity range should therefore be set to 0.5–1.
24. Make sure that the correct method is selected with the input parameters. *Method A* will perform the analysis faster than *Method B*.
25. The results of the measurement will be shown in a table on your screen. If a single image was analyzed (by selecting "active single image" in the tab for *Source data*), the results will be shown together with the original image in which each of the callose spots that was measured is labeled with the running number of the measurement shown in the table. Should the

measurements not accurately reflect the specific PD-associated callose staining signals, change the parameters to improve the measurement. If the measurements are accurate, you may apply the parameters for batch analysis of images in a stack or data folder.

Acknowledgments

We thank the Chinese Scholarship Council (CSC) for a PhD fellowship to C.H.

References

1. De Storme N, Geelen D (2014) Callose homeostasis at plasmodesmata: molecular regulators and developmental relevance. Front Plant Sci 5:138
2. Lim GH, Shine MB, de Lorenzo L et al (2016) Plasmodesmata Localizing proteins regulate transport and signaling during systemic acquired immunity in plants. Cell Host Microbe 19:541–549
3. Cui W, Lee JY (2016) Arabidopsis callose synthases CalS1/8 regulate plasmodesmal permeability during stress. Nat Plants 2:16034
4. Vaten A, Dettmer J, Wu S et al (2011) Callose biosynthesis regulates symplastic trafficking during root development. Dev Cell 21:1144–1155
5. Simpson C, Thomas C, Findlay K et al (2009) An Arabidopsis GPI-anchor plasmodesmal neck protein with callose binding activity and potential to regulate cell-to-cell trafficking. Plant Cell 21:581–594
6. Levy A, Erlanger M, Rosenthal M et al (2007) A plasmodesmata-associated beta-1,3-glucanase in Arabidopsis. Plant J 49:669–682
7. Zavaliev R, Ueki S, Epel BL et al (2011) Biology of callose (beta-1,3-glucan) turnover at plasmodesmata. Protoplasma 248:117–130
8. Benitez-Alfonso Y, Faulkner C, Pendle A et al (2013) Symplastic intercellular connectivity regulates lateral root patterning. Dev Cell 26: 136–147
9. Stahl Y, Faulkner C (2016) Receptor complex mediated regulation of symplastic traffic. Trends Plant Sci 21:450–459
10. Lee JY (2015) Plasmodesmata: a signaling hub at the cellular boundary. Curr Opin Plant Biol 27:133–140
11. Faulkner C, Petutschnig E, Benitez-Alfonso Y et al (2013) LYM2-dependent chitin perception limits molecular flux via plasmodesmata. Proc Natl Acad Sci U S A 110:9166–9170
12. Lee JY, Lu H (2011) Plasmodesmata: the battleground against intruders. Trends Plant Sci 16:201–210
13. Oparka KJ, Roberts AG, Boevink P et al (1999) Simple, but not branched, plasmodesmata allow the nonspecific trafficking of proteins in developing tobacco leaves. Cell 97:743–754
14. Brunkard JO, Zambryski P (2019) Plant cell-cell transport via plasmodesmata is regulated by light and the circadian clock. Plant Physiol 181: 1459–1467
15. Brunkard JO, Xu M, Scarpin MR et al (2020) TOR dynamically regulates plant cell-cell transport. Proc Natl Acad Sci U S A 117:5049–5058
16. Brunkard JO, Runkel AM, Zambryski PC (2013) Plasmodesmata dynamics are coordinated by intracellular signaling pathways. Curr Opin Plant Biol 16:614–620
17. Liang D (2018) A salutary role of reactive oxygen species in intercellular tunnel-mediated communication. Front Cell Dev Biol 6:2
18. Xu B, Cheval C, Laohavisit A et al (2017) A calmodulin-like protein regulates plasmodesmal closure during bacterial immune responses. New Phytol 215:77–84
19. Han X, Hyun TK, Zhang M et al (2014) Auxin-callose-mediated plasmodesmal gating is essential for tropic auxin gradient formation and signaling. Dev Cell 28:132–146
20. Wang X, Sager R, Cui W et al (2013) Salicylic acid regulates plasmodesmata closure during innate immune responses in Arabidopsis. Plant Cell 25:2315–2329
21. Rinne PL, van der Schoot C (1998) Symplasmic fields in the tunica of the shoot apical

meristem coordinate morphological events. Development 125:1477–1485

22. Rinne PL, Welling A, Vahala J et al (2011) Chilling of dormant buds hyperinduces FLOWERING LOCUS T and recruits GA-inducible 1,3-beta-glucanases to reopen signal conduits and release dormancy in Populus. Plant Cell 23:130–146
23. Ruan YL, Llewellyn DJ, Furbank RT (2001) The control of single-celled cotton fiber elongation by developmentally reversible gating of plasmodesmata and coordinated expression of sucrose and K^+ transporters and expansin. Plant Cell 13:47–60
24. Zavaliev R, Levy A, Gera A et al (2013) Subcellular dynamics and role of Arabidopsis beta-1,3-glucanases in cell-to-cell movement of tobamoviruses. Mol Plant-Microbe Interact 26:1016–1030
25. Gaudioso-Pedraza R, Beck M, Frances L et al (2018) Callose-regulated symplastic communication coordinates symbiotic root nodule development. Curr Biol 28:3562–3577.e6
26. Currier HB (1957) Callose substance in plant cells. Am J Bot 44:478–488
27. Thistlewhite P, Porter I, Evans N (1986) Photophysics of the aniline blue fluorophore: a fluorescent probe showing specificity toward (1->3)-beta-D-glycans. J Phys Chem 90:5058–5063
28. Smith MM, McCully ME (1978) A critical evaluation of the specificity of aniline blue induced fluorescence. Protoplasma 95:229–254
29. Zavaliev R, Epel BL (2015) Imaging callose at plasmodesmata using aniline blue: quantitative confocal microscopy. Methods Mol Biol 1217: 105–119

Chapter 10

Immunofluorescence Detection of Callose in Plant Tissue Sections

Sam Amsbury and Yoselin Benitez-Alfonso

Abstract

The accumulation of the cell wall component callose at plasmodesmata (PD) is crucial for the regulation of symplastic intercellular transport in plants. Here we describe protocols to fluorescently image callose in sectioned plant tissue using monoclonal antibodies. This protocol achieves high-resolution images by the fixation, embedding, and sectioning of plant material to expose internal cell walls. By using this protocol in combination with high-resolution confocal microscopy, we can detect PD callose in a variety of plant tissues and species.

Key words Callose, Immunofluorescence microscopy, Monoclonal antibody

1 Introduction

Cell-cell signalling is crucial to enable systemic coordination of growth and environmental responses in multicellular organisms. In plants, the presence of a cell wall spatially separates outer cell membranes posing a barrier to cell-cell signalling. One way to overcome this impediment is by the flux of signals through cell wall-embedded channels called plasmodesmata (PD) which connect neighboring cells cytoplasmically to form a continuous symplasm [1]. The flux of signals through PD is dynamically regulated by controlling the aperture of the channel, and one mechanism by which this is achieved is through the regulation of the cell wall composition at PD. It is well established that the dynamic turnover of the cell wall component callose, a β-(1,3)-glucan, plays an important role in the regulation of PD signalling [2]. PD cell wall microdomains are enriched in callose [3], and it has been shown that increasing callose restricts PD transport [4, 5]. On the other hand, when callose is reduced, for example, by the enhanced activity of

Yoselin Benitez-Alfonso and Manfred Heinlein (eds.), *Plasmodesmata: Methods and Protocols*,
Methods in Molecular Biology, vol. 2457, https://doi.org/10.1007/978-1-0716-2132-5_10,

degrading enzymes, intercellular transport increased [6]. Methods to detect callose are an important part of PD research.

Immunohistochemical labeling of plant cell walls using monoclonal antibodies against specific wall polysaccharides has been widely used in cell wall biology [7]. Monoclonal antibodies offer excellent specificity and reliable binding to epitopes of interest. The availability of antibodies which detect callose makes immunolabeling of callose possible in sectioned material [8]. There are a number of advantages to using fixed and sectioned material for the detection of callose in plants. These techniques are simple and require relatively little specialist equipment meaning they are widely accessible. Fixed material is stable, with long-term preservation of antigenicity, and sectioned material is very amenable to high-resolution microscopy with extremely high levels of detailed imaging being possible to achieve (Fig. 1). Antibody detection of callose is highly specific and extremely reproducible. We describe techniques for fixation, sectioning, and embedding of plant material and present optimized conditions for immunohistochemical detection of callose using monoclonal antibodies, as shown in Fig. 1. We discuss the merits of both wax and resin embedding which are suitable for a wide range of plant tissues.

2 Materials

2.1 Fixation of Plant Material

1. PEM buffer: 50 mM pipes, 5 mM EGTA, 5 mM $MgSO_4$, pH adjusted to 7.0 with KOH.
2. Formaldehyde solution: 16% preprepared commercially available ampules. Alternatively, formaldehyde fixative solution can be made fresh (*see* **Note 1**).
3. Fixative solution: 4% (v/v) formaldehyde in PEM buffer.

2.2 Embedding and Sectioning

1. Ethanol.
2. Steedman's wax: Mix sodium stearate and polyethylene glycol 400 distearate in a 1:9 (v/v) ratio at 65 °C until fully melted (*see* **Note 2**).
3. LR white resin (hard grade).
4. Disposable base molds (15 × 15 × 5 mm) with appropriately sized embedding cassettes for wax embedding.
5. Gelatine capsules for resin embedding.
6. Microscope slides: Poly-L-lysine-coated for wax embedded tissue, or Vectabond-coated glass slides for resin-embedded tissue (*see* **Note 3**).
7. Microtome: For resin sectioning an Ultramicrotome (such as a Reichert-Jung Ultracut) and for wax sections, a rotary

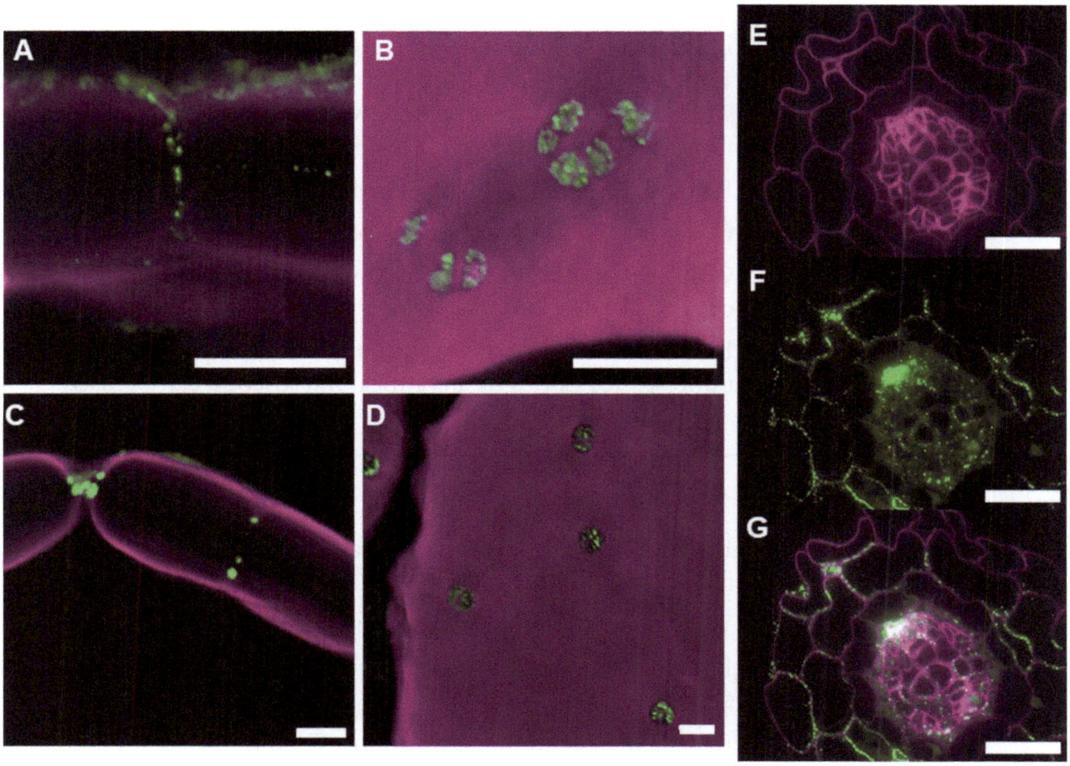

Fig. 1 Callose antibody labeling of sectioned material. Callose antibody signal shown in green and calcofluor counterstain shown in magenta. Images were captured using an Olympus BX51 microscope. Alexa488 was visualized using a filter set with 488 nm excitation and 525 nm emission, while calcofluor was imaged using excitation at 360 nm and emission at 430 nm. **a–d**: Sections of wax embedded tomato fruit pericarp showing pit fields. **a** and **c** show pit fields in transverse orientation. **b** and **d** show pit fields from the inner surface of the cell wall (longitudinal orientation). **e–g**: Sections of resin embedded *Arabidopsis* roots showing punctate callose binding indicative of PD. **e**, calcofluor signal; **f**, callose antibody signal; **g**, merge of both channels. Scale bars **a–d**, 10 μm; **e–g**, 100 μm

microtome with stainless steel blades (such as Leica-RM2125 RTS) should be used.

2.3 Enzymatic Pretreatment

1. 0.1 M sodium carbonate adjusted to pH 11.4 using 5 M NaOH.
2. Pectate lyase from *Cellvibrio japonicus* (Megazyme).
3. CAPS buffer: 50 mM N-cyclohexyl-3-aminopropanesulfonic acid, 2 mM $CaCl_2$, pH 10.

2.4 Immuno-Microscopies

1. Primary antibody. Callose monoclonal antibody from Biosupplies (www.biosupplies.com.au) (*see* **Note 4**).
2. Secondary antibody. Anti-mouse antibody conjugated to Alexa Fluor 488 or other fluorophores (*see* **Note 4**).

3. Phosphate-buffered saline (PBS): Prepare 10× stock containing 1.37 M NaCl, 27 mM KCl, 100 mM Na_2HPO_4, and 18 mM KH_2PO_4. Adjust to pH 7.4 with 4 M NaOH and sterilize by autoclaving. Dilute with water to 1× working solution before use.
4. Blocking buffer: 3% (w/v) dried milk in 1× PBS.
5. Citifluor AF1 antifadent solution.
6. Counterstain solutions. 0.25% (w/v) calcofluor white (fluorescent brightener 28) in dH_2O.
7. Fluorescence microscope. Either a microscope fitted with epifluorescence optics or a confocal microscope can be used.

3 Methods

3.1 Fixation of Plant Material

1. Excise samples from the plant and plunge them directly into fixative solution (*see* **Note 5**).
2. Vacuum infiltrate the samples in the fixative for 10 min.
3. Incubate infiltrated samples for at least 2 h and not more than 6 h at room temperature.
4. Repeat the vacuum cycle until the samples sink to the bottom of the tube (*see* **Note 6**).
5. Wash fixed samples in PEM buffer before proceeding to embedding in wax (Subheading 3.2) or in resin (Subheading 3.3).

3.2 Wax Embedding

Steedman's wax is a formulation of polyester wax and has good sectioning properties for firm and semi-firm plant materials (e.g., grass leaves, seeds, *Arabidopsis* inflorescence stems). The removal of wax by ethanol means that antigenicity is very well maintained. Very soft tissues, such as soft fruit or *Arabidopsis* leaves, are challenging to section using wax, and so resin embedding is recommended for these samples (*see* Subheading 3.3):

1. Dehydrate fixed samples using a series of ethanol solutions of ascending concentration (30, 50, 80, 100%, v/v). Incubate samples while rocking at 4 °C for 40 min with each ethanol solution.
2. Heat up wax to 37 °C.
3. Transfer the samples into wax by incubating in a molten wax/-ethanol mixture (1:1, v/v) at 37 °C overnight.
4. Transfer samples into 100% molten wax and incubate for 1 h at 37 °C.
5. Change the 100% wax twice, incubating 1 h per change at 37 °C.

6. Place wax molds on a 40 °C hot plate and fill them with molten wax.
7. Transfer samples from **step 3** into the molds filled with molten wax.
8. Transfer molds with samples to a cold plate (0 °C; *see* **Note 7**), and using fine forceps, working quickly before the wax solidifies, orientate the sample correctly for sectioning.
9. Once the wax is close to setting and is starting to become hazy, add the embedding cassette to the mold.
10. Leave the sample and embedding cassette in the mold on the cold block for 30 min and then at room temperature for at least 12 h. Wax blocks can be stored indefinitely if they are kept cool and dry.
11. Proceed to sectioning in Subheading 3.4.

3.3 Resin Embedding

LR white resin is a low viscosity acrylic resin which has excellent sectioning and light microscopy properties. LR white offers greater support than wax and so is better suited to soft tissues. In addition, LR white can be sectioned thinner than wax and so can offer better resolution at high magnification. Antigenicity is well preserved in LR white, but as the resin is not removed prior to immunolabeling, only the outer/sectioned surface is exposed to antibody binding:

1. Dehydrate fixed samples using an ethanol incubation series of ascending concentration (30, 50, 80, 90, 100% v/v) with 40-min incubation for each change. Incubations are done at 4 °C with samples kept rocking or rotating during this time.
2. Keep samples at 4 °C and rocking throughout the whole process from **steps 3** to **6**.
3. Infiltrate samples with LR white using an ascending resin series (10, 20, 30, 50, 70, 90% v/v resin in ethanol) incubating for 1 h per change.
4. Transfer the samples to 100% resin and incubate overnight.
5. Change the resin for fresh 100% resin and incubate for 8 h.
6. Change the resin again and leave incubating overnight.
7. Transfer the samples to gelatine capsules filled with resin and orientate using fine forceps.
8. Seal the gelatine capsules tightly to exclude oxygen (*see* **Note 8**) and incubate at 37 °C for at least 5 days until the resin is fully solidified.
9. Remove gelatine capsules and store at room temperature or proceed to sectioning in Subheading 3.5.

3.4 Sectioning of Wax Embedded Samples

These instructions are kept general as there are many microtome systems. This approach will work on most rotary microtomes with stainless steel blades such as the RM2125 RTS or the HM 325:

1. Cut wax sections, using a microtome, to a thickness of 12 μm as dry ribbons. Transfer these ribbons to paper using cool forceps or a fine brush and separate them gently.
2. Place sections onto a drop of water on poly-L-lysine slides and leave them to dry overnight at room temperature.
3. Dewax the sections by incubating the slides in 100% ethanol. Change the ethanol three times incubating for 10 min per change.
4. Rehydrate the sections by incubating the slides in 90% ethanol (v/v) for 10 min, 60% ethanol (v/v) for 10 min, and 30% ethanol (v/v) for 10 min followed by three changes of water (20 min per change).
5. Air-dry the slides and store at room temperature.
6. When ready, proceed to Subheading 3.6. If enzymatic pretreatment is not required, proceed to Subheading 3.7.

3.5 Sectioning of Resin Embedded Material

These instructions are for the Reichert-Jung Ultracut Ultramicrotome but would be equally applicable to other glass knife ultramicrotomes such as the EM UC7 (Leica):

1. Prepare glass knives as per the knifemaker's instructions (*see* **Note 9**).
2. Cut sections to a thickness of 2 μm onto water.
3. Use either a glass rod, single-hair brush, or section transfer loop to transfer sections to a drop of water on Vectabond-coated slides.
4. Dry the sections using a 20–30 °C hot plate.
5. Leave slides for at least 12 h in air to fully remove all remaining water after they appear visibly dry. Slides can be stored at room temperature indefinitely.
6. When ready, proceed to Subheading 3.6. If enzymatic pretreatment is not required, proceed to Subheading 3.7.

3.6 Enzymatic Pretreatment of Sections

Samples can be labeled without pretreatment, but in the event of signal being weak or variable, then enzymatic pretreatment may improve the imaging (*see* **Note 10**):

1. Incubate slides with sections, obtained from either Subheading 3.4 or Subheading 3.5, in 0.1 M sodium carbonate (pH 11.4) for 1 h at room temperature.
2. Wash four times with PBS (5 min per wash).

3. Incubate with pectate lyase (10 μg/ml) in CAPS buffer for 4 h at room temperature. Keep the samples in a humid chamber (*see* **Note 11**) during this incubation to prevent the wells from drying out.
4. Wash with three changes of PBS (5 min per change).
5. Proceed to immunolabeling as detailed in Subheading 3.7.

3.7 Immunolabeling of Wax and Resin Sections

1. If sections are not placed on multi-well slides, then isolate sections by drawing around them with a hydrophobic marker.
2. Incubate the sections in blocking solution for 30 min.
3. Remove blocking solution and wash samples twice in PBS (5 min per change).
4. Apply callose antibody diluted 1:80 in blocking solution (*see* **Note 12**). Fully cover the section with antibody solution; 40 μl per well on a multi-well slide is sufficient.
5. Incubate for 2 h at room temperature.
6. Wash the samples four times with PBS (5 min per change).
7. Apply anti-mouse Alexa Flour 488—conjugated secondary antibody diluted 1:100 in blocking solution. Fully cover the section with antibody solution; 40 μl per well on a multi-well slide is sufficient.
8. Incubate for 2 h at room temperature in the dark. From this step onward, samples should be protected from light.
9. Wash the samples four times with PBS (5 min per change).
10. Counterstain the tissue sections using either calcofluor white or pontamine red counterstain solution diluted in PBS 1:10. If using calcofluor white, incubate for 10 min at room temperature. If using pontamine red, incubate for 30 min at room temperature (*see* **Note 13**).
11. Wash the tissue sections four times with PBS (5 min per change) at room temperature.
12. Add a drop (approx. 10 μl) of Citifluor AF1 antifadent solution to each well/tissue section. Place a glass coverslip on top.
13. Visualize and collect images of the samples using an epifluorescence or confocal microscope.

4 Notes

1. To prepare fixative solution, a 8% (w/v) solution of paraformaldehyde in dH_2O is made. Heat to 60 °C and then add a small amount of 1 M NaOH to make the solution alkaline (approximately 10 drops/100 ml). Stir the solution and add further 1 M NaOH dropwise until dissolved, and then add an

equal volume of 2× PEM buffer to obtain a final concentration 4% formaldehyde in PEM. Adjust to pH 7 using 10% H_2SO_4. Do not use HCl to neutralize as this will form the carcinogen bis(chloromethyl) ether. Use pH strips to measure the pH as the fixative can degrade pH meter electrodes. Formaldehyde may cause cancer and so should always be used under the fume hood. Use only heatproof beakers on the hot plate, and do not use thick-walled glass bottles due to risk of cracking.

2. Steedman's wax is a common formulation which offers good sectioning properties and preserves tissue antigenicity well.
3. For resin sectioning, we usually use Vectabond-coated multi-well slides which have 8 or 16 wells with a hydrophobic coating between them (MP biomedicals); this makes it easy to apply multiple antibodies to a single slide. Alternatively, standard glass slides can be Vectabond-coated and then wells isolated by drawing around the sections using a PAP hydrophobic blocking pen. Vectabond coating is done according to manufacturer instructions.
4. The principles in this protocol will be applicable to any other callose-recognizing monoclonal antibodies. If applying this protocol to other antibodies, then a different secondary antibody may need to be used such as anti-rat depending on the hybridoma system used. When using a new antibody, always test a range of concentrations to determine the optimum concentration for labeling as this may differ from this protocol.
5. Excised tissues should ideally be smaller than 5 mm in thickness. Larger tissues may require longer incubation times during the fixing and embedding stages. Consider sample size when selecting capsules/molds; the sample should fit easily within the mold with space to be fully surrounded by resin or wax.
6. Some highly hydrophobic samples, such as rice leaves, may require a small drop of Tween 20 to be added to the fixative to break the surface tension and enable penetration. Use as little as possible and do not exceed 0.1% v/v Tween 20.
7. If a cold plate is not available, then a suitable alternative is a large petri dish (lid removed) pressed down into a tray of ice to create a cooled solid platform. Rapid cooling of the wax minimizes crystal formation and improves the sectioning.
8. The presence of oxygen within the capsules will inhibit the polymerization of the resin. For this reason, many plastic capsules are unsuitable for resin embedding as they allow too much oxygen into the sample; as such, we only recommend using gelatine capsules.

9. Glass knives should be clean and smooth. Poor-quality knives should be discarded as it will have strong adverse impacts on section quality.
10. In some samples, accessibility to callose can be impeded by other polysaccharides, and this can hinder antibody labeling. In these situations, digestion of cell wall pectin can aid antibody access to callose leading to stronger signal. Enzymatic digestion will often not be necessary and needs to be decided on a case-by-case basis. When using enzymatic pretreatment, it is important to include undigested controls.
11. To make a simple humid chamber, cover a square petri dish with aluminum foil, then place a piece of wet paper in the petri dish, and place your slides on the wet paper before closing the petri dish.
12. For most microscopy applications, 1:80–1:150 dilution is a good starting point, but some optimization might be required.
13. The choice of counterstain depends on user preference. Calcofluor white is a broad cell wall stain which binds widely to β-glycans including cellulose. Pontamine red is a specific stain which binds only to cellulose. Pontamine red can also be used to determine cellulose fibril orientation as it exhibits bifluorescence [9, 10]; this means that when viewed under polarized light, it fluoresces most strongly when aligned to the cellulose fibrils.

Acknowledgments

Financial support for this research was provided by the Leverhulme Trust Grant RPG-2016-136.

References

1. Amsbury S, Kirk P, Benitez-Alfonso Y (2018) Emerging models on the regulation of intercellular transport by plasmodesmata-associated callose. J Exp Bot 69:105–115. https://doi.org/10.1093/jxb/erx337
2. Cui W, Lee J-Y (2016) Arabidopsis callose synthases CalS1/8 regulate plasmodesmal permeability during stress. Nat Plants 2:16034. https://doi.org/10.1038/nplants.2016.34
3. Radford JE, Vesk M, Overall RL (1998) Callose deposition at plasmodesmata. Protoplasma 201:30–37. https://doi.org/10.1007/BF01280708
4. Li W, Zhao Y, Liu C, Yao G, Wu S, Hou C, Zhang M, Wang D (2012) Callose deposition at plasmodesmata is a critical factor in restricting the cell-to-cell movement of Soybean mosaic virus. Plant Cell Rep 31: 905–916. https://doi.org/10.1007/s00299-011-1211-y
5. Maule AJ, Gaudioso-Pedraza R, Benitez-Alfonso Y (2013) Callose deposition and symplastic connectivity are regulated prior to lateral root emergence. Commun Integr Biol 6: e26531. https://doi.org/10.4161/cib.26531
6. Benitez-Alfonso Y, Faulkner C, Pendle A, Miyashima S, Helariutta Y, Maule A (2013) Symplastic intercellular connectivity regulates lateral root patterning. Dev Cell 26:136–147. https://doi.org/10.1016/j.devcel.2013.06.010

7. Knox JP (2008) Revealing the structural and functional diversity of plant cell walls. Curr Opin Plant Biol 11:308–313. https://doi.org/10.1016/j.pbi.2008.03.001
8. Meikle PJ, Bonig I, Hoogenraad NJ, Clarke AE, Stone BA (1991) The location of (1→3)-β-glucans in the walls of pollen tubes of *Nicotiana alata* using a (1→3)-β-glucan-specific monoclonal antibody. Planta 185:1–8. https://doi.org/10.1007/BF00194507
9. Thomas J, Ingerfeld M, Nair H, Chauhan SS, Collings DA (2013) Pontamine fast scarlet 4B: a new fluorescent dye for visualising cell wall organisation in radiata pine tracheids. Wood Sci Technol 47:59–75. https://doi.org/10.1007/s00226-012-0483-x
10. Anderson CT, Carroll A, Akhmetova L, Somerville C (2010) Real-time imaging of cellulose reorientation during cell wall expansion in Arabidopsis roots. Plant Physiol 152:787–796. https://doi.org/10.1104/pp.109.150128

Chapter 11

Callose Detection and Quantification at Plasmodesmata in Bryophytes

Arthur Muller, Tomomichi Fujita, and Yoan Coudert

Abstract

In bryophytes (i.e., mosses, liverworts, and hornworts), extant representatives of early land plants, plasmodesmata have been described in a wide range of tissues. Although their contribution to bryophyte morphogenesis remains largely unexplored, several recent studies have suggested that the deposition of callose around plasmodesmata might regulate developmental and physiological responses in mosses. In this chapter, we provide a protocol to image and quantify callose levels in the filamentous body of the model moss *Physcomitrium* (*Physcomitrella*) *patens* and discuss possible alternatives and pitfalls. More generally, this protocol establishes a framework to explore the distribution of callose in other bryophytes.

Key words Aniline blue, Bryophytes, Callose, Moss, *Physcomitrella patens*, *Physcomitrium patens*, Plasmodesmata

1 Introduction

Bryophytes, which comprise mosses, liverworts, and hornworts, originated between 460 and 506 million years (My) ago [1] and constitute extant representatives of early land plants. Most recent phylogenomic studies support bryophytes as a monophyletic sister lineage to tracheophytes (vascular plants) [2–5]. Comparative ultrastructural studies have shown that extant bryophytes and tracheophytes possess similar plasmodesmata with desmotubules and spoke-like extensions that bridge the endoplasmic reticulum and the plasma membrane lining the intercellular cytoplasmic strand. These similarities suggest that the plasmodesmata of bryophytes and tracheophytes may be homologous, i.e., inherited from a shared ancestor [5–10]. In bryophytes, the structure of plasmodesmata has been characterized using transmission electron microscopy in various cell types and tissues, including apical meristematic cells, leptoids (food-conducting cells), phyllids (leaves), rhizoids and protonemata (filaments) in mosses, and placental cells at the

Yoselin Benitez-Alfonso and Manfred Heinlein (eds.), *Plasmodesmata: Methods and Protocols*,
Methods in Molecular Biology, vol. 2457, https://doi.org/10.1007/978-1-0716-2132-5_11,

gametophyte-sporophyte junction in hornworts and liverworts [11–15]. While plasmodesmata are a conserved structural feature of land plants, their contribution to the development and physiology of bryophytes remains poorly understood.

The reversible deposition of callose (β-1,3-glucan) at plasmodesmata provides a mechanism for dynamically controlling symplasmic trafficking in response to developmental or environmental cues in flowering plants [16–19]. Callose deposits can be visualized with fluorescence microscopy following aniline blue staining [20]. Using this method, callose has been observed around plasmodesmata connecting protonemal cells in *Funaria hygrometrica* and *Physcomitrium (Physcomitrella) patens* and in various tissues of the gametophore and the sporophyte in other moss species [21–24]. Exogenous treatment of *P. patens* with the phytohormone abscisic acid (ABA) triggers callose accumulation at cell-cell junctions within 72 h, when the intercellular movement of macromolecules such as the photoconvertible fluorescent protein Dendra2 is reduced [22, 23]. Inhibition of callose biosynthesis with the glucose analog, 2-deoxy-D-glucose, strongly reduces branching in *P. patens* gametophores, and this effect might be caused by an increase of auxin diffusion capacity through plasmodesmata in the stem [25]. Beyond these studies, the mechanisms and function of callose accumulation and turnover in bryophytes remain largely unexplored and deserve more attention.

Hereafter, we describe a method that uses aniline blue to stain callose deposits in protonemata of the model moss *P. patens* [26, 27]. We provide a procedure to image and quantify aniline blue fluorescence levels using confocal imaging and the ImageJ software. In Subheading 4, we highlight some of the pitfalls that we encountered when setting up this protocol. It must be noted that the main protocol described below has been established by Arthur Muller and Yoan Coudert to measure callose levels between adjacent protonemal cells. Alternative steps to this protocol (referred to as "Alternative steps" in the Subheadings 2 and 3) have been established independently by Tomomichi Fujita to detect changes in global callose levels and punctate signals in protonemal cell walls.

2 Materials

2.1 Plant Tissue Growth

1. Sterile 90 mm Petri dishes.
2. 250 mL autoclavable glass bottle.
3. Deionized water.
4. BCDAT growth medium stock solutions:

(a) Stock solution B (100×): 25 g/L $MgSO_2$, $7H_2O$.
(b) Stock solution C (100×): 25 g/L KH_2PO_4, adjust pH to 6.5 with 4 M KOH.
(c) Stock solution D (100×): 101 g/L KNO_3, 1.25 g/L $FeSO_4$, $7H_2O$.
(d) Stock solution AT (100×): 92 g/L $C_4H_{12}N_2O_6$.
(e) Stock solution TE (1000×): 614 mg/L H_3BO_3, 55 mg/L $CuSO_4$ $5H_2O$, 389 mg/L $MnCl_2$ $4H_2O$, 55 mg/L $CoCl_2$ $6H_2O$, 55 mg/L $ZnSO_4$ $7H_2O$, 28 mg/L KI, 25 mg/L $NaMoO_4$ $2H_2O$.
(f) Stock solution $CaCl_2$ (500×): 73.5 g/L $CaCl_2$.

5. Agar.
6. Micropore surgical tape (3 M).
7. Laminar flow hood for plant tissue culture.
8. Tweezers for plant tissue culture.
9. Glass bead sterilizer.
10. Growth cabinet.
11. *Physcomitrium (Physcomitrella) patens.*

2.2 Aniline Blue Solution Preparation

1. 0.1 mg/mL aniline blue fluorochrome solution (Biosupplies) in sterile deionized water. Solution must be stored at 4 °C in the dark.
2. 0.1 M phosphate buffer, pH 8.5: 97 mM K_2HPO_4, 3 mM KH_2PO_4.

2.3 Alternative Steps to Subheading 2.2 for Aniline Blue Preparation

1. Dissolve aniline blue (Wako) in 100 mM phosphate buffer (pH 8.7) to a concentration of 5% (w/v) and sterilize by filtration (pore size = 0.22 μm). This is the stock solution (blue color).
2. Dilute the stock solution in 100 mM phosphate buffer to a concentration of 1.6% (w/v). This is the working solution to be used within a year.
3. Incubate the working solution for 2 or more days at room temperature with exposure to air (*see* **Note 1**).

2.4 Biological Sample Preparation

1. Aluminum foil.
2. Microscope slides.
3. Microscope cover slips.
4. Thin forceps.
5. Sterile deionized water.
6. 200 μL automatic pipette.
7. 200 μL tips for automatic pipette.

8. Binocular microscope.
9. 24-well cell culture plates.

2.5 Biological Sample Observation and Data Analysis

1. Confocal microscope equipped with 10× dry and 40× water objectives, a 405 nm UV laser diode, a photomultiplier tube (PMT) spectral detector, and a UV light source were used to image aniline blue fluorescence.
2. The Leica Application Suite X software is used for image acquisition.
3. ImageJ (https://imagej.nih.gov/ij/download.html), an open source image processing software, is used for quantification of fluorescence intensity.
4. Microsoft Excel or a free equivalent (e.g., Apache OpenOffice), or R (https://www.r-project.org), is used to make plots.

3 Methods

3.1 Growth of Plant Material

1. Make up 100 mL of 1× BCDAT agar growth medium: Pour about 80 mL of sterile deionized water into a clean autoclavable bottle, add 1 mL of stock solution B, 1 mL of stock solution C, 1 mL of stock solution D, 1 mL of stock solution AT, 100 μL of stock solution TE, and 0.7 g of agar, top up with sterile deionized water to 100 mL, autoclave the bottle and cool it down in a laminar flow hood, and add 200 μL of sterile $CaCl_2$ stock solution while the medium is still warm.
2. Distribute warm autoclaved BCDAT agar growth medium into five sterile 90 mm Petri dishes (i.e., 20 mL per Petri dish).
3. Inoculate solid BCDAT agar medium with 1-week-old protonemata or sterile spores of the moss *P. patens* under sterile conditions.
4. Seal Petri dishes with Micropore tape.
5. Incubate Petri dishes for 2 weeks at 23 °C under a 16 h light/ 8 h dark cycle, at 50–150 $\mu mol/m^2/s$.

3.2 Alternative Steps to Subheading 3.1 for Growth of Plant Material

1. Prepare 100 mL of BCDAT medium containing 5 g/L glucose and 5 g/L gellan gum (BCDATG (gellan gum) medium) in a bottle. Autoclave and cool it down (*see* **Note 2**).
2. Melt the BCDATG (gellan gum) medium in a microwave (*see* **Note 3**).
3. Pour 500 μL of cooled BCDATG medium into each well of a four-well chamber slide and let it solidify.
4. Inoculate a small number of 3- to 5-day-old protonemata into each well.

5. Cover protonemata with 250 µL of cooled BCDATG.
6. Put on the lid and seal the chamber slide by wrapping parafilm around the chamber slide one and a half times (*see* **Note 4**).
7. Cultivate the protonemata for 10–15 days under continuous red light with the glass surface of the chamber slide up (*see* **Note 5**).

3.3 Sample Preparation

The protocol below has been optimized for moss filaments (*see* **Notes 6–11**).

1. Prepare a fresh 1:3 dilution of the aniline blue stock solution (i.e., 33 µg/mL final concentration) in 0.1 M phosphate buffer, pH 8.5.
2. Dispense 100 µL of diluted aniline blue solution into each well of a 24-well cell culture plate. Prepare one well per sample.
3. Harvest a few filaments from the edge of a well-grown moss colony. Make sure not to take any agar with the filaments.
4. Place the filaments in aniline blue solution.
5. Wrap 24-well cell culture plate in aluminum foil to keep the tissue in the dark.
6. Incubate for 30 min at room temperature.
7. Drop 30 µL of distilled water onto a microscope slide.
8. Transfer a few filaments into the water droplet.
9. Gently tease apart the filaments with thin forceps.
10. Place a cover slip onto the sample.
11. Check with a binocular microscope that filaments are individualized.

3.4 Alternative Steps to Subheading 3.3 for Sample Preparation

1. Add 150 µL of the aniline blue solution to the samples to a final concentration of 0.1% (w/v).
2. Keep the slide with the glass surface down for 12–24 h to let the aniline blue solution diffuse through the medium (*see* **Note 12**).
3. Incubate the sample for 12–24 h with the glass surface up (*see* **Note 13**).

3.5 Confocal Microscope Settings and Data Collection

After the filaments have been stained and mounted between slides and coverslips, they can be imaged directly with the confocal microscope. The following steps describe the image acquisition procedure with a Leica SP8 confocal microscope, but any equivalent imaging system can be used:

1. Set 405 nm UV laser diode power at 10%.
2. Set 440–520 nm (blue/cyan) band-pass emission filter for aniline blue.
3. Set scanning parameters as follows: frame scanning mode; frame size 1024 × 1024; pixel depth 12-bit; numerical zoom 1.5×; pinhole 1 A.U.; frame average 8; scan speed 8000 Hz.
4. Use the UV light (not the laser diode) and 10× dry lens to locate the aniline blue signal at the cell wall junction between the apical and subapical cell.
5. Select the 40× water immersion lens and focus on the cell wall.
6. Set Z-stack parameters. Select the beginning and end point of the volume to image; set number of steps to 30 and Z-step size to 0.7 μm.
7. Perform image acquisition.

3.6 Data Visualization

1. In the microscope software, perform a maximum intensity projection for each Z-stack and save resulting image as a TIFF file. Alternatively, maximum intensity projections can be performed in ImageJ, as described below.
2. Open .lif project (for Leica confocal microscopes) with ImageJ.
3. Tick *Open files individually*, click *OK*, and select the Z-series to analyze.
4. Perform maximum intensity projection: Go to *Image* > *Stack* > *Z-project*, and choose "Max Intensity" in "Projection type."
5. Select the aniline blue channel.
6. Normalize the image contrast (Fig. 1a): Go to *Image* > *Adjust* > *Contrast* > *Set*. Set the minimum displayed value to 0 and the maximum displayed value to 4095 for 12-bit pictures and to 63,353 if working with 16-bit pictures.
7. Apply false colors to the aniline blue fluorescence channel, and use the "Fire" look up table (LUT): Go to *Image* > *Lookup tables* > *Fire*. This will help visualize intensity gradients (Fig. 1b, c).

3.7 Alternative Steps to Subheading 3.5 for Confocal Microscope Settings and Data Collection

1. Detection of changes in global callose levels.

   Take images of five Z-stack images (1024 × 1024 pixels) with a 20× objective (zoom 4×) and a 405-nm diode laser using a confocal laser-scanning microscope. The pinhole size is set to 14.05 μm, which yields a 1.77-μm optical slice (*see* **Note 14**).
2. Detection of punctate signals in cell walls.

   To detect punctate callose signals, acquire images with a 100× objective (zoom 4×) and a 405-nm diode laser. The

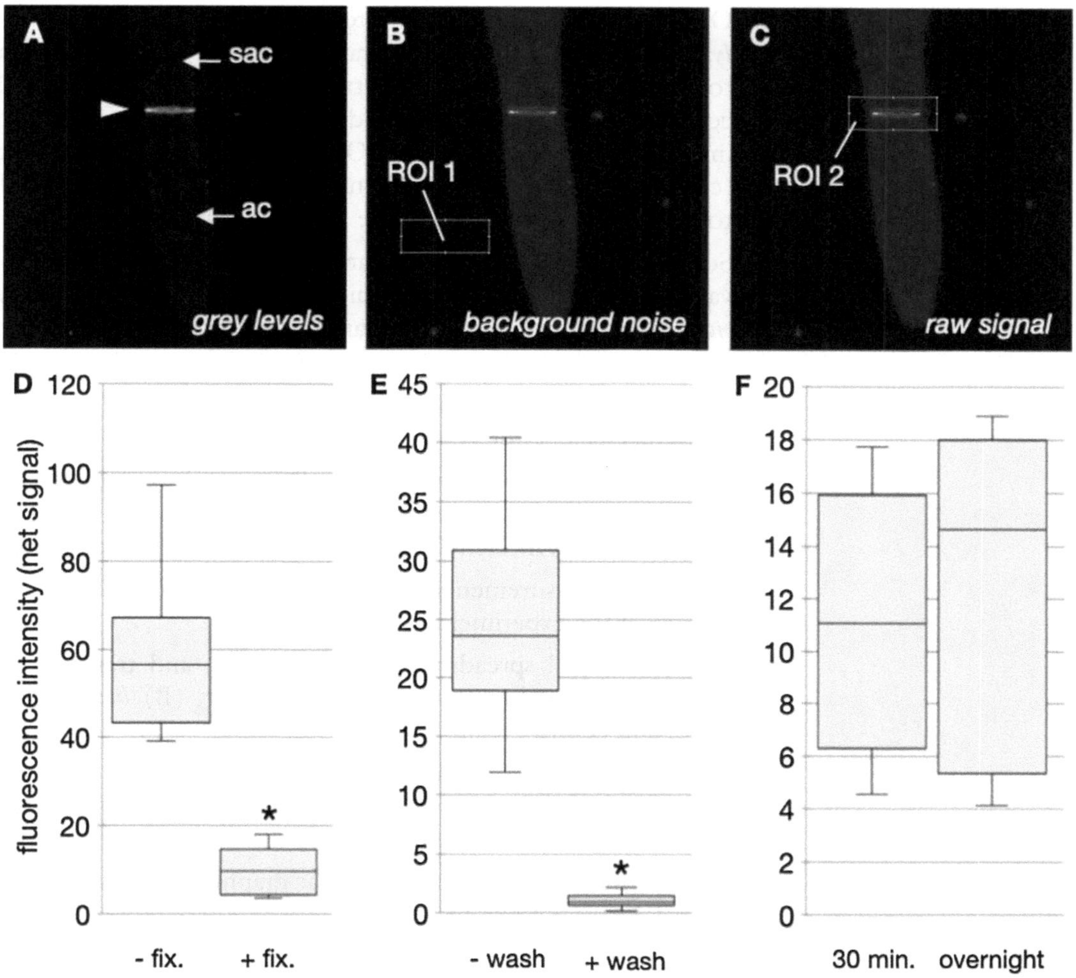

Fig. 1 Aniline blue staining of callose in *P. patens* protonemata. (**a**) Confocal image of a filament tip stained with aniline blue and shown in gray levels. Arrowhead indicates the junction between the subapical cell (sac) and the apical cell (ac). (**b**, **c**) Same images as **a** shown in "Fire" LUT of ImageJ. ROI 1 and ROI 2 indicate ROIs for quantification of background noise and raw signal, respectively. (**d**–**f**) Boxplots showing the effect of tissue fixation (**d**), tissue wash (**e**), and staining duration (**f**) on aniline blue fluorescence intensity (Wilcoxon signed-rank test different from control, *$p < 0.05$). Net signal values are indicated in millions. "fix." means FAA-based fixative

pinhole size is set to 40.87 μm, which yields a 0.31-μm optical slice (*see* **Notes 15–17**).

3.8 Signal Intensity Quantification

1. Open .lif project (for Leica confocal microscopes) with ImageJ.
2. Tick *Open files individually*, click *OK*, and select the Z-series to analyze.
3. Sum gray values for each pixel: go to *Image* > *Stack* > *Z-project*, chose "Sum slices" in "Projection type."
4. Select the aniline blue channel.

5. Open the ROI (region of interest) manager: Go to *Analyze > Tools > ROI manager*. Draw a rectangle around the area to analyze, such as the wall between the apical and subapical cell. The size of the ROI should be large enough to fit all the images analyzed. Name the ROI (click *Rename*) and save it (click *More > Save*) in the ROI manager menu. It is important to keep an ROI of constant size for all the images analyzed.
6. Set measurements to retrieve « area » and « integrated density » values from the ROI: in the main toolbar go to *Analyze > Set measurements*, and tick "*area*" and "*integrated density.*"
7. Place the ROI in the background and measure pixel intensity in the ROI: Go to *Analyze > Measure* (Fig. 1b). The measured "RawIntDens" value is considered as *background noise*.
8. Place the ROI around the cell wall between the apical and subapical cell and measure pixel intensity (Fig. 1c). The measured "RawIntDens" value is considered as the *raw signal*.
9. Save both measurements and repeat **steps** 7 and **8** for all the images of the experiment.
10. Open an Excel spreadsheet (or equivalent) and transfer the data to three columns: (A) *image number*, (B) *background noise*, and (C) *raw signal*.
11. In a fourth column (D), calculate the net fluorescence intensity value for each image as follows: *net signal (D)* = *raw signal (C)* – *background noise (B)*.
12. Represent data as boxplots or similar graphical representations (Fig. 1d–f) (*see* **Note 18**).

4 Notes

1. The color of the solution should change from blue to brownish yellow. Subsequent staining is carried out after this color change. This color change is observed with the pH 8.7 phosphate buffer, although it has previously been reported with K_3PO_4 (pH 12) [20]. Aniline blue dissolved in K_3PO_4 (pH 12) is unsuitable for time-lapse imaging due to the deleterious effect of pH on cells. pH 8.7 phosphate buffer alleviates damage permitting time-lapse imaging [23].
2. Application of glucose facilitates growth of *P. patens*. The high translucence of gellan gum improves imaging. BCD medium (BCDAT without ammonium tartrate) may alternatively be used to grow and observe gametophore tissues.
3. This medium can be used several times but should be discarded when it no longer solidifies.

4. Parafilm prevents drying of the medium but may delay protonemal growth as it limits gas exchange.
5. Red light suppresses initiation of side branching and makes observation of protonemata easier.
6. As aniline blue is expensive, we tried to dilute it further (e.g., 1: 30) but found that it did not yield sufficient staining.
7. Tissues treated in the same conditions but without aniline blue fluorochrome were used as a negative control.
8. Previous reports suggested that tissue fixation may help visualize callose deposition [20]. We tested the effect of tissue fixation on aniline blue staining intensity in moss and found that a 30-min treatment with an FAA-based fixative [5% acetic acid, 50% ethanol, 13.5% FAA (37% stock solution), 31.5% H_2O], prior to aniline blue staining, altered tissue color and significantly reduced fluorescence levels (Fig. 1d).
9. We tested the effect of washing tissues after aniline blue treatment and found that it significantly reduced fluorescence levels (Fig. 1e).
10. We tested different staining times and found that treating tissues with aniline blue solution more than 30 min (e.g., overnight) did not increase fluorescence levels (Fig. 1f).
11. It took about 5 min to prepare one slide and locate the sample to image under the microscope and about 5 min to perform a Z-stack. This must be considered when performing large-scale experiments.
12. Since the light is coming from above the sample, the protonemata tend to grow away from the glass surface during this period.
13. This step is important for the observation of the most apical cell walls because it will encourage the tips of the protonemata to grow on the glass surface, making them easier to observe.
14. The signal from callose stained with aniline blue was stable during time-lapse imaging over 180 min at 20-min intervals.
15. Manual adjustment of brightness and contrast are required to calculate density of punctate signals.
16. This method is unsuitable for time-lapse imaging to detect changes of global callose levels at lower magnification, because the higher laser power causes photobleaching.
17. Septa of *P. patens* in proximal regions of protonemata (i.e., closer to the colony center) are rich in callose, which makes it difficult to distinguish individual punctate signals.
18. Aniline blue fluorescence signal intensity was very variable between and within experiments, which could be attributable to biological stochasticity (age of sampled tissues, location of

the filaments within the colony). A minimum of 30 different cell walls per treatment ($n > 30$) was imaged to take variability into account in the measurements (Fig. 1d–f).

Acknowledgments

Y.C. acknowledges the CNRS (ATIP-Avenir fellowship) and the University of Lyon (Impulsion grant) for funding, the PLATIM team for technical assistance, Teva Vernoux for his continual support and Gwyneth Ingram for rereading this article. The work in TF's laboratory was supported by JSPS KAKENHI; Grant-in-Aid for Challenging Exploratory Research (25650089) and Grant-in-Aid for Scientific Research on Innovative Areas (18H04829).

References

1. Morris JL, Puttick MN, Clark JW et al (2018) The timescale of early land plant evolution. Proc Natl Acad Sci U S A 115:E2274–E2283. https://doi.org/10.1073/pnas.1719588115
2. Puttick MN, Morris JL, Williams TA et al (2018) The interrelationships of land plants and the nature of the ancestral embryophyte. Curr Biol 28:733–745.e2. https://doi.org/10.1016/j.cub.2018.01.063
3. Harris BJ, Harrison CJ, Hetherington AM, Williams TA (2020) Phylogenomic evidence for the monophyly of bryophytes and the reductive evolution of stomata. Curr Biol 30:2001–2012.e2. https://doi.org/10.1016/j.cub.2020.03.048
4. Leebens-Mack JH, Barker MS, Carpenter EJ et al (2019) One thousand plant transcriptomes and the phylogenomics of green plants. Nature 574:679–685. https://doi.org/10.1038/s41586-019-1693-2
5. Delaux P-M, Hetherington AJ, Coudert Y et al (2019) Reconstructing trait evolution in plant evo-devo studies. Curr Biol 29:R1110–R1118. https://doi.org/10.1016/j.cub.2019.09.044
6. Raven JA (1997) Miniview: multiple origins of plasmodesmata. Eur J Phycol 32:95–101
7. Nishiyama T, Sakayama H, de Vries J et al (2018) The Chara genome: secondary complexity and implications for plant terrestrialization. Cell 174:448–464.e24. https://doi.org/10.1016/j.cell.2018.06.033
8. Cook M, Graham L, Botha C, Lavin C (1997) Comparative ultrastructure of plasmodesmata of Chara and selected bryophytes: toward an elucidation of the evolutionary origin of plant plasmodesmata. Am J Bot 84:1169–1169
9. Cook ME, Graham LE (1999) Evolution of plasmodesmata. In: van Bel AJE, Van Kesteren WJP (eds) Plasmodesmata: structure, function, role in cell communication. Springer, Berlin, pp 101–117
10. Lucas W, van der Schoot C (1994) Plasmodesmata and the supracellular nature of plants. New Phytol 126:719–719. https://doi.org/10.1111/j.1469-8137.1994.tb02967.x
11. Brunkard JO, Zambryski PC (2017) Plasmodesmata enable multicellularity: new insights into their evolution, biogenesis, and functions in development and immunity. Curr Opin Plant Biol 35:76–83. https://doi.org/10.1016/j.pbi.2016.11.007
12. Pressel S, Ligrone R, Duckett JG (2006) Effects of de- and rehydration on food-conducting cells in the moss *Polytrichum formosum*: a cytological study. Ann Bot 98:67–76. https://doi.org/10.1093/aob/mcl092
13. Pressel S, Ligrone R, Duckett JG (2008) Cellular differentiation in moss protonemata: a morphological and experimental study. Ann Bot 102:227–245. https://doi.org/10.1093/aob/mcn080
14. Kitagawa M, Fujita T (2015) A model system for analyzing intercellular communication through plasmodesmata using moss protonemata and leaves. J Plant Res 128:63–72. https://doi.org/10.1007/s10265-014-0690-7
15. Gambardella R, Ligrone R (1987) The development of the placenta in the anthocerote

Phaeoceros laevis (L.) Prosk. Planta 172:439–447. https://doi.org/10.1007/BF00393859

16. Carafa A, Duckett JG, Ligrone R (2003) The placenta in *Monoclea forsteri* Hook. and *Treubia lacunosa* (Col.) Prosk: insights into placental evolution in liverworts. Ann Bot 92:299–307. https://doi.org/10.1093/aob/mcg140
17. Benitez-Alfonso Y, Faulkner C, Pendle A et al (2013) Symplastic intercellular connectivity regulates lateral root patterning. Dev Cell 26: 136–147. https://doi.org/10.1016/j.devcel.2013.06.010
18. Chen X-Y, Liu L, Lee E et al (2009) The Arabidopsis callose synthase gene GSL8 is required for cytokinesis and cell patterning. Plant Physiol 150:105–113
19. Vatén A, Dettmer J, Wu S et al (2011) Callose biosynthesis regulates symplastic trafficking during root development. Dev Cell 21:1144–1155. https://doi.org/10.1016/j.devcel.2011.10.006
20. Caillaud M-C, Wirthmueller L, Sklenar J et al (2014) The plasmodesmal protein PDLP1 localises to haustoria-associated membranes during downy mildew infection and regulates callose deposition. PLoS Pathog 10:e1004496. https://doi.org/10.1371/journal.ppat.1004496
21. Zavaliev R, Epel BL (2015) Imaging callose at plasmodesmata using aniline blue: quantitative confocal microscopy. Methods Mol Biol 1217: 105–119. https://doi.org/10.1007/978-1-4939-1523-1_7
22. Hébant C (1968) Observations en microscopie à fluorescence sur la répartition de la callose chez les mousses (Bryopsida). Rev Bryol Lichénol 36:111–113
23. Kitagawa M, Tomoi T, Fukushima T et al (2018) Abscisic acid acts as a regulator of molecular trafficking through plasmodesmata in the moss *Physcomitrella patens*. Plant Cell Physiol 60:738–751. https://doi.org/10.1093/pcp/pcy249
24. Tomoi T, Kawade K, Kitagawa M et al (2020) Quantitative imaging reveals distinct contributions of SnRK2 and ABI3 in plasmodesmatal permeability in *Physcomitrella patens*. Plant Cell Physiol 61:942–956. https://doi.org/10.1093/pcp/pcaa021
25. Schuette S, Wood A, Geisler M et al (2009) Novel localization of callose in the spores of *Physcomitrella patens* and phylogenomics of the callose synthase gene family. Ann Bot 103: 749–756
26. Coudert Y, Palubicki W, Ljung K et al (2015) Three ancient hormonal cues co-ordinate shoot branching in a moss. eLife 4:e06808. https://doi.org/10.7554/eLife.06808
27. Stefan A, Rensing B, Goffinet R, Meyberg S-Z, Wu M, Bezanilla (2020) The Plant Cell 32:1361–1376 https://doi.org/10.1105/tpc.19.00828

Chapter 12

Isolation of Plasmodesmata Membranes for Lipidomic and Proteomic Analysis

Laetitia Fouillen, Stéphane Claverol, Emmanuelle M. F. Bayer, and Magali S. Grison

Abstract

Plasmodesmata (PD) are membranous intercellular nanochannels crossing the plant cell wall to connect adjacent cells in plants. Our understanding of PD function heavily relies on the identification of their molecular components, these being proteins or lipids. In that regard, proteomic and lipidomic analyses of purified PD represent a crucial strategy in the field. Here we describe a simple two-step purification procedure that allows isolation of pure PD-derived membranes from *Arabidopsis* suspension cells suitable for "omic" approaches. The first step of this procedure consists on isolating pure cell walls containing intact PD, followed by a second step which involves an enzymatic degradation of the wall matrix to release PD membranes. The PD-enriched fraction can then serve to identify the lipid and protein composition of PD using lipidomic and proteomic approaches, which we also describe in this method article.

Key words Plasmodesmata, Membrane, Cell wall, Cellulase, Suspension cell, Proteomic, Lipidomic, Mass spectrometry, *Arabidopsis thaliana*

1 Introduction

In plants, cell-to-cell communication relies, to a large extent, on plasmodesmata (PD) channels. These are highly specialized structures spanning the plant cell wall creating direct cytosolic bridges between cells. PD are composed of segments of the plasma membrane (PM) and contain a strand of endoplasmic reticulum (ER) in their center. The two forms concentric cylinders tethered together by protein elements [1]. PD are considered as specialized ER-PM contact sites and contain a specific set of proteins and lipids [2–4]. This unique molecular signature likely defines the functional and structural feature organization of PD. The identification of PD molecular components is critical for understanding PD function, and, as such, tremendous efforts have been made to purify these structures along with proteomic and lipidomic approaches [5, 6].

Yoselin Benitez-Alfonso and Manfred Heinlein (eds.), *Plasmodesmata: Methods and Protocols*, Methods in Molecular Biology, vol. 2457, https://doi.org/10.1007/978-1-0716-2132-5_12,

Nevertheless, the purification of PD has been a long and tedious task. First, PD are firmly anchored to the cell wall and cannot be easily separated from the cellulose network. Second, because of their nanoscopic size, PD only represent a tiny fraction of the total plant tissue mass. Historically, PD purification has relied on subcellular fractionation and purification of cell wall fractions from various plant tissues. Due to their position embedded into the wall polymers, PD co-purify with the wall fragments. These approaches were successful in getting subcellular fractions enriched for PD and have led to the identification of several PD protein constituents [7–15]. Despite this achievement, the low proportion of PD compared to cell wall mass was a major shortcoming of this method. A major step forward was achieved with the use of cell wall-degrading enzymes which allows the release of the PD membranes from the cell wall matrix [3, 16]. Combined with proteomic analyses, this approach enables a giant step toward the identification of PD-associated proteins [3, 7, 8, 12, 17]. Later, through comparative lipidomic analyses, several groups were able to identify the lipid species specifically associated with the channels, highlighting the importance of membrane lipids in functionally defining PD [4, 18].

One crucial aspect of the PD purification protocol is to obtain highly pure cell wall fractions. Contrarily to plant tissues which are complex and often resistant to breakage due to the presence of lignified cell walls, *Arabidopsis thaliana* liquid cultured cells offer a friable population of large cells, from which clean wall fractions can be obtained [3, 4, 16]. The PD population within division walls is relatively homogenous and resembles to PD found in intact tissues. Cultured cells also represent a quick and easy way to generate a large amount of biological material from which large quantities of PD-derived membranes, suitable for biochemical analysis, can be produced.

Here, we will outline a detailed protocol for the isolation of PD from *A. thaliana* liquid cultured cells. Wall fragments containing intact PD are isolated by cell disruption and organelle release by N_2 decompression coupled to grinding in liquid nitrogen. The recovered wall fragments are then digested by cell wall-degrading enzymes enabling the release of PD membranes. PD are finally recovered by high-speed centrifugation. In this protocol, we also detail a procedure to identify the total fatty acid, sterol, and phospholipid content of isolated PD and a protocol for label-free comparative proteomic analysis of PD fractions.

2 Materials

Prepare all solutions with ultrapure water (sensitivity 18MΩ at 25 °C) and analytical grade reagents. All buffers and reagents should be kept at 4 °C. For longer preservation, buffers can be sterilized at 110 °C for 30 min before storage at 4 °C.

2.1 Plant Material

A. thaliana liquid cultured cells (ecotype *Landsberg erecta*) are grown in Murashige and Skoog medium containing vitamins and supplemented with 2.5 μM α-naphthaleneacetic acid (NAA), 0.25 μM kinetin, 87 mM sucrose, and 2.5 mM 2(N-morpholino)-ethane sulfonic acid (MES). Adjust the pH to 5.8 with 1 M KOH solution and sterilize at 110 °C for 30 min. Subculture the cells once a week (transfer 20 mL of the cultured cells into 200 mL fresh medium) under a laminar flow hood and keep on a shaker set to 120 rpm, at 24 °C, with constant light.

2.2 Purification of Cell Wall Fragments from Arabidopsis thaliana Suspension Cells

1. Cell wall preparation (CWP) buffer: 100 mM Tris-HCl, 100 mM KCl, 10 mM EDTA, 0.45 M D-mannitol and 10% (v/v) glycerol, pH 8 with KOH 1 M. Keep the solution at 4 °C. Before use, add one tablet of an EDTA-free protease inhibitor cocktail per 50 mL of buffer.
2. Cell wall washing (CWW) buffer: 10 mM Tris-HCl, 100 mM NaCl, 10 mM EDTA and 10% (v/v) glycerol, pH 8 with KOH 1 M. Keep the solution at 4 °C.
3. N_2 decompression-based cell disruption: cell disruption device, such as Parr cell disruption device. The total volume of the vessel should be about 900 mL; however, samples to be processed are adjusted to a volume of 50 mL. To reduce the volume capacity of the cell disruption vessel, a zinc cylinder with a hole that can contain a 50 mL falcon tube has to be engineered by a local workshop facility (Fig. 1a).
4. Refrigerated centrifuge allowing centrifugation of 50 mL tubes at low speed in a swinging-bucket rotor.
5. 50 mL screw cap tubes.
6. Sterile 25 mL pipettes and an electric pipette controller.
7. Ice and ice bucket.
8. Rough mortar (diameter 20 cm) and pestle.
9. N_2 liquid and N_2 liquid container.

2.3 Cell Wall Digestion with Cellulase

1. Digestion buffer: 10 mM MES-KOH, 240 mM D-mannitol, pH 5.5. Immediately before use, add 1 mM phenylmethylsulfonyl fluoride (PMSF) and 1 tablet of complete protease inhibitor cocktail.
2. Cellulase R-10 (Karlan) 1.4% in digestion buffer (w/v).
3. Tris-buffered saline buffer (TBS 1×): 20 mM Tris-HCl, 0.14 M NaCl, and 2.5 mM KCl, pH 7.4, containing 1 tablet per 25 mL protease inhibitor cocktail before use.
4. 50 mL screw top tubes.

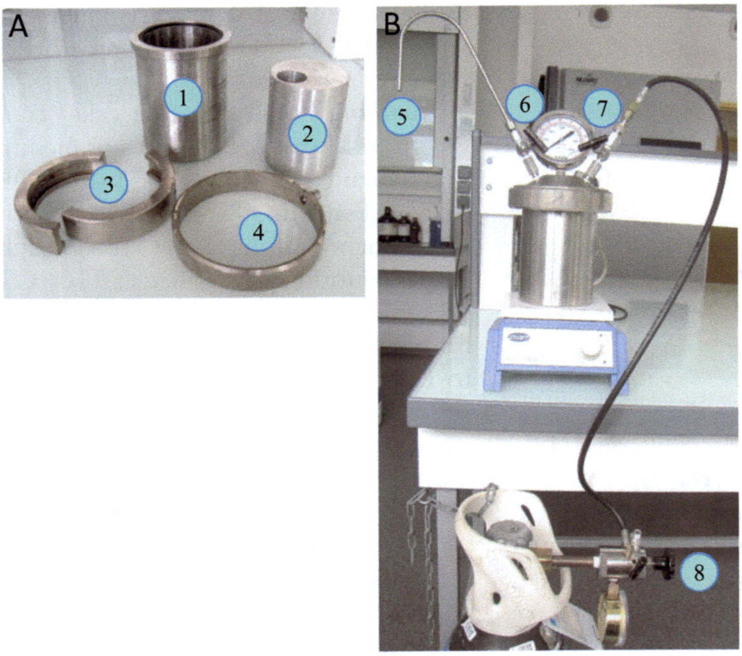

Fig. 1 Description of the N_2 decompression-based cell disruption device: (**a**) The parts of the disassembled cell disruption device. (**b**) Cell disruption device ready for use with loaded sample. Vessel (1), zinc cylinder which goes inside the vessel (2), split ring (3), cover ring (4), exit sample collection (5), discharging valve (6), charging valve (7), and valve connecting the cell disruption vessel with the gas cylinder (8)

5. 20 mL syringe without needle.
6. Syringe filter (0.2 μm) with support membrane.
7. Water bath.
8. Orbital shaking incubator (37 °C).
9. Refrigerated centrifuge for rotation of 30 mL glass centrifuge tubes at 6000 × *g*.
10. 30 mL glass centrifuge tubes, such as Corex tubes.
11. Ultraspeed refrigerated centrifuge for bucket rotation of 30 mL centrifuge tubes at 100,000 × *g* with fixe angle rotor.
12. 30 mL ultracentrifugation tubes.
13. Ultraspeed refrigerated centrifuge for bucket rotation of 4 mL centrifuge tubes at 100,000 × *g* with a fixe angle rotor.
14. 4 mL microfuge tubes.

2.4 Purity Assessment of the PD Fraction

2.4.1 Antibodies

Antibodies against PD proteins PDLP1 [9] and PDCB1 [15] and against potential contaminants, such as BiP (ER marker) [19], Membrin 11 (Golgi marker; we used an antibody provided by A. Hocquellet, L. Maneta-Peyret, and P. Moreau, Bordeaux, France), P16 (thylakoid marker) [20], PMA2 (PM marker) [21], Aquaporin PIP2;2 (PM marker) [22], the cellulose synthase subunits CesA3 and CesA6 (PM marker) [23], V-ATPase; e-subunit of tonoplast H+-ATPase (Agrisera), and ECHIDNA (*trans*-Golgi network marker) [24].

2.4.2 Protein Quantification

1. Bicinchoninic acid (BCA) protein assay kit.
2. 1.5 mL microfuge tubes.
3. Spectrophotometer capable of 562 nm.

2.4.3 SDS PAGE Protein Electrophoresis

1. Laemmli buffer 2×: 4% (w/v) sodium dodecyl sulfate (SDS), 16% (w/w) glycerol, 20 mM Tris-HCl, pH 6.8 and 0.02% (w/w) bromophenol blue.
2. Protein electrophoresis device.
3. Power supply capable of an output of 250 V, 3.0 A, and 300 W.
4. Fast cast gel kit, acrylamide kit 7.5% or 12%.
5. Running buffer: Tris-glycine-SDS buffer pH 8.5.
6. SDS-PAGE molecular weight standards.

2.4.4 Western Blot Analysis

1. Trans-Blot cell system.
2. Biorad Trans-Blot Turbo RTA transfer kit, nitrocellulose.
3. Washing buffer: 1× TBS 0.1% (v/v) Tween 20.
4. Blocking buffer: 5% (w/v) nonfat dry milk powder in 1× TBS.
5. Secondary antibodies, such as horseradish peroxidase-coupled goat anti-rabbit.
6. Enhanced chemiluminescence (ECL) system for the visualization of immunoreactive proteins.
7. Imaging system.

2.5 Lipidomic Analysis

2.5.1 Fatty Acid Profiling by GC-MS

1. Hydrolysis solution: 5% H_2SO_4 in methanol.
2. Hexane 99%.
3. 2.5% NaCl in H_2O.
4. 100 mM Tris-HCl, 0.09% NaCl, pH 8.
5. BSTFA-TMCS: N,O-bis(trimethylsilyl)trifluoroacetamide with 1% trimethylchlorosilane.
6. Heptadecanoic acid (C17:0).
7. 2-Hydroxytetradecanoic acid (h14:0).

2.5.2 Sterol Profiling by GC-MS

1. Chloroform/methanol 2/1 (v/v).
2. 0.09% NaCl solution.
3. Ethanol.
4. 11 N KOH.
5. Hexane 99%.
6. BSTFA-TMCS: N,O-bis(trimethylsilyl)trifluoroacetamide with 1% trimethylchlorosilane.
7. Cholestan-2-ol (also known as α-cholestanol).

2.5.3 Phospholipid Relative Quantification by LC-MS

1. Commercially available internal standards: phosphatidyl-choline PC 17:0/14:1, phosphatidyl-inositol PI 17:0/14:1, phosphatidyl-serine PS 17:0/17:0, phosphatidyl-ethanolamine PE 17:0/17:0 phosphatidyl-glycerol PG 17:0/17:0.
2. Eluent A: isopropanol/CH_3OH/H_2O (5/1/4 v/v/v) + 0.2% formic acid +0.028% NH_3.
3. Eluent B: isopropanol +0.2% formic acid +0.028% NH_3.

2.5.4 Instruments and Software for Lipidomic Analyses

1. Evaporator under N2 flow.
2. GC-MS (e.g., Agilent 7890A and MSD 5975 Agilent EI). An HP-5MS capillary column (5% phenyl-methyl-siloxane, 30 m, 250 mm, and 0.25 mm film thickness) is used.
3. MassHunter, version B.07.
4. Ultrahigh-performance LC 1290 Infinity II LC System with a Luna C8 150 × 1 mm column, with 100 Å pore size, 5 μm particles.
5. Qtrap 6500 Mass spectrometer.
6. Analyst software, version 1.7.1 and Analyst Device Driver for Agilent 1.3.
7. Multiquant software version 3.0.3.

2.6 Label-Free Proteomic Analysis

2.6.1 Gel Electrophoresis and Staining

1. Laemmli buffer 4×.
2. Coomassie Brilliant Blue G-250.
3. Destaining solution NH_4HCO_3 25 mM/ACN 50%.
4. Proteolysis buffer NH_4HCO_3 50 mM.

2.6.2 In-Gel Trypsin Digestion

1. Dithiothreitol 10 mM in NH_4HCO_3 50 mM.
2. Iodoacetamide 100 mM in NH_4HCO_3 50 mM.

2.6.3 Peptide Extraction

1. Extraction solution water/ACN/ formic acid 48:48:4 (v:v:v).
2. 100% ACN.
3. 0.1% trifluoroacetic acid (in MS-quality water).

2.6.4 NanoHPLC

1. 0.05% trifluoroacetic acid (in MS-quality water).
2. Solvent A: 0.1% formic acid (in MS-quality water).
3. Solvent B: 80% ACN, 0.1% formic acid (in MS-quality water).

2.6.5 Instruments and Software for Proteomic Analysis

1. GS-900 Densitometer.
2. Dionex Ultimate 3000 nanoHPLC system.
3. Orbitrap Fusion™ Lumos™ Tribrid™ Mass Spectrometer equipped with an EasySpray ion Source.
4. Image Lab™.
5. Proteome Discoverer.

3 Methods

PD purification takes place in two main steps (Fig. 2): (1) the purification of cell wall fragments containing PD followed by (2) the enzymatic digestion of the cell wall fraction allowing the release of "free" PD. Each step (about 4–5 h of work for each) can be performed independently during different days as during the first step the pelleted cell wall fraction can be stored at −20 °C after Subheading 3.1.1 **step 9** or **step 6**. All steps should be carried out at 4 °C unless specified otherwise.

3.1 PD Purification

The procedure starts with the disruption of *A. thaliana* suspension cells using a nitrogen decompression-based cell disruption device and recovery of the cell walls by low-speed centrifugation. In order to remove any trapped subcellular debris and organelles, the sample is then ground in liquid nitrogen using a rough mortar and pestle.

3.1.1 Cell Disruption Using a N_2 Decompression-Based Device

1. Divide 200 mL of 5- to 6-day-old cultured cells (*see* **Notes 1** and **2**) into 4× 50 mL screw-top tubes, and centrifuge the suspension cells at 400 × *g* for 5 min, at RT.
2. Meanwhile, place the cell disruption device (precooled at 4 °C) on a magnetic stirrer, and open the valve connecting the disruption vessel and the gas nitrogen cylinder so that the device is ready to use (Fig. 1b).
3. Discard the supernatant by inverting the tubes and resuspend the cell pellet (about 20 mL) with an equal volume of ice-cold CWP buffer containing protease inhibitors (total volume should be around 50 mL). Gently place a magnetic stirring bar into the screw-top tube.
4. Position the sample inside the cell disruption device; set the head on the vessel with the dip tube extending into the sample and attach the split ring cover clamp (Fig. 1a). Make sure the magnetic bar is stirring the sample before closing (*see* **Note 3**).

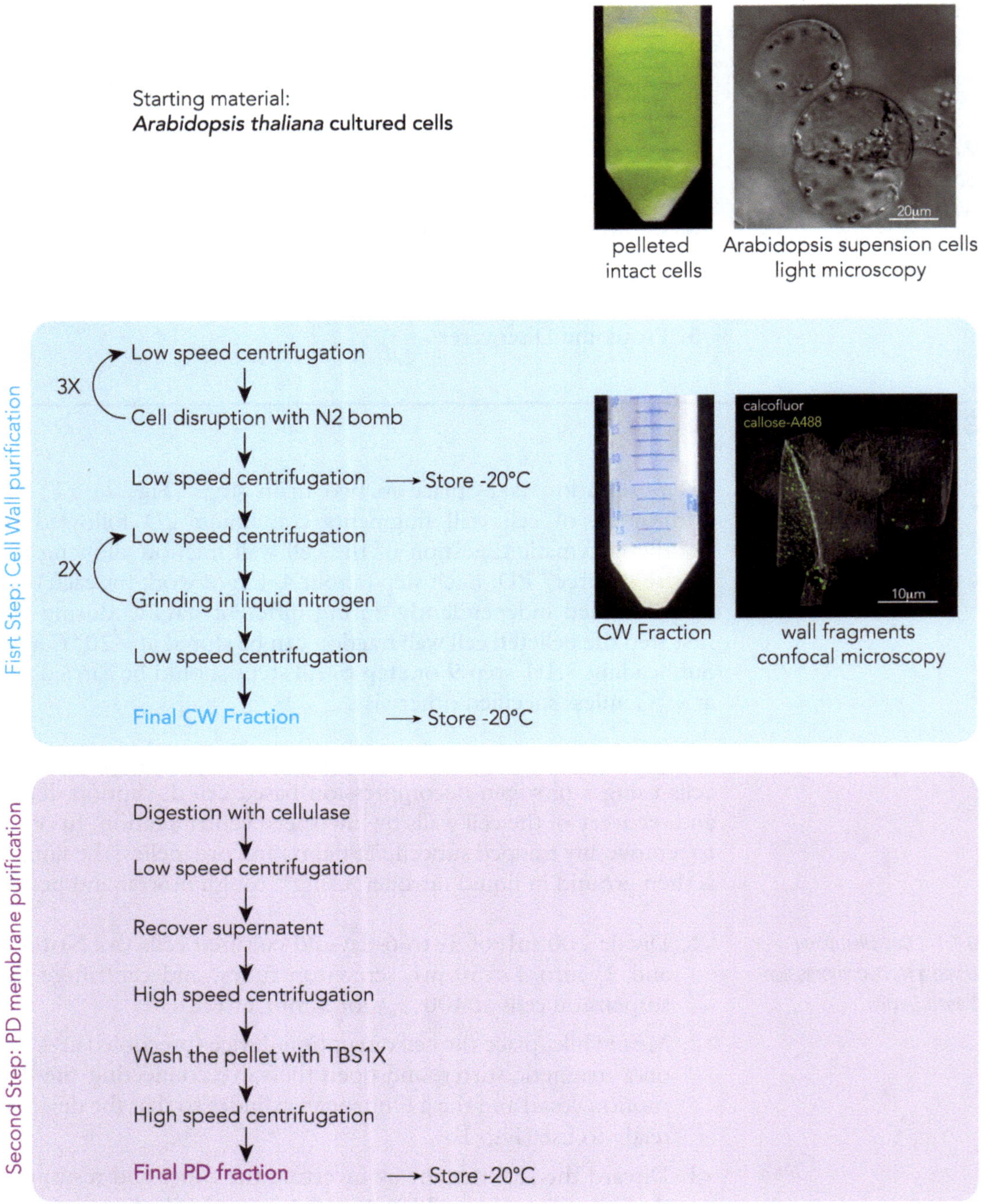

Fig. 2 PD purification procedure. This figure describes the workflow for PD membrane purification. The top of the figure describes the starting material: *Arabidopsis thaliana* cells, pelleted after low-speed centrifugation and live cells observed by light microscopy. The blue box describes the cell wall purification steps. Figure insets on the right show (1) the final purified cell wall fraction, pelleted after low-speed centrifugation, and (2) cell wall fragment in which callose is detected by immunolocalization (green) and cellulose stained using calcofluor white observed by confocal microscopy. The purple box describes the steps for PD membrane purification by digestion. "Store −20 °C" indicates steps where the purification process can be interrupted

Close the discharge valve (Fig. 1b) and gradually pressurize the vessel to 120 bar (*see* **Note 4**) by opening the charging valve (Fig. 1b). When the appropriate pressure is reached, close the charging valve.

5. To disrupt the cells, slowly open the releasing valve (Fig. 1b) to obtain a drip (*see* **Note 5**) and collect the disrupted cells in a 50 mL screw-top tube. Proceed until only gas exiting the cell disruption device (*see* **Note 6**). Leave the discharging valve opened until the pressure in the vessel is zero.
6. Collect an aliquot (total cell extract) for Western blot and proteomic/lipidomic analysis (*see* Subheading 3.3). Store at −20 °C.
7. Divide the solution containing disrupted cells (about 40–50 mL) into two new 50 mL screw-top tubes and adjust the volume to 40 mL with ice-cold CWW buffer. Centrifuge the solution at 400 × *g* for 5 min at 4 °C. Cell walls will sediment, while the cytosolic content stays in the supernatant.
8. Carefully remove the supernatant using 25 mL pipettes and resuspend the sedimented walls in ice-cold CWW buffer. Adjust the volume to 50 mL.
9. Repeat **steps 4–8** two more times (*omit* **step 6**).
10. The cell wall sediment can be stored at −20 °C.

3.1.2 Cell Wall Grinding in Liquid Nitrogen

1. Harvest the frozen pellet of cell walls from Subheading 3.1.1 using a spatula. Place it in a mortar and grind the sample with a pestle in liquid N_2 to a fine powder.
2. Use the spatula to load the frozen powder into a 50 mL screw-top tube and fill with ice-cold CWW buffer. Mix by vortexing.
3. Spin the suspension at 400 × *g* for 5 min at 4 °C.
4. Carefully remove the supernatant using a 25 mL pipette. Freeze the wall pellet by immersing the 50 mL screw-top tube into liquid N_2.
5. Repeat the grinding **steps 1–4**, two more times.
6. Wash the wall fragments three times with excess ice-cold CWW buffer by centrifugation at 400 × *g*, for 5 min at 4 °C.
7. The final cell wall fraction can be stored at −20 °C (*see* **Note 7**).

3.1.3 Cell Wall Digestion Using Cellulase

1. Thaw 10 mL of the cell wall fraction from Subheading 3.1.2 and resuspend with excess volume (~ 40 mL) of digestion buffer (without protease inhibitors). Centrifuge for 5 min at 400 × *g* at 4 °C in 50 mL screw-top tubes. Discard the supernatant and keep the 10 mL pellet on ice.
2. Activate the cellulase solution 5 min at 55 °C (*see* **Note 8**) and filter it using a 0.2 μm filter (*see* **Note 9**). Immediately before

use, add 1 tablet per 25 mL of protease inhibitor cocktail (*see* **Note 10**).

3. Mix one volume of the 1.4% (w/v) cellulase solution with 1 volume of the cell wall fraction from **step 1** (10 mL/10 mL). The final concentration of cellulase should be 0.7% (w/v).
4. Incubate the mixture at 37 °C for 1.5 h with gentle shaking in a rotating incubator (50–100 rpm).
5. Remove undigested wall fragments by centrifugation at 5850 × *g* for 5 min at 4 °C in 30 mL glass centrifuge tubes.
6. Collect the supernatant for ultracentrifugation at 110,000 × *g* for 40 min at 4 °C with a swing out rotor in 30 mL ultracentrifuge tubes.
7. Carefully discard the supernatant by inverting the tubes and resuspend the pellet, containing PD membranes, with an excess volume of cold 1× TBS.
8. Sediment the resuspended pellet again at 110,000 × *g* for 40 min at 4 °C.
9. Resuspend the final pellet (the PD fraction) with 100 μL of 1× TBS containing protease inhibitors.
10. Store the PD fraction at −20 °C.

3.1.4 Quality Assessment of the PD Fraction

The purity level of the PD fraction should be assessed by Western blot analysis and compared to total cell extract (Fig. 2; *see* **Note 11**):

1. Measure the protein concentration of the final PD fraction using a BCA Assay Kit (*see* **Note 12**).
2. Resuspend 1 volume of PD and total cell extract (*see* **step 6** in Subheading 3.1.1) with 1 volume of 1× Laemmli buffer and heat the samples for 30 min at 50 °C.
3. Load the same amount of proteins for each sample and separate proteins by 1D-SDS polyacrylamide gel electrophoresis. Subsequently, transfer to nitrocellulose membranes using standard protocols.
4. Perform immunoblotting using the appropriate antisera (*see* **Note 11**) and visualize specific binding using standard techniques.

3.2 Lipidomic Analysis

3.2.1 Fatty Acid Profiling Preparation

1. Transfer samples containing around 150 μg of protein equivalent in a screw glass tube, named tube A.
2. Add 1 mL of the hydrolysis solution containing 5 μg/mL of internal standard C17:0 and h14:0 (*see* **Note 13**).

3. Incubate overnight at 85 °C in a dry bath; be sure that the contents of tubes are not evaporating by retightening the screw caps every 5–10 min during the first 30 min. This step aims at releasing the FAs of different lipid classes from their respective backbone.
4. Remove the tubes from the dry bath and let them cool down.
5. Add 1 mL of 2.5% NaCl and then 1 mL of hexane.
6. Mix vigorously.
7. Centrifuge 5 min at 800 × *g* at room temperature to separate the phases.
8. Prepare new screw glass tubes named tube B containing 1 mL of 100 mM Tris-HCl, 0.09% NaCl, pH 8.
9. Collect the hexane upper phase and transfer it to the tube B prepared in previous step.
10. Add 1 mL of hexane to the initial tube A.
11. Mix vigorously.
12. Centrifuge 5 min at 800 × *g* at room temperature to separate the phases.
13. Collect the hexane upper phase and add it to tube B.
14. Mix vigorously the content of tube B.
15. Centrifuge 5 min at 800 × *g* at room temperature to separate the phases.
16. Recover gently the hexane upper phase taking care not to contaminate the sample with the lower aqueous phase and transfer to a new glass tube.
17. Evaporate the hexane through airflow evaporator.
18. Add 200 mL of BSTFA-TMCS.
19. Incubate 15 min at 110 °C in a dry bath.
20. Evaporate the BSTFA-TMCS through air flux evaporator.
21. Once BSTFA-TMCS is completely evaporated, add 100 μL of hexane and close the lid immediately to avoid hexane evaporation.
22. Mix by vortexing.
23. Transfer into GC vials containing a conical insert.
24. Run the sample in a GC-MS instrument using the FAMES methods.

3.2.2 Sterol Profiling Preparation

1. Transfer samples containing around 150 μg of protein equivalent in centrifuge glass tube with 1 mL of water.
2. Add 2 mL of chloroform/methanol 2/1 v/v.
3. Sonicate for 30 s at room temperature and mix vigorously.

4. Centrifuge 5 min at 800 × *g* at room temperature to separate organic and aqueous phases.
5. Collect organic lower phases and transfer to clean glass tubes.
6. Add 2 mL chloroform in the aqueous upper phase.
7. Sonicate for 30 s and mix vigorously.
8. Centrifuge 5 min at 800 × *g* at room temperature to separate organic and aqueous phases.
9. Collect the organic lower phase and combine with organic phase collected in **step 5**.
10. Add 2 mL of 0.09% NaCl solution in organic phases.
11. Sonicate for 30 s and mix vigorously.
12. Centrifuge 5 min at 800 × *g* at room temperature to separate organic and aqueous phases.
13. Collect the organic lower phase.
14. Use the evaporator to dry the combined organic phases.
15. Add 1 mL of ethanol containing 5 μg/mL of cholestan-2-ol and then 100 μL of 11 N KOH (*see* **Note 14**).
16. Incubate 1 h at 80 °C in a dry bath.
17. Let the samples cool.
18. Add 1 mL of hexane.
19. Mix using vortex.
20. Add 2 mL of distillated H_2O.
21. Mix vigorously.
22. Centrifuge 5 min at 800 × *g* at room temperature to separate the phases.
23. Recover gently the hexane upper phase taking care not to contaminate the sample with the lower aqueous phase (*see* **Note 15**).
24. Evaporate the hexane through airflow evaporator.
25. Prepare the sample as described in **steps 18–23** in Subheading 3.2.1.
26. Run the sample in GC-MS instrument using the sterol methods.

3.2.3 GC-MS Methods for FAMES and Sterol Profiling

1. Set the helium carrier gas at 2 mL/min.
2. Use the splitless mode for injection.
3. Set the temperatures of injector to 250 °C and the auxiliary detector to 325 °C; the temperature of MS Source is set to 230 °C and the MS Quad to 150 °C.

4. MS Analyzer is set in scan only with mass range 40–700 m/z in positive mode with an electron emission set to 70 eV.
5. For Fames profiling, the temperature program starts at 50 °C for 1 min, then 25 °C/min ramp from 50 °C to 150 °C (2-min hold), and a 10 °C/min ramp to 320 °C (6-min hold).
6. For Sterol profiling, the temperature program starts at 200 °C for 1 min, then 10 °C/min ramp from 200 °C to 305 °C (2.5-min hold), and a 15 °C/min ramp to 320 °C.

3.2.4 Mass Spectrometry Analysis Dedicated to Phospholipids

1. Phospholipids are extracted as described in **steps 1–14** in Subheading 3.2.2.
2. Resuspend samples in 50 μl of eluent A with 40 μg of each internal standard.
3. Transfer samples into glass vial suitable for injection.
4. Put the vials in the autosampler at 4 °C.
5. Separate phospholipid molecules species on an LC 1200 Infinity II system equipped with a Luna C8 150× 1 mm column with the following gradient:
   (a) 0 min start at 70% eluent A, 30% eluent B.
   (b) 0 to 5 min ramp at 50% eluent A, 50% eluent B.
   (c) 5 to 30 min ramp to 20% eluent A, 80% eluent B.
   (d) 31 to 41 min hold at 5% eluent A, 95% eluent B.
   (e) 42 to 52 min hold to 70% eluent A, 30% eluent B.
6. Subject HPLC elutes to coupled electrospray ionization in the negative (PE, PS, PI, and PG) and positive (PC) modes with fast polarity switching (50 ms); use nitrogen for the curtain gas (set to 15), gas 1 (set to 20), and gas 2 (set to 10). Set the needle voltage at −4500 or +5500 V without needle heating; use nitrogen as collision gas. Adjust the declustering potential at −50 V for PE, −35 V for PG, −50 V for PS, −70 V for PI, and +35 V for PC; set collision energy to −43 eV for phosphatidyl-ethanolamine (PE), −50 eV for phosphatidyl-glycerol (PG), −55 eV for phosphatidyl-serine (PS), −60 eV for phosphatidyl-inositol (PI), and +37 eV for phosphatidyl-choline (PC) (*see* **Note 16**).
7. Define in Multiquant the integration parameters: Integration Algorithm MQ4; Gaussian Smooth Width 1.0 points and Noise percentage: 40.0%.
8. Integrate all the sample using the Multiquant method and report the area under the peaks for each multiple reaction monitoring (MRM).

9. Sum the transitions corresponding to the same species, for example, for PE: PE 32:0 with PE32:0_14:0, PE32:0_16:0 and PE32:0_18:0.
10. Present the results as a ratio corresponding to the area of the phospholipid species, and use the corresponding internal standard for each phospholipids class as reference.

3.3 Label-Free Proteomic Analysis

3.3.1 Gel Electrophoresis and Staining

Samples are solubilized in Laemmli buffer and deposited onto a 10% acrylamide SDS–PAGE gel for concentration and cleaning purposes:

1. Depending on sample concentrations, take the volume V corresponding to 10 μg of proteins.
2. Add V/3 μL of Laemmli buffer 4× to the samples.
3. Heat samples at 99 °C for 2 min.
4. Load samples onto a 10% acrylamide SDS-PAGE.
5. Migrate the gel (150 V. max, 20 mA) until the migration front progresses by 5 mm.
6. Stop the migration, unmold the gel from the plate, and rinse it twice in ultrapure water for 3 min.
7. Incubate the gel overnight in 100 mL of Coomassie Brilliant Blue G-250.
8. Wash the gel in water until complete destaining of the background.

3.3.2 In-Gel Trypsin Digestion

1. Scan the gel on a densitometer and estimate protein quantities.
2. Cut each lane with a scalpel. Subsequently cut the gel pieces in 1 × 1 mm smaller cubes and transfer into a tube.
3. Wash with 300 μL ultrapure water for 15 min.
4. Wash with 300 μL (v:v) 25 mM NH_4HCO_3: 100% ACN for 4 h or until complete destaining.
5. Dehydrate with 300 μL 100% ACN for 10 min and remove the solution.
6. Dry the gel pieces at room temperature for 15 min.
7. Add 300 μL dithiothreitol 10 M and incubate for 30 min at 56 °C.
8. Remove the solution and add iodoacetamide 100 mM for 30 min at room temperature and in total darkness.
9. Remove solution and dehydrate with 300 μL 100% ACN for 10 min. Remove the solution.
10. Dry the gel pieces at room temperature for 15 min.
11. Rehydrate with 50 μL of 10 ng/μL trypsin in 50 mM NH_4HCO_3 for 10 min at 4 °C.

12. Add sufficient 50 mM NH_4HCO_3 to cover gel pieces.
13. Incubate overnight at 37 °C.

3.3.3 Peptide Extraction

1. Transfer supernatant to a new sample tube and add 150 μL of 50 mM NH_4HCO_3 and mix for 15 min.
2. Transfer supernatant to sample tube and add 150 μL extraction buffer water/ACN/formic acid 48:48:4 (v:v:v) and mix for 15 min.
3. Repeat **step 2**.
4. Lyophilize pooled sample in a vacuum to complete dryness.
5. Suspend the pellet in 100 μL 0.1% trifluoroacetic acid (*see* **Note 17**).

3.3.4 High-Performance Liquid Chromatography (HPLC) in Line with Mass Spectrometer

1. Trypsin-digested peptides are separated using a Dionex Ultimate 3000 nanoHPLC system.
2. Load 1 μg of the sample in 1% formic acid (v:v) on a 300-μm inner diameter × 5-mm C18 PepMap™ trap column (LC Packings) for 3 min at a constant flow rate of 10 μL/min (*see* **Note 17**).
3. After trapping and desalting on the C18 cartridge, peptides are eluted on an analytical PepMap C18 nanocolumn (75 μm × 50 cm) with a linear gradient of 4–40% solvent B over 108 min with a flow rate of 200 nL/min.

3.3.5 Mass Spectrometry Analysis

1. The mass spectrometer operates in positive ion mode at a 2-kV needle voltage.
2. Acquire data using Xcalibur software in a data-dependent mode. Record MS scans (at 375–1500 *m/z*) at a resolution of $R = 120{,}000$ (at 200 *m/z*) and collect AGC target of 4×10^5 within 50 ms. Set the dynamic exclusion to 60 s and perform top speed fragmentation in HCD mode over a 3-s cycle. Collect the MS/MS scans with a target value of 2×10^3 ions in the ion trap with a maximum fill time of 35 ms. Additionally, select only +2 to +7 charged ions for fragmentation. Additional settings include no sheath nor auxiliary gas flow, a 275 °C heated capillary temperature, a 35% normalized HCD collision energy, and a 1.6 *m/z* isolation width. Monoisotopic precursor selection (MIPS) is set to peptide, and an intensity threshold is set to 5×10^3 to trigger MS/MS.

3.3.6 Data Processing for Protein Identification and Quantitation

1. Search the data using SEQUEST HT through Proteome Discoverer interface against *Arabidopsis thaliana* protein database (https://www.arabidopsis.org/index.jsp).
2. Set parameters as follows:

(a) Reject spectra from peptides higher than 5000 Da or lower than 350 Da.

(b) Set up the mass accuracy of the monoisotopic peptide precursor and peptide fragments to 10 ppm and 0.6 Da, respectively.

(c) Consider only b- and y-ions for mass calculation.

(d) Consider oxidation of methionine (+16 Da) and protein N-terminal modifications (acetylation +42 Da, Met-loss −131 Da, Met-loss+Acetyl −89 Da) as variable modifications and carbamidomethylation of cysteines (+57 Da) as fixed modifications.

(e) Allow two missed trypsin cleavages.

(f) Perform peptide validation using Percolator algorithm [25] and consider only "high confidence" peptides corresponding to a 1% false positive rate at peptide level.

(g) Use the Minora algorithm embedded in Proteome Discoverer to detect and integrate the peaks.

(h) Quantify the proteins on unique peptide intensities. Normalize on total protein amount and calculate protein ratio as the median of all possible pairwise peptide ratios. Calculate a *t*-test based on background population of peptides or proteins. Then consider as quantitative data the proteins quantified by a minimum of two unique peptides and an adjusted *p*-value lower than 0.05.

3. Using a reference list of previously identified PD proteins, determine cutoff enrichment ratios. Select different cutoff scores for PD/PM, PD/microsome, 30 PD/TP, and PDCW to filter out the false positives.

4 Notes

1. *Arabidopsis* cells grow in clumps of about 30–40 cells. Optimal culture condition with the appropriate hormonal balance is important for cell wall purification. In the case of hormonal overdose, cells often divide faster, forming large clumps (liquid culture looks grainier). These clumps are more resilient to disruption with the cell disruption device and could result in a wall fraction of insufficient quality.

2. In *Arabidopsis thaliana* suspension cells, PD are mostly present on the division walls. Cell walls are prepared from cultured cells in the middle of the linear growth phase (about 5 days after subculturing) in order to maximize the proportion of dividing walls containing PD.

3. Stirring will help to maintain cells in a uniform suspension in the N_2 cell disruption vessel. This will guarantee maximum cell disruption and accelerate the equilibrium step.
4. A pressure lower than 10,000 kPa inside the vessel will reduce the efficiency of the disruption process and may lead to a contaminated cell wall fraction.
5. The actual disruption process does not take place while the cells are pressurized within the vessel. Cell disruption occurs at the instant of decompression as the sample goes from high-pressure environment to atmospheric pressure. Shearing of the cells as they pass through the discharging valve also participates in cell disruption. The valve is therefore opened gradually to leave as little space as possible for the liquid suspension cell to pass through.
6. Beware that the last 5 mL of solution may come at once and produce a splash.
7. The purity of the final cell wall fraction can be visually estimated, by looking at the color of the pelleted walls. The final pellet should be bright white and not yellowish. If it is not the case, additional grinding and washing steps may be necessary.
8. Do not exceed 5 min as the enzyme activity will diminish.
9. This step helps to remove any residual debris contained in the cellulase solution. It is recommended for electron microscopy studies.
10. It is necessary to use both PMSF and a complete inhibitor cocktail to avoid protein degradation during the incubation of the wall fragment with cellulase. PMSF is rapidly degraded upon light and water exposure. The stock solution should therefore be kept in a tube opaque to light. PMSF is very toxic and should be handled with appropriate care.
11. As PD membranes are continuous with the ER and the PM, these two compartments are likely to be a source of contamination and should therefore be tested for (using, for instance, PMA2 and BiP antisera; *see* Subheading 2.4.1). Contamination from any other subcellular compartments can also be investigated (for instance, chloroplast and Golgi). Enrichment in PD membranes is monitored through the use of intrinsic PD protein markers (PDCB and PDLP; *see* Subheading 2.4.1).
12. The PD purification yield (assessed by protein quantification) can be affected by a decrease of the cellulase activity. This effect has been observed with the use of "old" cellulase (older than a year).
13. The internal standard chosen for lipidomic classes is absent from plant membranes. The C17:0 standard, for GC analysis, is used to quantify the most common fatty acids (FAs) coming

from the glycerolipid and glycerophospholipid pool. The h14:0 standard is used to quantify the α-hydroxylated (2-hydroxy)-FAs coming exclusively from the sphingolipid pool. For LC-MS analyses, all internal standard contains a fatty acid with an odd number of carbons, i.e., 17 carbons.

14. This method based on saponification will allow the quantification of the sterol backbone from sterols, steryl esters, steryl glycosides, or acetylated steryl glycosides.
15. It is important to not collect any of the lower aqueous phase as this will result in misleading chromatogram results and may destroy the GC column. If some of the aqueous phase was accidentally collected, a white residue comes from the hydrolysis of trimethylsilyl (TMS) molecules that release silica atoms.
16. The MRM mode on the Qtrap 6500 is based on the loss of the fatty acid part for the negative analysis (PE, PS, PI, and PG) and on the loss of the polar head in positive mode for phosphatidyl-choline (PC) analysis. The PC class can be analyzed in negative ionization mode, but the intensity will be lower than in positive ionization mode.
17. Sample injection should be randomized in order to reduce experimental bias induced by instrument performance variation over the period of analysis.

Acknowledgments

This work was supported by the National Agency for Research (Grant ANR-18-CE13-0016 to E.M.B). Lipidomic analyses were performed on the Bordeaux Metabolome Facility supported by the grant MetaboHUB (ANR-11-INBS-0010).

References

1. Tilsner J, Amari K, Torrance L (2011) Plasmodesmata viewed as specialised membrane adhesion sites. Protoplasma 248:39–60
2. Faulkner C (2013) Receptor-mediated signaling at plasmodesmata. Front Plant Sci 4:521
3. Fernandez-Calvino L, Faulkner C, Walshaw J et al (2011) Arabidopsis plasmodesmal proteome. PLoS One 6:e18880
4. Grison MS, Brocard L, Fouillen L et al (2015) Specific membrane lipid composition is important for plasmodesmata function in Arabidopsis. Plant Cell 27:1228–1250
5. Faulkner C, Brandom J, Maule A et al (2005) Plasmodesmata 2004. Surfing the symplasm. Plant Physiol 137:607–610
6. Salmon MS, Bayer EMF (2013) Dissecting plasmodesmata molecular composition by mass spectrometry-based proteomics. Front Plant Sci 3:307
7. Benitez-Alfonso Y, Faulkner C, Pendle A et al (2013) Symplastic intercellular connectivity regulates lateral root patterning. Dev Cell 26:136–147
8. Faulkner C, Petutschnig E, Benitez-Alfonso Y et al (2013) LYM2-dependent chitin perception limits molecular flux via plasmodesmata. Proc Natl Acad Sci U S A 110:9166–9170
9. Jo Y, Cho WK, Rim Y et al (2011) Plasmodesmal receptor-like kinases identified through analysis of rice cell wall extracted proteins. Protoplasma 248:191–203

10. Lee JY, Taoka K, Yoo BC et al (2005) Plasmodesmal-associated protein kinase in tobacco and Arabidopsis recognizes a subset of non-cell-autonomous proteins. Plant Cell 17:2817–2831
11. Lee JY, Yoo BC, Rojas MR et al (2003) Selective trafficking of non-cell-autonomous proteins mediated by NtNCAPP1. Science 299: 392–396
12. Levy A, Erlanger M, Rosenthal M et al (2007) A plasmodesmata-associated beta-1,3-glucanase in Arabidopsis. Plant J 49:669–682
13. Sagi G, Katz A, Guenoune-Gelbart D et al (2005) Class 1 reversibly glycosylated polypeptides are plasmodesmal-associated proteins delivered to plasmodesmata via the golgi apparatus. Plant Cell 17:1788–1800
14. Simpson C, Thomas C, Findlay K et al (2009) An Arabidopsis GPI-anchor plasmodesmal neck protein with callose binding activity and potential to regulate cell-to-cell trafficking. Plant Cell 21:581–594
15. Thomas CL, Bayer EM, Ritzenthaler C et al (2008) Specific targeting of a plasmodesmal protein affecting cell-to-cell communication. PLoS Biol 6:180–190
16. Bayer E, Thomas CL, Maule AJ (2004) Plasmodesmata in Arabidopsis thaliana suspension cells. Protoplasma 223:93–102
17. Brault M, Petit J, Immel F et al (2019) Multiple C2 domains and transmembrane region proteins (MCTPs) tether membrane at plasmodesmata. EMBO Rep 20:e47182
18. Liu L, Zhang T, Liu Z et al (2019) Phytosphinganine affects plasmodesmata permeability via facilitating PDLP5-stimulated callose accumulation in Arabidopsis. Mol Plant 13: 128–143
19. Hofte H, Chrispeels MJ (1992) Protein sorting to the vacuolar membrane. Plant Cell 4: 995–1004
20. Vallon O, Wollman FA, Olive J (1986) Lateral distribution of the main protein complexes of the photosynthetic apparatus in Chlamydomonas reinhardtii and in spinach: an immunocytochemical study using intact thylakoid membranes and PS II enriched membrane preparation. J Photobiochem Photobiophys 12:203–220
21. Morsomme P, Dambly S, Maudoux O et al (1998) Single point mutations distributed in 10 soluble and membrane regions of the Nicotiana plumbaginifolia plasma membrane PMA2 H+-ATPase activate the enzyme and modify the structure of the C-terminal region. J Biol Chem 273:34837–34842
22. Santoni V, Vinh J, Pflieger D et al (2003) A proteomic study reveals novel insights into the diversity of aquaporin forms expressed in the plasma membrane of plant roots. Biochem J 373:289–296
23. Desprez T, Juraniec M, Crowell EF et al (2007) Organization of cellulose synthase complexes involved in primary cell wall synthesis in Arabidopsis thaliana. Proc Natl Acad Sci U S A 104:15572–15577
24. Gendre D, Oh J, Boutte Y et al (2011) Conserved Arabidopsis ECHIDNA protein mediates trans-Golgi-network trafficking and cell elongation. Proc Natl Acad Sci U S A 108: 8048–8053
25. Käll L, Canterbury JD, Weston J et al (2007) Semi-supervised learning for peptide identification from shotgun proteomics datasets. Nat Methods 4:923–925

Chapter 13

Methods for Detection of Protein Interactions with Plasmodesmata-Localized Reticulons

Verena Kriechbaumer and Stanley W. Botchway

Abstract

Plant reticulon family proteins (RTN) tubulate the ER by dimerization and oligomerization, creating localized ER membrane tensions that result in membrane curvature. Two RTN ER-shaping proteins have been found in the plasmodesmata (PD) proteome which could potentially contribute to the formation of the desmotubule, an ER-derived structure that crosses primary PD and physically connects the ER of two cells. Here we describe two methods used to identify partners of two PD-resident reticulon proteins, RTN3 and RTN6 that are located in primary PD at cytokinesis in tobacco (*Nicotiana tabacum*): immunoprecipitations using GFP-Trap®_A beads to find novel interaction partners and FRET-FLIM to test for and quantify direct protein-protein interactions *in planta*.

Key words Plasmodesmata, Reticulon proteins, Endoplasmic reticulum, Protein-protein interaction, Co-immunoprecipitation, GFP-trap, FRET-FLIM

1 Introduction

The cortical ER network of plants is involved in a plethora of functions and pathways including protein trafficking [1, 2] and pathogen responses (for review, *see* [3, 4]). In plants, the protein family of reticulon proteins (RTN) contributes significantly to tubulation of the ER [5–7]. RTN proteins are integral ER membrane proteins found ubiquitously in eukaryotes and have been linked to a number of biological functions including neurodegenerative diseases [8–10] in animals. RTNs feature a reticulon homology domain that comprises two hydrophobic regions flanking a hydrophilic loop [11]. In Arabidopsis, 21 reticulons have been identified so far, and overexpression of some of these induces a constriction phenotype to ER tubules and is capable of converting cisternae into tubules [5–7, 12]. Reticulons dimerize or oligomerize and so create ER membrane tension inducing membrane curvature [13]. This topology has been shown for the Arabidopsis

Yoselin Benitez-Alfonso and Manfred Heinlein (eds.), *Plasmodesmata: Methods and Protocols*,
Methods in Molecular Biology, vol. 2457, https://doi.org/10.1007/978-1-0716-2132-5_13,

reticulons 1, 2, 3, 4, and 13 [6, 13]. Arabidopsis coding for 21 reticulons in comparison to 4 mammalian ones raises question about the potential of different functionality. A subgroup of RTN's, RTN17-21, contain terminal extensions that might allow for specialized functions within the RTN family. RTN19 and 20 contain some N-terminal extension homologues to 3β-hydroxysteroid dehydrogenase and have a role in sterol regulation in Arabidopsis roots via a novel mechanism [14]. Both RTN19 and 20 are not able to constrict the ER [14]. RTN1 and RTN2 have recently been proposed to act as receptors for autophagy-mediated ER turnover [15]. The ability of RTN proteins to constrict membranes is of interest for cell plate development and the formation of primary plasmodesmata (PD) [16], the formation of which requires extensive remodeling of the cortical ER into tight tubules to form the desmotubule [17, 18]. The desmotubule is only 15 nm in diameter and hence one of the most constricted membrane structures found in nature [19]. How the ER is furled into desmotubules is poorly understood. RTN3 and RTN6 have been found in a PD proteomics approach [20] and are so far unique in labeling of the desmotubule within PD. Here they co-localize with viral movement proteins suggesting a role in cell-cell communication and plant immunity [16].

If two proteins are in close proximity (molecular distances of 1–10 nm) and therefore are most likely physically interacting, this can be measured using energy transfer processes described as Förster or fluorescence resonance energy transfer (FRET). This method was first described by Theodor Förster over 70 years ago [21] and is based on the non-radiative energy transfer from an excited fluorescent donor molecule to another non-excited fluorescent acceptor molecule in close proximity via dipole-dipole interactions. FRET will only take place if the donor emission spectrum has some overlap with the acceptor absorption spectrum with donor and acceptor dipoles displaying a mutual molecular orientation. The rate of the energy transfer k_T is calculated using Eq. (1):

$$k_T = \left(\frac{1}{\tau_D}\right)\left(\frac{R_0}{R}\right)^6 \quad (1)$$

where τ_D is the donor excited-state lifetime in the absence of the acceptor, R is the distance between D (donor) and A (acceptor), and R_0 is the Förster radius. At the Förster radius, 50% of the donor molecules emit fluorescence while the rest undergoes energy transfer. During FRET, the decay rate is reduced due to quenching that depletes the excited state of the donor fluorophore and leads to shortening of the donor fluorescence lifetime. By measuring changes in the excited-state lifetime of the donor at each pixel in the image, steady-state FRET is improved. This is described as FRET-fluorescence lifetime imaging microscopy or FRET-FLIM.

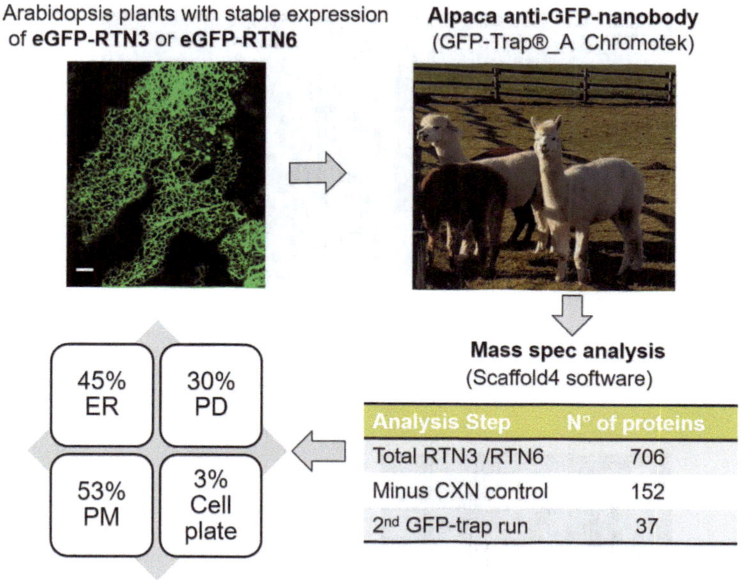

Fig. 1 Workflow for proteomic analysis. Arabidopsis plants expressing eGFP-RTN3 or eGFP-RTN6, respectively, in a stable manner are used for protein pulldowns with an anti-GFP-antibody (Chromotek). Interacting proteins are analyzed using mass spectrometry. GFP-Calnexin (CXN) plants are used as negative controls, and protein hits from CXN samples are deducted from the RTN samples. This is repeated at least twice. This resulted in 37 unique proteins interacting with the RTN proteins comprised of ER, PD, plasma membrane (PM), and cell plate proteins [22]. Size bar = 10 μm

We describe here two methods that have been used in protein–protein interaction studies for PD-localized reticulons. This work showed that RTN3 and RTN6 interact with a significant number of PD proteins as well as proteins associated with ER-PM contact sites, lipid raft proteins, and proteins that interact with viral movement proteins [22].

The methods described are:

1. Immunoprecipitations using GFP-Trap®_A beads to find novel interaction partners for PD-localized reticulon proteins (Fig. 1).
2. FRET-FLIM to test for and quantify real-time protein–protein interactions *in planta* (Fig. 2).

2 Materials

2.1 Immuno-precipitation

1. *Arabidopsis thaliana* seedlings expressing the protein of interest fused to a fluorescent tag.
2. Murashige and Skoog (MS) medium: 4.4 g/l Murashige and Skoog basal salts, pH 5.7, 10 g/l agar.

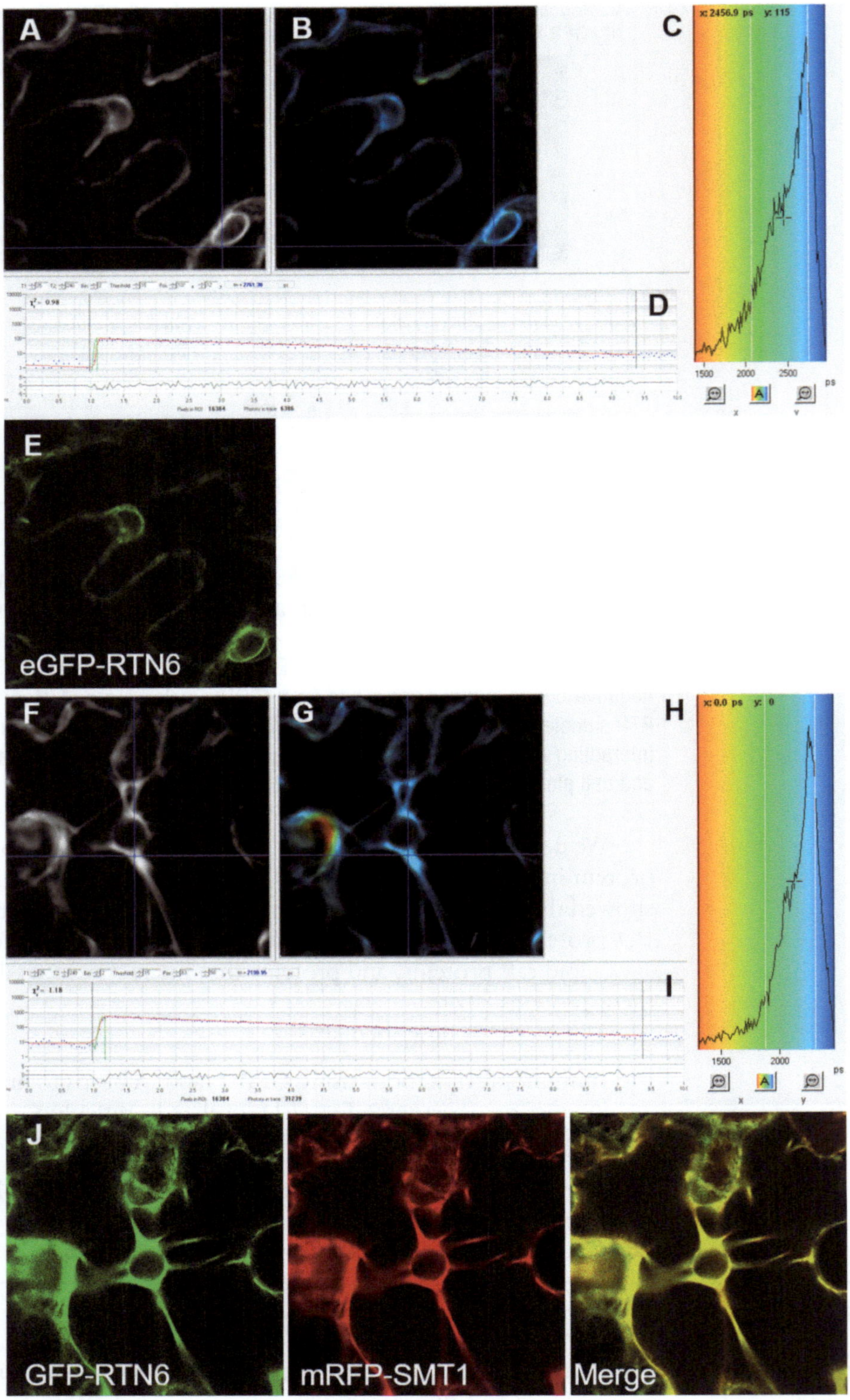

Fig. 2 FRET-FLIM analysis of eGFP-RTN6. eGFP-RTN6 without an interaction partner (**a–e**) or with mRFP-SMT1 (**f–j**). (**a**, **f**) Raw FRET-FLIM data; (**b**, **g**) pseudocolored lifetime maps showing the lifetime values for each point within the region of interest; (**c**, **h**) distribution of lifetimes across the image. Blue shades represent longer eGFP fluorescence lifetimes than green ones. (**d**, **i**) Representative decay curves of a single point with

3. Lysis buffer: 10 mM Tris–HCl, pH 7.5, 150 mM NaCl, 0.5 mM EDTA, 0.5% NP-40, 1 mM PMSF, protease inhibitor according to manufacturers' instructions.
4. Equilibration buffer: 10 mM Tris–HCl, pH 7.5, 150 mM NaCl, 0.5 mM EDTA.
5. GFP-Trap®_A beads.
6. Petri dishes (large, with lid $L \times W \times H$: 245 mm × 245 mm × 25 mm).
7. 50-ml Falcon tubes.
8. Glass bottles for buffers.
9. Cheese cloth.
10. Refrigerated tabletop centrifuge.

2.2 FRET-FLIM Using Tobacco Leaf Agrobacterium-Mediated Protein Expression

1. *Nicotiana tabacum* plants about 5–6 weeks old, grown at 21 °C, 14 h light, 10 h dark.
2. Plasmid constructs for Agrobacterium-mediated plant transformation fused to green fluorescent protein (here RTN proteins) and red fluorescent proteins (RFP) (interaction test proteins), respectively.
3. YEB medium: 5 g/l beef extract, 1 g/l yeast extract, 5 g/l sucrose, 0.5 g/l $MgSO_4 \cdot 7H_2O$.
4. 28 °C shaker for falcons.
5. Infiltration buffer: 50 mM MES, 2 mM Na_3PO_4, 0.1 mM acetosyringone, and 5 mg/ml glucose in 50-ml Falcon tube.
6. 15-ml Falcon tubes.
7. 2-ml Eppendorf tubes.
8. Tabletop centrifuge.
9. Spectrophotometer.
10. 1-ml syringes.
11. At least two-channel confocal and one-channel FLIM setup.
12. Becker and Hickl SPC 830 or SPC150 time correlated single-photon counting card.
13. SPCImage software version 5.1 or higher (Becker and Hickl).

Fig. 2 (continued) an optimal single exponential fit, where $\chi 2$ values from 0.9 to 1.2 were considered an excellent fit to the data points (a binning factor of 2 was applied). The confocal images for the analysis in (**e**, **j**) show the eGFP construct in green and the mRFP construct in red. This example of FRET-FLIM analysis shows eGFP-RTN6 alone as a negative control and mRFP-SMT1 for protein–protein interaction. The fluorescence lifetime values for eGFP-RTN6 + mRFP-SMT1 are 2.25 ± 0.64 ns and therefore statistically lower than the lifetime values for the eGFP-RTN6 fusion alone (2.71 ± 0.57 ns)

3 Methods

3.1 Immuno-precipitation with GFP-Trap®_A Beads

This part of the procedure requires between 4 and 5 h of time depending on sample size. This protocol uses GFP-Trap®_A beads (*see* **Note 1**), and the procedure is carried according to the company's protocol with slight modifications:

Arabidopsis thaliana seedlings expressing the protein of interest fused to a fluorescent tag in a stable manner are grown for 2 weeks on MS plates (*see* **Note 2**):

1. Grind 5 g of Arabidopsis whole seedling plant material in liquid nitrogen to a fine powder and then grind further in 5-ml lysis buffer until liquid (*see* **Note 3**). Distribute the extract into three 2-ml Eppendorf tubes.
2. Incubate the extracts on ice for 30 min and then centrifuge at 10,000 × *g* for 10 min at 4 °C.
3. Pour the supernatant (about 4 ml) into 50-ml Falcon tubes via two layers of cheese cloth.
4. For each plant sample, add 100 μl of GFP-Trap®_A agarose slurry to 500-μl equilibration buffer and centrifuge at 2500 × *g* for 2 min to pellet the beads. Discard the supernatant and repeat this wash twice.
5. Add the 100-μl washed agarose beads to the plant extract and shake the mixture on ice for 2 h with the tubes being placed horizontally in the ice.
6. Centrifuge the tubes at 2500 × *g* for 2 min at 4 °C and discard the supernatant carefully without disturbing the agarose pellet.
7. Wash the resulting agarose pellet twice with dilution buffer.
8. Use this pellet straight away for tryptic peptide digest and mass spectrometry analysis for containing proteins [22].

3.2 Real-Time Imaging of Protein–Protein Interactions Using FRET-FLIM

Potential interaction candidates resulting from the immunoprecipitation are verified *in planta* with the reticulon bait protein. The green fluorescent protein (GFP)-tagged RTN constructs are used, whereas candidate proteins are fused to a RFP. This procedure is done via FRET-FLIM using Agrobacterium-mediated transient expression in tobacco leaf epidermal cells [22, 23]:

1. Grow 5-ml overnight Agrobacterium cultures in 15-ml Falcon tubes at 28 °C with 200 rpm shaking in YEB with the appropriate antibiotics for constructs to be tested.
2. Transfer 1 ml per culture into separate labeled 2-ml Eppendorf tubes.
3. Spin for 5 min at 4000 rpm and remove the supernatant.
4. Add 1-ml infiltration buffer at RT and resuspend the pellet.

5. Spin for 5 min at 4000 rpm and remove the supernatant.
6. Resuspend the pellet in 1-ml infiltration buffer.
7. Measure the OD_{600}.
8. Prepare your various construct mixes (the green RTN construct with each of your red test constructs) in individual tubes, and in doing this dilute the resuspension with infiltration medium until the required OD is reached (*see* **Note 4**).
9. For the leaf infiltration, take up the cells in a 1-ml sterile syringe (no needle required).
10. Punch a small hole into the tobacco leaf avoiding using a P100 tip.
11. Place the syringe tip firmly against the underside of the leaf covering the hole and press the plunger down gently while exerting pressure against the other side of the leaf with your finger. The liquid will diffuse throughout the mesophyllar airspace.
12. Repeat for all your constructs in at least two tobacco plants.
13. Mark the leaf upper epidermis with a permanent marker to keep track of the transformation sites.
14. Place the infiltrated plants back into growth conditions (21 °C with 14 h light, 10 h dark) for at least 30 h to allow transformation and gene expression to occur (*see* **Note 5**).
15. Cut out epidermal samples of infiltrated tobacco leaves (~5 mm) with a scalpel, and perform both confocal/multiphoton FRET-FLIM data capture by microscope [24] (*see* **Notes 6** and 7).
16. Set laser light emission at a wavelength of 920 nm using a mode-locked titanium sapphire laser (Mira; Coherent Lasers), with 200 fs pulses at 76 MHz, pumped by a solid-state continuous wave 532-nm laser (Verdi V18; Coherent Laser).
17. To illuminate specimens on the microscope stage, focus the laser beam to a diffraction limited spot using a water-immersion objective (Nikon VC; numerical aperture of 1.2).
18. Collect fluorescence emission without pinhole (non-descan), bypassing the scanning system, but instead passing through a BG39 (Comar) filter to block the near-infrared laser light.
19. Generate the raw time-correlated single-photon (TCSPC) data by linking these via a time-correlated single-photon counting PC module SPC830 (Becker and Hickl). Analyze the pixel by pixel TCSPC data to generate a FLIM image or map (*see* **Note 8**).

20. Use a 633-nm interference filter to reduce chlorophyll autofluorescence that will otherwise strongly obscure the mRFP as well as eGFP emission.
21. Analyze data by obtaining excited-state lifetime values on a region of interest on the nucleus, and make calculations using SPCImage software version 5.1.
22. Generate the range of lifetime values (for each pixel) within a region of interest and display as a distribution curve.
23. Fit decay data to a single exponential parameter as $f(t) = ae^{-t/\tau}$ (*see* **Note 9**). To allow for optimal fit, consider only values with a chi-square (χ^2) between 0.9 and 1.4 for statistical analysis (*see* **Note 10**).
24. Produce the range of lifetimes per sample by taking into account the median lifetime and minimum and maximum values for one-quarter of the median lifetime values from the curve.
25. Analyze at least three nuclei from at least three independent biological samples per protein-protein combination, and take the average of the ranges.
26. Calculate the degree or percentage efficiency of energy transfer (E), from one protein to the other, using Eq. (2):

$$E\% = \left[1 - \left(\frac{\tau_{DA}}{\tau_D}\right)\right] \times 100 \quad (2)$$

here, τ_D and τ_{DA} are the measured excited-state lifetime of the donor and acceptor, respectively. Example data is shown in Fig. 2.

4 Notes

1. Gloves have to be worn for the whole procedure to reduce contamination with human proteins.
2. This protocol can also be carried out with tobacco transiently expressing the protein of interest but will result in reduced number of interacting proteins and will of course only show interactors in tobacco.
3. Tissue can also be separated, e.g., if there is an interest in root proteins.
4. The optimum OD_{600} will vary according to the construct; hence, different ODs have to be tested until sufficient expression levels are achieved. An OD_{600} of 0.1 is a good place to start with a new construct. After this, lower ODs (e.g., 0.03) or higher ODs (e.g., 0.2–0.3) can be tested.

5. This growth period depends on the construct and protein longevity, but 40–48 h is mostly required before fluorescence can be observed.
6. For the experiments here, a two-photon microscope built around a Nikon TE2000-U inverted microscope is used with a modified Nikon EC2 confocal scanning system to enable near-infrared laser wavelength for FLIM [25].
7. Best measurement results for the ER have been shown for the nuclear envelope. Latrunculin B can be applied to the leaf disks to depolymerize the actin cytoskeleton and therefore inhibit movement of the ER. This is not necessary given sufficient expression levels of the donor protein which result in shorter time for data capture, and therefore movement is less critical. FRET-FLIM measurements can also be taken on ER cisternae after application of latrunculin B.
8. Prior to FLIM data collection, eGFP and mRFP expression levels in the plant samples within the region of interest are confirmed using a Nikon EC2 confocal microscope with excitation at 488 and 543 nm, respectively.
9. The intensity data provides the fluorescence decay function (f) at time, t. Single, double, or triple exponential analysis of the decay yields the excited-state lifetime (τ). The amplitude of the exponential components, a, defines the contribution to each lifetime component.
10. χ^2 describes the goodness of data fitted to the exponential function, so that a value of 1 represents analysis that perfectly describes the decay data points.

Acknowledgments

This work was supported by the Science and Technology Facilities Council Program (grant no. 14230008) and by a Vice-Chancellor's research fellowship from Oxford Brookes University to VK.

References

1. Hawes C, Kiviniemi P, Kriechbaumer V (2015) The endoplasmic reticulum: a dynamic and well-connected organelle. J Integr Plant Biol 57:50–62
2. Kriechbaumer V, Brandizzi F (2020) The plant endoplasmic reticulum: an organized chaos of tubules and sheets with multiple functions. J Microsc 280:122–133
3. Pattison RJ, Amtmann A (2009) N-glycan production in the endoplasmic reticulum of plants. Trends Plant Sci 14:92–99
4. Beck M, Heard W, Mbengue M, Robatzek S (2012) The INs and OUTs of pattern recognition receptors at the cell surface. Curr Opin Plant Biol 15:367–374
5. Tolley N, Sparkes I, Hunter PR, Craddock CP, Nuttall J, Roberts LM, Hawes C, Pedrazzini E, Frigerio L (2008) Overexpression of a plant reticulon remodels the lumen of the cortical endoplasmic reticulum but does not perturb protein transport. Traffic 9:94–102

6. Tolley N, Sparkes I, Craddock CP, Eastmond PJ, Runions J, Hawes C, Frigerio L (2010) Transmembrane domain length is responsible for the ability of a plant reticulon to shape endoplasmic reticulum tubules in vivo. Plant J 64:411–418
7. Sparkes IA, Frigerio L, Tolley N, Hawes C (2009) The plant endoplasmic reticulum: a cell-wide web. Biochem J 423:145–155
8. Oertle T, Klinger M, Stuermer CAO, Schwab ME (2003) A reticular rhapsody: phylogenic evolution and nomenclature of the RTN/Nogo gene family. FASEB J 17: 1238–1247
9. Nziengui H, Bouhidel K, Pillon D, Der C, Marty F, Schoefs B (2007) Reticulon-like proteins in *Arabidopsis thaliana*: structural organization and ER localization. FEBS Lett 581: 3356–3362
10. Yang YS, Strittmatter SM (2007) The reticulons: a family of proteins with diverse functions. Genome Biol 8:234–244
11. Nziengui H, Schoefs B (2009) Functions of reticulons in plants: what we can learn from animals and yeasts. Cell Mol Life Sci 66: 584–595
12. Breeze E, Dzimitrowicz N, Kriechbaumer V, Brooks R, Botchway SW, Brady JP, Hawes C, Dixon AM, Schnell JR, Fricker MD, Frigerio L (2016) A C-terminal amphipathic helix is necessary for the in vivo tubule-shaping function of a plant reticulon. Proc Natl Acad Sci U S A 113:10902–10907
13. Sparkes I, Tolley N, Aller I, Svozil J, Osterrieder A, Botchway S, Mueller C, Frigerio L, Hawes C (2010) Five plant reticulon isoforms share ER localisation, topology, ER membrane shaping properties. Plant Cell 22:1333–1343
14. Kriechbaumer V, Maneta-Peyret L, Fouillen L, Botchway SW, Upson J, Hughes L, Richardson J, Kittelmann M, Moreau P, Hawes C (2018) The odd one out: Arabidopsis reticulon 20 does not bend ER membranes but has a role in lipid regulation. Sci Rep 8:2310
15. Zhang X, Ding X, Marshall RS, Paez-Valencia J, Lacey P, Vierstra RD, Otegui MS (2020) Reticulon proteins modulate autophagy of the endoplasmic reticulum in maize endosperm. eLife 9:e51918
16. Knox K, Wang P, Kriechbaumer V, Tilsner J, Frigerio L, Sparkes I, Hawes C, Oparka K (2015) Putting the squeeze on plasmodesmata: a role for reticulons. Plant Physiol 168: 1563–1572
17. Overall RL, Blackman LM (1996) A model of the macro-molecular structure of plasmodesmata. Trends Plant Sci 1:307–311
18. Ehlers K, Kollmann R (2001) Primary and secondary plasmodesmata: structure, origin, and functioning. Protoplasma 216:1–30
19. Tilsner J, Amari K, Torrance L (2011) Plasmodesmata viewed as specialised membrane adhesion sites. Protoplasma 248:39–60
20. Fernandez-Calvino L, Faulkner C, Walshaw J, Saalbach G, Bayer E, Benitez-Alfonso Y, Maule A (2011) Arabidopsis plasmodesmal proteome. PLoS One 6:e18880
21. Förster T (1948) Zwischenmolekulare Energiewanderung und Fluoreszenz. Ann Phys 437:55–75
22. Kriechbaumer V, Botchway SW, Slade SE, Knox K, Frigerio L, Oparka K, Hawes C (2015) Reticulomics: protein–protein interaction studies with two plasmodesmata-localized reticulon family proteins identify binding partners enriched at plasmodesmata, endoplasmic reticulum, and the plasma membrane. Plant Physiol 169:1933–1945
23. Sparkes IA, Runions J, Kearns A, Hawes C (2006) Rapid, transient expression of fluorescent fusion proteins in tobacco plants and generation of stably transformed plants. Nat Protoc 1:2019–2025
24. Schoberer J, Botchway SW (2014) Investigating protein-protein interactions in the plant endomembrane system using multiphoton-induced FRET-FLIM. Methods Mol Biol 1209:81–95
25. Botchway SW, Scherer KM, Hook S, Stubbs CD, Weston E, Bisby RH, Parker AW (2015) A series of flexible design adaptations to the Nikon E-C1 and E-C2 confocal microscope systems for UV, multiphoton and FLIM imaging. J Microsc 258:68–78

Chapter 14

Studying Protein–Protein Interactions at Plasmodesmata by Measuring Förster Resonance Energy Transfer

Patrick Blümke, Vicky Howe, and Rüdiger Simon

Abstract

Plasmodesmata (PD) provide interconnectivity between plant cells to enable the intercellular transport and communication that is requisite to multicellularity. Being at the interface of the apoplast, plasma membrane (PM), endoplasmic reticulum (ER), and symplast, PD are uniquely positioned to integrate exogenously and endogenously derived signals with plant developmental and physiological responses. The distinct membrane curvature and composition of PD allow them to function as microdomains to facilitate dynamic protein–protein interactions. Förster resonance energy transfer (FRET) combined with fluorescence lifetime imaging microscopy (FLIM) and fluorescence anisotropic decay measurements provides valuable tools to analyze these interactions in vivo and *in planta*. Here we describe a detailed methodology to perform FRET-FLIM and fluorescence anisotropy measurements to analyze protein–protein interactions at PD in a transient expression system using *Nicotiana benthamiana*; however this can be adapted to other plant species and subcellular compartments.

Key words Plasmodesmata, Protein–protein interaction, Confocal fluorescence microscopy, Live-cell imaging, FRET, FLIM, Anisotropy, Fluorescence lifetime

1 Introduction

Plasmodesmata (PD) are plasma membrane (PM)-lined channels connecting adjacent plant cells, thereby enabling intercellular transport and communication (*see* [1] for an overview of PD). PD-localized proteins regulate trafficking through and signaling at PD (*see* [2, 3] for reviews). Because of their small size (typically 20–40 nm wide and 100–400 nm long) and the fact that they are an integral part of the cell wall, PD and their protein composition are notoriously difficult to study [2].

To capture the temporal and spatial dynamics of PD-associated protein complexes, it is important to measure protein–protein

Patrick Blümke and Vicky Howe contributed equally to this work.

Yoselin Benitez-Alfonso and Manfred Heinlein (eds.), *Plasmodesmata: Methods and Protocols*,
Methods in Molecular Biology, vol. 2457, https://doi.org/10.1007/978-1-0716-2132-5_14,

interactions in intact, living cells. Förster resonance energy transfer (FRET)-based state-of-the-art confocal fluorescence microscopy techniques offer the possibility to observe protein proximity, potentially reflecting interaction, in vivo and *in planta*. FRET describes an emission-free energy transfer between FRET pairs from a so-called donor fluorophore, which can be excited by a laser pulse, to an acceptor fluorophore. This results in the excitation and fluorescence of the acceptor fluorophore while the donor fluorophore's fluorescence is quenched [4]. For FRET to occur, the emission spectrum of a FRET donor must overlap sufficiently with the excitation spectrum of the FRET acceptor. The fluorophores must also be in close proximity (<10 nm), where FRET efficiency increases the closer the distance between FRET pairs, and the dipole moment of the donor fluorophore's emission must be parallel to the acceptor fluorophore's excitation dipole moment (i.e., parallel polarization) [5, 6].

Since FRET is distance-dependent, it can be utilized to determine whether two proteins are in close physical proximity and thus potentially form functional complexes. In FRET interaction studies, one protein of interest (POI) is fused to a FRET donor and the other to a FRET acceptor. Traditionally, the FRET pair eGFP and mCherry have been used in plants; however a number of other suitable FRET partners also work well in plant cells [7]. In contrast to other fluorescence-based interaction assays such as bimolecular fluorescence complementation, the two fluorophores used in FRET experiments have no innate affinity toward one another. FRET can only occur if the two fluorophores are brought into close proximity by the association of the two POIs, suggesting their interaction [8] (Fig. 1).

FRET can be measured in a number of ways, but here we will focus on fluorescence lifetime imaging microscopy (FLIM) and fluorescence anisotropy. Both FRET-FLIM and fluorescence anisotropy can be measured noninvasively, requiring only a low laser power. This avoids photobleaching of fluorophores and/or damage to the plant cell; therefore measurements can be taken over a period of minutes or even hours in live, intact cells, allowing dynamic processes to be observed [9]. FRET-FLIM and anisotropy measurement can be taken concurrently, referred to as multiparameter fluorescence image spectroscopy (MFIS), and can provide valuable information on the nature of protein-protein interactions in FRET fusion complexes [9].

In FRET-FLIM, the change in fluorescence lifetime of a sample is measured over time. Fluorescence lifetime (denoted as τ, tau) is a measurement of the time that a fluorophore remains in the excited state after excitation by a short laser pulse before returning to the ground state through emission of a photon (i.e., fluorescence) [4]. Fluorescence lifetime is specific for each fluorophore and is dependent on its FRET state [10]. When FRET occurs between a

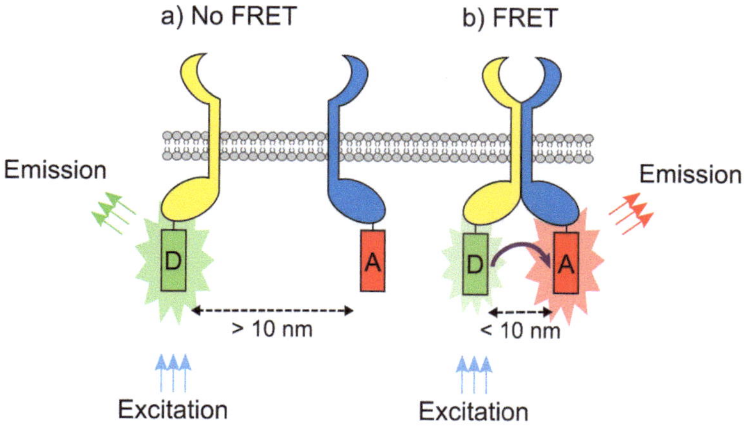

Fig. 1 Schematic of FRET. A donor fluorophore (D) is excited by a short laser pulse close to its absorption maximum (indicated by parallel blue arrows), causing it to fluoresce. FRET can only occur when the donor and acceptor (A) fluorophores are brought into close proximity (<10 nm) and the dipole moments are aligned. If these conditions are not met, as in (**a**), there is no FRET, and fluorescence is only detected in the donor (indicated by parallel green arrows). However, if the donor and acceptor fluorophores are brought into close proximity through the interaction of the two proteins of interest (shown as yellow and blue membrane receptors), as in (**b**), energy can be transferred from the donor to the acceptor fluorophore (indicated by purple arrow). This causes fluorescence of the acceptor fluorophore (indicated by parallel red arrows) and corresponding quenching of fluorescence in the donor fluorophore (indicated by smaller and fainter glow around donor)

donor and acceptor fluorophore, i.e., "hetero-FRET," the average fluorescence lifetime of the donor fluorophore in a FLIM measurement is decreased as the acceptor quenches the donor's fluorescence. In a practical context, one can detect this by comparing the average fluorescence lifetimes of FRET samples with a donor-only sample, where a decreased fluorescence lifetime in the FRET sample compared with the donor-only sample indicates heteromeric protein–protein interactions.

In addition to hetero-FRET, fluorophores can also undergo "homo-FRET" through homomerization of identical FRET fusion proteins. Unlike hetero-FRET, homo-FRET does not affect fluorescence lifetime but instead affects the polarization of the light emitted through fluorescence, i.e., the fluorescence anisotropy. Fluorescence anisotropy (also known as fluorescence polarization, denoted r) refers to the property whereby light is emitted at different intensities along a fluorophore's planes of polarization. For a fluorophore to be excited, it must have the same dipole moment (i.e., parallel polarization) as the laser beam [4]. If the fluorophore changes orientation in the time between absorption and emission of a photon, the light becomes depolarized, and the amount of depolarization gives a measure of fluorescence anisotropy decay (Fig. 2).

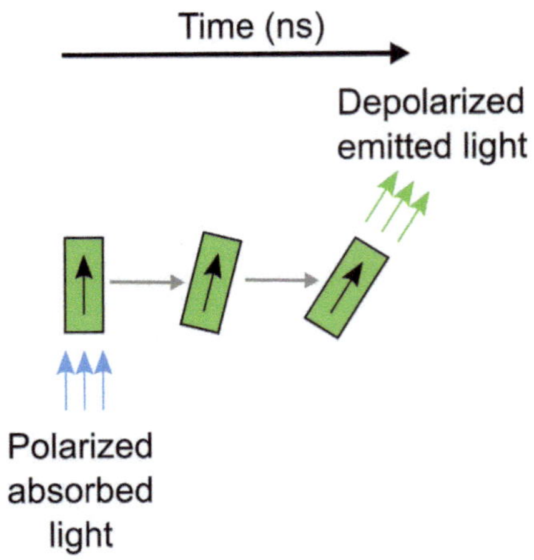

Fig. 2 Schematic of fluorescence anisotropy. Polarized light from the laser source (shown as blue parallel arrows) is absorbed by a fluorophore (shown as a green rectangle) with a parallel polarization (indicated by the black arrow). If the fluorophore rotates in solution in the time between absorption of the polarized light and emission of fluorescence (indicated by green parallel arrows), the light becomes depolarized, and anisotropy is decreased

Fluorescence anisotropy is dependent on the rotational freedom of the fluorophore. Large protein fluorophores such as GFP have relatively low rotational freedom compared to a small dye molecule that can rotate rapidly, and thus there is little depolarization of light over the fluorophore's fluorescence lifetime [11]. In heteromeric complexes of FRET fusion proteins, the rotational freedoms of the fluorophores are restricted compared to noninteracting fusion proteins; thus the fluorescence anisotropy (i.e., polarization) is increased. However, if FRET fusion proteins undergo homo-FRET, energy from an excited fluorophore with parallel polarization to the laser can be transferred to a fluorophore with a nonparallel dipole moment. This depolarizes the signal, decreasing the total anisotropy of the fluorophore, but does not affect the fluorescence lifetime [11]. FLIM and anisotropy measurements can be visualized by plotting fluorescence lifetime against fluorescence anisotropy in a so-called 2D MFIS plot [11, 12] to determine whether homomeric, heteromeric, or mixed complexes are occurring in a FRET sample (Fig. 3).

Furthermore, heterogeneity of fluorescence lifetimes in a FRET-FLIM sample can provide valuable information on the stoichiometry or spatial distribution of FRET complexes. As fluorescence lifetime is dependent on the orientation and distance between FRET pairs, each complex will have a unique fluorescence lifetime [7, 10]. Low heterogeneity of fluorescence lifetimes indicates that the same types of complexes are formed throughout the FLIM dataset; high heterogeneity suggests that distinct complexes are formed from the same proteins. For example, the stem

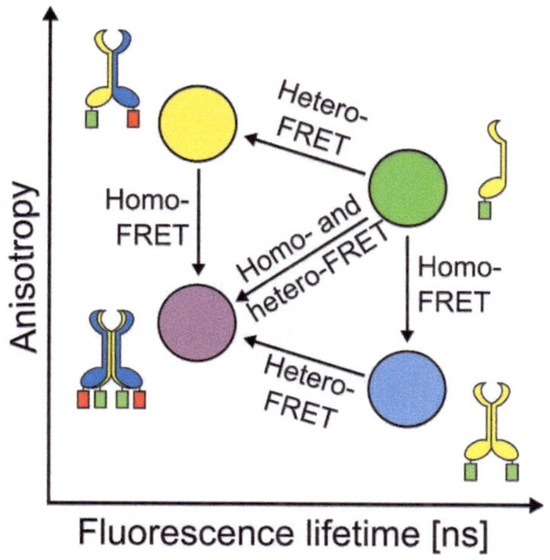

Fig. 3 Schematic of an ideal 2D MFIS plot. Non-interacting FRET fusion proteins (e.g., donor only, shown as a green rectangle, fused to a yellow receptor protein) will have relatively high fluorescence lifetimes (indicated by green circle). Hetero-FRET through heteromerization of FRET fusion proteins (shown as yellow and blue receptor proteins fused to donor, green rectangle, and acceptor, red rectangle, fluorophores) decreases fluorescence lifetime and increases fluorescence anisotropy (indicated by a yellow circle). Homo-FRET through homomerization of FRET fusion proteins decreases fluorescence anisotropy without affecting fluorescence lifetime (indicated by blue circle). If both homo-FRET and hetero-FRET are occurring, both the fluorescence lifetime and fluorescence anisotropy will decrease (shown by the purple circle)

cell-regulating RLKs ARABIDOPSIS CRINKLY 4 and CLAVATA1 appear to form complexes with differing stoichiometries at PD compared to at the bulk PM [11, 12]. Measuring changes in fluorescence lifetime heterogeneity over time can also indicate whether the composition or spatial distribution of membrane complexes changes upon signal perception. This is particularly useful when observing protein–protein interactions at PD, for example, where ligand perception can cause redistribution of PM receptor to PD, forming clusters of receptor complexes at PD [11, 13, 14].

Combining FRET-FLIM and fluorescence anisotropic decay and fluorescence lifetime heterogeneity measurements can therefore give insight into the nature, composition, distribution, and diversity of complexes formed by the fluorophore-fused POIs in near-endogenous conditions, which is particularly useful when observing protein–protein interactions at PD. Here we provide a detailed guide to perform FRET-FLIM and fluorescence anisotropy measurements to observe protein–protein interactions at PD in a transient expression system using *Nicotiana benthamiana*. This protocol can also be adapted to other plant species and subcellular compartments.

2 Materials

1. Vectors: Plant expression plasmids (*see* **Note 1**) harboring gene of interest fusions to coding sequences of appropriate fluorophores under the control of suitable promotors (*see* Subheading 3.1) and the p19 gene silencing suppressor for efficient transgene expression [15].
2. Plant material: 3–4-week-old *N. benthamiana* plants grown under greenhouse conditions (*see* **Note 2**).
3. Bacteria: *Agrobacterium tumefaciens* GV3101 pMP90 (or alternative strains).
4. Selective antibiotics (stock solutions in deionized water and sterile filtrated): 50 mg/mL rifampicin, 50 mg/mL gentamycin, and 50 mg/mL kanamycin and the antibiotic corresponding to the resistance of the expression plasmid (*see* **Note 3**).
5. *A. tumefaciens* growth medium ("double Yeast Tryptone," dYT): 1.6% w/v tryptone, 1% w/v yeast extract, and 0.5% w/v NaCl, solved in deionized water and autoclaved.
6. Infiltration medium (IM): 5% w/v sucrose, 150 μM acetosyringone (150 mM stock in DMSO), and 0.01% v/v Silwet, solved in deionized water.
7. Confocal laser scanning microscope (CLSM) equipped with pulsed lasers and precise photon counting electronics comprising:

   Pulsed laser source (e.g., Laser Diode Head (LDH)-Series from PicoQuant). Parallel and perpendicular detectors with single photon sensitivity (e.g., single photon avalanche diodes (τ-SPADs) from PicoQuant).

   Time-Correlated Single Photon Counting (TCSPC) unit (e.g., Hydra Harp 400 from PicoQuant).
8. Erythrosine B solution (according to PicoQuant instructions):

   1 mL of saturated aqueous solution of potassium iodide (KI).

   +0.17 mL of saturated aqueous solution of erythrosine B.

   +0.03 mL of 0.004 M KOH.
9. Microscope slides and coverslips (thickness #1.5).
10. Laser power meter (e.g., FieldMaxII from Coherent).
11. Optional:

    Induction solution: 10 μM ß-estradiol (10 mM stock in DMSO) and 0.1% v/v Tween-20.

    Aniline blue (water-soluble powder, use 0.01% w/v in 0.01 M K_3PO_4, pH 12 [16]).

3 Methods

3.1 Selection of FRET Pairs

1. Choose suitable FRET fluorophores for the POI to be analyzed (*see* **Notes 4** and **5**).
2. Test the in vivo functionality of a FRET fluorophore when fused to the POI (*see* **Note 6**).
3. Make sure that the subcellular localization of the chimeric protein is as expected before utilizing a POI-fluorophore fusion in FRET experiments. Colocalization with an aniline blue staining (*see* Subheading 3.3, **step 8**) can help to identify PD localization of a POI. Ideally, stable expression of the fusion construct should complement a mutant phenotype.
4. Include negative controls:
   (a) Fluorescence lifetime (hetero-FRET)—To control for specificity of protein–protein interactions, include a non-specific acceptor fluorophore fusion protein (*see* **Note 7**). The fluorescence lifetime of the POI-donor fusion should not be affected by the presence of the control-acceptor fusion.
   (b) Anisotropy (homo-FRET)—To determine the intrinsic anisotropic value of a FRET fusion protein, i.e., with no homo-FRET occurring, dilute the donor fusion protein with untagged POI by over-expressing the untagged POI (*see* **Note 8**). Alternatively, if the POI is not solely found at PD, the same POI-fluorophore fusion in another subcellular region, like the PM, can provide a reference value for anisotropy (*see* [12]).
5. Adjust protein levels by controlling the timing of gene expression and protein maturation. Even when driven by the same promotors, expression levels of different fusion constructs can vary. To synchronize two POI-fluorophore fusions, inducible promoter systems, e.g., estradiol induction [17], can be used.

3.2 Preparation of Infiltration Cultures

1. Transform *A. tumefaciens* GV3101 pMP90 with the p19 silencing suppressor plasmid and the appropriate plant expression vector. Select for positive transformants by plating on dYT plates containing the appropriate antibiotics and incubating at 28 °C for 48 h until distinct colonies form.
2. Select single colonies to inoculate 5 mL dYT supplemented with 50 μg/mL antibiotics and incubate overnight with shaking at 28 °C. Use 100 μL of this pre-culture to inoculate 50 mL dYT supplemented with antibiotics and again incubate overnight shaking at 28 °C (*see* **Note 9**).

3. Spin down the *A. tumefaciens* culture (10 min, 4000 × *g*, 4 °C) and remove supernatant by decanting and removing remainder with a pipette. Resuspend the pellet in 5 mL IM. Keep on ice.
4. Determine the optical density of the suspension at 600 nm (OD_{600}) using a spectrophotometer. The final OD_{600} of the suspension for infiltration should be 0.3 (*see* **Note 10**).
5. Incubate the *A. tumefaciens* cultures for 2–3 h at 4 °C.

3.3 Infiltration of N. benthamiana Leaves and Preparation of Microscope Slide

1. Select plants with leaf diameters of ~2–3 cm and strong venation. Plants should be well watered and kept at high humidity for several hours prior to infiltration to induce stomata opening.
2. Label each leaf to be infiltrated by attaching a label with thread. Up to two leaves per plant may be infiltrated.
3. Infiltrate the abaxial side of *N. benthamiana* leaves with the *A. tumefaciens* culture using a 1 mL syringe (with no needle) by pressing the syringe against the abaxial side and applying pressure against a gloved finger on the adaxial side of the leaf (*see* **Note 11**).
4. Incubate *N. benthamiana* plants for 2–5 days (21 °C, continuous light). The optimal maturation time of the individual protein fusions should be accessed prior to this to optimize the time between infiltration and analysis (*see* **Note 12**).
5. Dissect a small area of an infiltrated leaf (<1 cm^2 either using a razor or a leaf punch), avoiding veins to ensure the section is flat. Place the sample abaxial side up on a microscope slide mounted with deionized water and cover with a coverslip. Carefully press the sample to remove the air in the intercellular spaces of the leaf.
6. Optional: Counterstain callose of PD by infiltrating a 0.01% w/v aniline blue solution (dissolved in deionized water) into the leaf. After 30–60 min, dissect a leaf disc as described above, excite the sample at 405 nm, and detect the emission between 410 and 480 nm (*see* **Note 13**).

3.4 Measuring Hetero-FRET Via Lifetime Imaging

Here we describe how to measure and analyze fluorescence lifetime and anisotropy with instruments and software from PicoQuant that are integrated in the setup of the Zeiss CLSM 780. The measurements are controlled via both ZEN software from Zeiss and SymPhoTime64 software from PicoQuant. Prior to lifetime acquisition, the microscope settings need to be adjusted for the selected FRET pair. Furthermore, the system needs to be calibrated before starting a new experiment.

1. Choose the C-Apochromat 40×/1.20 water objective.
2. Remove the differential interference contrast (DIC) filter if it is plugged into the objective.
3. Select the excitation wavelength of the donor fluorophore in the SymPhoTime64 software (e.g., 485 nm for GFP/YFP derivates), choose a relative laser power of 60%, and deselect "cw," which allows pulsed instead of continuous wavelength (cw). Set the pulse frequency to 32 MHz.
4. Select an appropriate filter system, including a beam splitter, e.g., LP560 for GFP/YFP-derived donors with RFP-derived acceptors. This separates the emission light to two separate detectors (*see* **Note 14**). The fractionated emission light is further separated by band pass filters, e.g., for GFP derivate detection, use a 520/35 filter that allows wavelengths spread 35 nm around the peak value at 520 nm to pass.
5. Calibrate the attenuation of the pulsed laser (e.g., PicoQuant LDH-D-C-485 for GFP/YFP-derived donors) via adjustment wheels to reach a laser power of 1 μW. Use a laser power meter to determine the laser output of the system at the objective.
6. The instrument response function (IRF) is required to factor in the intrinsic error of the setup and can be estimated by measuring the short lifetime of an erythrosine B sample (*see* **Note 15**). Create a new experiment folder and a file for the IRF measurement in SymPhoTime64. Pipette 30 μL erythrosine B solution onto a coverslip and place on the objective, focus on the center of the drop, and perform a 120-s point measurement in the TCSPC mode. The peak of the resulting decay should not exceed 10^4 counts.

   For *in planta* lifetime measurements, select a ZEN configuration that enables the use of the external pulsed laser sources and the τ-SPAD detectors. Choose an appropriate major beam splitter, a picture size of 256 × 256 pixels, and a time-series acquisition mode (with 40–80 frames; *see* **Note 16**). We recommend a total $n = 30$ for each experimental condition acquired within three independent experiments (10 measurements per biological repetition).
7. Prepare a leaf sample (Subheading 3.3, **step 7**) and search for a cell area with moderate expression of the donor and (if co-expressed) acceptor at PD. Choose a region of interest (ROI), applying the same zoom factor during the entire experiment (we recommend 8× for PD measurements; *see* **Note 17**).
8. In SymPhoTime64, create a new file in the existing experiment folder. Switch from "point" to "image" measurement in the TCSPC mode and start the experiment. This activates the external detectors, which rest in a steady state until excitation is initiated in the next step.

9. Start the time series in ZEN, thereby triggering the pulsed laser source. Photon collection in the τ-SPAD detectors is monitored in SymPhoTime64 displaying, for example, count rates, pixel-wise lifetime previews, and lifetime decays. To avoid a pile-up effect, which is photon oversaturation of the detectors, the count rate should not exceed 1.6 mega counts/s (Mcps) at a pulse frequency of 32 MHz.
10. After the time series is completed, a control measurement of the acceptor is recommended to estimate its expression level in the corresponding cell. Since these data are not included in the lifetime analysis, a pulsed laser source is not necessary. A setting using a suitable internal Zeiss laser and a time series of few frames (e.g., 10) is sufficient.
11. Start again from **step 7** until the desired amounts of repetitions are reached. Each leaf sample can be used for ~5 single measurements in different areas.

3.5 Measuring Homo-FRET Via Anisotropy

To analyze the anisotropy of a sample, the parallel and perpendicular detectors of the POI-fluorophore fusion emission are compared. Photon acquisition is obtained as above (*see* Subheading 3.4), and the same data can be used for lifetime and anisotropy analysis. Anisotropy calculation via SymPhoTime64 (see below) requires values for the G-factor and the correction factors L1 and L2. These parameters are characterized by the imaging setup and thus need to be determined individually. They can be experimentally accessed as described previously [12, 18].

3.6 Data Analysis with SymPhoTime64

Both lifetime and anisotropy of the acquired raw data can be analyzed with the SymPhoTime64 software. Load the workspace with the individual measurements, choose one for the calculation, and select "Lifetime FRET Image" or "Anisotropy Image," respectively, in the "Analysis/Imaging" tab.

1. For lifetime calculation, select channels 1 and 2, which correspond to the parallel and perpendicular detector of donor emission.
2. Set a threshold of 100 counts/pixel and use the freehand ROI picker to select only PD. In a second analysis, you may compare ROIs of the same image excluding PD, if your POI displays differential subcellular localization.
3. Press "calculate fast FLIM" to update the ROI.
4. Load the IRF measurement.
5. Select the "Multi Exponential Donor" fitting model, the decay of your ROI, and the number of exponents of the fitting curve (depends on the donor fluorophore) and conduct an initial fit.

6. The initial fit must result in positive amplitudes and should be characterized by a Chi^2 close to 1.
7. Press "Fit" to apply the model to your data and export the results, including the intensity weighted lifetime given by τ_{AvInt}.
8. Repeat the analysis for all measurements and compare τ_{AvInt} values of different samples.
9. For anisotropy calculation, first assign the parallel and the perpendicular detectors, respectively, to the correct channels of the measurement.
10. Perform **step 2** also in this analysis workflow.
11. Insert the setup specific values for the G-factor and the correction factors L1 and L2 (*see* Subheading 3.5).
12. Press "Calc. Aniso" and save results.

4 Notes

1. We use GreenGate destination vectors [19], which comprise six modules that generally represent a promoter, N-terminal tag, coding sequence (i.e., the gene of interest), C-terminal tag, plant terminator, and plant resistance cassette flanked by left border (LB) and right border (RB) T-DNA elements in a plant transformation vector backbone. The plasmid contains the spectinomycin resistance gene and origins of replication for *E. coli* and *A. tumefaciens* for plasmid amplification and plant transformation, respectively. The *A. tumefaciens* strain used should additionally carry the pSOUP helper plasmid, which provides the replicase needed for replication in *A. tumefaciens.* The T-DNA fragment between the LB and RB will be inserted into the plant genome by *A. tumefaciens.* When using the GreenGate system, the two POIs can either be expressed in expression cassettes on separate T-DNAs and co-infiltrated into *N. benthamiana*, or the expression cassettes can be combined into one T-DNA [19].
2. Alternatively, any other plant tissue expressing the desired combination of fluorophore fusions at sufficient levels at PD can be used.
3. The *A. tumefaciens* GV3101 pMP90 strain carries resistance genes against rifampicin and gentamycin, while the p90 plasmid confers kanamycin resistance. If using the GreenGate system, the pSOUP helper plasmid confers tetracycline resistance.
4. The frequent description of novel fluorophore variants with improved features continuously increases the number of potential fluorophore combinations forming suitable FRET partners.

Online databases like the FPbase (www.fpbase.org) provide details of the physical properties of fluorophores that can help to estimate the general FRET efficiency between a donor and an acceptor fluorophore.

5. Select fluorophores that match the requirements of the microscope setup you are planning to use. For example, the available lasers, filters, and detectors should be able to provide the correct excitation wavelength for the donor and detect and differentiate fluorescence from both the donor and acceptor fluorophores.
6. Besides their overlapping emission (donor) and excitation (acceptor) ranges, many factors affecting energy transfer between two FRET fluorophores depend on the exact cellular and molecular context. This includes the position of the fluorophore tag on the POI (N- or C-terminal). PD-localized proteins are often PM integral proteins such as receptor-like kinases, i.e., they feature an extracellular and a cytosolic protein terminus. In these cases, the cytosolic side should be selected for fluorophore tagging because the cytosol provides a more stable environment for fluorophore activity. Furthermore, the length of a linker sequence, which can determine the flexibility of the fluorescent tag, and the relative size of the fluorophore in comparison to the POI must be considered. If the fluorophore has a relatively large molecular mass compared to the POI, it is likely that it will affect the POI's localization, activity, and interaction with other proteins.
7. When analyzing interactions between two POIs at PD, this could be a fusion protein of the acceptor fluorophore and a PD-localized protein, like PDLP1a [20].
8. The untagged POI outcompetes the POI-donor fusion protein for homomeric interactions, minimizing the probability that two POI-donor fusion proteins will interact via homo-FRET.
9. Alternatively, a fresh dYT agar plate supplemented with appropriate antibiotics can be inoculated with transformed *A. tumefaciens*, incubated for 48 h at 28 °C, and bacteria can be harvested by scraping off the cells with a spatula. This *A. tumefaciens* bacteria material can be resuspended in IM.
10. If the genes of interest are being expressed on a single T-DNA or only one construct is being transformed (e.g., for donor-only control), the suspension is adjusted to an OD_{600} of 0.3 with IM. When co-infiltrating with two different *A. tumefaciens* strains, adjust each suspension to an OD_{600} of 0.6 and mix 1:1 to obtain a final OD_{600} of 0.3.
11. Ensure the entire leaf is infiltrated to make it easier to select regions that have been infiltrated for microscopy. If resistance prevents infiltration of the leaf, move the syringe to a different

section. It is highly recommended to wear safety glasses for infiltration to avoid eye contamination with *A. tumefaciens*, as resistance can cause the suspension to spray back toward you. Use a new pair of gloves for each construct to avoid cross-contamination.

12. Inducible expression cassettes, like the estradiol-inducible XVE system [17], can help to precisely control timing of target protein availability. *N. benthamiana* leaves transformed with an estradiol-inducible promoter system must be sprayed with induction solution 6–16 h before imaging. Optimal induction time depends on the maturation rate of the analyzed protein and should be determined individually for each protein.
13. The spectral requirements of aniline blue detection overlap with the filter settings of blue fluorophores; thus, samples with derivates of cyan fluorescent protein (CFP) cannot be combined with aniline blue staining.
14. The long pass (LP) filter passes the fraction of the emission light above the indicated wavelength to one detector and reflects the wavelength below to another detector.
15. For blue donors, i.e., when using excitation of 440 nm, perform the IRF measurement with Atto425 dye diluted 1:5 in saturated KI.
16. More details of software settings for lifetime acquisition via ZEN/SymPhoTime64 have been described in a previous "Methods in Molecular Biology" article [8].
17. Avoid the presences of chloroplasts in the ROI, because their autofluorescence is detected in the lifetime measurements. Chloroplasts can be spotted by their emission in the far-red wavelength area (600–800 nm) after excitation at 561 nm.

Acknowledgments

This work was supported by the DFG through CRC1208, the IMPRS "Understanding complex plant traits using computational and evolutionary approaches," and CEPLAS (EXC2048). Patrick Blümke and Vicky Howe contributed equally to this work.

References

1. Sager RE, Lee J-Y (2018) Plasmodesmata at a glance. J Cell Sci 131:jcs209346. https://doi.org/10.1242/jcs.209346
2. Han X, Huang LJ, Feng D, Jiang W, Miu W, Li N (2019) Plasmodesmata-related structural and functional proteins: the long sought-after secrets of a cytoplasmic channel in plant cell walls. Int J Mol Sci 20(12):2946. https://doi.org/10.3390/ijms20122946
3. Petit JD, Li ZP, Nicolas WJ, Grison MS, Bayer EM (2020) Dare to change, the dynamics behind plasmodesmata-mediated cell-to-cell communication. Curr Opin Plant Biol 53:

80–89. https://doi.org/10.1016/j.pbi.2019.10.009

4. Förster T (1946) Energiewanderung und Fluoreszenz. Naturwissenschaften 33: 166–175. https://doi.org/10.1007/BF00585226
5. Förster T (1948) Zwischenmolekulare Energiewanderung und Fluoreszenz. Ann Phys 437:55–75. https://doi.org/10.1002/andp.19484370105
6. Clegg RM (1995) Fluorescence resonance energy transfer. Curr Opin Biotechnol 6: 103–110. https://doi.org/10.1016/0958-1669(95)80016-6
7. Denay G, Schultz P, Hänsch S, Weidtkamp-Peters S, Simon R (2019) Over the rainbow: a practical guide for fluorescent protein selection in plant FRET experiments. Plant Direct 3: 1–14. https://doi.org/10.1002/pld3.189
8. Somssich M, Simon R (2017) Studying protein-protein interactions in planta using advanced fluorescence microscopy. Methods Mol Biol 1610:267–285. https://doi.org/10.1007/978-1-4939-7003-2_17
9. Weidtkamp-Peters S, Felekyan S, Bleckmann A, Simon R, Becker W, Kühnemuth R, Seidel CAM (2009) Multiparameter fluorescence image spectroscopy to study molecular interactions. Photochem Photobiol Sci 8:470–480. https://doi.org/10.1039/b903245m
10. Bajar B, Wang E, Zhang S, Lin M, Chu J (2016) A guide to fluorescent protein FRET pairs. Sensors 16:1488. https://doi.org/10.3390/s16091488
11. Somssich M, Ma Q, Weidtkamp-Peters S, Stahl Y, Felekyan S, Bleckmann A, Seidel CAM, Simon R (2015) Real-time dynamics of peptide ligand-dependent receptor complex formation in planta. Sci Signal 8(388):ra76. https://doi.org/10.1126/scisignal.aab0598
12. Stahl Y, Grabowski S, Bleckmann A, Kühnemuth R, Weidtkamp-Peters S, Pinto KG, Kirschner GK, Schmid JB, Wink RH, Hülsewede A, Felekyan S, Seidel CAM, Simon R (2013) Moderation of arabidopsis root stemness by CLAVATA1 and ARABIDOPSIS CRINKLY4 receptor kinase complexes. Curr Biol 23:362–371. https://doi.org/10.1016/j.cub.2013.01.045
13. Grison MS, Kirk P, Brault ML, Wu XN, Schulze WX, Benitez-Alfonso Y, Immel F, Bayer EM (2019) Plasma membrane-associated receptor-like kinases relocalize to plasmodesmata in response to osmotic stress. Plant Physiol 181:142–160. https://doi.org/10.1104/pp.19.00473
14. Hunter K, Kimura S, Rokka A, Tran HC, Toyota M, Kukkonen JP, Wrzaczek M (2019) CRK2 enhances salt tolerance by regulating callose deposition in connection with PLDα1. Plant Physiol 180:2004–2021. https://doi.org/10.1104/pp.19.00560
15. Qu F, Morris TJ (2002) Efficient infection of Nicotiana benthamiana by tomato bushy stunt virus is facilitated by the coat protein and maintained by p19 through suppression of gene silencing. Mol Plant-Microbe Interact 15: 193–202. https://doi.org/10.1094/MPMI.2002.15.3.193
16. Zavaliev R, Epel BL (2015) Imaging Callose at Plasmodesmata using aniline blue: quantitative confocal microscopy. Methods Mol Biol 1217: 105–119. https://doi.org/10.1007/978-1-4939-1523-1_7
17. Zuo J, Niu QW, Chua NH (2000) An estrogen receptor-based transactivator XVE mediates highly inducible gene expression in transgenic plants. Plant J 24:265–273. https://doi.org/10.1046/j.1365-313X.2000.00868.x
18. Koshioka M, Sasaki K, Masuhara H (1995) Time-dependent fluorescence depolarization analysis in three-dimensional microspectroscopy. Appl Spectrosc 49:224–228
19. Lampropoulos A, Sutikovic Z, Wenzl C, Maegele I, Lohmann JU, Forner J (2013) GreenGate - a novel, versatile, and efficient cloning system for plant transgenesis. PLoS One 8(12):e83043. https://doi.org/10.1371/journal.pone.0083043
20. Thomas CL, Bayer EM, Ritzenthaler C, Fernandez-Calvino L, Maule AJ (2008) Specific targeting of a plasmodesmal protein affecting cell-to-cell communication. PLoS Biol 6: 0180–0190. https://doi.org/10.1371/journal.pbio.0060007

Chapter 15

Quantifying the Organization and Dynamics of the Plant Plasma Membrane Across Scales Using Light Microscopy

Joseph F. McKenna

Abstract

The plant cell surface continuum is composed of the cell wall, plasma membrane, and cytoskeleton. Plasmodesmata are specialized channels in the cell wall allowing intercellular communication and resource distribution. Proteins within these organelles play fundamental roles in development, perception of the external environment, and resource acquisition. Therefore, an understanding of protein dynamics and organization within the membrane and plasmodesmata is of fundamental importance to understanding both how plants develop as well as perceive the myriad of external stimuli they experience and initiate appropriate downstream responses. In this chapter, I will describe protocols for quantifying the dynamics and organization of the plasma membrane and plasmodesmata proteins across scales. The protocols described below allow researchers to determine bulk protein mobility within the membrane using fluorescence recovery after photobleaching (FRAP), imaging, and quantification of nanodomain size (with Airyscan confocal microscopy) and determining the dynamics of these nanodomains at the single particle level using total internal reflection (TIRF) single particle imaging.

Key words Plasma membrane (PM), Nanodomains, Fluorescence recovery after photobleaching (FRAP), Airyscan, Total internal reflection fluorescence (TIRF)

1 Introduction

The plasma membrane (PM) forms a barrier between the external surroundings and the cell, allowing the formation of an environment which is energetically favorable to biochemical processes. In addition to this, it is also the first point of contact of the cell with extracellular signals and stimuli. Proteins within the PM have fundamental roles in development via the perception and transport of developmental cues, perception of external stimuli, and resource acquisition. In plants, the PM exists as part of the cell surface continuum with the plant cell wall and the actin cytoskeleton [1]. Lipids and proteins confined within two-dimensional plane of the membrane are able to laterally diffuse via thermal agitation

Yoselin Benitez-Alfonso and Manfred Heinlein (eds.), *Plasmodesmata: Methods and Protocols*,
Methods in Molecular Biology, vol. 2457, https://doi.org/10.1007/978-1-0716-2132-5_15,

[2]. It was hypothesized that the PM was organized as a homogenous mix of proteins and lipids by the fluid mosaic model [3]. However, over the last few decades, this view has been challenged, and heterogeneous distribution of PM resident proteins has been described using a number of different imaging techniques in mammalian [4] and plant cells [5–9]. To describe protein dynamics and organization in the PM, a number of techniques spanning the scale of PM organization from bulk lateral diffusion to imaging and tracking individual nanodomains are required. These techniques can also be used to study plasmodesmata protein dynamics.

Different proteins within the PM show different lateral mobilities within the membrane [10] as shown by fluorescence recovery after photobleaching (FRAP). FRAP is a relatively simple but invaluable technique for determining the bulk lateral mobility of a fluorescently tagged protein within the membrane. Moreover, it is possible to perform FRAP experiments on all commercial point scanning confocal microscope systems without specialist technical knowledge or equipment. It allows determination of the mobile fraction of a protein, how much of the protein is able to diffuse freely laterally, and the speed at which the protein diffuses through the membrane. FRAP is also extremely useful if domains of proteins have been modified in order to determine their effect on lateral mobility. For example, FORMIN1 in *A. thaliana* shows increased lateral mobility when its cell wall binding domain is removed and the cytoskeleton is depolymerized, as determined by FRAP [11]. Therefore, it can be useful for determining how protein function affects lateral mobility. However, while FRAP is a very useful tool for determining the bulk mobility of a protein in the membrane, it does not allow us to determine the size and dynamics of individual protein domains within the PM.

Protein heterogeneity in larger PM microdomains has been demonstrated by the PIN-FORMED (PIN) [12] and Breaking of Symmetry in the Stomatal Lineage (BASL) [13] proteins, which show polar localizations within plant cells in microdomains, and their localization is important for normal plant development. Moreover, the Casparian strip membrane domain proteins (CASP) family of proteins show highly specific polar localization at the site of the soon to be formed Casparian strip in the root endodermis [14]. A microdomain is defined as a PM protein domain which is 1 μm or larger, and hence, these protein microdomains can be visualized using standard confocal microscopy [15, 16].

In addition to microdomains, several studies over the last decade have demonstrated the existence of PM nanodomains in plants. These are submicron protein assemblies are 20 nm to 1 μm in size [15, 16]. The best characterized nanodomain proteins are the remorin (REM) family which exist in non-overlapping membrane nanodomains and are resolvable using conventional confocal microscopy [7, 8]. Remorins interact with the PM via their unconventional lipid-binding C-terminal anchor domain which confers nanodomain organization [17]. During host cell infection in Medicago, SYMBIOTIC REMORIN 1 (SYMREM1) segregates the receptor LYSINE MOTIF KINASE 3 (LYK3) into nanodomains, demonstrating a molecular mechanism where REM nanodomains support signalling of extracellular stimuli [18].

The diffraction limit of microscopy for GFP is ~250 nm as described 100 years ago in the resolution equation by Ernst Abbe [19]. In order to image nanodomains smaller than this diffraction limit, super-resolution imaging techniques have been employed by a range of groups. High spatial resolution imaging of nanodomain organization within the membrane can be performed; however for some techniques, the frame rate of these is insufficient to determine the dynamics of nanodomains. Therefore, other techniques can focus on very quick image acquisition and increased temporal resolution.

Spatial organization of nanodomains can be determined using techniques based on confocal microscopy, which point scan over an area to generate an image. Stimulated emission depletion (STED) imaging has demonstrated that PIN proteins localize to nanodomains within their polarized membrane microdomain [9]. In addition, Airyscan imaging has recently been used to image the proteins BRASSINOSTEROID INSENSITIVE 1 (BRI1), PLASMA MEMBRANE INTRINSIC PROTEIN 2A (PIP2A), FLAGELLIN-SENSITIVE 2 (FLS2), and PIN3 and shows that these form differently sized nanodomains within the PM [5]. Airyscan sub-diffraction limited imaging has a lateral resolution (X, Y axis) of 120 nm and allows easy live-cell imaging with all confocal-compatible fluorescently tagged proteins without the need to use high depletion lasers or specialist fluorophores [20–22]. Previously, the frame rate (temporal resolution) of Airyscan was insufficient to allow tracking of nanodomain dynamics. However, the release of Airyscan fast and Airyscan2 allows significantly faster X, Y scanning and opens up the possibility for quantification of nanodomain dynamics in the future [23].

In order to determine the dynamics of nanodomains or single molecules in the membrane, the most readily used technique is total internal reflection fluorescence (TIRF) microscopy or variable-angle epifluorescence microscopy (VAEM) or variations on these techniques [24]. TIRF allows significant faster acquisition rates than point scanning techniques, and as such, it is widely used for determining protein dynamics in the membrane. TIRF microscopy has been show to function on a range of organelles in plants [25]. TIRF has been used to demonstrate that perturbation of the cell wall and actin cytoskeleton can alter the dynamics of the PM proteins FLS2 (which perceives pathogens) and PIN3 (which transports an important plant hormone) [5, 10]. In addition, the pathogen receptor FLS2 and steroid receptor BRI1 exist within distinct remorin nanodomains as demonstrated via VARIABLE ANGLE EPIFLUORESCENCE MICROSCOPY (VAEM) and confocal microscopy [6]. This might explain how different signalling receptors achieve distinct downstream signals when sharing common components.

Single particle tracking photoactivated localization microscopy (sptPALM) can be performed with TIRF and allows sub-diffraction limited imaging. sptPALM has been used to demonstrate that RHO-RELATED GTPASES FROM PLANTS 6 (ROP6) is stabilized into nanodomains which are required for auxin signalling [26]. Moreover, sptPALM has more recently been used to demonstrate that during hyperosmotic stimuli, ROP6 is recruited to nanodomains and interacts with two NADPH oxidases which generate ROS [27].

In the chapter, methods for good sample mounting will be detailed. This will ensure reduced thermal drift of samples being imaged and hence higher-quality data being produced for all the subsequent imaging methods described. In addition, FRAP determination of the bulk mobility of proteins and sub-diffraction limited Airyscan imaging and quantification of individual nanodomains will be described. Finally, a method is provided for determining the single particle dynamics of nanodomains using TIRF microscopy. The techniques described will allow a detailed determination of protein dynamics and organization across scales within the PM. Moreover, these techniques (with the exception of single particle tracking TIRF microscopy) can be also used to determine plasmodesmata protein organization and dynamics.

2 Materials

2.1 Equipment

A confocal microscope capable of FRAP and Airyscan and a TIRF-equipped inverted microscope (*see* **Note 1**).

2.2 Microscopy Consumables

1. Standard glass microscope slides that fit the stage used.
2. No. 1.5 coverslips (*see* **Note 2**).
3. 3M Micropore tape.
4. Micropipettes set and plastic tips.

2.3 Image Analysis Tools

1. FIJI implementation of ImageJ (v1.53g) [28].
2. GraphPad Prism (version 9 used here, but earlier versions perform curve fitting just as well).
3. Zen Black and/or Blue with Airyscan processing enabled dongle.

4. Imperial College London Facility for Imaging by Light Microscopy FIJI Macro package (https://www.imperial.ac.uk/medicine/facility-for-imaging-by-light-microscopy/software/fiji/).
5. Line profile and FWHM Macro by John Lim, IMB (http://imagej.1557.x6.nabble.com/FWHM-on-line-selection-td5004777.html).
6. Trackmate (built into FIJI) [29].
7. TrajClassifier for classification of single particle tracks [30].

2.4 Plant Samples

1. *Arabidopsis* seedlings were grown on 0.5× Murashige and Skoog medium with 4-Morpholineethanesulfonic acid, 2-(*N*-Morpholino)ethanesulfonic acid (MES), and 0.8% Phytoagar in 16:8 light/dark cycle for 5 days.
2. *Arabidopsis* 5-day-old seedlings stably expressing a plasma membrane protein GFP fusion were used for all example data shown here. The stable lines used here (p35S::GFP-Lti6B and pFLS2::FLS2-GFP) have been previously described [31, 32]. However, identical experiments have been performed for transient expression of proteins tagged with numerous fluorescent proteins in both *N. tabacum* and *N. benthamiana* and over a range of plant ages.

3 Methods

3.1 Sample Mounting

For all the described techniques which follow, good mounting of the sample is fundamentally important irrespective of the age or complexity of the equipment being used. Sample and thermal drift are extremely detrimental to good data collection for PM proteins, especially when performing high-resolution time lapse imaging. Therefore, there are a few notes which help reduce sample and thermal drift of the microscope system. This procedure was developed from hundreds of trial and error tests and is the best found (Fig. 1). While there are some methods in the literature which correct for linear fluorescence excitation bleaching, significant *X*, *Y*, or *Z* sample drift will make the experimental data useless, and if this occurs, it is best end the current data acquisition and move to a new area/sample.

1. Before mounting the sample, ensure that the air-conditioning in the microscope room has been on for at least an hour (*see* **Note 3**).
2. Add 100 μl of dH_2O (or growth medium without Phytoagar) to the microscope slide using a p200 pipette.

Fig. 1 Sample mounting. (**a**) One side of the coverslip is placed flat on the microscope slide and fixed from moving by holding one finger toward it (white arrowhead). The other side of the coverslip is raised by the forceps so the coverslip and sample are not touching. By sliding out the forceps (in the direction shown, black arrowhead), the coverslip is gently lowered until it touches the sample and the forceps are removed. (**b**) 3M Micropore tap is wrapped over the coverslip and microscope slide, pulled tight, and wrapped around the back of the slide

3. Use fine forceps to remove the *Arabidopsis* seedling from the growing medium and place it onto the microscope slide within the dH_2O droplet.
4. Add a small bead of water to the center of the coverslip. Place one side of the coverslip on the slide and the other against the forceps to keep it elevated (Fig. 1a). Slide out slowly the forceps to allow the raised side of the coverslip to gently lower onto the sample to reduce air bubbles. If air bubbles occur, slide the forceps back underneath the coverslip and repeat.
5. Wrap micropore tape over the coverslip tightly around the back of the slide on both sides (Fig. 1b); this significantly reduces sample drift (*see* **Note 4**). However, be careful not to crush the sample.
6. Place the sample on the stage holder of the microscope, raise the oil immersion lens, and focus. If time permits, leave the sample on the microscope for 5–10 min before imaging to reduce sample drift further.

3.2 Fluorescence Recovery After Photobleaching (FRAP)

In order to determine the bulk mobility of proteins within the plasma membrane, fluorescence recovery after photobleaching (FRAP) is a readily available technique which can be performed on all commercially available point scanning confocal systems. The technique relies on bleaching all fluorophores in a defined region and then determining the recovery of fluorescence over a subsequent time series as freely diffusing proteins move into the

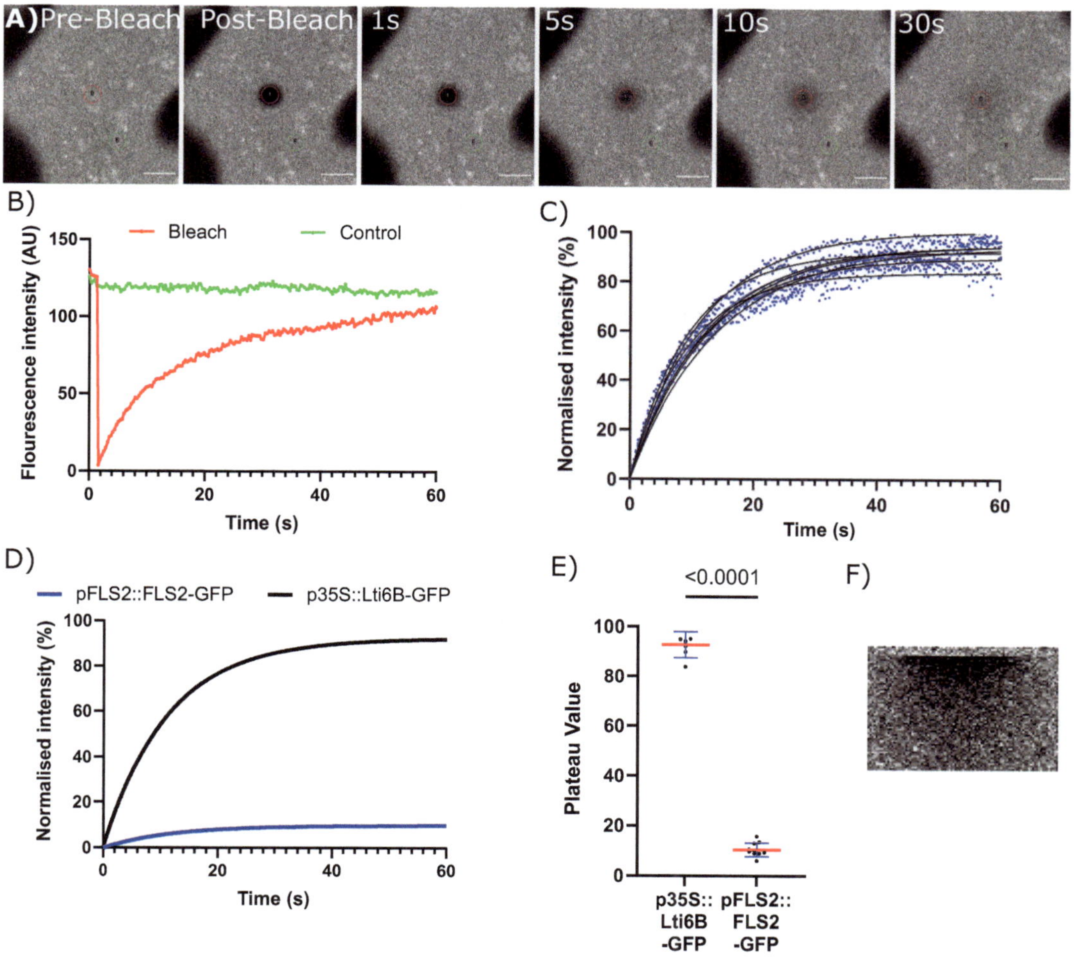

Fig. 2 Fluorescence recovery after photobleaching (FRAP). (**a**) Confocal images from FRAP time course of p*35S*::GFP-Lti6B. The bleached region is in red and control is in green. Representative images post- and pre-bleach, 1, 5, 10, and 30 s after bleaching. Scale bar denotes 5 μm. (**b**) Representative raw ROI fluorescence intensity data from FRAP of p*35S*::GFP-Lti6B shown in (**a**). Red line shows bleached region and green control. Note little change in control intensity which indicates that fluorophore bleaching, sample, and thermal drift are all low during imaging. (**c**) Normalized intensity plots of one-phase association curve fits of p*35S*::GFP-Lti6B. Black lines show curve fits from individual bleaches; blue dots show cloud of data points. (**d**) FRAP recovery curves from all data sets for p*FLS2*::FLS2-GFP (blue) and p35S::Lti6B-GFP (green). (**e**) Box plot of plateau values for p*FLS2*::FLS2-GFP and p*35S*::GFP-Lti6B showing that Lti6B has higher mobile fraction and FLS2 higher immobile fraction. Red lines show mean and blue error bars (standard deviation). Statistical significance determined by students T-test. (**f**) Example kymograph of p*35S*::GFP-Lti6b showing recovery post-bleach

bleach region (Fig. 2). The monitoring of the diffusion of proteins into the bleached region tells us about the bulk movement of proteins within the membrane.

1. Use 488 nm laser excitation and collect emission from 500 to 550 nm with a pinhole of 1AU for GFP imaging (*see* **Note 5**).

If imaging samples with multiple fluorophores, check that no fluorescence emission bleed-through occurs.

2. Set the digital zoom to 5×, frame size to 512 × 512 pixels, and a scan time of 0.316 s (3 frames per second) (*see* **Notes 6** and 7).
3. Define a circular region of interest in the center of the image (Fig. 2a) keeping consistent the size between experiments; in Fig. 2 a circle of 50 pixel diameter is used (*see* **Note 8**).
4. Set the bleaching experiment as follows: five pre-bleach scans to determine the pre-bleach average and a time course for 240 frames which can be trimmed to allow recovery 60 s post-bleach (*see* **Note 9**). Bleach the ROI using the 488 nm laser set to 100% intensity for 10 iterations.
5. Use FIJI (ImageJ) (*see* **Note 10**) to extract fluorescence recovery data; an Region of Interest (ROI) is selected based on the bleach and control areas (Fig. 2a and b) (*see* **Note 11**). The intensity is plotted over time using the ROI intensity time plot macro (*see* **Note 12**) and saved into Microsoft Excel.
6. Before curve fitting, normalize the fluorescence intensity data using the equation: $I_n = [(I_t - I_{min})/(I_{max} - I_{min})] \times 100$ in Microsoft Excel (*see* **Note 13**).

   where I_n is the normalized intensity, I_t is the intensity at any time point, I_{min} is the minimum (initial post-bleach) intensity, and I_{max} is the average pre-bleach maximum intensity (determined from five pre-bleach intervals taken).
7. Copy normalized data into GraphPad Prism and perform non-linear regression curve fitting using one- or two-phase association. Select Analyze > *XY* Analysis > Non-linear regression (curve fit). In model select one-phase (or two-phase) association. Individual curves for each FRAP experiment (Fig. 2c) and a single combined curve of all experiments (Fig. 2d) should be produced (*see* **Note 14**). In constrain set Y0 constant equal to 0. For individual curves, leave all other constraints. To generate a combined curve (Fig. 2d), select shared value for all datasets for all values (except Y0).
8. Key curve characteristics from the initial non-constrained individual FRAP curve fits can then be extracted. Click on the results sheet in GraphPad. These include the plateau values, $t_{1/2}$, and R^2 fit of the curves (Fig. 2e) and can be used for subsequent statistical analysis (*see* **Note 15**).
9. In order to visualize FRAP recovery from an example dataset, produce a kymograph (Fig. 2f). In FIJI draw a line profile over the bleached region and then Analyze > Multi Kymograph > Multi Kymograph (*see* **Note 16**).

3.3 Airyscan Sub-Diffraction Limited Imaging and Quantification of Nanodomain Size

While FRAP determination of bulk protein remodelling in a membrane is an extremely useful measure of protein dynamics, more recently proteins within the PM have been shown to exist in nanodomains. While some of the nanodomains exist in sizes above the optical diffraction limit (for GFP >250 nm) such as the remorins, a large number exist in nanodomains smaller than the diffraction limit of light, and therefore super-resolution microscopy techniques are required to image them [5]. A large number of techniques have been used, such as STED and STORM, but here I focus on Airyscan, a technique developed by Carl Zeiss [20]. Airyscan imaging functions by not using a standard confocal pinhole and single detector but allowing fluorescence emission to pass onto a specialized 32-channel circular arranged detector. Each detector element has a fixed pinhole of 0.2 Airy units, but the combined detector (sum of all elements) is 1.25 AU. Collected data is then processed with a Wiener filter-based deconvolution and pixel reassignment [20, 22]. This culminates in sub-diffraction limited imaging with increased signal to noise when compared to standard confocal. This technique has a lateral resolution limit (X, Y) of 120 nm. It is compatible with live-cell imaging of any confocal-compatible fluorescent protein line already produced and imaged.

1. For imaging GFP, use the 488 nm laser and a 495–-550BP + 570 nm LP duel filter preceding the Airyscan detector and use the SP620 filter preceding the band-pass filter (*see* **Note 17**).
2. Use the highest numerical aperture lens available (*see* **Note 18**). Set the optical zoom to be above 1.8, the pinhole open above 1AU, and the frame size set to optimal which will depend on the objective lens and optical zoom used (*see* **Note 19**).
3. Ensure the detector is not saturated by using the range indicator (located on the display tab) and setting up appropriate master gain and laser levels (*see* **Note 20**).
4. Focus the sample on a flat membrane sheet with a 5× optical zoom (Fig. 3a for comparison of confocal and Airyscan imaging on the same area using the same acquisition excitation and emission) (*see* **Note 21**).
5. Always monitor the detector view when performing Airyscan imaging. Perfect data will show highest intensity in the central detector and then decreasing intensity radiating outward; if this is not observed, increase the gain or laser power or level the stage (*see* **Note 22**).
6. Airyscan images must be processed pre-analysis, using Airyscan processing on the process tab (*see* **Note 23**).
7. Open processed Airyscan files in FIJI. There are many more nanodomains that can be quantified using the approach

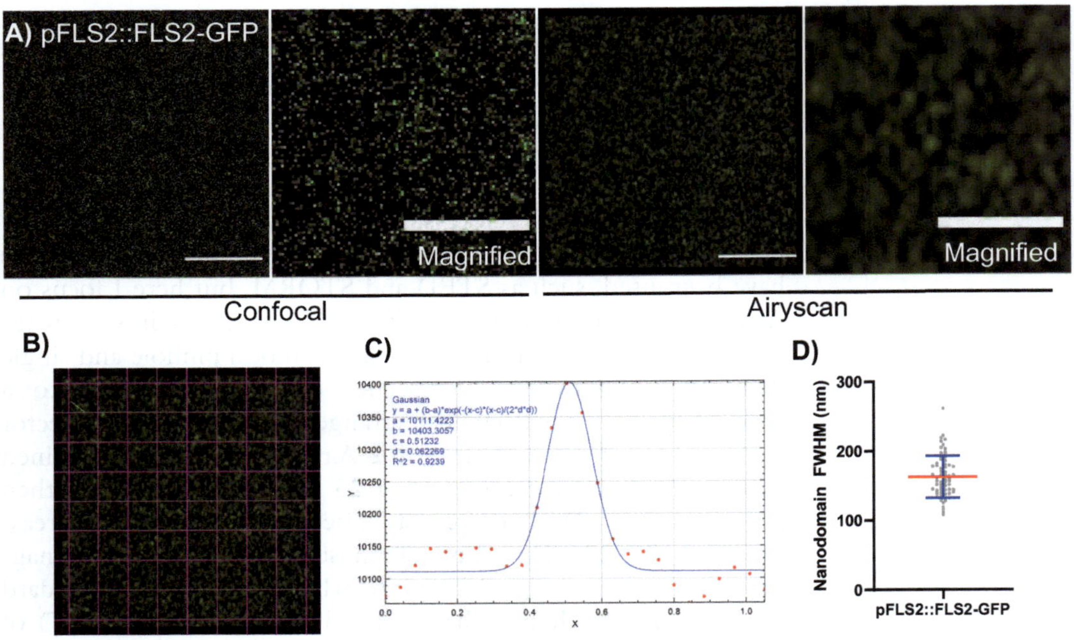

Fig. 3 Airyscan imaging of PM nanodomains and size quantification. (**a**) Comparison of confocal and Airyscan imaging of p*FLS2*::FLS2-GFP. Nanodomains clearly resolvable with Airyscan vs. confocal which shows high levels of noise and no resolvable objects. Note that confocal and Airyscan data are taken at the same time, with same laser power and gain settings on the same sample, therefore demonstrating capacity of the Airyscan system. Scale bars = 5 μm, magnified imaged scale bars = 2 μm. (**b**) Example grid over Airyscan data. Green line in first purple cell is an example 1 μm line over nanodomain used for subsequent quantification. (**c**) Gaussian curve fitted to nanodomain line profiles shown in (**b**). (**d**) Box plot of FWHM of multiple p*FLS2*::FLS2-GFP nanodomains shown in (**b**). Redline = mean (163.3 nm), blue error bars = standard deviation (30.59 nm), $N = 64$ for example dataset

described; therefore draw a grid (Analyze > Tools > Grid), and one full width half maximum (FWHM) line profile of a nanodomain from each square in the grid is performed (Fig. 3b).

8. Produce a full width half maximum (FWHM) measurement over a single nanodomain using FIJI. A line (1 μm in length) is drawn over the nanodomain, its profile plotted, and a Gaussian curve fitted to it; use Analyze > Plot profile (or Ctrl+K). The distance and intensity values should be copied. A Gaussian curve can then be fitted using Analyze > Tools > Curve fitting, selecting Gaussian and inputting the plot values obtained from plot profile. This can then be used to manually determine the FWHM (Fig. 3c) (*see* **Note 24**).

9. Alternatively, use available ImageJ plugins and macros, which can automate the process in **step 8**. The one used here is from John Lim of the IMB (Fig. 2c). This automatically plots the line and curve fits and determines the FWHM in a separate pop-up screen (*see* **Note 25**).

10. Process at least 64 full width half maximum measurements per Airyscan image and 30 images in total across three experimental repeats. Collate FWHM measurements in Excel and convert into nm (divided by 1000), copy into GraphPad, and produce a scatter dot plot showing nanodomain size for a given protein (Fig. 3d).

3.4 Total Internal Reflection Fluorescence Imaging of Particle Dynamics

Total internal reflection fluorescence (TIRF) is a technique which can produce significantly high temporal resolution of PM proteins with enhanced signal to noise ratio. To perform TIRF, a laser is reflected off a coverslip at a critical angle; a small amount of laser energy produces an evanescent wave which passes about 100 nm into the sample, exciting fluorescent proteins only within this wave. Therefore, this technique has significant axial (Z) resolution perfectly adapted for imaging PM proteins. Because of the high signal to noise ratio, these particles can be tracked using several different methods. This is coupled with extremely fast acquisition using EM CCD cameras. These features allow single particle tracking to be performed (Fig. 4).

1. Mount samples as described in Subheading 3.1 and place onto the TIRF microscope.
2. In order to achieve TIRF, adjust the angle of the laser away from epifluorescence until the critical angle is reached. This is at the point when the signal ratio is highest to the background (*see* **Note 26**).
3. Focus on the PM, and blinking molecules will be observed (*see* **Note 27**, Fig. 4). Ensure the laser is not too powerful and rapid bleaching is not occurring. The high signal to noise ratio is required for subsequent tracking of molecules (*see* **Note 28**).
4. Acquire a time course, set gain values for the camera to maximum and at least 20 frames per second (0.05 s) acquisition is required for 60 s (Fig. 4).
5. In FIJI, select Plugins > Tracking > Trackmate. Trackmate which is built into FIJI is an excellent option for image analysis (*see* **Note 29**) [29].
6. Select LoG detector to identify spots; estimated blob diameter is determined by increasing slightly the FWHM size determined previously (in Airyscan imaging), so here 0.2 μm (200 nm). Thresholding will enhance the quality of spots identified and is set to 50. The identified spots can then be previewed. Thresholding of spots should be performed based on their "quality" (deviation for input spot size).
7. For tracking, use the hyperstack displayer with no filter on spots selected. To track spots over time, select a Simple LAP tracker with a linking max distance of 0.3 μm and no gap

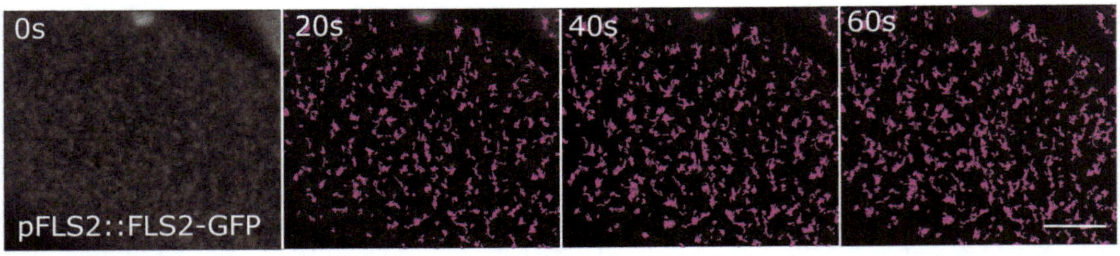

Fig. 4 Example of p*FLS2*::FLS2-GFP single particle movement acquisition and tracking. p*FLS2*::FLS2-GFP nanodomain movement collected by TIRF imaging over 0, 20, 40, and 60 s. Purple lines indicate particle tracks over time. Scale bar denotes 5 μm

closing. Once tracked, filter tracks by duration of tracks or number of spots. These should be set to 0.5 s or 10 s, respectively, and ensure that only spots tracked over 10 time points are taken further.

8. For visualizing tracks, set track display mode to show local tracks, backward, and set color by duration of track. To extract individual track data, click on analysis. On the final screen, an overlay video can be made; the Trackmate file should be saved (this will be a large file). For subsequent analysis, export tracks in an .xml file from the final page of the plugin.
9. After tracking is performed in Trackmate, use TrajClassifier [30] to classify tracks into different diffusion models and provide overlay videos. Open the exported tracks XML file (not the Trackmate saved file). Overview data tables and a video will be produced.

4 Notes

1. For the examples listed here, a confocal microscope is required. The data presented was captured on a Zeiss LSM880 system with Airyscan. However, all commercially available point scanning confocal microscope systems will be able to perform FRAP. For super-resolution, we describe acquisition using the Zeiss Airyscan system. However, the analysis of nanodomain size can be used on data collected from several sub-diffraction limited imaging techniques. A high numerical aperture (at least 1.4) objective of 63× or 100× is required for high-resolution imaging described. The data presented was collected on a Zeiss αPlan-Apochromat 100× 1.46NA DIC M27 Elyra lens, which is preferred for high-resolution imaging. The TIRF microscope was configured using an AxioObserver stand (Zeiss) equipped with 100x objective (Zeiss αPlan-Apochromat 100× 1.46NA DIC M27 Elyra lens) and a manual TIRF slider (Zeiss) with a 488 nm laser for excitation. Fluorescence emission was

collected using a Chroma HQ525/50 filter, and an EMCCD camera was used (iXon+, Andor).

2. Any number 1.5 coverslip can be used; however larger coverslips (24 × 50 mm) seem to reduce sample drift more than smaller. This is presumed to be due to increased surface tension of the surface area of the coverslip and water with the microscope slide.
3. Changes in air temperature when microscope systems are first used can significantly increase thermal drift of the microscope. If an on/off air-conditioning is used, leave on overnight before imaging to ensure the room is at a stable temperature. As a minimum, the microscope stand can take an hour to equilibrate to a different room temperature. A temperature chamber on the stage of the microscope is extremely beneficial to reduce thermal drift. It insulates the stage from air currents and if possible should be included on future microscope purchases if high-resolution time-lapse imaging is to be performed. The temperature can be set to that of the growth chamber (22 °C).
4. 3 M Micropore tape was found to be the easiest, quickest, and most versatile tape to hold coverslips and microscope slides together tightly and reduce sample drift. However, if longer-term imaging is required or thicker samples are to be used, double-sided tape can be used to stick coverslips down in slides. The benefit of double-sided tape is that its thickness protects the sample from mechanical stress. Also, if long-term imaging is required, Valap can be coated around the edge of the slide, significantly reducing water evaporation and sample drift. Valap is produced by melting equal volumes of Vaseline, lanolin, and paraffin together. Once made it can be reheated many times. To reheat it is best to use a hot plate. To apply melted Valap to slides, the back of a small spatula can be dragged along the seams of the coverslip, and it sets within 10 s.
5. A pinhole of 1AU is optimal for most imaging experiments; however with FRAP we are not concerned with spatial resolution but more temporal. Therefore, a larger pinhole size can be used if weak fluorescence occurs.
6. Example data collected here was generated from cotyledon cells; flat membrane sheets are imaged by adjusting the focus until only the top membrane of cells is visible and then using a 5× optical zoom (Fig. 2a). If imaging *A. thaliana* cotyledon cells, pavement cells immediately above the petioles appear more stable and show less sample drift; presumably the underlying tissue is thicker here.
7. For FRAP experiments, temporal resolution trumps spatial. Therefore, the frame size can be reduced (e.g., 256 × 256) to increase scan speed, and bidirectional scanning can be used.

The aim is not to generate a pretty picture but an accurate graph.

8. Use the pixel size from the image setup (look it up in the image metadata using the vendor's software if away from the microscope), to determine the area in μm of the bleached region.
9. All bleach settings (laser iterations, ROI, etc.) should be determined before the experiment begins. Try performing a time course on an area without bleaching to ensure there is no sample drift/the laser is not bleaching the sample during normal acquisition (Fig. 2b, green line). A smaller ROI can reduce the bleach energy and sample drift. The laser intensity required for bleaching should also be determined pre-experiment. It is best to perform as few iterations as possible with 100% laser intensity. If using a white light laser, multiple bleaching wavelengths can be used in close wavelength proximity to each other to provide enough energy for bleaching to occur. The time course you set up will be dependent on the dynamics of the protein you are imaging. If imaging a fast diffusing protein such as Lti6B, constant imaging can be used. However, if imaging a slower diffusing protein such as AGP4, an interval of 1.5 s or longer can be added. This will reduce the laser power being put into the sample and hence sample/thermal drift over long time periods.
10. If using Zeiss Zen software, or other manufacturers' software, this can be readily done by copying the intensity ROI data directly from the acquisition software.
11. ROIs can then be added to the ROI manager (Analyze > Tools > ROI manager), and the ROI can be saved so it can easily be used for analysis of subsequent data files. An additional ROI of the same size can be used as a control in a non-bleached region.
12. The Imperial College Facility for Imaging by Light Microscopy Multi ROI Intensity time plot macro (located in the Intensity suite of tools) is then used to plot the intensity in the bleached/control ROI over time (Fig. 2b). This also produces a data table which can be saved as a .csv file and opened in Excel for subsequent analysis. Irrespective of the method used to extract the ROI intensity data over time, the outcome should be an Excel file containing the acquisition frame interval (in seconds) in the first column and each experimental repeat bleached ROI intensity data in each subsequent column.
13. If analyzed in an Excel file with the original ROI intensity data in a separate tab, data normalization for all datasets can be rapidly performed by dragging the equation over multiple datasets from the linked sheets after ensuring that appropriate row values are locked using the F4 key in Excel. To get the correct time series (in seconds), the time required for the first

five pre-bleach scans must be subtracted from all subsequent time points. Therefore, the first postbleach scan (time point 6) is 0 s.

14. Once in prism, a one- or two-phase association curve fitting can be performed with the Y0 value constrained to 0 (as we have normalized all data to an initial % fluorescence recovery value of 0). The type of curve fitting performed depends on the type of data collected. A one-phase association was used here (Fig. 2). The curve fits for each time course can then be drawn as separate curves on a graph. It can be useful to additionally include the individual data points as a "cloud" around the curves to show the spread of the combined datasets (Fig. 2c, blue dots). In order to produce a single curve representative of all datasets, perform another nonlinear regression curve fit with Y0 set to 0 and all other values set to shared value for all datasets in the constrain tab. This will produce the single standard FRAP recovery curve for all of the datasets (Fig. 2d).
15. These can be copied from the results tab in GraphPad and pasted into a new data table and compared to other experimental conditions (Fig. 2e). The plateau values indicate how mobile the protein is in the PM; a higher plateau demonstrates that more protein has diffused into the bleach ROI.
16. A kymograph is a line profile taken through all time points in the time series and can be used to visualize FRAP recovery of the line over time. Remember, if doing a lot of these, save the line profiles as an ROI in ROI manager.
17. This ensures that GFP emission passes through but can also be used for line switching if dual GFP/RFP imaging is being performed. Always use the highest numerical aperture objective available (either 63 or 100×). The 620SP filter will block chlorophyll autofluorescence from entering the Airyscan detector and should always be used.
18. I have had excellent results with αPlan-Apochromat 100× 1.46NA DIC M27 Elyra lens (420792-9800-720). Ensure that any DIC prisms below the objective are removed.
19. The optimal pixel size is linked to the area which is being imaged; therefore if acquisition rate is important, optical zooming as much as possible will reduce the frame size required allowing faster acquisition. If fast temporal acquisition is required, use a letterbox frame size (reduce the Y frame by half, e.g., 512 × 256).
20. It should be noted that while the Airyscan system can still produce impressive results with low signal, remember the golden rule "photons are the currency of microscopy," and as such the strongest (non-saturated) signal will produce the best results.

21. It can help accuracy to use a piezo stage when performing high-resolution imaging.
22. Ensure that the Airyscan detector is set to continually calibrate in the Zen settings. If a bright sample is being used and the detector still does not look aligned, try calibrating the stage to ensure it is perpendicular to the objective.
23. Either this can be done individually or if many data files have been collected, a batch process can be set up and run at the end of the acquisition. For LSM880 and LSM800, the standard Airyscan processing has a resolution limit of 140 nm. To get to 120 nm, process in Zen Blue and tick the 2D-SR mode. Also note, the wiener filter processing strength can be increased above the automatic level. This can increase *X*, *Y* resolution but should be performed with care; too strong processing creates substantial image artifacts [22]. I always image with the automatic processing strength.
24. Per grid for nanodomain quantification, it is best to draw a line profile on the most clearly defined nanodomain. After analysis, drawing the line into the data (Ctrl+D) can help identify which nanodomain was quantified later. It is always best to save this modified copy separate to keep the original as well.
25. Once the FWHM macro is installed, setting a shortcut key (such as F4) allows rapid determination of nanodomain size when processing many images.
26. While the initial adjustment of TIRF angle at the start of a session might take some time. Subsequent samples should be near TIRF when placed on the microscope, and hence imaging will be much quicker.
27. It is worth noting that the cell wall does appear to show some autofluorescence as does any dust on the coverslip or unidentified fluorescing objects (UFOs) in the mounting solution (dH_2O or liquid growth media). Due to the self-thresholding nature of CCD cameras, this can be troublesome when the PM protein is expressed at a low level. Therefore it is important to ensure you are focused on the PM. A good tip is if the dots are blinking and moving.
28. Very high expression levels of fluorescently tagged proteins can perturb subsequent particle tracking due to crowding. This can be solved by either using a promoter which expresses at a moderate level (e.g., pFLS2 or pPIN3) or performing overexpression (with p35S) but using a paGFP fluorescent tag instead of regular GFP. A small population of paGFP is activated and fluorescing without 405 nm laser activation. Therefore, this can be observed and tracked without saturating the camera or crowding subsequent analysis.

29. I perform TIRF imaging at the Central Laser Facility at the Rutherford Appleton laboratories in Harwell, and they have an in-house analysis system designed for data acquired on site [33]. However, if TIRF data is acquired elsewhere, use Trackmate. If using Trackmate, it is a good idea to calibrate it appropriately by cropping TIRF data and having a small area to initially track with only a few hundred frames. This can help you determine the appropriate spot size, threshold, and track length to use without using large amounts of computer time processing large datasets before optimum quantification parameters have been determined. But for every type of different microscope technique used or different protein tracked, the parameters described below will need to be modified. Settings presented here are a good starting point. A benefit of Trackmate is you can easily go backward and forward between steps with ease, modifying parameters until you get appropriate identification of spots and tracks.

Acknowledgments

I would like to acknowledge Oxford Brookes University for funding the research fellowship which enabled this work to be produced. In particular, I would like to thank David Evans and John Runions of Oxford Brookes University for tireless support during my time at Oxford Brookes. The Bioimaging Center at Oxford Brookes generously provided the time on the confocal system to gather the examples presented. I would also like to acknowledge the STFC-funded Central Laser Facility at Harwell and Dan Rolfe for help with the TIRF imaging.

References

1. McKenna JF, Tolmie AF, Runions J (2014) Across the great divide: the plant cell surface continuum. Curr Opin Plant Biol 22:132–140. https://doi.org/10.1016/j.pbi.2014.11.004
2. Martinière A, Runions J (2013) Protein diffusion in plant cell plasma membranes: the cell-wall corral. Front Plant Sci 4:515. https://doi.org/10.3389/fpls.2013.00515
3. Singer SJ, Nicolson GL (1972) The fluid mosaic model of the structure of cell membranes. Science 175:720–731. https://doi.org/10.1126/science.175.4023.720
4. Bernardino de la Serna J, Schütz GJ, Eggeling C, Cebecauer M (2016) There is no simple model of the plasma membrane organization. Front Cell Dev Biol 4:106. https://doi.org/10.3389/fcell.2016.00106
5. McKenna JF, Rolfe DJ, Webb SED et al (2019) The cell wall regulates dynamics and size of plasma-membrane nanodomains in *Arabidopsis*. Proc Natl Acad Sci U S A 116:12857–12862. https://doi.org/10.1073/pnas.1819077116
6. Bücherl CA, Jarsch IK, Schudoma C et al (2017) Plant immune and growth receptors share common signalling components but localise to distinct plasma membrane nanodomains. eLife 6:e25114. https://doi.org/10.7554/eLife.25114
7. Jarsch IK, Konrad SSA, Stratil TF et al (2014) Plasma membranes are subcompartmentalized into a plethora of coexisting and diverse microdomains in *Arabidopsis* and *Nicotiana benthamiana*. Plant Cell 26:1698–1711. https://doi.org/10.1105/tpc.114.124446

8. Raffaele S, Bayer E, Lafarge D et al (2009) Remorin, a solanaceae protein resident in membrane rafts and plasmodesmata, impairs *potato virus x* movement. Plant Cell 21:1541–1555. https://doi.org/10.1105/tpc.108.064279
9. Kleine-Vehn J, Wabnik K, Martinière A et al (2011) Recycling, clustering, and endocytosis jointly maintain PIN auxin carrier polarity at the plasma membrane. Mol Syst Biol 7:540. https://doi.org/10.1038/msb.2011.72
10. Martiniere A, Lavagi I, Nageswaran G et al (2012) Cell wall constrains lateral diffusion of plant plasma-membrane proteins. Proc Natl Acad Sci U S A 109:12805–12810. https://doi.org/10.1073/pnas.1202040109
11. Martinière A, Gayral P, Hawes C, Runions J (2011) Building bridges: formin1 of Arabidopsis forms a connection between the cell wall and the actin cytoskeleton: AtFH1 links the cell wall and actin cytoskeleton. Plant J 66:354–365. https://doi.org/10.1111/j.1365-313X.2011.04497.x
12. Galweiler L, Guan C, Muller A et al (1998) Regulation of polar auxin transport by AtPIN1 in Arabidopsis vascular tissue. Science 282:2226–2230. https://doi.org/10.1126/science.282.5397.2226
13. Dong J, MacAlister CA, Bergmann DC (2009) BASL controls asymmetric cell division in Arabidopsis. Cell 137:1320–1330. https://doi.org/10.1016/j.cell.2009.04.018
14. Roppolo D, De Rybel B, Tendon VD et al (2011) A novel protein family mediates Casparian strip formation in the endodermis. Nature 473:380–383. https://doi.org/10.1038/nature10070
15. Ott T (2017) Membrane nanodomains and microdomains in plant–microbe interactions. Curr Opin Plant Biol 40:82–88. https://doi.org/10.1016/j.pbi.2017.08.008
16. Jaillais Y, Ott T (2020) The nanoscale Organization of the Plasma Membrane and its Importance in signaling: a proteolipid perspective. Plant Physiol 182:1682–1696. https://doi.org/10.1104/pp.19.01349
17. Gronnier J, Crowet J-M, Habenstein B et al (2017) Structural basis for plant plasma membrane protein dynamics and organization into functional nanodomains. eLife 6:e26404. https://doi.org/10.7554/eLife.26404
18. Liang P, Stratil TF, Popp C et al (2018) Symbiotic root infections in *Medicago truncatula* require remorin-mediated receptor stabilization in membrane nanodomains. Proc Natl Acad Sci U S A 115:5289–5294. https://doi.org/10.1073/pnas.1721868115
19. Abbe E (1873) Beiträge zur Theorie des Mikroskops und der mikroskopischen Wahrnehmung. Arch Mikr Anat 9:413–468. https://doi.org/10.1007/BF02956173
20. Huff J (2015) The Airyscan detector from ZEISS: confocal imaging with improved signal-to-noise ratio and super-resolution. Nat Methods 12:i–ii. https://doi.org/10.1038/nmeth.f.388
21. Huff J, Bergter A, Birkenbeil J et al (2017) The new 2D Superresolution mode for ZEISS Airyscan. Nat Methods 14:1223–1223. https://doi.org/10.1038/nmeth.f.404
22. Korobchevskaya K, Lagerholm B, Colin-York H, Fritzsche M (2017) Exploring the potential of Airyscan microscopy for live cell imaging. Photo Dermatol 4:41. https://doi.org/10.3390/photonics4030041
23. Huff J (2016) The fast mode for ZEISS LSM 880 with Airyscan: high-speed confocal imaging with super-resolution and improved signal-to-noise ratio. Nat Methods 13:i–ii. https://doi.org/10.1038/nmeth.f.398
24. Martin-Fernandez ML, Tynan CJ, Webb SED (2013) A 'pocket guide' to total internal reflection fluorescence. J Microsc 252:16–22. https://doi.org/10.1111/jmi.12070
25. Vizcay-Barrena G, Webb SED, Martin-Fernandez ML, Wilson ZA (2011) Subcellular and single-molecule imaging of plant fluorescent proteins using total internal reflection fluorescence microscopy (TIRFM). J Exp Bot 62:5419–5428. https://doi.org/10.1093/jxb/err212
26. Platre MP, Bayle V, Armengot L et al (2019) Developmental control of plant rho GTPase nano-organization by the lipid phosphatidylserine. Science 364:57–62. https://doi.org/10.1126/science.aav9959
27. Smokvarska M, Francis C, Platre MP et al (2020) A plasma membrane nanodomain ensures signal specificity during osmotic signaling in plants. Curr Biol 30:4654–4664.e4. https://doi.org/10.1016/j.cub.2020.09.013
28. Schindelin J, Arganda-Carreras I, Frise E et al (2012) Fiji: an open-source platform for biological-image analysis. Nat Methods 9:676–682. https://doi.org/10.1038/nmeth.2019
29. Tinevez J-Y, Perry N, Schindelin J et al (2017) TrackMate: an open and extensible platform for single-particle tracking. Methods 115:80–90. https://doi.org/10.1016/j.ymeth.2016.09.016
30. Wagner T, Kroll A, Haramagatti CR et al (2017) Classification and segmentation of nanoparticle diffusion trajectories in cellular

micro environments. PLoS One 12:e0170165. https://doi.org/10.1371/journal.pone.0170165

31. Robatzek S, Chinchilla D, Boller T (2006) Ligand-induced endocytosis of the pattern recognition receptor FLS2 in Arabidopsis. Genes Dev 20:537–542. https://doi.org/10.1101/gad.366506
32. Kurup S, Runions J, Köhler U et al (2005) Marking cell lineages in living tissues. Plant J 42:444–453. https://doi.org/10.1111/j.1365-313X.2005.02386.x
33. Rolfe DJ, McLachlan CI, Hirsch M et al (2011) Automated multidimensional single molecule fluorescence microscopy feature detection and tracking. Eur Biophys J 40:1167–1186. https://doi.org/10.1007/s00249-011-0747-7

Chapter 16

Using Steady-State Fluorescence Anisotropy to Study Protein Clustering

Ajeet Chaudhary and Kay Schneitz

Abstract

Signaling pathways rely on the precise control of protein-protein interactions. Therefore, it is essential to be able to investigate such interactions with spatiotemporal resolution and in live cells. Here we describe a microscope-based fluorescence spectrometry technique to investigate homotypic interactions between GFP-labeled fusion proteins in a rapid and reproducible fashion using fluorescence anisotropy. This method is of great value for the study of protein complexes in live tissue with subcellular resolution.

Key words *Arabidopsis*, Cell wall, Confocal laser scanning microscopy, Fluorescence anisotropy, Methods, Microscopy, Plasmodesmata, Protein–protein interaction, Receptor kinase

1 Introduction

Successful deciphering of signaling processes requires the ability to investigate protein–protein interactions in live cells and with spatiotemporal resolution. Förster resonance energy transfer (FRET) between two functional proteins carrying a distinct fluorophore (such as the fluorescent proteins GFP and mCherry) has long been the method of choice, also in plants [1–6]. Indeed, a growing number of FRET-based studies addressing interactions of proteins present at the plasma membrane and at plasmodesmata (PD) can be found in the literature [7–10]. A central requirement for FRET to occur is that the two fusion proteins are separated by 10 nm or less. In addition, the emission and absorption transition dipole moments of donor and acceptor should not be perpendicular to each other, and their emission and absorption spectrum should significantly overlap. The required distance of less than 10 nm essentially indicates direct physical interaction. FRET is usually measured between different compatible donor and acceptor probes (hetero-FRET). GFP and mCherry have turned out to represent a

Yoselin Benitez-Alfonso and Manfred Heinlein (eds.), *Plasmodesmata: Methods and Protocols*,
Methods in Molecular Biology, vol. 2457, https://doi.org/10.1007/978-1-0716-2132-5_16,

robust pair of probes, but other combinations of fluorescent proteins are in use [11].

Interestingly, FRET can also occur between identical fluorophores (homo-FRET) [12]. Investigating homo-FRET can reveal interesting insight into homo-oligomerization properties of the protein under study. The decay of fluorescence anisotropy (FA) of a fluorophore in a time-resolved manner represents a straightforward way to detect homo-FRET [12, 13]. Conveniently, FA is largely independent of the concentration of the fluorophore, making FA detection a robust technique to investigate protein clustering. Moreover, measuring FA values using microscopy allows the investigation of the oligomerization properties of a fusion protein of interest at the subcellular level. Although complex physics underlies FA, the basic principle of FA detection is relatively simple. In a nutshell, a population of fluorophores of different orientations is excited by plane-polarized excitation light with the help of a polarizer. In a microscope setup, the emitted light is typically detected using a dual channel setup for parallel and perpendicular detection of polarization. Comparing the number of photons collected by the two differentially oriented detectors allows assessment of FA. Only fluorophores that have their absorption transition dipole moment aligned in parallel to the direction of polarization of the excitation light will be excited. Initially, this results in a high level of FA of the emitted light. In principle, FA drops over time due to rotation of the excited molecule and/or energy transfer. In the case of large fluorophores, such as GFP or GFP-based fusion proteins, the rotational aspect can be largely neglected. However, excited GFP molecules present in a protein cluster can transfer energy to closely neighboring GFPs that initially were not excited as their dipole moments were not aligned in parallel to the direction of polarization of the exciting light. Thus, homo-FRET between clustering GFP moieties of different orientations results in depolarization that can be detected as a correspondingly low FA value. In such instances, it is not strictly necessary to use time-resolved approaches. Due to the phenomenon of photoselection, the excitation with polarized light results in a population of excited GFP fluorophores that are partially oriented along the vertical polarization axis. When there are no processes resulting in depolarization, the maximum anisotropy value is $r = 0.4$ [12, 14]. Free cytosolic GFP was measured to have a FA value of 0.332 [7]. For comparison, assessment of the FA of GFP fusions of the receptor kinase ARABIDOPSIS CRINKLY 4 (ACR4) revealed values of 0.300 for ACR4:GFP at the plasma membrane and 0.263 for PD-localized ACR4:GFP [7]. The latter result indicates increased clustering of ACR4:GFP at PD.

Here, we describe the instrumentation and methodology that is commonly used in our lab to perform steady-state FA experiments. As an example, we employ a C-terminal fusion of EGFP to

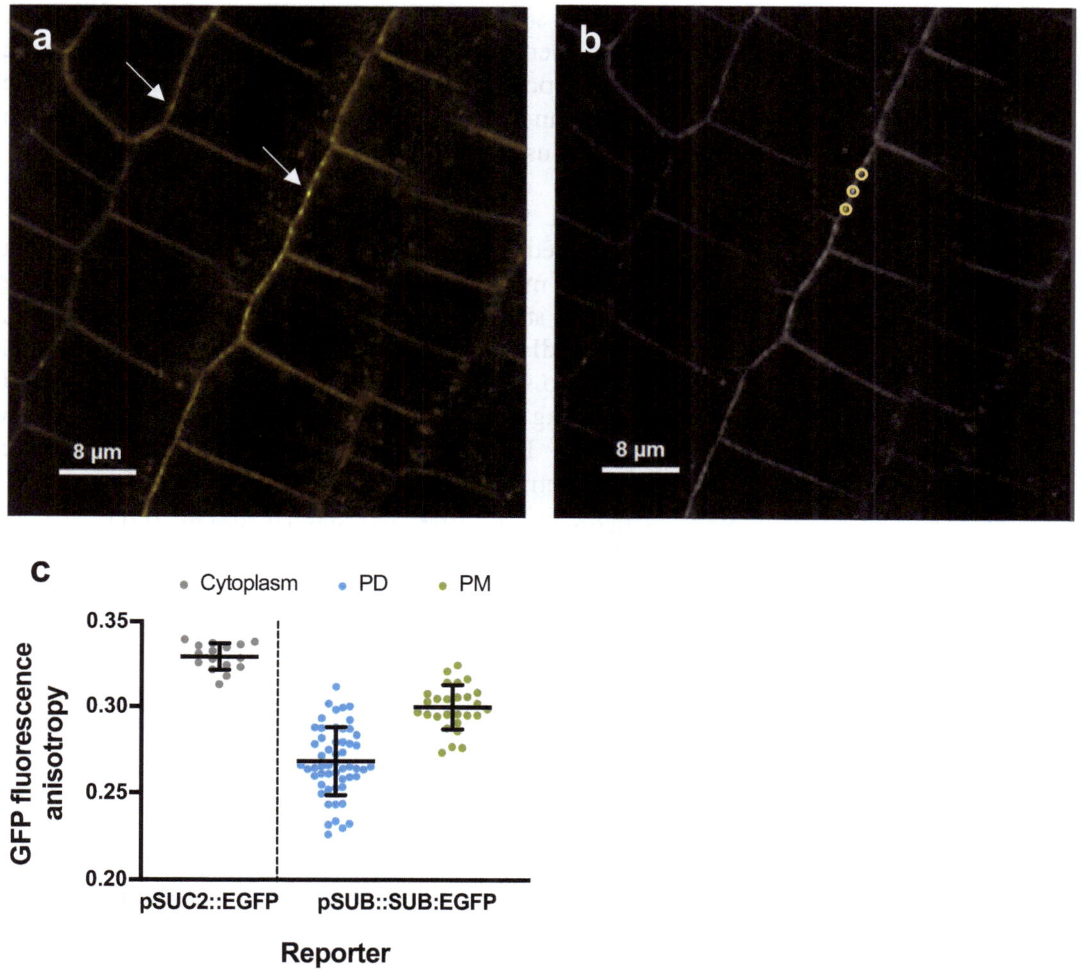

Fig. 1 Fluorescence anisotropy image of *pSUB::SUB:EGFP* signal present in epidermal cells of the root meristem from a 7-day-old seedling. (**a**) The arrows indicate reporter signal at PD puncta localized at the lateral side of cells. (**b**) Individual PD puncta are selected as ROIs by precisely drawing circles over PD using the ROI tool. (**c**) FA measurement. Free cytoplasmic EGFP was driven by the SUC2 promoter [19]. Mean $\pm$ SD are indicated. Cytoplasmic EGFP: 0.336 ($n = 16$); SUB:EGFP/PD: 0.269 ($n = 54$); SUB:EGFP/PM: 0.300 ($n = 27$). All means differ significantly from each other (one-way ANOVA, $p < 0.0001$)

STRUBBELIG (SUB). SUB is a plasma membrane-localized receptor kinase that is enriched at PD [8, 15, 16]. The measurements indicate SUB:EGFP to exhibit increased clustering at PD in epidermal cells of the root meristem (Fig. 1).

2 Materials

FA analysis requires stably or transiently transformed lines expressing GFP-tagged fusion proteins. To exemplify this protocol, we use *pSUB::SUB:EGFP*, a well-characterized reporter for the receptor kinase SUB (*see* **Note 1**).

Prepare all the solutions in double distilled ultrapure water (ddH_2O) unless mentioned otherwise and use analytical grade reagents. Store all stock solutions at the recommended temperature (4 °C or −20 °C) and equilibrate the working solutions to room temperature before using it.

2.1 Plant Growth and Mounting

1. Murashige and Skoog basal plant growth medium (half-strength MS medium): Dissolve 2.2 g MS medium and 10 g sucrose in 900 ml dH_2O for a final volume of 1000 ml (*see* **Note 3**). While stirring, adjust the pH to 5.8 using 37% HCl. Add additional dH_2O to bring the medium to the final volume. Add 9 g agar to solution. Sterilize the medium in a validated autoclave at 1 kg/cm^2 (15 psi). The medium should attain a temperature of 121 °C for at least 15 min. Pour culture medium into petri plate and let it settle.
2. Sodium hypochlorite: 20% commercial bleach, 0.02% Triton X-100.
3. 22 × 22 mm glass coverslips of 0.17 mm thickness (no. 1.5H) (*see* **Note 2**).

2.2 Imaging and Analysis

1. Confocal laser scanning microscope (e.g., FV3000 Olympus Europa GmbH, Hamburg, Germany) equipped with a 60× water objective (Olympus UPlanSApo, NA: 1.2) and a single photon counting device with picosecond time resolution (LSM upgrade kit, PicoQuant, Berlin, Germany). The microscope must have a 485 nm laser with a linearly polarized, pulsed (40 MHz) diode laser (LDH-D-C-485, PicoQuant) for GFP excitation and a PMA Hybrid 40s detector (PicoQuant) for detection.
2. FA analysis software: SymPhoTime 64 software (PicoQuant).
3. Statistics software: GraphPad Prism 9 (GraphPad Software, San Diego, USA).

3 Methods

3.1 Sample Preparation

1. Surface sterilize *Arabidopsis* seeds with sodium hypochlorite for 20 min and wash 5× with ddH_2O.
2. Place at 4 °C for 48 h (*see* **Note 4**).
3. Spread single seeds on half-strength MS agar plate in a straight line.
4. Place the plate in a growth chamber at vertically inclined position (80°).
5. Collect 7-day-old seedlings for mounting and imaging (*see* **Note 5**).

3.2 Image Acquisition

1. For live-cell imaging, mount tissue sample on a slide in half-strength liquid MS medium.
2. First, rapidly scan the sample with a 485 nm laser to find the region of interest (ROI) for FA analysis using a 60× water objective (*see* **Note 6**).
3. Excite GFP with a 485 nm laser line with a linearly polarized, pulsed (40 MHz) diode laser using a 60× water immersion objective.
4. Collect the emitted light in the same objective and separate the collected light into perpendicular and parallel polarization with respect to excitation polarization.
5. Detect the GFP fluorescence using a PMA Hybrid 40s detector in a narrow range within its emission spectrum (BP520/35).
6. To obtain the FA image, collect a minimum of 25 photons per pixel using time-correlated single photon counting with a resolution of 25 ps per ROI (512 px × 512 px, 0.103 μM photons/px) (Fig. 1a) (*see* **Notes** 7 and **8**).

3.3 FA Counting

1. Calculate pixel-wise anisotropy with the help of SymPhoTime 64 software from PicoQuant using instrument correction factors L1 (0.035) and L2 (0.031) as well as the G-factor (1.018) (*see* **Note 9**).
2. Select individual PD puncta as ROIs as described in Fig. 1b and calculate FA values with prefixed software settings (*see* **Note 10**).

3.4 Data Analysis

1. Analyze all collected data points for statistical significance (Fig. 1c) (*see* **Note 11**).

4 Notes

1. The pSUB::SUB:EGFP reporter line employed in this protocol has been previously well characterized [8, 17]. For an unknown protein, it is best to first analyze its localization and expression pattern. GFP is the recommended fluorophore for FA analysis.
2. The thickness of the coverslip is important to avoid optical aberration. Most confocal microscope objective lenses are designed to be used with a cover glass of 0.17 mm thickness and a refractive index of 1.515. When the numerical aperture (NA) is 0.4 or less, cover glass thickness has less of an effect on image formation. However, in case of high numerical aperture dry objectives (NA of 0.8 or greater), variation in cover glass thickness (even by few micrometers) leads to dramatic image distortion due to aberration.

3. In our hands, the composition and strength of the MS medium greatly affect seedling growth as well as the FA value of SUB: EGFP. We obtain the best and consistent results using half-strength MS supplemented with 1% sucrose.
4. Higher concentrations of sodium hypochlorite or longer washing time than recommended significantly reduces seed germination.
5. In this protocol, we analyze the FA of SUB:EGFP in epidermal cells from the root meristem, but in principle any other tissue of choice can be analyzed in similar fashion.
6. Longer exposure to laser causes photo damage and bleaching of the fluorophore. To avoid cytotoxicity caused by the laser, the tissue is first scanned with very low laser power (in most of CSLM setups, laser power output is set to 70% power only; in scanning mode, use 1–2% power for a PMT detector or 0.1–1% for a GaAsP detector of total 70% laser power output), and the ROI is adjusted into the scanning frame for final FA image acquisition.
7. Photon counting per pixel depends on the expression level of the GFP-tagged fusion protein. The higher the expression of GFP, the faster will be photon counting. However, the number of emitted photons will greatly diminish over time because of photobleaching. A minimum of 25 photons per pixel is recommended for FA measurements. Although higher photon counting is in principle possible, longer laser exposure creates tissue damage and heterogenous photon counting over the image frame. An image frame of 512 px × 512 px with 0.103 μM photons/px is optimal. Using a bigger frame for acquiring FA images takes more time, and thus the sample is prone to suffer from laser-induced damage.
8. To avoid sample movement, minimal amount of mounting solution should be used. Additionally, extra mounting solution could be removed from the edges of the cover slip using tissue paper (Kimwipes) before mounting on the imaging platform.
9. The L1, L2, and G factors are adjusted by PicoQuant technicians during installation and depend on the CLSM setup. The L1/L2 factors relate to the numerical aperture. The G factor relates to the ratio of the sensitivity of detector 1 (antiparallel or horizontal) over the sensitivity of detector 2 (parallel or vertical). The factors can also be obtained at the beginning of experiment following the instrument manufacturer's instruction.

10. Once a FA image is obtained, individual PD puncta can be selected as ROI for FA measurement. Multiple ROI selection is also possible.
11. More details about data interpretation can be found in [18].

Acknowledgments

Our work is currently supported by grants SFB 924 (TPA2) and FOR2581 (TP7) from the German Research Council (DFG) to KS. We thank Ramón A. Torres-Ruiz for help with FA imaging and comments on the manuscript. We also thank Wanney Walja and Sebastian Scholz for sharing data. We further acknowledge support by the Center for Advanced Light Microscopy (CALM) of the TUM School of Life Sciences.

References

1. Harter K, Meixner AJ, Schleifenbaum F (2012) Spectro-microscopy of living plant cells. Mol Plant 5:14–26
2. Bücherl CA, Bader A, Westphal AH, Laptenok SP, Borst JW (2014) FRET-FLIM applications in plant systems. Protoplasma 251:383–394
3. Long Y, Stahl Y, Weidtkamp-Peters S, Postma M, Zhou W, Goedhart J et al (2017) In vivo FRET-FLIM reveals cell-type-specific protein interactions in Arabidopsis roots. Nature 548:97–102
4. Weidtkamp-Peters S, Stahl Y (2017) The use of FRET/FLIM to study proteins interacting with plant receptor kinases. Methods Mol Biol 1621:163–175
5. Shah K, Gadella TW, van Erp H, Hecht V, de Vries SC (2001) Subcellular localization and oligomerization of the Arabidopsis thaliana somatic embryogenesis receptor kinase 1 protein. J Mol Biol 309:641–655
6. Somssich M, Ma Q, Weidtkamp-Peters S, Stahl Y, Felekyan S, Bleckmann A et al (2015) Real-time dynamics of peptide ligand-dependent receptor complex formation in planta. Sci Signal 8:ra76
7. Stahl Y, Grabowski S, Bleckmann A, Kuhnemuth R, Weidtkamp-Peters S, Pinto KG et al (2013) Moderation of Arabidopsis root stemness by CLAVATA1 and ARABIDOPSIS CRINKLY4 receptor kinase complexes. Curr Biol 23:362–371
8. Vaddepalli P, Herrmann A, Fulton L, Oelschner M, Hillmer S, Stratil TF et al (2014) The C2-domain protein QUIRKY and the receptor-like kinase STRUBBELIG localize to plasmodesmata and mediate tissue morphogenesis in *Arabidopsis thaliana*. Development 141:4139–4148
9. Rosas-Diaz T, Zhang D, Fan P, Wang L, Ding X, Jiang Y et al (2018) A virus-targeted plant receptor-like kinase promotes cell-to-cell spread of RNAi. Proc Natl Acad Sci U S A 115: 1388–1393
10. Cheval C, Samwald S, Johnston MG, de Keijzer J, Breakspear A, Liu X et al (2020) Chitin perception in plasmodesmata characterizes submembrane immune-signaling specificity in plants. Proc Natl Acad Sci U S A 117: 9621–9629
11. Denay G, Schultz P, Hänsch S, Weidtkamp-Peters S, Simon R (2019) Over the rainbow: a practical guide for fluorescent protein selection in plant FRET experiments. Plant Direct 3: e00189
12. Bader AN, Hoetzl S, Hofman EG, Voortman J et al (2011) Homo-FRET imaging as a tool to quantify protein and lipid clustering. ChemPhysChem 12:475–483
13. Jares-Erijman EA, Jovin TM (2003) FRET imaging. Nat Biotechnol 21:1387–1395
14. JR L (2006) Principles of fluorescence spectroscopy. Springer Science and Business Media, New York
15. Chevalier D, Batoux M, Fulton L, Pfister K, Yadav RK, Schellenberg M et al (2005) *STRUBBELIG* defines a receptor kinase-mediated signaling pathway regulating organ development in *Arabidopsis*. Proc Natl Acad Sci U S A 102:9074–9079

16. Kwak SH, Shen R, Schiefelbein J (2005) Positional signaling mediated by a receptor-like kinase in Arabidopsis. Science 307:1111–1113
17. Vaddepalli P, Fulton L, Batoux M, Yadav RK, Schneitz K (2011) Structure-function analysis of STRUBBELIG, an Arabidopsis atypical receptor-like kinase involved in tissue morphogenesis. PLoS One 6:e19730
18. Chaudhary A, Chen X, Leśniewska B, Gao J, Wolf S, Schneitz K (2020) Cell wall damage impairs root hair cell patterning and tissue morphogenesis mediated by the Arabidopsis receptor kinase STRUBBELIG bioRxiv 2020.11.04.368241
19. Imlau A, Truernit E, Sauer N (1999) Cell-to-cell and long-distance trafficking of the green fluorescent protein in the phloem and symplastic unloading of the protein into sink tissues. Plant Cell 11:309–322

Part IV

Functional Analysis of Plasmodesmata

Chapter 17

Quantification of Cell-to-Cell Connectivity Using Particle Bombardment

Estee E. Tee, Sebastian Samwald, and Christine Faulkner

Abstract

Plant cells are connected by cytoplasmic bridges called plasmodesmata. Plasmodesmata are lined by the plasma membrane, essentially forming tunnels that directly connect the cytoplasm of adjacent cells through which soluble molecules can move from cell to cell. This cell-to-cell mobility is underpinned by cytoplasmic advection and diffusion in a manner dependent on molecular size. This movement of molecules is regulated by the aperture of plasmodesmata. GREEN FLUORESCENT PROTEIN (GFP) is a 27 kDa soluble protein that can move passively between cells via plasmodesmata. Thus, it serves as an ideal probe to assess plasmodesmal aperture. GFP can be transgenically produced in single cells by microprojectile bombardment-mediated transformation, and its cell-to-cell mobility can be measured by live-cell imaging and counting the number of cells (or cell layers) to which it has moved. Thus, the number of cells in which GFP is visible serves as a measure of plasmodesmal aperture and functional cell-to-cell connectivity. Here we present methods for microprojectile bombardment of GFP into leaf epidermal cells and statistical analysis of resulting data.

Key words Plasmodesmata, Cell-to-cell communication, Cell-to-cell mobility, GFP, Microprojectile bombardment

1 Introduction

Plants have an interconnected cytoplasm, termed the symplast, that allows molecules to move from cell to cell and tissue to tissue. The symplast is critical for key processes such as the allocation of photo-assimilates and allows for signaling at both local (cell-to-cell) and distant (organ-to-organ) scales [1–3]. The cytoplasm of adjacent cells is connected via structures called plasmodesmata, which in essence are membrane-lined tunnels that cross the cell wall. The aperture of plasmodesmata can be regulated by the deposition of callose in the cell wall that surrounds the plasmodesmal tunnel; callose causes constriction of the plasmodesma and thus restricts molecular passage between cells. Thus, via the process of callose

Yoselin Benitez-Alfonso and Manfred Heinlein (eds.), *Plasmodesmata: Methods and Protocols*,
Methods in Molecular Biology, vol. 2457, https://doi.org/10.1007/978-1-0716-2132-5_17,

synthesis and hydrolysis, plasmodesmal connections are closed and opened allowing for dynamic connectivity within and between tissues.

Callose regulation of plasmodesmata occurs in response to a variety of stresses and developmental transitions. For example, callose-mediated plasmodesmal closure occurs in response to perception of pathogen-derived molecules [4], the defense hormone salicylic acid [5], and hydrogen peroxide [6]. This implicates cellular isolation as a critical component of stress responses, presumably relating to a need to retain cellular signals or resources in stressed cells as part of the initial response phase. Plasmodesmal closure is also associated with developmental transitions such as lateral root formation [7] and processes corresponding to tissue sink-to-source transition [8], suggesting that information and resources associated with cellular reprogramming need to be contained within relevant cells.

Small soluble molecules can pass from cell to cell if they are freely available in the cytosol, exemplified by the easy movement of non-endogenous fluorescent molecules such as carboxyfluorescein (376 Da) and green fluorescent protein (GFP, 27 kDa). Such probes can represent the capacity for movement of endogenous molecules under given conditions, spreading to more cells if the plasmodesmal aperture is open and to fewer cells if the plasmodesmal aperture is constricted or closed. Thus, such molecules can be exploited to assay for plasmodesmal aperture or functional cell-to-cell connectivity.

Cell-to-cell mobility assays require a clear point of synthesis or delivery of the fluorescent probe to allow measurement of the distance it spreads. A range of approaches have been used to deliver non-endogenous fluorescent molecules to a localized site or cell including microinjection [9], localized dye application [5], uncaging of dyes [9], microprojectile bombardment [4, 10], genetically manipulated expression of fluorescent protein genes [7, 11], or photoactivation or switching of fluorescent proteins [9, 12]. Each method has advantages and constraints and requires a moderate to high level of technical and analytical dexterity. While dye loading, microinjection, and dye uncaging primarily deal with small fluorescent molecules, microprojectile bombardment, localized gene expression, and photoactivation rely on synthesis and movement of larger fluorescent proteins such as GFP. Thus, users must consider both the resolution of analysis required for their experiment (which relates to the size change in plasmodesmal aperture expected in their system) and the method of delivery most suited for their biological question and system in choosing an approach to assay for plasmodesmal function and cell-to-cell connectivity.

Microprojectile bombardment involves shooting DNA-coated metal particles at high velocity into plant tissues [13]. For transformation to occur, the gold particle must hit the nucleus of the cell; a combination of factors that relate to the gold concentration, the pressure of the shot, and the distance of the plant material beneath the source allows for separated single cell transformation with minimal wounding. If the metal particles are coated with plasmid DNA that encodes for high expression of a fluorescent protein (FP), then transformed cells synthesize the fluorescent protein which then moves cell to cell in a manner dependent on the functional state and aperture of the plasmodesmata. Live-cell imaging of bombarded tissues allows identification of bombardment sites and quantification of the spread of the fluorescent protein.

Different metrics can be chosen for measuring cell-to-cell spread of a fluorescent protein: the number of cells in which the FP can be detected or the number of cell layers in which the FP can be detected. In our experience, cell-to-cell mobility data has a high variance and is heteroskedastic, demanding appropriate statistical analysis. For this, we recently developed a bootstrap analysis that accommodates such data [14] and recommend its use to assay for a change in the median (or mean) of FP movement.

Here we present detailed methods for microprojectile bombardment of Arabidopsis leaves. These methods can be used for a variety of plants, but experimental parameters are dependent on the physical properties of the tissue. Thus, the method requires empirical optimization for individual circumstances. Here we present a strategy for such optimization, as well as analysis of collected data sets. As we routinely use this method to investigate cell-to-cell connectivity in response to treatments such as immune elicitors, we have included this as an optional method step.

2 Materials

2.1 Plant Material

For bombardment of *Arabidopsis thaliana* leaves, ecotype Columbia-0 plants are grown on soil in short day growth conditions (8 h light: 16 h dark) at 23 °C (*see* **Note 1**).

2.2 Microcarrier Preparation

1. Microcarriers: for Arabidopsis leaves, we use 1 nm gold particles.
2. Water bath sonicator (Ultrasonic Cleaning Device).
3. Vortex.
4. Microcentrifuge.
5. DNA constructs:

   For synthesis of a soluble fluorescent protein (FP), use a plasmid that encodes high expression of the *FP* gene such as from the *35S* promoter. We frequently use pB7WG2.0.GFP [10] for this purpose to produce GFP. For a cell transformation marker, use a plasmid (*see* **Note 2**) that expresses an immobile, cell-anchored FP such as ER-targeted RFP (*35S::SP-RFP-HDEL*, pB7WG2.0.RFP_{ER} [10]).

6. Sterile dH_2O.
7. 2.5 M $CaCl_2$.
8. 0.1 M spermidine.
9. 100% ethanol.

2.3 Bombardment

1. Gene gun:

   PDS-1000/He Particle Delivery System (Bio-Rad, https://www.bio-rad.com/webroot/web/pdf/lsr/literature/10000070900.pdf), tank of pressurized helium, and vacuum pump.
2. Rupture discs, macrocarriers, and stopping screens (Fig. 1).

   For expanded Arabidopsis leaves, we use 1100 psi or 1350 psi rupture discs.
3. 1× Murashige-Skoog (MS) medium or water agar (0.6%) plates (*see* **Note 3**).
4. 1 mL syringe for post-bombardment treatment.

2.4 Imaging

1. Microscope:

   We use a confocal laser scanning microscope with 20× or 25× water dipping lens.
2. Slides and mounting:

   For mature leaves, we mount whole leaves on slides using Vaseline, double-sided tape, or medical adhesive to secure the leaf.

3 Methods

Cell-to-cell mobility assay by microprojectile bombardment takes place in two main steps: (Subheading 3.1) the bombardment followed by (Subheading 3.2) imaging. The timing of these steps can be adjusted depending on the rate of production of the FP (usually minimally 6–8 h) and the requirement to introduce a post-bombardment treatment to assay for plasmodesmal responses if applicable.

3.1 Bombardment

3.1.1 Microcarrier Preparation

1. Make a suspension of 50 mg microcarriers (e.g., 1.0 μm gold) in 1 mL sterile dH_2O.
2. Vortex the suspension vigorously for 1–2 min.
3. Centrifuge the suspension at 13,000 × *g* for 1 min.
4. Discard supernatant.

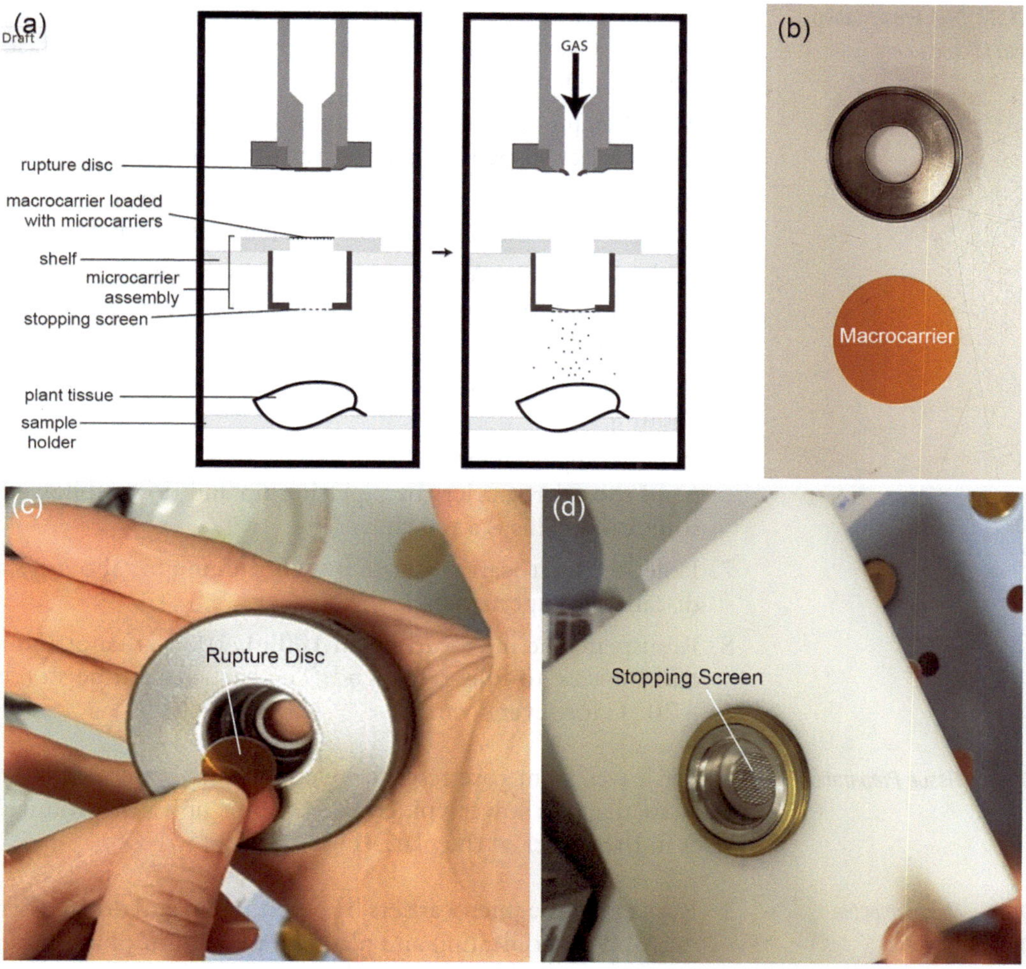

Fig. 1 The PDS-1000/He Biolistic Particle Delivery System. (**a**) A cartoon of the gene gun components and mechanics of particle bombardment. (**b–d**) PDS-1000/He consumables: (**b**) macrocarriers are mounted in microcarrier holders (top); (**c**) rupture discs are mounted into the retaining cap; (**d**) stopping screens are secured in the microcarrier launch assembly

5. Wash the microcarriers three times with 100% ethanol by vortexing followed by centrifuging.
6. Wash once with sterile dH_2O – vortex, centrifuge, and discard the supernatant.
7. Resuspend microcarriers in 1-mL sterile dH_2O and aliquot 25 μL per 1.5 mL tube. Continuously vortex the suspension as you aliquot to ensure equal volumes of microcarriers in each tube.
8. Store tubes at 4 °C.

3.1.2 DNA Precipitation onto Microcarriers

1. Resuspend a 25 μL aliquot of microcarriers by briefly vortexing or with several pulses in a sonicator (a water bath Ultrasonic Cleaning Device).
2. Add 2–5 μg (up to 5 μL volume) of plasmid DNA to the microcarrier suspension and vortex or sonicate briefly again.
3. Add 25 μL 2.5 M $CaCl_2$ and pipette up and down 5–10 times (do not cause bubbles) or vortex briefly (*see* **Note 4**).
4. Add 10 μL of 0.1 M spermidine, mix by pipetting, and vortex at low speed for 2 min. Incubate the tube on ice for 5–30 min (*see* **Note 4**).
5. Pellet the DNA-coated microcarriers by micro-centrifuging for about 5 s.
6. Remove supernatant and add 180 μL of 100% ethanol. Resuspend the microcarriers by pipetting up and down followed by sonicating. This is a washing step.
7. Pellet the microcarriers again by centrifuging for about 5 s and discard the supernatant.
8. Resuspend the microcarriers in 100 μL 100% ethanol and mix and sonicate to fully resuspend. The particles can be stored at −20 °C for several weeks.

3.1.3 Tissue Preparation

Dissect leaves from rosettes and invert on MS/water agar (0.6%) plates so that the lower epidermis is facing upward. Mount 2–3 leaves in the center of the plate (Fig. 2).

3.1.4 Microprojectile Bombardment

1. Resuspend the microcarriers by pipetting up and down, vortexing, and/or sonicating and place a 5 μL aliquot of your microcarrier preparation on the downward facing surface of the macrocarrier.
2. Assemble the microcarrier launch assembly, placing a stopping screen at the base. Place the microcarrier holder in the brass cylinder, position the assembled microcarrier launch in the top, and position the shelf in the top position.
3. Open the main valve on the helium cylinder and adjust the pressure as required.
4. Place a rupture disk (e.g., 1100 psi for Arabidopsis leaves) in the rupture disc retaining cap and screw into position and tighten the cap with the torque wrench.
5. Place the sample (on the MS/water agar plate) on the target plate and slot into the desired position.
6. Close the chamber and turn on the vacuum pump (press and hold VAC). When the appropriate vacuum (we use the maximum vacuum) is reached, flick the switch to HOLD.

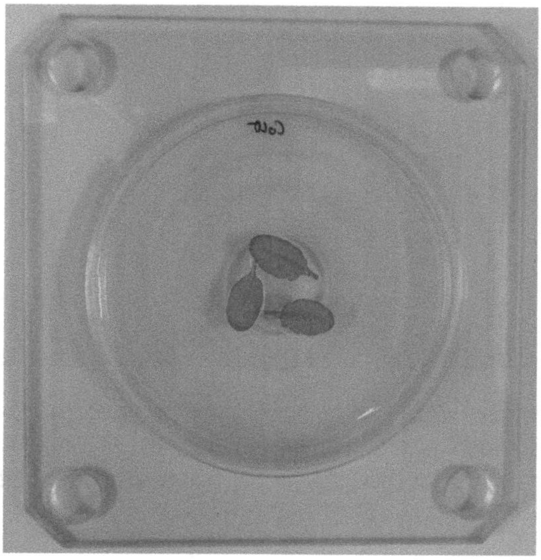

Fig. 2 Arabidopsis leaves mounted on an MS plate for bombardment with abaxial surface facing upward

7. Press and hold the fire button and watch the pressure gauge on the top of the gun rise. When the pressure reaches the appropriate psi, the rupture disk will burst and launch the microcarriers.
8. After bombardment is completed, release the vacuum from the chamber by pushing the vacuum button to VENT and remove the samples when the vacuum is released.
9. Keep bombarded tissue well hydrated until imaging (*see* **Note 5**).

3.1.5 Treatment

Syringe infiltrate a solution of the treatment of interest (e.g., 500 μg/mL chitin, 100 nM flg22, 100 μM salicylic acid), into the abaxial side of the leaf with a 1 mL syringe (*see* **Note 6**). This can be done 2–4 h post-bombardment.

3.2 Live-Cell Imaging

3.2.1 Tissue Mounting

1. Mount the whole leaf (or leaf pieces) on a microscope slide (abaxial epidermis facing up) using Vaseline, medical adhesive, or double-sided tape to secure the leaf to the slide.

3.2.2 Imaging and Data Collection

1. Using a confocal laser scanning microscope (*see* **Note 7**) to image the bombarded tissue, systematically survey the full length of the leaf to identify and image bombarded cells. Bombarded cells can be identified by the expression of the cell transformation marker (e.g., RFP_{ER}).

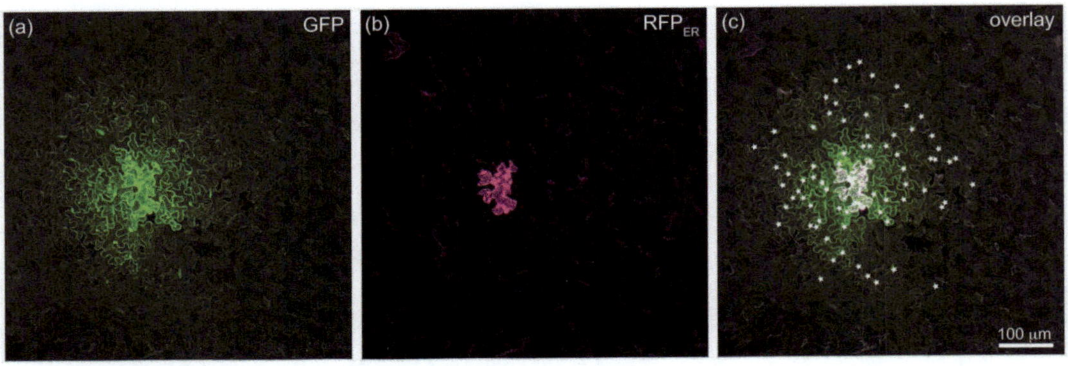

Fig. 3 Bombardment site of an Arabidopsis leaf showing cell-to-cell movement of GFP. Bombardment site shows (**a**) GFP in many cells and (**b**) the transformed cell with RFP_{ER}. (**c**) Shows the overlay of the images and the cells counted for data collection for this bombardment site are marked by stars

2. Collect images (usually z-stacks) of each bombardment site, ensuring that sufficient planes are collected such that all cells that show the FP are imaged.
3. Count the number of cells in each bombardment site that show FP fluorescence (Fig. 3, *see* **Notes 8** and **9**).

3.2.3 Data Analysis

Bombardment data is usually non-normal, and data from different treatments can display different variances. Bootstrapping is a suitable method to statistically analyze such data sets. The medianbootstrap.R script is available for this (https://github.com/faulknerfalcons/Johnston-2020-Bootstrap). We recommend performing analysis on the median of the data because the data distributions are often skewed with a small number of large data points (>50 cells).

4 Notes

1. The success of microprojectile bombardment depends on the physical features of the tissue that are defined by the cuticle thickness and turgor pressure of the epidermis. Thus, the method is empirical and requires optimization for any new plant growth conditions. This is also true for different tissues, e.g., young and old leaves of the same plant can require different pressures or different distances from the rupture disc. Most commonly we use mature expanded leaves of *Arabidopsis*, but we also bombard seedling cotyledons and young leaves. Further, we have bombarded *Nicotiana benthamiana* leaves, wheat leaves, and leek hypocotyls.
2. In our experience, the concentration of the DNA with which you coat the microcarriers is a significant factor in determining

the frequency of transformation. The higher the concentration (within the range, 500–1000 ng/μL), the higher the frequency of transformation. Thus we recommend doing a plasmid DNA maxiprep to obtain higher concentrations of DNA for the DNA precipitation Subheading 3.1.2.

3. We use 90-mm diameter petri dishes to fit in the gene gun chamber. Altering the concentration of agar may help secure the sample better depending on the tissue type (*see* **Note 1**). For Arabidopsis leaves, we use 0.6% agar.
4. $CaCl_2$ precipitates the DNA onto the microcarriers. Spermidine is used to protect the DNA from cellular DNAses.
5. It is important to keep the tissue hydrated during bombardment and until imaging is complete. For Arabidopsis, we use MS plates (0.6% agar) and keep the lids on the plates at all times during incubation periods. Keeping the petioles inserted into the MS agar can also prevent desiccation. Other possibilities are plates lined with wet paper towel, or for larger *Nicotiana benthamiana* leaves, we have kept the leaves with their cut petioles in water.
6. We primarily use syringe infiltration for application of various treatments to the bombarded tissue. If stomata are closed, infiltration can be improved by punching a small hole in the tissue with the tip of a white or yellow pipette tip (10–200 μL) and placing the syringe over the hole. Adding 0.01% Silwet to the infiltration solution can ensure even infiltration. For control samples, infiltrate water or an appropriate buffer to control for wounding effects. Depending on your experiment, treatments can be applied as appropriate, and there is potential to use a variety of methods (e.g., using a droplet, spraying a fine mist, etc.) as long as the treatment passes the cuticle and is perceived by the bombarded epidermal cells.
7. A confocal microscope is ideal for data collection of FP mobility assays because of the possibility to collect z-stacks to cover the whole bombardment site. We use 20× or 25× water dipping lens on a confocal laser scanning microscope to image bombarded tissue.

   However, it is also possible to observe and count cells using an epi-fluorescence microscope. In this scenario, it can be difficult to archive image data that covers the full bombardment site, but an experienced user can easily recognize and count fluorescent cells.
8. We find identification of fluorescence in nuclei a good indicator of cells containing FP. We count cells at the boundary for visible movement of FP around a transformation site. It is important to be consistent in your definition of visible or detectable fluorescence within an experiment. This can be

guided by a single experimenter performing a complete experiment and/or defining fluorescence above background levels with fixed parameters.

9. Bombardment carries the potential to cause significant damage to the tissue. This is often visible as holes on the tissue or many dead cells, both detectable by autofluorescence. If tissue is damaged, do not use it for quantification of mobility of an FP; cells might be transformed, but movement is rarely observed.

Acknowledgments

Research in the Faulkner lab is supported by the European Research Council (725459, "INTERCELLAR") and the Biotechnology and Biological Research Council (BBS/E/J/000PR9796). SS is supported by the John Innes Foundation.

References

1. Liesche J (2017) Sucrose transporters and plasmodesmal regulation in passive phloem loading. J Integr Plant Biol 59:311–321
2. Rennie EA, Turgeon R (2009) A comprehensive picture of phloem loading strategies. Proc Natl Acad Sci U S A 106:14162
3. Vatén A, Dettmer J, Wu S, Stierhof Y-D, Miyashima S, Yadav Shri R, Roberts Christina J, Campilho A, Bulone V, Lichtenberger R, Lehesranta S, Mähönen Ari P, Kim J-Y, Jokitalo E, Sauer N, Scheres B, Nakajima K, Carlsbecker A, Gallagher Kimberly L, Helariutta Y (2011) Callose biosynthesis regulates symplastic trafficking during root development. Dev Cell 21: 1144–1155
4. Faulkner C, Petutschnig E, Benitez-Alfonso Y, Beck M, Robatzek S, Lipka V, Maule AJ (2013) LYM2-dependent chitin perception limits molecular flux via plasmodesmata. Proc Natl Acad Sci U S A 110:9166–9170
5. Wang X, Sager R, Cui W, Zhang C, Lu H, Lee JY (2013) Salicylic acid regulates Plasmodesmata closure during innate immune responses in Arabidopsis. Plant Cell 25:2315–2329
6. Cui W, Lee JY (2016) Arabidopsis callose synthases CalS1/8 regulate plasmodesmal permeability during stress. Nat Plants 2:16034
7. Benitez-Alfonso Y, Faulkner C, Pendle A, Miyashima S, Helariutta Y, Maule A (2013) Symplastic intercellular connectivity regulates lateral root patterning. Dev Cell 26(2): 136–147
8. Brunkard JO, Xu M, Scarpin MR, Chatterjee S, Shemyakina EA, Goodman HM, Zambryski P (2020) TOR dynamically regulates plant cell–-cell transport. Proc Natl Acad Sci U S A 117: 5049–5058
9. Christensen NM, Faulkner C, Oparka K (2009) Evidence for unidirectional flow through plasmodesmata. Plant Physiol 150: 96–104
10. Thomas CL, Bayer EM, Ritzenthaler C, Fernandez-Calvino L, Maule AJ (2008) Specific targeting of a plasmodesmal protein affecting cell-to-cell communication. PLoS Biol 6(1):e7
11. Gaudioso-Pedraza R, Beck M, Frances L, Kirk P, Ripodas C, Niebel A, Oldroyd GED, Benitez-Alfonso Y, de Carvalho-Niebel F (2018) Callose-regulated symplastic communication coordinates symbiotic root nodule development. Curr Biol 28:3562–3577.e3566
12. Kitagawa M, Fujita T (2015) A model system for analyzing intercellular communication through plasmodesmata using moss protonemata and leaves. J Plant Res 128:63–72
13. Sanford JC (1988) The biolistic process. Trends Biotechnol 6:299–302
14. Johnston MG, Faulkner C (2021) A bootstrap approach is a superior statistical method for the comparison of non-normal data with differing variances. New Phytol 230:23–26

Chapter 18

Investigating Plasmodesmata Function in *Arabidopsis Thaliana* Using a Low-Pressure Bombardment System and GFP Movement Assay

Jessica C. Fernandez and Tessa M. Burch-Smith

Abstract

Plasmodesmata are nanopores in the plant cell wall that allow direct cell-to-cell communication. They are key for plant growth, development, and defense. However, studying these pores is challenging due to their small size, with diameters of 30–50 nm and lengths that match cell wall thickness. One particular challenge is measuring how much cell-to-cell trafficking is facilitated by the plasmodesmata in a tissue or between particular cells. Here, we present an approach for studying plasmodesmata-mediated trafficking in the model plant *Arabidopsis thaliana* by using an easy-to-build and affordable low-pressure particle bombardment apparatus. Using low-pressure particle bombardment at around 60 psi, we are able to transform individual cells in the leaf epidermis and study by confocal fluorescence microscopy the subsequent cell-to-cell trafficking of the diffusible molecule green fluorescent protein (GFP). The technique and equipment could be used by any plant biologist to measure intercellular trafficking through plasmodesmata under varying growth conditions including exposure to different stresses, light conditions, chemical treatments, or in various mutant backgrounds.

Key words Plasmodesmata, Green fluorescent protein, Particle bombardment, Gene gun, Movement assay, Intercellular trafficking, *Arabidopsis*, Agroinfiltration

1 Introduction

Plasmodesmata (PD) are cell wall pores that connect adjacent plant cells, thus allowing direct cell-to-cell communication and intercellular trafficking of molecules including metabolites, signaling molecules, and ions. Many small molecules, such as metabolites and small ions, move through plasmodesmata via simple diffusion. Some proteins, like the green fluorescent protein (GFP), also traverse plasmodesmata by diffusion [1, 2]. However, other proteins, exemplified by the WUSCHEL (WUS) transcription factor and its mRNA, move in a directional manner via plasmodesmata from the

Yoselin Benitez-Alfonso and Manfred Heinlein (eds.), *Plasmodesmata: Methods and Protocols*,
Methods in Molecular Biology, vol. 2457, https://doi.org/10.1007/978-1-0716-2132-5_18,

"organizing center" to the outer layer of the meristem (layer 1) indicating a more controlled and targeted movement [3].

Determining plasmodesmal flux is challenging because of the need to introduce a labeled molecule whose movement can be easily tracked between cells without disrupting normal tissue and plasmodesmata function (reviewed in [4]). One of the earliest techniques used to measure intercellular trafficking was microinjection of fluorescent small molecules and dyes into plant cells [5–7]. Microinjection presented a number of problems including requiring a great deal of skill to master the technique. Further, the large vacuoles of plant cells rendered the technique extremely difficult since the dye needed to be delivered to the cytoplasm as this is the path for cell-to-cell diffusion, but often the dye was delivered instead to the large central vacuole. A recently developed technique, the "Drop-ANd-See" or DANS assay, utilizes carboxyfluorescein diacetate (CFDA) as a proxy to measure plasmodesmata permeability by measuring the spread of the dye [8]. CFDA is colorless until it enters cells where esterases cleave the diacetate from the carboxyfluorescein (CF) moiety, yielding a CF molecule that fluoresces after excitation by ultraviolet light [9]. The dye is applied to the adaxial side of the leaf, and the size of the cell cluster containing CF on the abaxial side of the leaf is measured a short time later. An overall assessment of plasmodesmata-mediated trafficking could be quantified using this assay. In the DANS assay, the overall diameter of the CF-containing cell cluster is quantified, and therefore, cell-to-cell spread quantification is not possible. CFDA can also be used to measure the import of the dye into the vascular tissue and measure the movement and unloading into sink tissues although this gives estimates of systemic spread and is more challenging for determining cell-to-cell trafficking [10, 11].

Other plasmodesmal flux studies have utilized free cytoplasmic GFP as a proxy for cell-to-cell spread. In *Arabidopsis* roots, GFP expressed from the *SUC2* promoter in vascular cells is commonly used to study unloading from the phloem into companion cells via plasmodesmata [4, 12]. While this model for measuring intercellular trafficking in roots is robust, studying plasmodesmata in *Arabidopsis* leaves remains challenging. Therefore, measuring the diffusion of GFP between cells in leaves would provide a good understanding of plasmodesmal function in leaf tissues. Monitoring intercellular trafficking by measuring the cell-to-cell spread of fluorescent proteins has been used with much success in the *Nicotiana benthamiana* model system [13–16]. *Agrobacterium* carrying a *35S:GFP* construct can be infiltrated into leaves at very low concentrations, resulting in transformation of individual cells [17]. This approach has been coupled with virus-induced gene silencing (VIGS), drug treatments, and varying growth conditions to understand how genetic and environmental perturbations impact intercellular trafficking [15, 16, 18]. Unfortunately, the

N. benthamiana genome is not yet well annotated, and genetic manipulation outside VIGS through the generation of stable transgenic lines is fairly uncommon.

In this chapter, we present a method to measure plasmodesmata-mediated trafficking using an easily constructed low-pressure bombardment system in the well-studied *Arabidopsis* model system. In this system, small metal particles are coated with plasmid DNA encoding GFP expressed from the *35S* promoter and then shot into plant tissues under low pressure. Transformed cells express GFP that then diffuses into neighboring cells allowing quantification of intercellular trafficking, similar to the method used for trafficking assays in *N. benthamiana*. The large number of mutant lines available allows any researcher to study how a mutation of interest may affect intercellular trafficking. Commercial particle bombardment systems are expensive, and the typical bombardment for plant transformation is performed around 1000 psi. However, these high pressures can inhibit plasmodesmata-mediated trafficking by forcing closure of the pores [19]. Low-pressure bombardment ensures that plasmodesmata are not shut down [20], and therefore is an excellent method to quantifying cell-to-cell trafficking in systems like *Arabidopsis* that are resistant to transient expression by *Agrobacterium* [21]. The method we describe here is based on [20]. Another advantage is that the low-pressure bombardment system uses common lab materials and supplies readily available from university machine shops or facilities departments. Using this low-pressure bombardment system, a variety of mutants, treatments, and growth conditions can also be easily studied for their effects on plasmodesmata and intercellular trafficking. For example, the effects of various metabolites or hormones on plasmodesmata-mediated trafficking could be measured.

Here we offer methods for a single experiment with one wild-type *Arabidopsis* ecotype and a mutant in the same ecotype. The experiment should be replicated at least three times and should identify and count a minimum of 15 foci (individual transformed cells and surrounding cells into which GFP has diffused) per replicate.

2 Materials

2.1 Plant Growth

1. *Arabidopsis* accessions *Columbia* (*Col-0*), *Landsberg erecta* (*Ler*), or *Nossen* (No-0) wild-type ecotypes.
2. Soil (any soil mix which provides adequate drainage and aeration).
3. Small 5 cm × 5 cm inserts for flats.
4. Cleaned and sanitized flats.

5. Clear plastic humidity domes.
6. Plant growth chambers.

2.2 Constructs for Particle Bombardment and Preparation of DNA-Coated Metal

1. *E. coli* competent cells for transformation.
2. *Plasmid* pRTL2rsGFP [20] (*see* **Note 1**).
3. Luria-Bertani broth and plates containing 100 μg/ml carbenicillin (from stock 100 mg/ml) for the growth of *E. coli* containing pRTL2rsGFP.
4. 0.5× MS + 0.1% sucrose plant agar plates (*see* **Note 2**).
5. 2.5 M $CaCl_2$.
6. 0.1 M spermidine filtered sterilized with syringe 0.22 μm filter.
7. 70% ethanol.
8. 100% ethanol.
9. Autoclaved ddH_2O.

2.3 Low-Pressure Particle Bombardment

1. Modified plastic vacuum desiccator (e.g., EW-06514-30, Bel-Art).
2. 1/8 in. npt 10–32 reducer bushing.
3. Pipe bulkhead.
4. 1/16 in. Tygon tubing.
5. 1/16 in. × 10–32 hose barb.
6. Momentary switch.
7. Attenuator.
8. Refrigerated vapor trap system (e.g., RVT400-115, Thermo Fisher).
9. Two lengths vacuum rubber hose.
10. Laboratory jack (e.g., 14-673-50, Fisher).
11. Two types of washers: flat washers and raised round silicone washers. Flat washers with 9 mm inner diameter and a 1 mm width. Round washer is 1 mm in diameter.
12. Mesh grid (1.3 cm diameter).
13. Aluminum disks (1.2–1.3 cm diameter) (*see* **Note 3**).
14. Metal microcarriers for bombardment (tungsten is discontinued; 1.0 μm gold microcarriers from Bio-Rad are recommended).

2.4 Microscopy

1. A confocal laser scanning microscope with 25× lens. GFP excitation wavelength of 488 nm and emission between wavelengths 500 nm and 530 nm.
2. Glass microscope slides and cover slips.
3. Single-edged razor blades.

4. Leica LasX software (https://www.leica-microsystems.com/products/microscope-software/p/leica-las-x-ls/) or ImageJ (https://imagej.nih.gov/ij/index.html) [22].

3 Methods

3.1 Construction of Low-Pressure Bombardment System (Fig. 1)

1. Modify a vacuum desiccator by attaching a refrigerated vapor trap system and a vacuum pump to create the low-pressure bombardment apparatus (*see* **Note 4**).
2. Retool the lid of the vacuum desiccator (Fig. 1f) to include a DNA holder that consists of washers for holding the aluminum disks and mesh grids (Fig. 1c). Make a hole on the top to fit a 1/8 in. npt 10–32 reducer bushing secured with a pipe bulkhead. Connect the gun to the bushing via a length of 1/16 in. Tygon tubing and a 1/16 in. × 10–32 hose barb.

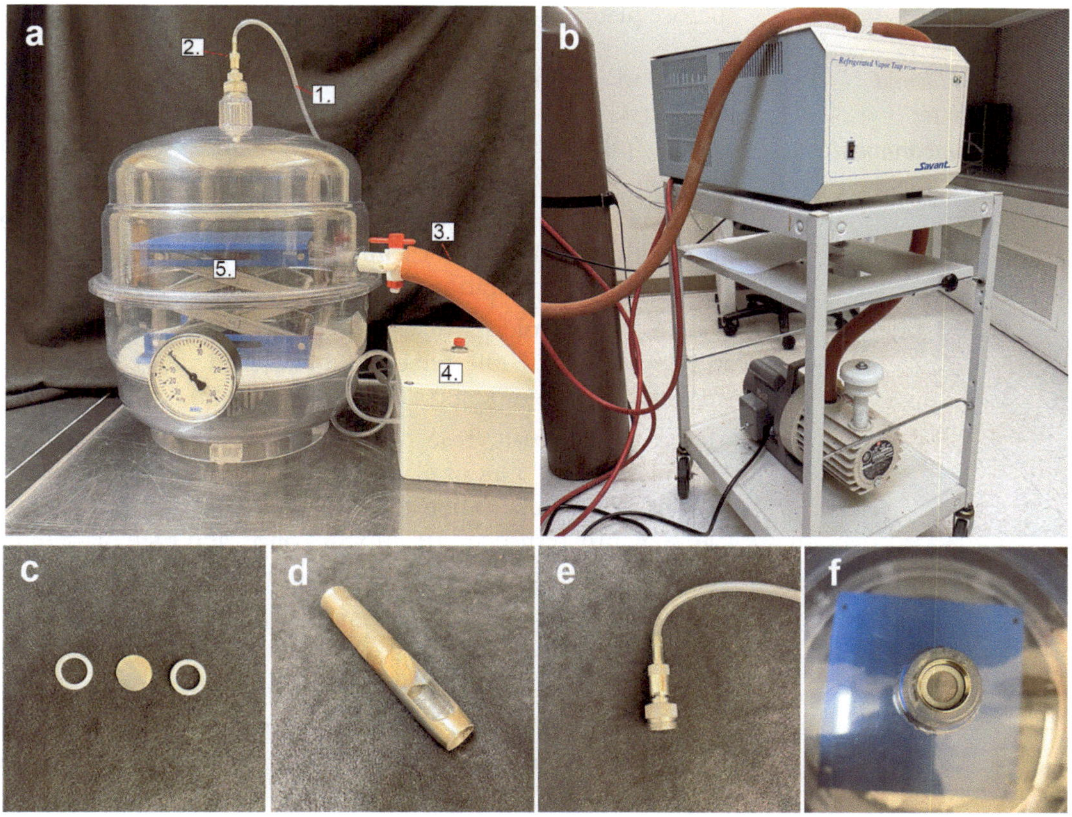

Fig. 1 Bombardment apparatus. (**a**) Modified vacuum desiccator with (1) 1/16 in. Tygon tubing, (2) connection of Tygon tubing and a 1/16 in. × 10–32 hose barb, 1/8 in. npt 10–32 bushing secured with a pipe bulkhead, (3) rubber vacuum rubber hose, (4) gun switch (box), (5) laboratory jack. (**b**) Refrigerated vapor trap and vacuum. (**c**) Flat silicone washer, mesh grid, and washer (from left to right). (**d**) Hole punch for bombardment aluminum disk. (**e**) Exit gun attachment. (**f**) Top-down view of bombardment apparatus

3. To assemble the gun, use a timer circuit in which a momentary switch and attenuator activate the pressured air release into the Tygon tubing for 1 s.
4. Connect the vacuum to a refrigerated vapor trap system via a vacuum rubber hose. Attach a second hose from the vapor trap to the vacuum desiccator.
5. Place lab jack inside the vacuum desiccator and adjust to desired height.

3.2 Plant Growth

1. Plant *Arabidopsis* seeds directly to the soil and allow them to stratify at 4 °C for 2 days.
2. Move flats to a plant growth chamber under standard conditions of 16 h day 22 °C, 8 h night 20 °C (light intensity of 120 μmol m^{-2} s^{-1}). Water from the bottom until the soil is wet and then discard excess water.
3. Once plants are 3 weeks old, take leaves 5 and 6 from five plants (*see* **Note 5**) and carefully place them on 0.5× MS + 0.1% sucrose plate. Place 5–6 leaves per plate.

3.3 DNA Plasmid Preparation for Particle Bombardment and Tungsten Preparation

1. Grow an overnight culture of *E. coli* containing the pRTL2rsGFP plasmid in Luria-Bertani (LB broth) and 100 μg/ml carbenicillin. Isolate the plasmid using any plasmid isolation kit or protocol (e.g., QIAGEN).
2. Generate large amounts of plasmid and ensure a concentration greater than 300 ng/ul. Vacuum concentrate, if necessary, to achieve this concentration.
3. Wash 5 mg of gold particles three times with 100 μl ddH_2O and centrifuge at 1000 rpm on a tabletop centrifuge such as the Eppendorf 5424 model (or 94 × *g*) for 1 min.
4. After the last wash, resuspend in 80 μl ddH_2O and make 12.5 μl aliquots. 5 μg of plasmid pRTL2rsGFP DNA is added to the washed metal. Then add 12.5 μl of filter-sterilized 2.5 M $CaCl_2$, gently resuspend the mixture once, and then place onto a vortex shaker (level 2–3 on Fisher brand Vortex Genie).
5. Quickly add 5 μl of 0.1 M spermidine (free base, freshly made, and filter-sterilized), wait 30 s, and finger vortex. Shake the suspension in the vortex shaker for 10 min (*see* **Note 6**).
6. After the 10 min, remove the tubes and let them settle for 3–4 min. Centrifuge the tubes for 15 s at 5000 rpm on a tabletop centrifuge (or 2352 × *g*).
7. Remove the supernatant and resuspend the metal in 100 μl 70% ethanol by finger vortexing (*see* **Note 7**).
8. Allow the tubes to settle (3–5 min) and centrifuge for 15 s at 5000 rpm on a tabletop centrifuge (or 2352 × *g*).

9. Finally, remove the supernatant and resuspend by finger vortexing with 40 μl 100% ethanol.

3.4 Macro-carrier Assembly and Low-Pressure Bombardment

1. Use a sharpened metal hole punch (Fig. 1d) to make bombardment disks to a final diameter of 1.3 cm. Make the bombardment disks by layering aluminum foil and a thin paper towel to three layers total. Hammer the hole punch in clean strokes and then separate the disks and autoclave them on a dry cycle (20 min sterilization, 50 minspiepr146 dry).
2. Place the washers and grid in the DNA holder in the following order: flat washer, grid, and top washer (Fig. 1c). Rest the aluminum bombardment disk (with the DNA-coated microcarriers) on top with the DNA-coated side face down (Fig. 2a inset).
3. Load each carrier with 3–5 μl of DNA-coated tungsten, and once the ethanol evaporates, it is placed DNA side down onto the bombardment system.
4. Place the plate containing the leaves 10.5 cm from the exit of gun (Fig. 2a) and align with the gun and then adjust the distance from the gun using the lab jack (Fig. 1a; *see* **Note 8**).
5. Connect the vacuum hose to the attachment (Fig. 1a) and turn the vacuum on until the gauge reaches 25 in. Hg.
6. Once this pressure is reached, activate the particle gun twice (press button twice) per macro-carrier. Bombard each plate twice (with two macro-carriers) and shift the plate slightly from the starting position between bombardments to ensure transformation across different leaves (*see* **Note 9**).
7. Allow the leaves to recover for 1 h after bombardment on the same plate and then transfer to new plates as appropriate.
8. Place the plates into the growth chamber and image 24 h post-bombardment (*see* **Note 10**).

3.5 Measuring Intercellular Trafficking by Confocal Fluorescence Microscopy

1. Take a leaf from the post-bombardment plate and remove the midrib with a single-edged razor blade. Place the remaining leaf tissue between a glass slide and coverslip and ensure the tissue is fully submerged in water.
2. Image the fluorescent foci on the confocal fluorescence microscope with the 25× objective and with 488 nm excitation wavelength and collecting emission from 500 to 530 nm.
3. Identify single transformed epidermal cells expressing GFP by their bright fluorescence (*see* **Note 11**) (Fig. 2c, d). Once the primary GFP-expressing cell is identified, move up and down the z-plane and locate contiguous and neighboring cells containing GFP.

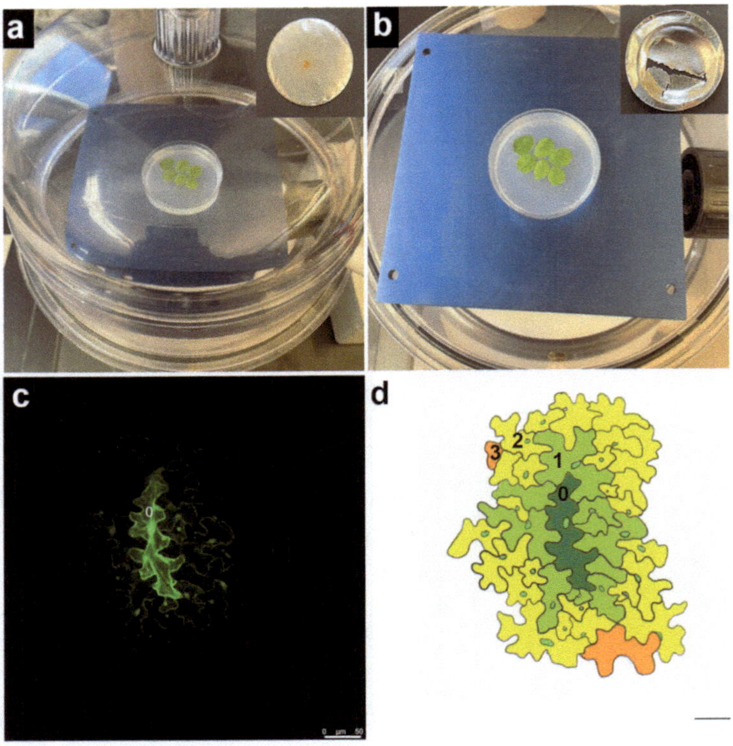

Fig. 2 Bombardment of leaves on a plate and GFP movement assay (**a**) MS plate with *Arabidopsis* leaves is aligned with the exit of the gun. Inset: pre-bombardment aluminum disk with gold particles. (**b**) After bombardment, the plate should not appear much different than prior to bombardment. Inset: disk ruptured by bombardment. (**c**) Primary transformed cell, labeled 0, has the brightest fluorescence in a focus as seen under the confocal fluorescence microscope. The GFP has trafficked into the neighboring cells with weaker fluorescence signal. (**d**) Cartoon depicting the three different cell layers surrounding the primary transformed cell in dark green (0). Layer 1 in lime green, layer 2 in yellow, and layer 3 in orange. Nuclei are in bright green. Scale bars are 50 μm

4. Collect successive images in the z-plane to generate a z-stack containing all the GFP-expressing cells in a cluster (*see* **Note 12**).

3.6 Analysis of the Data

1. Use the Leica LasX software or ImageJ to identify the primary transformed cell within a stack of images and count the layers of cells to which the GFP spread, taking caution to identify layers (Fig. 2d). Count cell layers as they represent discrete rounds of trafficking through PD. Also count the total number of cells containing GFP from the primary transformed cell.

4 Notes

1. The plasmid concentration for bombardment needs to be above 350 ng/μl for adequate transformation efficiency. If the concentrations are below 350 ng/μl after the isolation procedure, the DNA can be concentrated by using a vacuum dryer to remove water.
2. If the bombarded leaves are to be transferred to media with selective chemical or hormone treatments, fresh media should be made the day of bombardment, unless the hormone or chemical does not degrade.
3. If the aluminum disks are made in house, use regular aluminum foil and not heavy duty aluminum foil. Heavy duty aluminum foil will not burst during the bombardment event and therefore not transform the leaves. The aluminum foil is purchased from a regular supermarket or local store.
4. Pressure is generated in the chamber with helium, and the pressure is brought to 60 psi. Machine shops at universities can easily modify the vacuum desiccator.
5. For reproducibility and consistency, leaves of similar age and size should always be used. This will ensure that any variability in movement is not due to developmental differences between leaves.
6. During the 10-min incubation, the refrigerated vapor trap should be turned on and allowed to run for a total of about 30 min prior to bombardment.
7. At this point, the metal should have a very silky appearance, and there should be no clumping. If there are clumps, do not use the metal as this will cause severe damage to bombarded tissue and will not give reliable movement data. Finger vortexing the sample ensures that the metal will not aggregate but instead fully mix.
8. Prior to performing experiments, it is highly recommended to optimize the system. For example, the height of the jack inside the bombardment apparatus may have to be adjusted to be closer or further away from the bombardment "gun" attachment and the height in which sufficient transformations should be used for experiments. Other parameters that can be adjusted (although not as common) are the amount of vacuum applied to the system.
9. The bombardment is performed under vacuum with 60 psi (helium). The leaves should not drastically differ from their pre-bombardment appearance, with only faint marks from the microcarriers being visible (Fig. 2b), while the bombardment aluminum macro-carrier is ruptured after the bombardment event (Fig. 2b inset).

10. The bombardment and the following imaging should be performed at the same point in the diurnal cycle for each replicate.
11. It is critical to find isolated transformed cells. One easy way to identify the transformed cell is to observe the fluorescence. Typically, a single transformed cell will consist of one cell brightly expressing GFP (Fig. 2c labeled "0"), and the GFP fills the nucleus and cytoplasm. Neighboring cells into which the GFP has diffused will be dimmer with less fluorescence in the cytoplasm and nucleus. In some cases, the GFP has spread more to the mesophyll layers than the neighboring epidermal layers, and this could be quantified as well, as it represents an increase in trafficking through secondary plasmodesmata (those plasmodesmata formed across established cell walls). Further, some cells extend further than others, and counting layers needs to be done more cautiously (Fig. 2d labels "2 and 3"). In addition to layers, one can quantify the numbers of cells in which the GFP diffused as a metric for quantification of spread.
12. The typical z-stack consists of 10–15 slices; however, due to the undulating nature of the leaf, the thickness of the z-stack will vary. This should be optimized for each focus to ensure enough detail in the images collected of the GFP spread.

Acknowledgments

We thank George Butler, an undergraduate researcher in the lab, for excellent technical assistance. The Staff of the Biological Service Facility at the University of Tennessee, Knoxville, are acknowledged for assembling the bombardment apparatus and providing technical details for this article. Research reported in this publication was supported by the National Institute of General Medical Sciences of the National Institutes of Health under Award Number F31GM131671. The content is solely the responsibility of the authors and does not necessarily represent the official views of the National Institutes of Health. Additional support was from the National Science Foundation through grant MCB 1846245.

References

1. Schonknecht G, Brown JE, Verchot-Lubicz J (2008) Plasmodesmata transport of GFP alone or fused to potato virus X TGBp1 is diffusion driven. Protoplasma 232:143–152
2. Liarzi O, Epel BL (2005) Development of a quantitative tool for measuring changes in the coefficient of conductivity of plasmodesmata induced by developmental, biotic, and abiotic signals. Protoplasma 225:67–76. https://doi.org/10.1007/s00709-004-0079-x
3. Daum G, Medzihradszky A, Suzaki T, Lohmann JU (2014) A mechanistic framework for noncell autonomous stem cell induction in Arabidopsis. Proc Natl Acad Sci U S A 111: 14619–14624. https://doi.org/10.1073/pnas.1406446111
4. Stadler R, Wright KM, Lauterbach C, Amon G, Gahrtz M, Feuerstein A, Oparka KJ, Sauer N (2005) Expression of GFP-fusions in Arabidopsis companion cells reveals non-specific

protein trafficking into sieve elements and identifies a novel post-phloem domain in roots. Plant J 41:319–331. https://doi.org/10.1111/j.1365-313X.2004.02298.x

5. Kragler F (2015) Analysis of the conductivity of plasmodesmata by microinjection. Methods Mol Biol 1217:173–184. https://doi.org/10.1007/978-1-4939-1523-1_12
6. Lucas WJ, Bouche-Pillon S, Jackson DP, Nguyen L, Baker L, Ding B, Hake S (1995) Selective trafficking of KNOTTED1 homeodomain protein and its mRNA through plasmodesmata. Science 270:1980–1983
7. Storms MMH, van der Schoot C, Prins M, Kormelink R, van Lent JW, Goldbach RW (1998) A comparison of two methods of microinjection for assessing altered plasmodesmal gating in tissues expressing viral movement proteins. Plant J 13:131–140
8. Cui W, Wang X, Lee JY (2015) Drop-ANd-See: a simple, real-time, and noninvasive technique for assaying plasmodesmal permeability. Methods Mol Biol 1217:149–156. https://doi.org/10.1007/978-1-4939-1523-1_10
9. Wright KM, Oparka KJ (1996) The fluorescent probe HPTS as a phloem-mobile, symplastic tracer: an evaluation using confocal laser scanning microscopy. J Exp Bot 47:439–445
10. Ross-Elliott TJ, Jensen KH, Haaning KS, Wager BM, Knoblauch J, Howell AH, Mullendore DL, Monteith AG, Paultre D, Yan D, Otero S, Bourdon M, Sager R, Lee JY, Helariutta Y, Knoblauch M, Oparka KJ (2017) Phloem unloading in Arabidopsis roots is convective and regulated by the phloem-pole pericycle. eLife 6:e4125. https://doi.org/10.7554/eLife.24125
11. Peron T, Candat A, Montiel G, Veronesi C, Macherel D, Delavault P, Simier P (2016) New insights into phloem unloading and expression of sucrose transporters in vegetative sinks of the parasitic plant *Phelipanche ramosa* L. (Pomel). Front. Plant Sci 7:2048. https://doi.org/10.3389/fpls.2016.02048
12. Fitzgibbon J, Beck M, Zhou J, Faulkner C, Robatzek S, Oparka K (2013) A developmental framework for complex plasmodesmata formation revealed by large-scale imaging of the Arabidopsis leaf epidermis. Plant Cell 25:57–70. https://doi.org/10.1105/tpc.112.105890
13. Burch-Smith TM, Zambryski PC (2010) Loss of increased size exclusion limit (ISE)1 or ISE2 increases the formation of secondary plasmodesmata. Curr Biol 20:989–993. https://doi.org/10.1016/j.cub.2010.03.064
14. Bobik K, Fernandez JC, Hardin SR, Ernest B, Ganusova EE, Staton ME, Burch-Smith TM (2019) The essential chloroplast ribosomal protein uL15c interacts with the chloroplast RNA helicase ISE2 and affects intercellular trafficking through plasmodesmata. New Phytol 221:850–865. https://doi.org/10.1111/nph.15427
15. Brunkard JO, Xu M, Scarpin MR, Chatterjee S, Shemyakina EA, Goodman HM, Zambryski P (2020) TOR dynamically regulates plant cell-cell transport. Proc Natl Acad Sci U S A 117:5049–5058. https://doi.org/10.1073/pnas.1919196117
16. Ganusova EE, Reagan BC, Fernandez JC, Azim MF, Sankoh AF, Freeman KM, McCray TN, Patterson K, Kim C, Burch-Smith TM (2020) Chloroplast-to-nucleus retrograde signalling controls intercellular trafficking via plasmodesmata formation. Philos Trans R Soc Lond Ser B Biol Sci 375:20190408. https://doi.org/10.1098/rstb.2019.0408
17. Brunkard JO, Burch-Smith TM, Runkel AM, Zambryski P (2015) Investigating plasmodesmata genetics with virus-induced gene silencing and an agrobacterium-mediated GFP movement assay. Methods Mol Biol 1217:185–198. https://doi.org/10.1007/978-1-4939-1523-1_13
18. Brunkard JO, Zambryski P (2019) Plant cell-cell transport via plasmodesmata is regulated by light and the circadian clock. Plant Physiol 181:1459–1467. https://doi.org/10.1104/pp.19.00460
19. Oparka KJ, Prior DAM (1992) Direct evidence for pressure-generated closure of plasmodesmata. Plant J 2:741–750
20. Crawford KM, Zambryski PC (2000) Subcellular localization determines the availability of non-targeted proteins to plasmodesmatal transport. Curr Biol 10:1032–1040
21. Stonebloom S, Brunkard JO, Cheung AC, Jiang K, Feldman L, Zambryski P (2012) Redox states of plastids and mitochondria differentially regulate intercellular transport via plasmodesmata. Plant Physiol 158:190–199. https://doi.org/10.1104/pp.111.186130
22. Schindelin J, Arganda-Carreras I, Frise E, Kaynig V, Longair M, Pietzsch T, Preibisch S, Rueden C, Saalfeld S, Schmid B, Tinevez JY, White DJ, Hartenstein V, Eliceiri K, Tomancak P, Cardona A (2012) Fiji: an open-source platform for biological-image analysis. Nat Methods 9:676–682. https://doi.org/10.1038/nmeth.2019

Chapter 19

Quantifying Plasmodesmatal Transport with an Improved GFP Movement Assay

Wilson Horner and Jacob O. Brunkard

Abstract

Plasmodesmata (PD) are membrane-lined channels that cross the cell wall to connect the cytosol of adjacent plant cells, permitting diverse cytosolic molecules to move between cells. PD are essential for plant multicellularity, and the regulation of PD transport contributes to metabolism, developmental patterning, abiotic stress responses, and pathogen defenses, which has sparked broad interest in PD among diverse plant biologists. Here, we present a straightforward method to reproducibly quantify changes in the rate of PD transport in leaves. Individual cells are transformed with *Agrobacterium* to express fluorescent proteins, which then move beyond the transformed cell via PD. Forty-eight to 72 h later, the extent of GFP movement is monitored by confocal fluorescence microscopy. This assay is versatile and may be combined with transient gene overexpression, virus-induced gene silencing, physiological treatments, or pharmaceutical treatments to test how PD transport responds to specific conditions. We expect that this improved method for monitoring PD transport in leaves will be broadly useful for plant biologists working in diverse fields.

Key words Plasmodesmata, *Agrobacterium*, Virus-induced gene silencing, TARGET OF RAPAMYCIN, Plant development, *Nicotiana benthamiana*, Protein trafficking, GFP, Cell signaling

1 Introduction

Plasmodesmata (PD) are nanoscopic, membrane-bound channels that cross the cell wall, connecting the cytosol of neighboring cells. Cytosolic molecules may freely pass through PD if they are small enough: water, ions, metabolites (e.g., sugars), and phytohormones can readily move through PD, as can proteins up to ~80 kDa and small RNA duplexes (e.g., RNAi signals) [1, 2]. Viruses hijack PD to promote their own trafficking between cells [3], and other pathogens may coopt or even modify PD to spread virulence effectors [4–7]. PD transport is absolutely essential for plant life: no mutants lacking PD have ever been isolated, PD connect all cells with only highly specialized exceptions (e.g.,

Yoselin Benitez-Alfonso and Manfred Heinlein (eds.), *Plasmodesmata: Methods and Protocols*,
Methods in Molecular Biology, vol. 2457, https://doi.org/10.1007/978-1-0716-2132-5_19,

stomatal guard cells), and any severe disruption to PD transport causes pleiotropic deleterious phenotypes. PD or PD-like structures evolved repeatedly, in parallel, in every complex multicellular lineage of eukaryotes that make cellulosic cell walls, including multiple green algal lineages, distantly related brown algae, and land plants [5, 8]. The homoplasy of PD-like structures in organisms with cellulosic cell walls suggests that PD were necessary adaptations for complex multicellularity, i.e., in these lineages, PD are required for three-dimensional multicellularity such that some cells are not in direct contact with the external environment [5, 9].

Broadly, PD transport can be divided into two categories: targeted and non-targeted movement [10, 11]. Some molecules may be targeted or delivered to PD to facilitate their movement between cells, such as viral movement proteins [3, 12–14] and the SHORTROOT transcription factor [15–17]. In general, however, any molecule in the cytosol is potentially capable of moving through PD, provided it is small enough to pass through a PD channel [2, 10]. The method presented here only addresses non-targeted PD trafficking, the predominant functional role for PD. Various fluorescent tracers have been used to monitor rates of PD transport, but GFP has several advantages over many other fluorophores. First, GFP is synthesized within the cytosol, is not membrane permeable, and does not exit the cytosol through membrane transporters. Many chemical fluorescent tracers must first be imported by cells, and they can then be sequestered in the vacuoles or apoplast [18, 19], confounding efforts to distinguish between rates of intercellular trafficking and rates of tracer uptake or export from the cytosol. Second, GFP is a heterologous molecule that does not have endogenous interacting partners in plant cells once it is folded, so GFP trafficking provides an unbiased measure of protein movement between cells. Third, because GFP is synthesized by the host cell, rather than directly injected to the cell, this method is relatively noninvasive. Notoriously, microinjection and high-pressure biolistic bombardment, by contrast, can induce severe PD closure. Finally, because GFP movement can be tracked by confocal microscopy at the cellular level, this method counts how many cell wall barriers GFP has crossed and is not biased by incidental developmental phenotypes (e.g., differences among plants in cell size, cell number, or leaf thickness).

PD transport rates are remarkably variable and dynamically regulated by myriad cues, including pathogen infection [20], phytohormones [21, 22], light and the circadian clock [23], redox status [24, 25], chloroplast retrograde signals [26–31], and metabolic status [32]. To test how these and other cues impact PD transport, here, we describe three genetic approaches that can be combined with GFP movement quantification: virus-induced gene silencing (VIGS), transient gene overexpression, and chemical genetics. Virus-induced gene silencing takes advantage of the plant's antiviral RNAi system to silence host genes. The *Tobacco rattle virus* (TRV) genome is modified to include a noncoding fragment, ~300 to 700 nucleotide (nt) long, cloned from a host gene's mRNA; the plant is infected with this modified TRV, and as the TRV infection spreads, the host generates 21–22 nt long siRNAs derived from the viral RNA genome that then target both the virus and the host gene for silencing by ARGONAUTE-siRNA RNAi silencing complexes [33, 34]. TRV has a broad host range, so TRV-based VIGS can be used in species including *Nicotiana benthamiana* (as described here) and *Arabidopsis thaliana*. The transient gene overexpression method described here uses *Agrobacterium tumefaciens* to transform leaf cells with any T-DNA construct, driving high levels of gene expression from ~24 to 72 h post-infiltration. For chemical genetics, leaves may be treated with pharmaceuticals that target the function of specific gene products (e.g., selective protein kinase inhibitors or phytohormones). Depending on the experimental design, these three approaches may be used in tandem or in parallel to dissect genetic pathways that regulate PD transport.

To quantify changes in PD trafficking induced by the previously described genetic approaches, plants can be infiltrated with a low inoculum of *Agrobacterium* carrying a binary vector to transform cells to express GFP driven by the *cauliflower mosaic virus* (CaMV) 35S promoter. This low level of inoculation results in only several isolated cells being transformed in each leaf. Two days after infiltration with this inoculum, leaves can be imaged with confocal fluorescence microscopy. Transformed cells will glow brightly with GFP fluorescence, and varying numbers of surrounding cells will also fluoresce depending on the distance GFP has moved through PD from the transformed cell. Figure 1 shows an example VIGS experiment: plants are grown for 3 weeks, infiltrated with VIGS constructs to induce silencing, and subsequently infiltrated with the GFP inoculum 2 weeks later. Two to 3 days after this agroinfiltration, leaves are imaged by confocal microscopy, and the total number of GFP-containing cells is quantified to assess effects on the rate of plasmodesmatal trafficking [35].

2 Materials

2.1 Plants

1. *Nicotiana benthamiana* accession Nb-1 seeds [36] (*see* **Note 1**).
2. Soil providing good drainage and aeration (such as Sunshine Mix, or similar). Autoclave and fully cool soil if planting for VIGS experiments.
3. 4″ pots (clean and sterilized).
4. Flats (clean and sterilized).

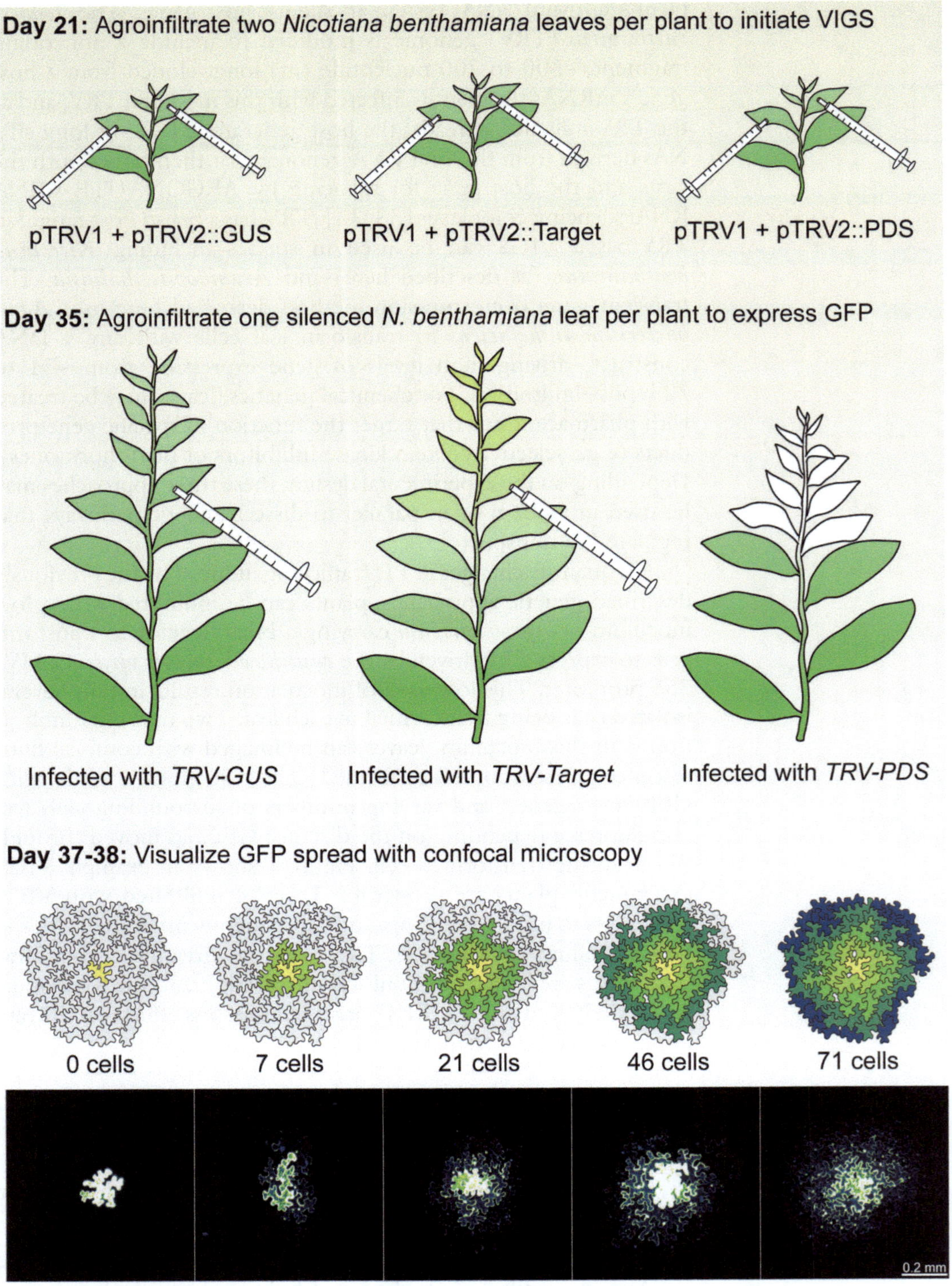

Fig. 1 Overview of methodology to quantify PD transport, using VIGS experiments (Subheading 3.3) as an example. *N. benthamiana* seeds are sown under sterile conditions and left to grow for 3 weeks. At 21 days old, two expanded leaves are infiltrated with *Agrobacterium* to express pTRV1 and pTRV2 carrying either a mock silencing trigger (*GUS*), a trigger that silences a gene of interest (*Target*), or a positive control silencing trigger (*PDS*). Silencing will begin to appear within ~7 to 10 days after agroinfiltration; this can be tracked by monitoring for photobleached (white) leaves in plants infected with *TRV-PDS*. Fourteen days after

5. Light source (photosynthetic flux density ~100 μmol photons/ m^2 s).
6. Controlled release 14-14-14 N/P/K fertilizer.

2.2 Materials for *Agrobacterium*-Mediated Transformation

1. *Agrobacterium tumefaciens* cells (*see* **Notes 2** and **3**) competent for transformation.
2. For VIGS: pTRV1, pTRV2::*PDS*, pTRV2::*GUS*, and pTRV2:: *[Target]* (*see* **Notes 4** and **5**).
3. Binary vector for transient gene overexpression and the mock control binary vector, pEarlyGate::*GUS*.
4. Plasmid containing CaMV 35S promoter driving expression of GFP for the GFP movement assay.

2.3 Media

1. Infiltration medium: 10 mM MES (2-(*N*-morpholino)ethanesulfonic acid), 10 mM $MgCl_2$, and 200 μM acetosyringone, pH adjusted to 5.5–5.6 with KOH (*see* **Note 6**).
2. LB agar plates (10 g tryptone, 5.0 g yeast extract, 10 g NaCl, and 15 g agar in 1.0 L H_2O; autoclave, cool to 65 °C, and pour 25 mL LB agar per 100 mm diameter sterile plate) containing appropriate antibiotics (e.g., 50 mg/L kanamycin, 50 mg/L rifampicin, and 50 mg/L gentamicin; *see* **Note 7**).
3. LB (10 g tryptone, 5.0 g yeast extract, 10 g NaCl, and 15 g agar in 1.0 L H_2O; autoclave and cool to room temperature before using) containing appropriate antibiotics (e.g., 50 mg/L kanamycin, 50 mg/L rifampicin, and 50 mg/L gentamicin; *see* **Note 7**).

2.4 Agroinfiltration

1. *Agrobacterium* cultures, recently streaked onto fresh LB agar plates (*see* **Note 8**).
2. Spectrophotometer capable of measuring optical density (OD) at 600 nm.
3. Needleless 1.0 mL syringes.

2.5 Microscopy

1. Confocal laser scanning microscope equipped with 488 nm Argon laser to excite GFP fluorescence and appropriate emission filters (to detect emissions from ~500 to ~530 nm).

Fig. 1 (continued) agroinfiltration (day 35), one expanded leaf is infiltrated with a very low inoculum of *Agrobacterium* (<1 mL media with OD_{600nm} ~ 10^{-5}) carrying a plasmid to express GFP driven by the *cauliflower mosaic virus* (CaMV) 35S promoter. Two or 3 days later (day 37 or 38), GFP movement is visualized using confocal laser scanning microscopy. Example results are shown in the bottom panels, ranging from extremely limited movement (GFP restricted to the transformed cell) to widespread movement (GFP moved to 71 cells from the transformed cell). GFP fluorescence is false colored in blue-green-yellow-white according to intensity.

2. Needleless syringe (5.0 mL).
3. Glass microscope slides.
4. Glass coverslips (#1.5).
5. Single-edged razor blades.
6. Image processing software (such as ImageJ, http://rsbweb.nih.gov/ij/).

3 Methods

3.1 Plant Growth

1. Place 4″ pots into planting flat, about 18 pots per flat. Fill each pot halfway with sunshine mix (or equivalent) soil, pack soil, and add small amount (~5 to 8 pellets) of 14-14-14 fertilizer. Finish filling pots with soil and water each pot, applying 1.5–2 L per flat.
2. Spread 5–10 *N. benthamiana* seeds in each pot filled with wet soil, and lightly cover in vermiculite.
3. Grow plants under a regular diurnal cycle (e.g., 16 h light/8 h dark) in a growth cart or greenhouse. Water plants from the bottom of the flat every 2 or 3 days, keeping the soil damp.
4. Once the seedlings are ~14 days old, prune out unhealthy or malformed seedlings, leaving one healthy seedling per pot.
5. Continue watering plants from the bottom (in the flat) throughout the experiment or from the top if the seedlings are large enough to be undamaged by the impact of water.

3.2 Preparing Agrobacterium for Infiltration into N. benthamiana

1. From stocks (such as frozen glycerol stocks or plates that are ~1 week old), streak *Agrobacterium* strains containing relevant binary vectors on LB agar plates containing appropriate antibiotics for selection (*see* **Note 7**) and grow in a dark incubator at 28 °C for 2 days.
2. Inoculate LB containing appropriate antibiotics for selection (*see* **Note 7**) with *Agrobacterium* and grow in a dark 28 °C rotary shaker at 250 rounds per minute (RPM) overnight. Recommended culture volumes are specified for each protocol (VIGS, transient expression, and the fluorescent tracer assay, *see* Subheadings 3.3, 3.4, and 3.5, respectively).
3. In the morning, centrifuge *Agrobacterium* culture at 700 × *g* for 10 min at room temperature.
4. In the morning, prepare fresh infiltration medium (*see* **Note 6**).
5. Resuspend the *Agrobacterium* pellet in infiltration medium and bring to desired optical density at 600 nm (OD_{600}; *see* below, Subheading 3.3, **step 1**, Subheading 3.4, **step 1**, and Subheading 3.5, **step 1**) measured with a spectrophotometer. Leave

cultures in a rotator at a low speed just sufficient to prevent cells from settling and at room temperature (~22 to 25 °C) for 2–4 h to induce virulence, i.e., to induce expression of the virulence genes required for T-DNA transfer.

3.3 Virus-Induced Gene Silencing

VIGS is a highly efficient method to knock down gene expression in diverse plant species, including *N. benthamiana* and *A. thaliana*. To infect plants for VIGS, we use *Agrobacterium* carrying T-DNAs that encode the viral RNA genome that has been engineered to include a silencing trigger. The RNA virus then spreads from the agroinfiltrated leaf to young, growing tissues in the shoot apex and root, where the host antiviral RNAi system knocks down expression of the target gene.

1. Prepare *Agrobacterium* for infiltration as in Subheading 3.2. For a minimal experiment with 24 plants, prepare the following *Agrobacterium* cultures: 12 mL carrying pTRV1, 5 mL carrying pTRV2::[*Target*], 5 mL carrying pTRV2::*GUS* (negative control), and 2 mL carrying pTRV2::*PDS* (positive control for silencing; *see* **Notes 4** and **5**). All resuspended bacteria should be adjusted to a final $OD_{600} = 1.0$ in fresh infiltration medium.
2. TRV replicates from a bipartite RNA genome composed of two RNAs, TRV1 and TRV2, both of which must be introduced to the plant for VIGS. Combine equal volumes of virulence-induced *Agrobacterium* inocula from Subheading 3.3, **step 1** ($OD_{600} = 1.0$) carrying pTRV1 and pTRV2 into a 15 mL tube immediately before infiltrating leaves.
3. Infiltrate the induced *Agrobacterium* inoculum through the abaxial surface of the two youngest fully expanded leaves of 3-week-old plants using a needleless 1 mL syringe without damaging the leaves. For a minimal standard experiment with 24 plants, infiltrate 10 plants with a 1:1 mixture of pTRV1 and pTRV2::*[Target]*, 10 plants with a 1:1 mixture of pTRV1 and pTRV2::*GUS*, and 4 plants with a 1:1 mixture of pTRV1 and pTRV2::*PDS*. To test additional silencing triggers, include an additional ten plants for each pTRV2::*[Target]* plasmid.
4. Gene expression should be silenced in the upper leaves of TRV-infected plants about 7–10 days after agroinfiltration (Fig. 1). Monitor plants infected with TRV2::*PDS* to track when gene silencing begins; upper leaves will begin to appear very pale green or white when *PDS* expression is silenced. After silencing is detected, movement assays should only be conducted in comparable, newly emerged leaves infected with TRV2::*[Target]* or TRV2::*GUS* (*see* **Note 5**).

3.4 Transient Gene Overexpression

Agrobacterium transforms plant cells to express genes encoded by transferred T-DNA within 24 h after agroinfiltration. The transient gene overexpression continues until approximately 72–96 h post-agroinfiltration, when T-DNA expression is silenced. In general, we recommend observing GFP movement precisely at 72 h post-agroinfiltration to maximize the effect of transient gene overexpression on observed PD transport rates.

1. Prepare *Agrobacterium* for infiltration as in Subheading 3.2. For a minimal experiment with 20 plants, prepare the following *Agrobacterium* cultures: 10 mL pEarlyGate::*GUS* and 10 mL binary vector for target gene overexpression (as described in Subheading 2.2, **item 3**). Each *Agrobacterium* inoculum should be adjusted to a final $OD_{600} = 0.1$ in fresh infiltration medium.
2. Infiltrate the induced *Agrobacterium* inoculum through the abaxial surface of the fourth leaf of 4–5-week-old plants (*see* **Note 9**) using a needleless 1 mL syringe without damaging the leaves.

3.5 Transformation with Fluorescent Tracer

The leaf GFP movement assay can be conducted in diverse contexts, including after VIGS (typically 14 days post-agroinfiltration with the binary vectors that encode the TRV VIGS constructs), alongside transient gene overexpression, or after or alongside chemical treatments.

1. Prepare *Agrobacterium* carrying the GFP binary vector (e.g., pWH001) for agroinfiltration as in Subheading 3.2, growing a 5 mL overnight culture. In the morning, first resuspend and dilute bacteria in infiltration medium to $OD_{600} = 0.1$. Then, dilute the bacteria at least 1:10,000 in infiltration medium (*see* **Note 10**), so that the final OD_{600} of the inoculum is 10^{-5} (*see* **Note 11**).
2. Infiltrate the induced *Agrobacterium* culture through the abaxial surface of the leaves using a needleless 1 mL syringe. Be careful not to severely wound the leaf; if a leaf is severely wounded, discard that plant and do not include it in further analysis.
3. Two to 3 days after agroinfiltration, image GFP movement using confocal microscopy as described in Subheading 3.6 (*see* **Note 12**).

3.6 Assaying Transport: Microscopy

1. Forty-eight to 72 h post-infiltration, infiltrate each leaf that was transformed to express GFP with water to remove air bubbles that could interfere with microscopy.
2. Excise a small portion of the leaf (approximately a 1 cm × 1 cm section), then mount on glass slide with a coverslip, and image with a confocal laser scanning microscope, abaxial surface facing up.

3. To visualize GFP movement, excite fluorescence with a 488 nm argon laser and collect emitted light with a GFP-compatible filter (detecting emissions from ~500 to 530 nm). Collect a z-stack using a wide pinhole aperture so that all fluorescing cells surrounding the transformed (brightest) cell are visualized; the z-stack will typically be ~80 to 160 μm deep, taking an image every 10 μm.
4. GFP movement can then be scored by two methods: counting the total number of GFP-containing cells or counting rings of GFP-containing cells.

In the first method (total GFP-containing cells), all cells fluorescing with GFP are counted per transformation event. GFP-containing cells will be in a rough circle around the primary transformed cell (*see* Fig. 1); users should take care to adjust the OD_{600nm} of their infiltration media that contains their GFP-expressing plasmid to ensure that there are few enough transformation events per leaf to discern individual groups of GFP-containing cells. Typically, $OD_{600nm} = 10^{-5}$ will be ideal, but this may be adjusted as necessary.

In the second method (rings of GFP-containing cells), the distance in rings of cells to which GFP has moved from the original transformed cell can be scored. No movement is scored as zero, since GFP remains only in the transformed cell; movement into one or all cells immediately touching the original transformed cell but none beyond can be scored as one ring and so on.

Both scoring methods produce biologically relevant results [35]. In one example VIGS experiment investigating the influence of the cytokinin signaling pathway on plasmodesmatal trafficking, GFP movement was scored using the second, ring-counting method [35]. GFP was observed to move 1.61 ± 0.07 rings of cells from the transformed cell in mock-silenced (TRV-GUS) plants, whereas GFP movement in *ARABIDOPSIS HISTIDINE KINASE (AHK) 3*-silenced plants was reduced to 1.31 ± 0.07 rings of cells and to 1.35 ± 0.06 rings of cells in *AHK4*-silenced plants [35].

4 Notes

1. The GFP movement assay can also be performed in *Arabidopsis thaliana* leaves. The EFR leucine-rich repeat receptor-like kinase is proposed to detect *Agrobacterium* infections and initiate host pathogen defenses [37], which could interfere with some experiments; thus, the GFP movement assay in *A. thaliana* may be more reliable in lines carrying null *efr* mutations. *A. thaliana* of most genotypes, including Col-0,

are susceptible to VIGS using somewhat modified methods [38].

2. Any lab strain of *A. tumefaciens* may be used for this assay. Our lab routinely uses GV3101, a C58 strain carrying a disarmed (ΔT-DNA) derivative of the nopaline C58 tumor-inducing (Ti) plasmid, pMP90 [39]. GV3101 is efficient at transforming *N. benthamiana* and *A. thaliana*.

3. We use the freeze-thaw method to transform *A. tumefaciens*. Briefly, agrobacteria are streaked from glycerol stocks onto an LB agar plate with antibiotic selection (for GV3101, rifampin and gentamicin) and grown for 2 days in a dark incubator at 28 °C. Bacteria from this plate are used to inoculate 1 mL LB with antibiotic selection (for GV3101, rifampin and gentamicin) and grown overnight in a dark incubator at 28 °C shaking at 250 RPM. This 2 mL starter culture is added to a large flask containing 200 mL LB with antibiotic selection (for GV3101, rifampin and gentamicin) in the morning, and this culture is allowed to grow in a dark incubator at 28 °C shaking at 250 RPM until it reaches log phase (OD_{600} is at least 0.5 but no more than 0.8; this requires between 5 and 8 h). The culture is pelleted at 1000 × *g* for 10 min at room temperature, the pellet is washed with sterile Tris EDTA, and then the pellet is resuspended in a total of 20 mL LB. The resuspended agrobacteria are aliquoted in 250 μL fractions in microcentrifuge tubes, snap-frozen in liquid nitrogen, and stored at −70 °C for future transformation. To transform agrobacteria, one tube of cells is thawed on ice. A 1.0 μg of plasmid is added and gently mixed by stirring. The tube is kept on ice for 10 min before being frozen in liquid nitrogen for 5 additional minutes. The tube is then thawed in a 37 °C water bath for 5 min. A 1.0 mL of antibiotic-free LB is then added to the tube, which is left to shake or rotate gently (no more than 60 RPM) at room temperature for 2–4 h. Finally, cells are collected by centrifugation at 700 × *g* for 2 min, most supernatant is removed, and cells are then plated onto LB plates with appropriate selection antibiotics (rifampin and gentamicin for GV3101 and typically kanamycin for the inserted plasmid).

4. *Tobacco rattle virus* (TRV) is a positive-sense single-stranded RNA virus in the genus *Tobravirus*. The TRV genome is bipartite, composed of two RNA molecules, TRV1 and TRV2. Both TRV1 and TRV2 are required for the virus to replicate, spread, and induce systemic silencing. In this VIGS system [40], TRV1 and TRV2 were cloned and inserted in two separate *Agrobacterium* binary vectors, previously called pYL192 and pYL156; for simplicity, here, we call them pTRV1 and pTRV2, respectively. The silencing trigger is cloned into a multiple cloning

site that was introduced into pTRV2. Agroinfiltration of both pTRV1 and pTRV2 is required for VIGS.

5. At least three pTRV2-derived vectors are used in every silencing experiment: pTRV2::*NbPDS*, pTRV2::*GUS*, and pTRV2::*[Target]*. pTRV2::*NbPDS* triggers silencing of *PHYTOENE DESATURASE (PDS)*, which encodes an enzyme required for carotenoid biosynthesis in leaves. Beginning at around 7–10 days after agroinfiltration of pTRV1 and pTRV2::*NbPDS*, newly emerged leaves are white due to photobleaching in carotenoid-depleted photosynthetic tissue. This serves as a positive control for VIGS efficiency.
6. pTRV2::*GUS* contains a silencing trigger against a bacterial gene, β-*glucuronidase* (GUS), that is not present in the *N. benthamiana* genome. Therefore, pTRV2::*GUS* should not trigger any VIGS phenotypes but serves as an important negative control to account for any effects of agroinfiltration and TRV infection. It should be noted, however, that under our growth conditions, we have not observed any difference in PD transport between uninfected plants and plants infected using pTRV2::*GUS*.
7. To silence a gene of interest, a short portion (~300 to 700 bp) of the target mRNA is cloned from cDNA into pTRV2. A draft genome of *N. benthamiana* is available at the Sol Genomics Network (solgenomics.net) to facilitate design of silencing triggers and cloning. This database also hosts a "VIGS tool" to help design optimal VIGS triggers (vigs.solgenomics.net); we recommend altering the default settings, which are very relaxed, so that "n-mer size" is 20, "fragment length" is 500, and mismatches are "1" or "2," to best avoid any off-target silencing. We also recommend cloning two nonoverlapping silencing triggers against a single gene, when possible, to validate that the effects of the trigger are specific to the target gene, rather than caused by unpredicted off-target silencing.
8. Acetosyringone (3′,5′-dimethoxy-4′-hydroxyacetophenone), which induces *Agrobacterium* virulence [41], is not stable in aqueous solution, so the infiltration medium should always be prepared on the day of agroinfiltration. Acetosyringone can be stored in small aliquots as a 1000× stock solution (200 mM) in DMSO at −20 °C for at most a few months.
9. These antibiotics are appropriate for GV3101(pMP90) (resistant to rifampicin and gentamicin) carrying pTRV1 or pTRV2 (conferring resistance to kanamycin); adjust the antibiotic selection to match the strain of *Agrobacterium* and binary vectors being used.

10. *Agrobacterium* should always be used from fresh cultures on plates no more than ~10 days old that have not been stored in the cold.
11. The precise age of the plants is not critical, but we conduct the assay before flowering, which typically begins about 6 weeks after germination. Most critically, PD transport changes drastically as leaves and plants age [23, 32], so the leaf and plant ages must be kept consistent across all experiments to avoid introducing error.
12. For transient gene overexpression with the aim of affecting plasmodesmatal trafficking, the infiltration medium may be supplemented with the *Agrobacterium* carrying a binary vector at an appropriate OD_{600} (typically 0.1). Alternatively, the infiltration medium may be supplemented with chemicals suspected to influence plasmodesmatal trafficking (e.g., kinase inhibitors or phytohormones) immediately before agroinfiltration of the leaf. Be sure to include all necessary positive or negative control treatments in the experimental design, such as *Agrobacterium* carrying a binary vector that overexpresses a biologically inert protein (e.g., β-glucuronidase (GUS)) or chemical solvents.
13. In an overexpression experiment where GFP movement is being used to track plasmodesmatal trafficking, the agroinfiltration inoculum should have a higher concentration of the target gene-expressing agrobacteria for widespread cell transformation and a much lower concentration of GFP-expressing agrobacteria to ensure that individual transformation events are sufficiently spaced out for proper quantification. An example of the inoculum dilution is as follows: 1 μL GFP solution $OD_{600} = 0.1$ into 10 mL GUS or overexpression solution $OD_{600} = 0.1$, resulting in $OD_{600} = 0.1$ for GUS/overexpression and $OD_{600} = 10^{-5}$ for GFP. Depending on the GFP vector being used, final dilution may be desired at $OD_{600} = 10^{-4}$ to 10^{-5}.
14. Fluorophore movement can be assayed anytime between 24 h post-agroinfiltration and 72 h post-agroinfiltration. After 72 h, the fluorophore has typically moved too far beyond the original transformed cell to accurately assay by confocal microscopy, and transient expression may begin to be silenced by endogenous cellular mechanisms. In general, 48 h post-agroinfiltration is the most sensitive time point to detect differences in intercellular spread of the fluorophore via PD.

Acknowledgments

This work was supported by NIH grant DP5-OD023072 to J.O.B. We thank Dr. Tessa M. Burch-Smith, Dr. Anne M. Runkel, and Dr. Patricia Zambryski for their contributions to a previous version of this method [42] and Snigdha Chatterjee for constructive comments on the manuscript.

References

1. Brunkard JO, Runkel AM, Zambryski PC (2015) The cytosol must flow: intercellular transport through plasmodesmata. Curr Opin Cell Biol 35:13–20
2. Paultre DSG, Gustin MP, Molnar A, Oparka KJ (2016) Lost in transit: long-distance trafficking and phloem unloading of protein signals in arabidopsis homografts. Plant Cell 28: 2016–2025
3. Reagan BC, Burch-Smith TM (2020) Viruses reveal the secrets of plasmodesmal cell biology. Mol Plant-Microbe Interact 33:26–39
4. Khang CH, Berruyer R, Giraldo MC et al (2010) Translocation of *Magnaporthe oryzae* effectors into rice cells and their subsequent cell-to-cell movement. Plant Cell 22: 1388–1403
5. Brunkard JO, Zambryski PC (2017) Plasmodesmata enable multicellularity: new insights into their evolution, biogenesis, and functions in development and immunity. Curr Opin Plant Biol 35:76–83
6. Cheval C, Faulkner C (2018) Plasmodesmal regulation during plant–pathogen interactions. New Phytol 217:62–67
7. Aung K, Kim P, Li Z et al (2020) Pathogenic bacteria target plant plasmodesmata to colonize and invade surrounding tissues. Plant Cell 32:595–611
8. Raven J (2005) Evolution of plasmodesmata. In: Oparka KJ (ed) Plasmodesmata. Blackwell, Oxford, pp 33–52
9. Niklas KJ (2014) The evolutionary-developmental origins of multicellularity. Am J Bot 101:6–25
10. Crawford KM, Zambryski PC (2000) Subcellular localization determines the availability of non-targeted proteins to plasmodesmatal transport. Curr Biol 10:1032–1040
11. Crawford KM, Zambryski PC (2001) Non-targeted and targeted protein movement through plasmodesmata in leaves in different developmental and physiological states. Plant Physiol 125:1802–1812
12. Waigmann E, Zambryski P (1995) *Tobacco mosaic virus* movement protein-mediated protein transport between trichome cells. Plant Cell 7:2069–2079
13. Atkins D, Hull R, Wells B et al (1991) The tobacco mosaic virus 30K movement protein in transgenic tobacco plants is localized to plasmodesmata. J Gen Virol 72:209–211
14. Yuan C, Lazarowitz SG, Citovsky V (2016) Identification of a functional plasmodesmal localization signal in a plant viral cell-to-cell-movement protein. MBio 7:e02052–e02015
15. Gallagher KL, Paquette AJ, Nakajima K, Benfey PN (2004) Mechanisms regulating SHORT-ROOT intercellular movement. Curr Biol 14:1847–1851
16. Koizumi K, Wu S, MacRae-Crerar A, Gallagher KL (2011) An essential protein that interacts with endosomes and promotes movement of the SHORT-ROOT transcription factor. Curr Biol 21:1559–1564
17. Koizumi K, Hayashi T, Wu S, Gallagher KL (2012) The SHORT-ROOT protein acts as a mobile, dose-dependent signal in patterning the ground tissue. Proc Natl Acad Sci U S A 109:13010–13015
18. Tucker JE, Mauzerall D, Tucker EB (1989) Symplastic transport of carboxyfluorescein in staminal hairs of *Setcreasea purpurea* is diffusive and includes loss to the vacuole. Plant Physiol 90:1143–1147
19. Wright KM, Oparka KJ (1996) The fluorescent probe HPTS as a phloem-mobile, symplastic tracer: an evaluation using confocal laser scanning microscopy. J Exp Bot 47:439–445
20. Faulkner C, Petutschnig E, Benitez-Alfonso Y et al (2013) LYM2-dependent chitin perception limits molecular flux via plasmodesmata. Proc Natl Acad Sci U S A 110:9166–9170
21. Maule AJ, Gaudioso-Pedraza R, Benitez-Alfonso Y (2013) Callose deposition and symplastic connectivity are regulated prior to lateral root emergence. Commun Integr Biol 6: e26531

22. Wang X, Sager R, Cui W et al (2013) Salicylic acid regulates plasmodesmata closure during innate immune responses in Arabidopsis. Plant Cell 25:2315–2329
23. Brunkard JO, Zambryski P (2019) Plant cell-cell transport via plasmodesmata is regulated by light and the circadian CLOCK. Plant Physiol 181:1459–1467
24. Benitez-Alfonso Y, Cilia M, San Roman A et al (2009) Control of Arabidopsis meristem development by thioredoxin-dependent regulation of intercellular transport. Proc Natl Acad Sci U S A 106:3615–3620
25. Stonebloom S, Brunkard JO, Cheung AC et al (2012) Redox states of plastids and mitochondria differentially regulate intercellular transport via plasmodesmata. Plant Physiol 158: 190–199
26. Burch-Smith TM, Brunkard JO, Choi YG, Zambryski PC (2011) Organelle-nucleus cross-talk regulates plant intercellular communication via plasmodesmata. Proc Natl Acad Sci U S A 108:E1451–E1460
27. Brunkard JO, Runkel AM, Zambryski PC (2013) Plasmodesmata dynamics are coordinated by intracellular signaling pathways. Curr Opin Plant Biol 16:614–620
28. Brunkard JO, Burch-Smith TM (2018) Ties that bind: the integration of plastid signalling pathways in plant cell metabolism. Essays Biochem 62:95–107
29. Ganusova EE, Reagan BC, Fernandez JC et al (2020) Chloroplast-to-nucleus retrograde signalling controls intercellular trafficking via plasmodesmata formation. Philos Trans R Soc B Biol Sci 375:20190408
30. Bobik K, Fernandez JC, Hardin SR et al (2019) The essential chloroplast ribosomal protein uL15c interacts with the chloroplast RNA helicase ISE2 and affects intercellular trafficking through plasmodesmata. New Phytol 221: 850–865
31. Azim MF, Burch-Smith TM (2020) Organelles-nucleus-plasmodesmata signaling (ONPS): an update on its roles in plant physiology, metabolism and stress responses. Curr Opin Plant Biol 58:48–59
32. Brunkard JO, Xu M, Regina Scarpin M et al (2020) TOR dynamically regulates plant cell-cell transport. Proc Natl Acad Sci U S A 117: 5049–5058
33. Kumagai MH, Donson J, Della-Cioppa G et al (1995) Cytoplasmic inhibition of carotenoid biosynthesis with virus-derived RNA. Proc Natl Acad Sci U S A 92:1679–1683
34. Ratcliff F, Martin-Hernandez AM, Baulcombe DC (2001) *Tobacco rattle virus* as a vector for analysis of gene function by silencing. Plant J 25:237–245
35. Horner W, Brunkard JO (2021) Cytokinins stimulate plasmodesmatal transport in leaves. Front Plant Sci 12:674128
36. Bombarely A, Rosli HG, Vrebalov J et al (2012) A draft genome sequence of Nicotiana benthamiana to enhance molecular plant-microbe biology research. Mol Plant-Microbe Interact 25:1523–1530
37. Zipfel C, Kunze G, Chinchilla D et al (2006) Perception of the bacterial PAMP EF-Tu by the receptor EFR restricts Agrobacterium-mediated transformation. Cell 125:749–760
38. Burch-Smith TM, Schiff M, Liu Y, Dinesh-Kumar SP (2006) Efficient virus-induced gene silencing in Arabidopsis. Plant Physiol 142: 21–27
39. Hellens R, Mullineaux P, Klee H (2000) A guide to Agrobacterium binary Ti vectors. Trends Plant Sci 5:446–451
40. Liu Y, Schiff M, Marathe R, Dinesh-Kumar SP (2002) Tobacco Rar1, EDS1 and NPR1/NIM1 like genes are required for N-mediated resistance to tobacco mosaic virus. Plant J 30: 415–429
41. Stachel SE, Messens E, Van Montagu M, Zambryski P (1985) Identification of the signal molecules produced by wounded plant cells that activate T-DNA transfer in Agrobacterium tumefaciens. Nature 318:624–629
42. Brunkard JO, Burch-Smith TM, Runkel AM, Zambryski P (2015) Investigating plasmodesmata genetics with virus-induced gene silencing and an agrobacterium-mediated GFP movement assay. In: Heinlein M (ed) Plasmodesmata, Methods in molecular biology (Methods and protocols), vol 1217. Humana Press, New York

Chapter 20

An *Arabidopsis* Callus Grafting Method to Test Cell-to-Cell Mobility of Proteins

Frank Machin, Yağmur Hasbioğlu, and Friedrich Kragler

Abstract

A callus is a semi-disorganized tissue that can be induced to develop from diverse tissues by the addition of exogenous hormones. The fast growth and ease of propagation have made callus cultures useful for creating a wide variety of different experimental systems.

Here, we describe a detailed and simple procedure by which different, non-clonal calli from transgenic and wild-type *A. thaliana* plants can be co-cultured such that they form symplasmic connections via plasmodesmata (PD). We show that callus cultures can be used to study both PD formation and transport of macromolecules between non-clonal cells via PD in a tissue lacking a vasculature. Further, we include a simple protocol for a method by which calli can be sectioned to image cells and PD by confocal laser scanning microscopy.

Key words Callus, Plant cell culture, Graft, Intercellular macromolecular transport, Plasmodesmata, *Arabidopsis*

1 Introduction

Plant tissue culture is a technique that is commonly used for plant propagation and transformation but seldom used for cell biology studies. In contrast, tissue culture is frequently used in mammalian biology in a wide range of applications. Unlike mammalian tissue culture, plant tissue culture has not yet been used for tissue engineering or growth on artificial structures. We propose here that using plant tissue culture techniques to construct tissue chimeras can be a powerful tool to study intercellular connectivity and transport. In this chapter, we outline some simple protocols for plant tissue engineering specifically for the study of de novo secondary PD formation between non-clonal cells and the study of intercellular movement of macromolecules via PD.

Grafting is a powerful technique for the study and identification of mobile signals as well as the genetic requirement for

Yoselin Benitez-Alfonso and Manfred Heinlein (eds.), *Plasmodesmata: Methods and Protocols*,
Methods in Molecular Biology, vol. 2457, https://doi.org/10.1007/978-1-0716-2132-5_20,

mobility in plants, since two genotypes can be combined into an individual plant and the exchange of material can be monitored [1–3]. Stem and hypocotyl grafting are the most popular grafting methods due to the simplicity of the technique and the applicability to a large number of plant species. In general, a successful stem or hypocotyl graft relies on reestablishing a connection between plant organs via the phloem and xylem vasculature, and therefore alternative methods are needed to study the capacity of proteins and RNAs to be transported via PD to neighboring cells. In this respect, callus grafting, the combination of genetically different calli into one symplasmically connected chimeric callus tissue, has great potential as an experimental platform.

In this technique, two or more types of genetically distinct calli are co-cultured in a way that maximizes the formation of symplasmic connections between genetically distinct plant cells (Fig. 1a). By co-cultivating a line producing a mobile fluorescent protein (e.g., fused to yellow florescent protein, YFP) with a line producing an immobile fluorescent (e.g., tagged with the red fluorescent protein Cherry), protein mobility can be detected by observing their co-occurrence in cells expressing the immobile fluorescent protein variant. Therefore, by cloning and transforming different fluorescently tagged proteins into wild-type and mutant *Arabidopsis* plants, it is possible to test (a) the mobility of different proteins across secondary PD and (b) to identify factors involved in their transport. The chimerism displayed by the grafted calli allows for detailed dissection of the genetic components of mobility by using different genetic backgrounds on either side of the graft to elucidate both the components required to facilitate movement and also to differentiate whether they are necessary in the donor cell or the receiver cell.

In the protocols we present here, we use transgenic *Arabidopsis thaliana* lines expressing free (cytosolic) YFP which is known to diffuse between cells via PD [4] and two fluorescent fusion proteins unable to diffuse via PD (TUA5-GFP and nuclear localized H2B-Cherry [5]). In a callus grafting experiment in which these lines are combined, we can expect to detect free YFP protein, but not TUA5-GFP, in H2B-Cherry tagged cells. The presence of free YFP in H2B-Cherry expressing cells indicates zones of symplasmic connectivity (Fig. 2c), whereas the TUA5-GFP fusion found associated to microtubules serves as a control as it is unable to move between cells (Fig. 2d).

In addition, to enhance the rate of successful callus graft union formation, we established a method by which callus grafts are placed inside gel structures to limit their growth area and to enhance cellular connectivity [6]. This results in the formation of grafted calli that are restrained in a close, touching position that increases the efficiency of the method. In this protocol, we present

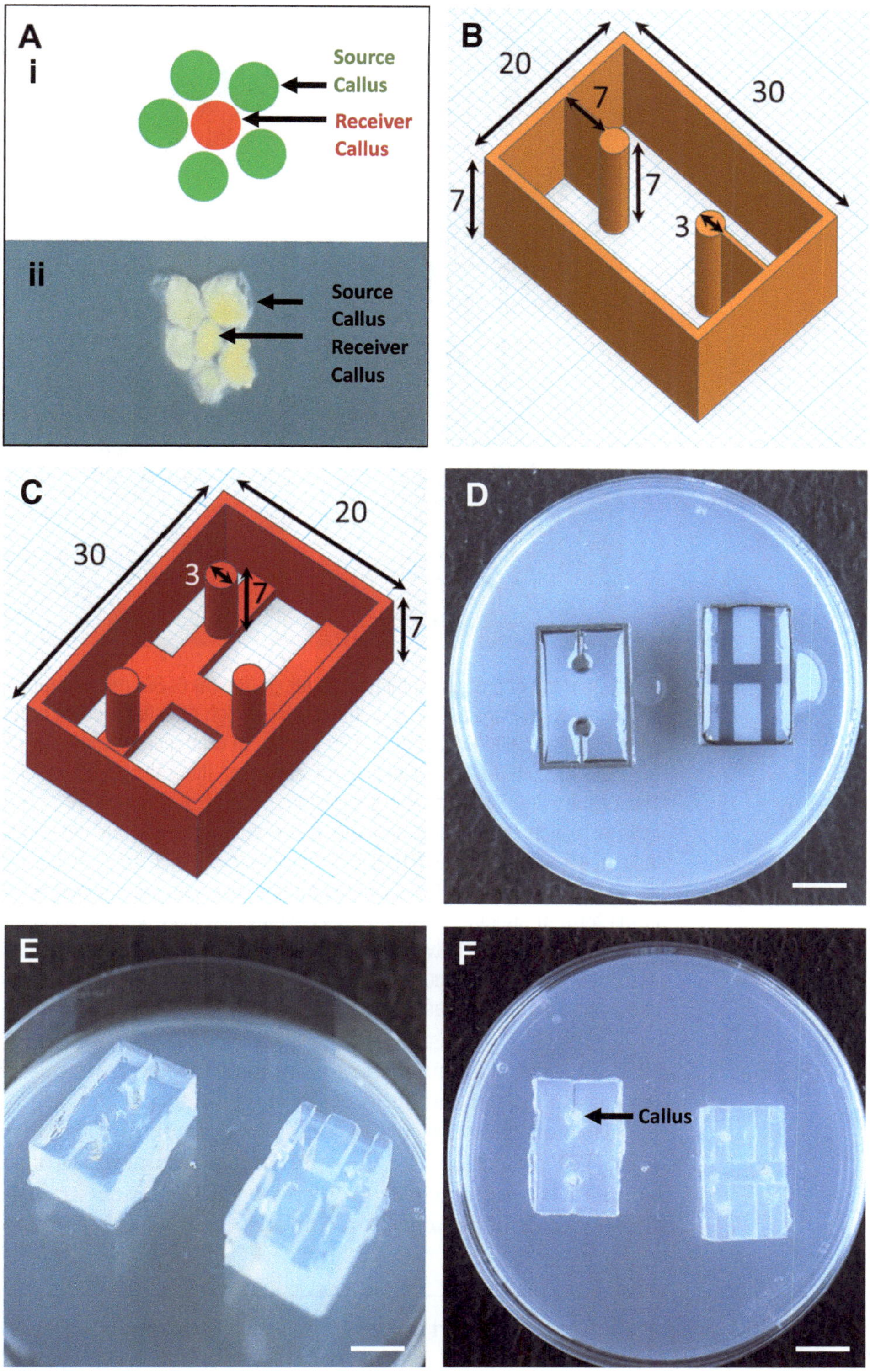

Fig. 1 Assembly of callus grafts. (**a**) (i) Scheme of callus placement in assembly of a callus graft. Source callus refers to the calli expressing the candidate mobile fluorescent protein, and receiver callus refers to the callus expressing the nonmobile fluorescent marker protein. (ii) Photograph of a callus graft arrangement after manual assembly on a callus induction medium (Subheading 2.2, **item 1**) plate. Scale bar = 1 mm. (**b**, **c**)

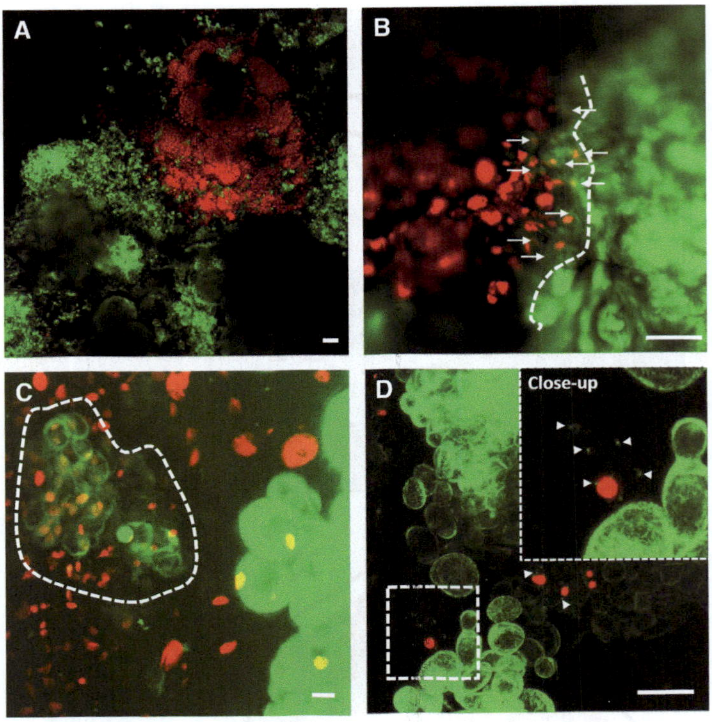

Fig. 2 Imaging of protein mobility in callus grafts. (**a**) CLSM micrograph of a *35S:YFP* and *35S:XVE::H2B-Cherry* callus graft in which both fluorescent proteins are seen in callus cells. Scale bar = 100 μm. (**b**, **c**) Confocal micrographs of a *35S:YFP/35S:XVE::H2B-Cherry* callus graft depicting the callus graft union region. The dashed line indicates the approximate site of graft union. Arrows indicate cells in which both green and red fluorescent proteins are detected, which indicates transport of YFP to H2B-Cherry expressing cells. (**b**) Scale bar = 50 μm; (**c**) Scale bar = 10 μm. (**d**) CLSM micrograph of green autofluorescent artifacts in the form of punctae sometimes observed when high-gain CLSM settings are used. This is a graft between nonmobile *35S::TUA5-GFP* and nonmobile *35S:XVE::H2B-Cherry*. The fluorescent objects detected in the imaged cell are most likely autofluorescent plastids. Scale bar = 50 μm

two 3D-printed gel support mold designs for gel structures, one for two grafts per gel and another for three grafts per gel (Fig. 1b–f).

To make best use of the callus graft method for studying de novo PD formation, we also present an optimized protocol for callus cryosectioning that preserves both the tissue structure and the fluorescent proteins for confocal laser scanning microscopy

Fig. 1 (continued) Diagrams of 3D-printed gel support molds (Subheading 2.5, **item 6b** and Subheading 2.5, **item 6c**). Dimensions given are in mm. (**d**) Gel support molds filled with callus graft gel on callus induction medium (Subheading 2.2, **item 1**) plates. Scale bar = 10 mm. (**e**) Gel supports fully formed after the molds have been removed. Scale bar = 10 mm. (**f**) The same gel supports as shown in (**e**) photographed from below and with completed callus grafts inside the callus chambers. Scale bar = 10 mm

(CSLM) imaging. Because handling the callus graft presents a significant risk of breaking the graft union, we have further adapted the protocol to avoid disturbing the callus tissue structure. Here, a grafted callus is fixed in place using 3D-printed plastic molds that create a chamber (Fig. 3a) in which fixation, washing, and gel encapsulation steps are performed before CLSM imaging. Using the cryosectioning technique, PD tagged by the viral movement protein 17-green fluorescent protein (GFP) (MP17-GFP) [7] can be imaged in calli cells with high resolution using a CLSM (Fig. 3e, f).

2 Materials

2.1 Growth of Arabidopsis *Seedlings*

1. Culture medium for *Arabidopsis* seedlings: 0.5× MS with vitamins (Duchefa, M0222), 1% microagar, and 0.5% sucrose. pH adjusted to 5.8 with 1 M KOH. Autoclave.
2. Seeds of transgenic *A. thaliana* Col-0 line expressing *35S:YFP* (available upon request).
3. Seeds of transgenic *A. thaliana* Col-0 line expressing *35S: XVE::H2B-Cherry* (available upon request).
4. Seeds of transgenic *A. thaliana* Col-0 line expressing *35S: MP17-GFP* (available upon request).
5. Seeds of transgenic *A. thaliana* Col-0 line expressing *35S:-α-tubulin 5 isoform (TUA5)-GFP* (available on request).
6. 15-cm square petri dishes (Greiner).
7. 70% ethanol solution.
8. 12% sodium hypochlorite solution.
9. Sterile distilled water.
10. Sterile 0.1% microagar solution.
11. Parafilm.

2.2 Culture of Callus and Callus Grafts

1. Callus induction medium: 0.68% microagar (Duchefa Biochemie) and 0.5× MS with vitamins (Duchefa, M0222). Autoclave and add 1.6% (w/v) glucose, 0.5 mg/L 2,4-D (Sigma), and 0.05 mg/L Kinetin (Sigma) after autoclaving and melting.
2. Callus induction medium for gel supports: 7% microagar (Duchefa Biochemie) and 0.5× MS with vitamins (Duchefa, M0222). Autoclave. Add 1.6% (w/v) glucose, 0.5 mg/L 2,4-D (Sigma), and 0.05 mg/L Kinetin after autoclaving and melting (*see* **Note 1**).
3. Round 94 × 16 mm petri dishes (Greiner).
4. Fine-tipped forceps.
5. No. 22 scalpel (Swann-Morton).

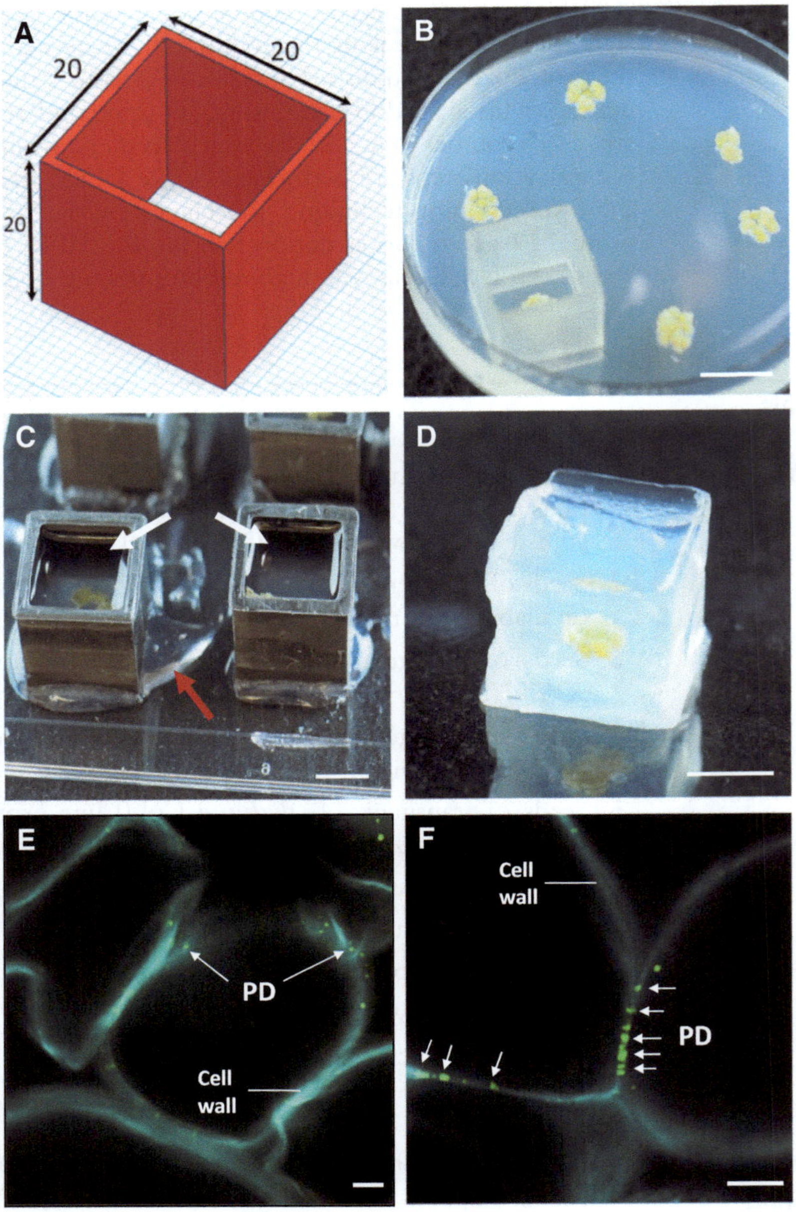

Fig. 3 Cryosectioning of callus grafts to image fluorescent PD. (**a**) Diagram of callus cryosectioning mold for 3D printing (Subheading 2.5, **item 6a**). Dimensions given are in mm. (**b**) Callus cryosectioning mold surrounding a 3-week-old callus graft. Note that the mold is pushed into the medium so as to be sealed tightly, and it can be removed from the plate together with the medium. Scale bar = 10 mm. (**c**) Callus cryosectioning molds containing intact callus grafts and filled with fixation solution (Subheading 2.3, **item 2**) (white arrows). Red arrow indicates the 7% agarose used to seal the bottom of the mold against the plate to prevent liquid leaking. Scale bar = 10 mm. (**d**) Completed gel-encapsulated intact cryopreserved callus graft. Scale bar = 10 mm. (**e**) CLSM micrograph of MP17-GFP expressing cell showing labeled PD in grafted callus cells. Blue represents Calcofluor-stained cell walls. Scale bar = 5 μm. (**f**) CLSM micrograph of MP17-GFP labeled PD at the boundary of two clonal callus cells. Blue: Calcofluor-stained cell walls. Scale bar = 5 μm

6. 100% ethanol.
7. Bunsen burner/ethanol lamp.
8. Parafilm.
9. Lightproof box. Cardboard box is sufficient provided it can be closed securely.

2.3 Callus Cryosectioning

1. Phosphate-buffered saline (10× PBS): 1.37 M NaCl, 27 mM KCl, 80 mM Na_2HPO_4, and 20 mM KH_2PO_4. Adjust pH to 7.4 with HCl.
2. 2.5% paraformaldehyde fixation solution: 1× PBS (Subheading 2.3, **item 1**), 2.5% paraformaldehyde (Sigma P-6148), 1% DMSO, and 0.1% Calcofluor White (Sigma). Adjust pH with 1 M NaOH to 7.4.
3. Paraformaldehyde wash solution: 1× PBS (Subheading 2.3, **item 1**) and 2.5% paraformaldehyde. Adjust pH with 1 M NaOH to 7.4.
4. 10% sucrose paraformaldehyde wash solution: 1× PBS (Subheading 2.3, **item 1**), 2.5% paraformaldehyde, and 10% sucrose. Adjust pH with 1 M NaOH to 7.4.
5. 20% sucrose paraformaldehyde wash solution: 1× PBS (Subheading 2.3, **item 1**), 2.5% paraformaldehyde, and 20% sucrose. Adjust pH with 1 M NaOH to 7.4.
6. 30% sucrose paraformaldehyde wash solution: 1× PBS (Subheading 2.3, **item 1**), 2.5% paraformaldehyde, and 30% sucrose. Adjust pH with 1 M NaOH to 7.4.
7. Cryosectioning gel: 1× PBS (Subheading 2.3, **item 1**) and 7% (w/v) TopVision Low Melting Point Agarose (Thermo R0801). Microwave to dissolve (*see* **Note 1**).
8. 3D-printed supports (Subheadings 2.5 and 3.4).
9. Leica 2000 microtome with a freezing facility.
10. Tissue-Tek® O.C.T Compound (e.g., Hartenstein).
11. Dry ice.

2.4 Confocal Laser Scanning Microscopy (CLSM)

1. Confocal laser scanning microscope (e.g., Leica SP8).
2. Glass slides (25 × 75 × 1 mm, e.g., Menzel-Glaser).
3. 22 × 50 mm coverslips (e.g., Carl Roth).
4. Perfluorodecalin (e.g., F2 Chemicals).

2.5 3D Printing

1. Formlabs Form 3 3D Printer.
2. FormLabs UV Cure Chamber.
3. Formlabs Black Resin (*see* **Note 2**).
4. Formlabs Clear Resin.

5. TinkerCAD Software (Autodesk). Freely available as a browser tool at http://www.tinkerrcad.com.
6. 3D print files (.stl) to create molds (downloaded from https://doi.org/10.5281/zenodo.4446896 [6]):
   (a) Callus_Cryosectioning_Mould.stl; print file to create mold for encapsulating callus grafts for cryosectioning.
   (b) Callus_Mould_Type1.stl; print file to create type 1 gel supports to aid in callus grafting.
   (c) Callus_Mould_Type2.stl; print file to create type 2 gel supports to aid in callus grafting.
7. Weck glass jars.

3 Methods

3.1 Growth of Arabidopsis *Seedlings*

1. Sterilize transgenic *A. thaliana* seeds (*35S:YFP*, *35S:XVE::H2B-Cherry* [5], and *35S:MP17-GFP* [7]) by washing in 70% ethanol for 15 min with gentle agitation.
2. Wash seeds in 12% bleach for 5 min (*see* **Note 3**).
3. Wash seeds twice with sterile water and sow on culture medium for *Arabidopsis* seedlings (Subheading 2.1, **item 1**) in 15-cm square petri dishes, approximately 20–30 seeds per plate.
4. Stratify the seeds for 2 days by placing the plates at 4 °C (e.g., in a cold room).
5. Transfer the plates to a controlled environmental chamber under long-day conditions (16 h 22 °C/8 h 18 °C) and grow vertically for 7 days.

3.2 Growth of Callus for Callus Grafting

All stages of this experiment are to be performed in a sterile flow hood.

1. Sterilize a scalpel and forceps by submerging in ethanol and then passing through a flame (*see* **Note 4**).
2. Cut approximately 5-mm long explants from the primary roots of the 7-day-old seedlings grown as described in Subheading 2.1 (*see* **Note 5**).
3. Transfer these explants to fresh callus induction medium on round petri dishes (Subheading 2.2, **item 1**) and seal with parafilm (*see* **Note 6**).
4. Move the plates to the same culture chamber in which the seedlings were grown (Subheading 2.1, **item 2**) and keep them in a lightproof box for approximately 10 days until callus initiation occurs (*see* **Note 7**).

5. When the calli are approximately 3–4 mm in diameter, use a sterile scalpel to remove the callus from the explant and transfer to a new callus induction medium plate (*see* **Note 8**).
6. After 10–14 days, the calli are ready to be used for grafting.

3.3 Callus Grafting Without Gel Support

All stages of this experiment are to be performed in a sterile flow hood.

1. Take calli as cultured in Subheading 3.2 and fresh callus induction medium in round plates (Subheading 2.2, **item 1**) (*see* **Note 9**).
2. Sterilize a scalpel and forceps by submerging in ethanol and then passing through a flame (*see* **Note 4**).
3. With sterile forceps, break off small pieces (around 1 mm diameter) of callus and arrange them on the plate as shown in the image in Fig. 1a (*see* **Note 10**).
4. Seal the plate with parafilm and return to the culture chamber for 21 days in a lightproof box (*see* **Notes 7, 11,** and **12**).

3.4 3D-Printed Molds

1. Import 3D print files (.stl files) for gel support molds (Subheading 2.5, **item 6b** or Subheading 2.5, **item 6c**) and cryosectioning supports (Subheading 2.5, **item 6a**).
2. Print objects using Form 3 Printer and Black Resin.
3. Remove objects from the printer carefully, and remove any scaffold structures (*see* **Note 13**).
4. Wash objects in isopropanol for 15 min.
5. Rinse objects in fresh isopropanol for 5 min with agitation.
6. Wash objects again in isopropanol for 15 min.
7. Leave objects to dry completely.
8. Cure in UV Cure Chamber for 1 h at 60 °C.
9. Wash objects with dish soap and rinse thoroughly with water.
10. Place objects in Weck glass jars and cover with water. Autoclave.

3.5 Callus Grafting with Gel Support

All stages of this experiment are to be performed in a sterile flow hood.

1. Take calli as cultured in Subheading 3.2.
2. Pour fresh callus induction medium in round plates (Subheading 2.2, **item 1**) (*see* **Note 14**).
3. Carefully place autoclaved gel support molds onto the surface of the callus induction medium.
4. Pour callus induction medium for gel supports (Subheading 2.2, **item 2**) into the gel support mold and allow to set solid.

5. Remove the gels from the molds as carefully as possible, both to avoid damaging the gels and to avoid contamination (Fig. 1e) (*see* **Note 15**).
6. Sterilize forceps by submerging in ethanol and then passing through a flame (*see* **Note 4**).
7. With sterile forceps, break off small callus pieces (around 1 mm diameter) and arrange them in the callus graft gel as shown in the scheme in Fig. 1a (*see* **Note 16**).
8. Seal the plate with parafilm and return to the culture chamber for 21 days in a lightproof box (*see* **Note** 7).

3.6 Callus Graft Preparation for Confocal Laser Scanning Microscopy (CLSM)

1. Take 21-day-old callus grafts as made in Subheading 3.4 or 3.5 (*see* **Notes 17** and **18**).
2. Keeping the grafted callus as intact as possible, place it upside down on a microscope slide (*see* **Notes 19** and **20**).
3. Immerse the callus in ~1 mL perfluorodecalin and place a microscope coverslip on the top. Do not press the coverslip flat.
4. Use low magnification (e.g., 5×) at first to record the position of the attached calli before proceeding with higher magnification images (20× or 40×) to detect fluorescent protein transport between cells (Fig. 2c) indicated by the presence of both fluorescent protein markers in the cells (*see* **Note 21**).
5. For imaging YFP and YFP-fusion proteins, use a 514 nm laser and filter for fluorescence emission between 524 and 550 nm. For the red fluorescent H2B-Cherry fusion protein, use the 561 nm laser and filter for fluorescence emission between 595 and 624 nm. Keep the gain within the range of 400–800 V for both fluorescent proteins (*see* **Note 22**).

3.7 Cryosectioning Callus Grafts for CLSM

1. Take 21-day-old callus grafts made according to Subheading 3.3 and 3D-printed molds as described in Subheading 3.4 for cryosectioning (Fig. 3a).
2. Push a mold into the agar in the plate so that it is surrounding the callus graft.
3. Lift the mold out of the plate together with the agar and callus inside it (*see* **Note 23**).
4. Spread some of the cryosectioning gel (Subheading 2.2, **item** 7) (*see* **Note 24**) onto the bottom of another clean, empty petri dish.
5. Before the gel solidifies, quickly transfer the mold from the empty petri dish on top of the cryosectioning gel so that the bottom of the mold is sealed and liquid cannot escape.

6. Fill the mold with fixation solution (Subheading 2.3, **item 2**) slowly in order to not to disrupt callus grafts and place at 4 °C overnight (>12 h) (*see* **Notes 25** and **26**).
7. The next day, keeping the callus graft at 4 °C (e.g., cold room), carefully remove the fixation solution and add paraformaldehyde wash solution (Subheading 2.3, **item 3**).
8. After 1 h, replace with the 10% wash solution (Subheading 2.3, **item 4**).
9. After 1 h, replace with 20% wash solution (Subheading 2.3, **item 5**).
10. After 1 h, replace with 30% wash solution (Subheading 2.3, **item 6**) for a final hour and then remove the solution slowly by using a 1 mL Gilson pipette.
11. Fill the mold with cryosectioning gel (Subheading 2.3, **item 7**) and allow to set solid. This will take approximately 5 min at 4 °C.
12. Remove the solidified gel from the mold by gentle finger pressure (Fig. 3d).
13. Fix the gels to the stage using Tissue-Tek. Place the gels on a piece of dry ice to freeze it.
14. Cut the gel into 80–100 μm sections using a microtome with freezing (*see* **Note 27**).
15. Place the sections on microscope slides and cover the sample with ~1 mL perfluorodecalin and a coverslip.
16. For imaging of Calcofluor white stain (added as a component of the paraformaldehyde fixation solution (Subheading 2.3, **item 2**) during the fixation step (Subheading 3.7, **step 6**)), use the 405 nm laser with power set to 15–20% and fluorescence emission filtering set to 420–460 nm. Set gain to around 600 V.
17. For imaging of MP17-GFP, use the 488 nm laser with power set to 20% and detect fluorescence emission at 510–550 nm (Fig. 3e, f). Keep the gain at around 650–700 V.

4 Notes

1. Mix gently by rolling the tube; do not shake, as this will introduce unwanted bubbles.
2. Ensure that the resin/plastic used for the molds is able to be autoclaved without melting or deforming, as sterile conditions are essential for the success of this method.

3. To minimize the risk of contamination of explants derived from these seedlings, seeds should be bleach-sterilized. Alcohol sterilization alone is not sufficient.
4. Tools for callus grafting (scalpel, forceps) should be flame-sterilized frequently to reduce the risk of contamination.
5. Callus tissue cuttings should be done with a very sharp scalpel to have clean cutting site to minimize damage to the tissue which might lead to tissue necrosis.
6. Take care not to overcrowd the plate, and leave approximately 1 cm gaps at minimum between individual explants.
7. It is important to keep the callus tissue in the dark to reduce chlorophyll production, which might lead to unwanted autofluorescence during imaging in later steps.
8. This step ensures that none of the original plant tissue will remain in the final callus graft, as the presence of vascular tissues may have unpredictable effects upon the success of the callus graft.
9. At this stage, it is advisable to screen a few individual calli to check whether the fluorescence is still visible. If not, or only some are fluorescent, it may be a sign that the line is undergoing silencing or the calli are not viable. It is imperative that only friable, actively dividing calli be used for the experiment.
10. It is important to push them as close together as possible, but be aware that it is common for calli to drift slightly apart due to the presence of water on the surface of the medium. This can be seen in Fig. 1a as the small gaps between individual calli. This does not matter much as the success of this method is determined by the growth of the calli into each other to form a graft union and does not rely solely upon the initial positioning of the calli.
11. It is important to seal the plate with parafilm as other methods (tape, cling film) have resulted in slight dehydration of the plates and a reduction in callus graft success.
12. Callus grafts should not be disturbed; if possible, do not move them after returning them to the growth chamber until the imaging step.
13. The steps required will differ according to each type of 3D printer and polymer, but it is important to ensure that no contaminants are present on the finished printed molds.
14. Be aware that, in this experiment, the gel supports have a height of 7 mm (Fig. 1b, c), and it is important that the gels do not press against the lid of the plate as this might limit airflow. Therefore, it is necessary that the medium in the plates

be poured to no higher than half way (approximately 25 mL for a plate with a diameter of 10 cm).

15. Gentle pressure from a sterile tip has been found to work well. It may be necessary to cut away some gel from the mold with a sterile scalpel in order to extract the gel from the mold. It may also be necessary to move the gels to new plates if the initial plate becomes damaged in the attempt.
16. This is more difficult than doing it without a gel support but with practice can be done very quickly (Fig. 1f). We have found that beginning with the central callus and then subsequently adding the remaining calli around it is the optimal strategy.
17. Twenty-one-day-old callus grafts should be taken for imaging experiments, as this has been shown to be a time at which macromolecular mobility is best detected between grafted calli.
18. The calli can dry out quickly, so it is important that they are kept inside the plate with the lid on as much as possible.
19. Inverting the calli helps to improve the observation of graft junctions, as the top-most regions of calli are mostly made up of dividing cells, which do not form graft junctions.
20. Callus grafts grown in growth restricting gel supports (Fig. 1e, f) can be most simply removed for imaging by first tearing away the gel with forceps. The callus grafts can then be separated using forceps and mounted on a slide as described in step in Subheading 3.6, **step 3**.
21. The most common sites of symplasmic connectivity can be found where the heterologous cells have been attached for the longest time period. One should consider that it is uncommon to find the callus graft junction intact as it breaks easily apart by this handling.
22. Note that at high-gain settings, autofluorescent plastids appear as yellow/green background dots (example shown in Fig. 2d).
23. It should be possible to remove the callus and agar together with the mold held together by the agar sticking inside the mold. If this is not possible, support the agar from below with a spatula during transfer to a new empty petri dish until the next step.
24. Make the 7% agarose solutions at least a day in advance and keep in an incubator at 60 °C. This will help the bubbles to disperse, as this takes a very long time in such a thick gel.
25. When adding or removing solutions from the cryosectioning mold, do so gently, dribbling solutions down the inside of the mold and pipetting slowly. All care must be taken to avoid disturbing the structure of the callus graft.

26. If adapting this protocol for larger tissues or larger calli, it may be necessary to vacuum infiltrate at this stage. It is not necessary, however, for calli of the type and size generated by this protocol.
27. Good quality slices for PD imaging can also be achieved without cryosectioning by cutting the gels with a razor blade at room temperature.

Acknowledgments

We would like to thank Stephanie Weber and Cindy Hauptvogel for their assistance. We thank the Oparka laboratory members for kindly donating *35S:MP17-GFP* seeds. The presented work was supported by the European Research Council (ERC) Synergy grant ERCSyG 2018_810131–PLAMORF funds awarded to Friedrich Kragler.

References

1. Thieme CJ, Rojas-Triana M, Stecyk E et al (2015) Endogenous Arabidopsis messenger RNAs transported to distant tissues. Nat Plants 1:15025
2. Turnbull CGN, Booker JP, Ottoline Leyser HM (2002) Micrografting techniques for testing long-distance signalling in Arabidopsis. Plant J 32:255–262
3. Yang L, Perrera V, Saplaoura E et al (2019) m5C methylation guides systemic transport of messenger RNA over graft junctions in plants. Curr Biol 29:2465–2476
4. Kim I, Cho E, Crawford K et al (2005) Cell-to-cell movement of GFP during embryogenesis and early seedling development in Arabidopsis. PNAS 102:2227–2231
5. Machin FQ, Beckers M, Tian X et al (2019) Inducible reporter/driver lines for the Arabidopsis root with intrinsic reporting of activity state. Plant J 98:153–164
6. Machin F, Hasbioğlu Y, Kragler F (2021) 3D printable moulds for Arabidopsis tissue co-cultivation. Zenodo. https://doi.org/10.5281/zenodo.4446896
7. Hofius D, Herbers K, Melzer M, Omid A et al (2001) Evidence for expression level-dependent modulation of carbohydrate status and viral resistance by the potato leafroll virus movement protein in transgenic tobacco plants. Plant J 28:529–543

Chapter 21

Quantification of Plasmodesmata Permeability in *Arabidopsis* Leaves by Tracing the Movement of GFP

Min Diao and Shanjin Huang

Abstract

Plasmodesmata (PD) play an important role in plant growth and development and defense. The permeability of PD is strictly regulated. Here, we describe an assay for measuring the permeability of PD in *Arabidopsis thaliana* leaves, which relies on tracing intercellular movement of green fluorescent protein (GFP) upon transient expression of the protein-encoding plasmid delivered by particle bombardment. The method allows to assess GFP movement at single-cell resolution.

Key words Plasmodesmata, PD permeability, GFP, Particle bombardment

1 Introduction

Plasmodesmata (PD) connect adjacent cells through cell walls and allow the transport of molecules between cells. Small molecules, such as ions and hormones, and also some large molecules, such as proteins, RNAs, and pathogens, can move through PD [1–4]. PD-mediated intercellular communication is crucial for plant growth and development as well as for the coordination of cellular and tissue responses triggered by the interaction of plants with the surrounding environment [5]. Dependent on cell type and tissue maturity, PD can have a simple or branched structure and different protein trafficking functions [6]. To study the mechanism that controls the function of PD, development of assays to quantitatively measure the permeability of PD is critical.

To date, several methods are used to measure the permeability of PD in plant tissues. One of the methods is to visualize the diffusion of fluorescent dextran or other fluorescent probes from microinjected cells to neighboring cells [7]. The advantage of this method is that it allows scientists to choose different sizes of fluorescent dextran, facilitating the study of the regulation of size

Yoselin Benitez-Alfonso and Manfred Heinlein (eds.), *Plasmodesmata: Methods and Protocols*,
Methods in Molecular Biology, vol. 2457, https://doi.org/10.1007/978-1-0716-2132-5_21,

exclusion limit (SEL) of PD under different conditions. In addition, this method allows co-injection of specific drugs or proteins to determine the effects of these drugs or proteins on PD permeability. However, this method is actually no longer used because it is intrusive and may itself cause alterations in PD permeability. Another method to assay PD permeability is Drop-ANd-See (DANS), which uses 5(6)-carboxyfluorescein diacetate (CFDA) dripped onto the adaxial leaf surface. Upon absorption, the dye is cleaved by cellular esterases and becomes fluorescent and at the same time membrane impermeable. Following diffusion through the mesophyll, the dye reaches the abaxial epidermis where the extent of dye spreading can be imaged and quantified by confocal microscopy [8]. Although DANS is a simple and fast method, it is difficult to track the spread of CFDA at single-cell resolution. A third method uses green fluorescent protein (GFP) and viral movement proteins (MPs) (e.g., the MP of cucumber mosaic virus (CMV)) fused to GFP (e.g., CMV MP-GFP) as probes to assay PD permeability and macromolecular movement in plant tissues [9–11]. This method enables the scientists to visualize the passive diffusion of GFP from GFP-expressing cells into adjacent cells and to compare this movement with that of active, PD-targeted movement of MP at single-cell resolution.

Here, we describe a method for transient GFP expression in single cells of *Arabidopsis* leaves using tissue microparticle bombardment [12] and measuring GFP cell-to-cell transport to deduce PD permeability. This method is adapted from our previously published methods [13, 14]. To identify the bombarded cell, the plasmid pHDEL-mCherry-pdGN was used for co-bombardment [13]. HDEL-mCherry is targeted to ER and cannot move between cells. HDEL-mCherry thus allows to identify the bombarded cells expressing GFP.

2 Materials

2.1 Biological Materials

1. *Arabidopsis* plants grown under 16 h/8 h light/dark cycle for 3–4 weeks.
2. *Nicotiana benthamiana* plants grown under the 16 h/8 h light/dark cycle for 4–6 weeks.
3. Plasmid pGFP-pdGN containing a 35S promoter driving a *GFP* gene [15].
4. Plasmid pHDEL-mCherry-pdGN [13].

2.2 Biolistic Bombardment

1. Petri dish (9 cm diameter) containing 20 mL of MS medium with 0.8% agar.

Dissolve MS powder in Milli-Q H_2O according to the instructions, adjust the pH to 5.8, add 0.8% agar, and then sterilize at 121 °C for 20 min.

2. Plasmid purification kit.
3. Ophthalmic scissors and tweezers.
4. 1.0 μm gold microcarriers (Bio-Rad, catalog number: #165-2263).
5. Gold suspension solution: add 30 mg gold microcarriers to 1 mL of 70% ethanol in a 1.5 mL microcentrifuge tube and vortex for 3 min. Allow the mixture to set for 15 min and centrifuge for 5 s at 3000 × *g*. Remove the supernatant and add 1 mL sterile water. Repeat the previous step two times. Finally, remove the supernatant and add 500 μL 50% sterile glycerin.
6. 0.1 M spermidine dissolved in Milli-Q-purified water. Prepare fresh 0.1 M spermidine from a 1 M stock (*see* **Note 1**).
7. 2.5 M $CaCl_2$ dissolved in Milli-Q-purified water.
8. 100% ethanol.
9. 70% ethanol.
10. Macrocarriers (Bio-Rad, catalog number: 1652335).
11. 1100 psi rupture discs (Bio-Rad, catalog number: 1652329).
12. Stopping screens (Bio-Rad, catalog number: 1652336).
13. PDS-1000/HeTM Biolistic Particle Delivery System (Bio-Rad, model: 1652257).

2.3 Microscopy and Image Acquisition

1. Zeiss LSM 510 META or any other equivalent confocal microscope.
2. Coverslips.
3. Glass slides.

2.4 Data Analysis Software

1. ImageJ (https://imagej.nih.gov/ij).
2. IBM SPSS Statistics (https://www.ibm.com/cn-zh/products/spss-statistics) or other software for statistical data analysis.

3 Methods

3.1 Biolistic Bombardment of the Abaxial Leaf Surface

1. Extract plasmid pGFP-pdGN and pHDEL-mCherry-pdGN using a plasmid purification kit from *Escherichia coli*. Ensure that the concentration of the purified plasmids is higher than 1 μg/μL (*see* **Note 2**).

Fig. 1 Preparation of the leaves of 4-week-old *Arabidopsis* plants and side view of the PDS-1000/HeTM Biolistic Particle Delivery System. (**a**, **b**) show the leaves of 4-week-old *Arabidopsis* plants for particle bombardment. (**c**) Side view of the PDS-1000/HeTM Biolistic Particle Delivery System. (1) and (2) highlight specific components that are shown enlarged and with individual subcomponents on the right. a, 1.100 psi rupture disk; b, microparticles carrier disk carrying 1.0 μm gold microcarriers; c, stopping screen. (Images are adapted from Diao et al. [14])

2. Select rosette leaves of healthy *Arabidopsis* plants that have not yet bolted (Fig. 1a). Use an ophthalmic scissor to cut leaves and place them into the petri dish with the abaxial side facing upward (Fig. 1b) (*see* **Note 3**).

3. Add 2 μg each of plasmid DNA (pGFP-pdGN together with pHDEL-mCherry-pdGN) into 10 μL gold suspension solution and vortex for 1 min (*see* **Note 4**).
4. Add 20 μL fresh 0.1 M spermidine and vortex again for 1 min.
5. Add 50 μL of 2.5 M $CaCl_2$, continue vortexing for another 3 min, and allow to set on ice for 5 min.
6. Centrifuge at 2348 × *g* for 10 s.
7. Wash the pellet containing the plasmid-coated gold particles in 1 mL of 70% ethanol and then in 1 mL of 100% ethanol and, finally, resuspend them in 15 μL of 100% ethanol.
8. Vortex the particles gently and transfer them to a macrocarrier using a 10–200 μL tip pipette.
9. Air-dry the macrocarrier and assemble all of the parts necessary for performing a particle bombardment according to the manufacturer's instructions (Fig. 1c).
10. Set the pressure to 1100 psi and use 1550 psi rupture discs.
11. Place the MS plate with *Arabidopsis* leaves 9 cm below the macrocarrier (level 3) in the bombardment chamber (Fig. 1c).
12. Set the vacuum level in the bombardment chamber to 28 psi and bombard.

3.2 Image Acquisition

1. Incubate the bombarded leaves for 24 and 48 h at 23 °C in the dark.
2. Mount the bombarded leaves on a glass slide with a drop of water, covered with a coverslip.
3. Image the leaves with a Zeiss LSM 510 META or an equivalent confocal microscope. Apply appropriate wavelengths and filtering for GFP excitation (use 488 nm laser) and emission (505–545 nm). Use a 20×/0.25 Fluar objective lens to scan the GFP cell clusters. Apply 20% laser intensity to image every fluorescent cell clearly. Identify the bombarded cells via applying appropriate wavelengths and filtering for mCherry excitation (use 543 nm laser) and emission (590–625 nm) (*see* **Note 5**).

3.3 Image Analysis

1. Open image files with ImageJ to analyze the clusters of GFP-containing cells. A GFP-containing cell cluster is the result of cell-to-cell diffusion of GFP out of a locally GFP-expressing cell. Usually, the local GFP-expressing cell is the brightest cell in a cell cluster [13, 14]. When both pGFP-pdGN and pHDEL-mCherry-pdGN plasmids were delivered into *Arabidopsis* leave cells (as in this protocol), verify the local GFP-expressing cell by the presence of the HDEL-mCherry marker. Define the GFP-containing cells connected to the GFP-expressing cell as the first layer, the GFP-containing cells

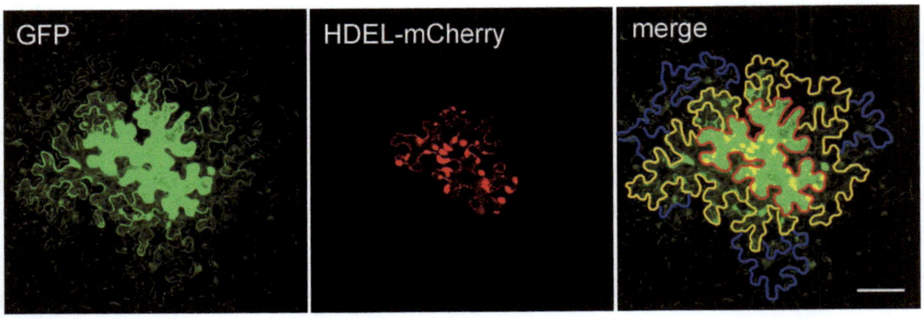

Fig. 2 *Arabidopsis* leaf epidermal cells expressing mobile GFP together with nonmobile HDEL-mCherry upon microparticle bombardment delivery of the respective expression plasmids. The transfected, GFP-expressing cell is marked by a red line. The edges of the layer 1 cells and layer 2 cells are marked by orange and blue lines, respectively. Normally, the cells with green signal accumulated in the nucleus and on plasma membrane were counted as GFP-expressing cells. Scale bar = 50 μm. (Images are adapted from Diao et al. [14])

connected to the first layer as the second layer, and so on (Fig. 2).

2. For each cluster, count the number of GFP-containing layers to obtain data sets. Evaluate numerous clusters from images taken from different plants; the bombardment experiments were repeated at least three times to obtain a sufficiently high number of GFP-expressing cell clusters for statistical analysis (*see* **Note 6**).
3. Analyze the data from experimental group (data sets from the mutant or transgene leaves) and control groups (data sets from the wild-type leaves) that were obtained in the previous step using IBM SPSS Statistics or similar software. The data are usually not normally distributed (*see* **Note 7**). Verify by using a Shapiro-Wilk test [16] before applying the Mann-Whitney *U*-test [17] to determine *p* values. If all the data sets are normally distributed, the student *t*-test will be performed.

4 Notes

1. The 1 M spermidine stock can be stored at −20 °C for at least 3 months.
2. The exact concentration of the plasmid is important for the next step. Prepare the plasmids in advance to ensure that the concentration is greater than 1 μg/μL.
3. Use young leaves.
4. The maximum total volume of plasmid DNA should be ≤10 μL. When both pGFP-pdGN plasmids and pHDEL-mCherry-pdGN plasmids need to be expressed simultaneously,

2 μg of each plasmid are mixed together with the gold particles in a single test tube.

5. Almost all GFP cell clusters contain a HDEL-mCherry expressing cell, and this red cell is always the one with the brightest green fluorescence in the cell cluster.
6. We usually take at least 60 pictures from leaves derived from 6 different plants for statistical analysis. Samples that express GFP but show serious damage resulting from the bombardment should not be used for analysis.
7. We usually compare the average values for diffusion layers from two or more data sets.

Acknowledgments

This work was supported by a grant from National Natural Science Foundation of China (31970180) and the funding from Beijing Advanced Innovation Center for Structural Biology.

Availability Statement of Materials *The plasmids pGFP-pdGN and pHDEL-mCherry-pdGN prepared during the current study are available from the corresponding author on reasonable request.*

References

1. Ding B (1998) Intercellular protein trafficking through plasmodesmata. Plant Mol Biol 38: 279–310
2. Ghoshroy S, Lartey R, Sheng J et al (1997) Transport of proteins and nucleic acids through plasmodesmata. Annu Rev Plant Physiol Plant Mol Biol 48:27–50
3. Lucas WJ, Lee JY (2004) Plasmodesmata as a supracellular control network in plants. Nat Rev Mol Cell Biol 5:712–726
4. Mezitt LA, Lucas WJ (1996) Plasmodesmal cell-to-cell transport of proteins and nucleic acids. Plant Mol Biol 32:251–273
5. Cheval C, Faulkner C (2018) Plasmodesmal regulation during plant-pathogen interactions. New Phytol 217:62–67
6. Ding B, Itaya A, Woo YM (1999) Plasmodesmata and cell-to-cell communication in plants. Int Rev Cytol 190:251–262
7. Ding B, Kwon M, Warnberg L (1996) Evidence that actin filaments are involved in controlling the permeability of plasmodesmata in tobacco mesophyll. Plant J 10:8
8. Cui W, Wang X, Lee JY (2015) Drop-ANd-See: a simple, real-time, and noninvasive technique for assaying plasmodesmal permeability. Methods Mol Biol 1217:149–156
9. Crawford KM, Zambryski PC (2001) Non-targeted and targeted protein movement through plasmodesmata in leaves in different developmental and physiological states. Plant Physiol 125:1802–1812
10. Iglesias VA, Meins F (2010) Movement of plant viruses is delayed in a beta-1,3-glucanase-deficient mutant showing a reduced plasmodesmatal size exclusion limit and enhanced callose deposition. Plant J 21:157–166
11. Liarzi O, Epel BL (2005) Development of a quantitative tool for measuring changes in the coefficient of conductivity of plasmodesmata induced by developmental, biotic, and abiotic signals. Protoplasma 225:67–76
12. Batoko H, Zheng HQ, Hawes C et al (2000) A Rab1 GTPase is required for transport between the endoplasmic reticulum and Golgi apparatus and for normal Golgi movement in plants. Plant Cell 12:2201–2217
13. Diao M, Ren S, Wang Q et al (2018) Arabidopsis formin 2 regulates cell-to-cell trafficking by capping and stabilizing actin filaments at plasmodesmata. elife 7:e36316

14. Diao M, Wang Q, Huang S (2019) Quantitative plasmodesmata permeability assay for pavement cells of *Arabidopsis* leaves. Bio Protoc 9(7):e3206
15. Lee JY, Taoka K, Yoo BC et al (2005) Plasmodesmal-associated protein kinase in tobacco and Arabidopsis recognizes a subset of non-cell-autonomous proteins. Plant Cell 17:2817–2831
16. Shapiro SS, Wilk MB (1965) An analysis of variance test for normality (complete samples). Biometrika 52(3/4):21
17. Fay MP, Proschan MA (2010) Wilcoxon–Mann–Whitney or t-test? On assumptions for hypothesis test and multiple interpretations of decision rules. Stats Surveys 4:1–39

Chapter 22

Tracking Intercellular Movement of Fluorescent Proteins in Bryophytes

Takumi Tomoi, Yoan Coudert, and Tomomichi Fujita

Abstract

An important approach to investigate intercellular connectivity via plasmodesmata is to visualize and track the movement of fluorescent proteins between cells. The intercellular connectivity is largely controlled by the size exclusion limit of the pores. Over the past few decades, the technique to observe and analyze intercellular movement of a fluorescent protein has been developed mainly in angiosperms such as *Arabidopsis thaliana*. We recently applied the corresponding system to track the intercellular movement of the fluorescent protein Dendra2 in the moss *Physcomitrium* (*Physcomitrella*) *patens*. The protonemal tissues are particularly suited for observation of the intercellular movement due to the simple organization. Here, we describe a protocol suitable for the analysis of Dendra2 movement between cells in *P. patens*.

Key words Confocal microscopy, Dendra2, ImageJ, Photoconversion, *Physcomitrium patens*, Plasmodesmata

1 Introduction

Fluorescent proteins (FPs) are widely used as reporters of intercellular movement through plasmodesmata (PD). Unless FPs are fused with endogenous proteins, FP movement between cells is nonselective and determined by the molecular size of the FP and the given PD aperture [1–5]. Some endogenous proteins fused with FPs can migrate to adjacent cells in a mode of selective transport, where an additional component likely alters PD aperture and/or the endogenous molecules to modulate PD permeability [6–10].

To determine intercellular transport, stably fluorescent FPs such as green fluorescent protein (GFP) must be expressed in single cells to observe their transport into adjacent cells. Thus, introduction by microinjection, microparticle bombardment, or expression under a tissue-specific promoter has been used for tracking intercellular movement [1–5, 7, 11–13]. Such analyses provided

Yoselin Benitez-Alfonso and Manfred Heinlein (eds.), *Plasmodesmata: Methods and Protocols*,
Methods in Molecular Biology, vol. 2457, https://doi.org/10.1007/978-1-0716-2132-5_22,

evidence that developmental and environmental cues alter intercellular connectivity through PD [2–5, 11–13]. Additional techniques, such as fluorescence recovery after photobleaching (FRAP), enabled to quantitatively describe the mode of GFP movement, for example, with a mathematical model [14]. In addition, photoactivatable FPs (PAFPs), photoconvertible FPs (PCFPs), or photoswitchable FPs (PSFPs) became available that allow to determine directional movement of a pool of the FPs by time-lapse imaging without influence of de novo FP synthesis [15–18]. Moreover, the observation of intercellular movement of FPs was further advanced by new microscopic techniques. For example, while conventional confocal microscopy limits the observations of intercellular transport to surface cell layers such as the epidermis, the use of two-photon microscopy in combination with photoswitching of PSFPs [17] or potentially with FRAP allows visualization of intercellular movement also in deeper tissue layers. Furthermore, photoconversion of PCFP through two stages with dual-wavelength (blue and red/near-infrared lasers) illumination, called primed conversion, may allow spatially restricted visualization and tracking of the intercellular movement of PCFP in deeper tissue layers of plants [19, 20].

We previously established an experimental system to track PD-mediated transport of the PCFP Dendra2 (≈26 kDa) in the moss *Physcomitrium* (*Physcomitrella*) *patens* [16, 21]. *P. patens* forms filamentous tissues (protonemata, consisting of two types of tissues, chloronemata and caulonemata) and a leaf-like structure (phyllid) composed of a single cell layer [22]. This simple structure makes *P. patens* highly suitable for observation of intercellular movement of FPs. Here, we describe a protocol to track intercellular movement of Dendra2 between protonemal cells and give an account of how to measure photoconverted Dendra2 intensity using the open-source imaging analysis software Fiji (ImageJ) [16, 23, 24].

Analyses of intercellular movement through PD in bryophytes have a relatively short history. We hope that this protocol applied to bryophytes will help researchers to investigate the regulatory mechanisms and function of intercellular movement of endogenous proteins and gain a better understanding of cell-cell communication via PD in plants.

2 Materials

1. Plant material: *P. patens* constitutively expressing *Dendra2* under the promoter of *ELONGATION FACTOR 1α* (Pp3c2_6690V3) (*ProEF1α:D2*) [16] (*see* **Notes 1–3**). Grow *P. patens* on solid BCDAT medium at 25 °C under continuous light [25].

2. Glass bottom dish. We used four-well chamber slides (Iwaki).
3. Stock B: 100 mM $MgSO_4{\cdot}7H_2O$ in water (autoclaved).
4. Stock C: 184 mM K_2PO_4 (adjusted to pH 6.5 with KOH) in water (autoclaved).
5. Stock D: 1 M KNO_3 and 4.5 mM $FeSO_4{\cdot}7H_2O$ in water (not autoclaved).
6. 500 mM ammonium tartrate (dibasic) in water (autoclaved).
7. Alternative TES: 0.22 mM $CuSO_4{\cdot}5H_2O$, 10 mM H_3BO_3, 0.23 mM $CoCl_2{\cdot}6H_2O$, 0.1 mM $Na_2MoO_4{\cdot}2H_2O$, 0.19 mM $ZnSO_4{\cdot}7H_2O$, 2 mM $MnCl_2{\cdot}4H_2O$, and 0.17 mM KI.
8. 50 mM $CaCl_2{\cdot}2H_2O$ in water (autoclaved).
9. BCD medium: 10% (v/v) of stock B, 10% (v/v) of stock C, 10% (v/v) of stock D, 1% (v/v) of alternative TES, 20% (v/v) of 50 mM $CaCl_2$, and filled up with water to the specific total volume (autoclaved).
10. BCDAT medium: BCD medium also containing 10% (v/v) of 500 mM ammonium tartrate.
11. BCDG: BCD medium containing 0.5% (w/v) glucose solidified with 0.5% (w/v) gellan gum (autoclaved) [25] (*see* **Notes 4** and **5**).
12. BCTATG: BCDAT medium containing 0.5% (w/v) glucose solidified with 0.5% (w/v) gellan gum (autoclaved) [25] (*see* **Notes 4** and **5**).
13. Pipette.
14. Thin forceps.
15. Parafilm.
16. Black container with no lid.
17. Red acrylic filter (Mitsubishi Rayon). This filter is needed to culture *P. patens* under red light. The use of a red light-emitting diode (LED) may be an alternative.
18. Confocal microscope. To observe Dendra2 movement between cells, use 405 and 561 nm lasers for the photoconversion and excitation of the photoconverted (red) Dendra2, respectively. A confocal microscope Nikon A1 (Nikon) equipped with a 405 nm diode laser (36 mW) and a 561 nm solid laser (10 mW, 2% output) and emission filter set to 570–620 nm for the detection of photoconverted Dendra2 was used for this protocol (*see* **Note 6**).
19. Image analysis software. We use the open-source software Fiji (ImageJ; https://imagej.net/Fiji).

3 Methods

3.1 Sample Preparation

1. Melt the solidified BCDATG or BCDG medium in a microwave.
2. Allow the medium to cool so that the glass of a four-well chamber slide is not broken in the subsequent step.
3. Pour 500 μL of the medium into each well of a four-well chamber slide and let it solidify.
4. Pick up and embed a small amount of 3–5-day-old protonemal tissues into the medium of each well with thin forceps (*see* **Note 7**).
5. Overlay the embedded tissue with additional 250 μL of the medium. The medium should be warm and liquid but not too hot to avoid any damage to protonemal tissues by heat.
6. Put on the lid and seal the chamber slide with parafilm (*see* **Note 8**).
7. Cultivate the protonemal tissues for 10–15 days under continuous red light with the glass surface of the chamber slide oriented upward (*see* **Note 9**). For red light conditions, place the four-well chamber slide into a black container with no lid, and place the lid of the red acrylic filter on the box for a red light culture (Fig. 1). We use a fluorescent lamp as the light source (*see* **Notes 10** and **11**).

3.2 Photoconversion

Perform photoconversion of Dendra2 with the 405 nm diode laser (10.0% laser output, 36 mW) with 3 iterations (8 s/frame, zoom 4×) (*see* **Notes 12–16**).

3.3 Time-Lapse Imaging

Apply the time-lapse settings to image intercellular movement of Dendra2 (Fig. 2a) (*see* **Notes 17–19**).

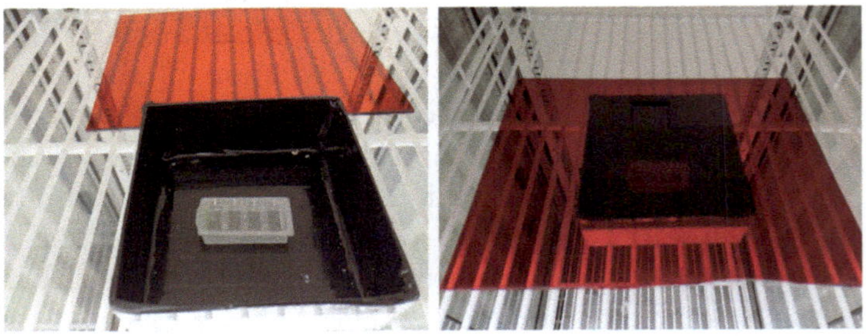

Fig. 1 A Black box with red filter used to culture the transgenic *P. patens* line *ProEF1α:D2*. A four-well chamber slide is placed into a box below the red acrylic filter. The left and right panels show the box without and with the filter, respectively

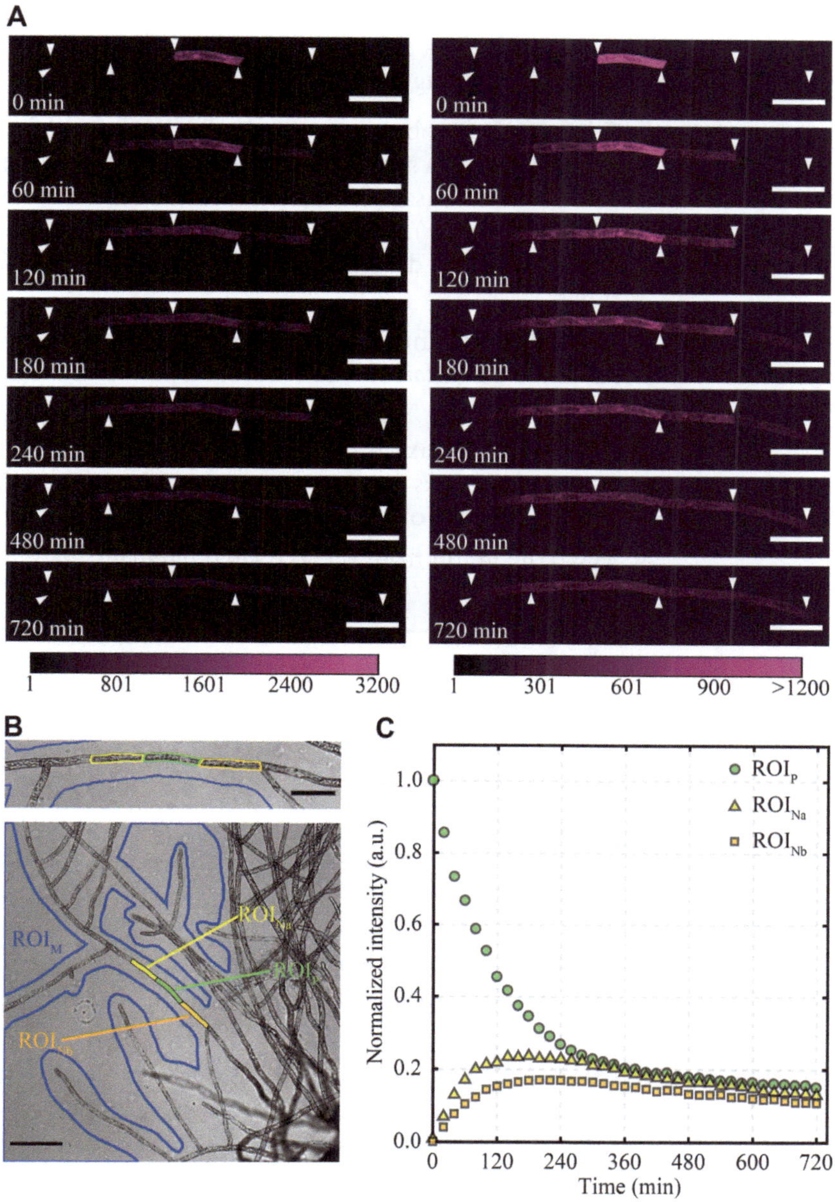

Fig. 2 An example of measurement of red Dendra2 intensity. (**a**) Red Dendra2 fluorescence (magenta) images of protonemal cells of the *ProEF1α:D2* line are shown. Times after the onset of the time-lapse imaging are indicated. The images on the right are enhancements of the images on the left. Scale bars = 100 μm. Bars shown below the images indicate ranges of displayed pixel value. (**b**) A transmission image of *ProEF1α:D2* protonemal cells. The upper image shows the cells analyzed by fluorescence in (**a**), and this image has been cropped from the bottom image. The ROI_P, ROI_{Na}, ROI_{Nb}, and ROI_M are indicated by green, yellow, orange, and blue lines, respectively. Scale bars = 100 μm (upper) and 200 μm (bottom), respectively. (**c**) Normalized red Dendra2 intensity of the protonemal cells in (**a**, **b**). The exponential decrease in the intensity of red Dendra2 at the photoconverted cell indicates a simple diffusion phenomenon [24]. Thus, the intensity change is seemingly obvious until around 240 min while less visible after 240 min from the pictures (**a**). In contrast, the slight intensity change is graphically illustrated even after 240 min of the onset of the time-lapse imaging (**c**)

3.4 Analysis

Analyze the Dendra2 movement between protonemal cells. Here, we limit the protocol to the analysis to Dendra2 intensity measurement using Fiji (ImageJ) software.

1. Select ROIs for photoconverted cells (ROI_P) and for neighboring cells on both sides of the photoconverted cell, i.e., in apical (ROI_{Na}) and basal (ROI_{Nb}) directions, and for the medium (ROI_M) (Fig. 2b). ROI_M is used to determine the background, which may be different between samples and observation times.
2. Measure mean intensity of the ROIs (*Analyze* > *Tools* > *ROI Manager...* > *More* ≫ *Multi Measure*) to obtain it as "Results" (*see* **Note 20**).
3. Pick up the max intensity of the mean values among the 5 z-stack images at the observation time from the data of "Results" (*see* **Note 21**).
4. Make the graph from the extracted data (Fig. 2c).

4 Notes

1. The *Dendra2* sequence in *ProEF1α:D2* is optimized to the human codon usage (Evrogen). We could not obtain transgenic lines possessing Dendra2 fluorescence with the *Dendra2* sequence optimized to *A. thaliana* codon usage (Evrogen).
2. Check beforehand whether cells of your interest show fluorescence levels sufficient for imaging analysis. In this regard, you may need to consider to test a choice of promoters. We observed Dendra2 fluorescence in different tissues such as protonemata and gametophores with the *EF1α* promoter (Fig. 3).
3. Instead of Dendra2 for visual tracking of the intercellular movement [16], other monomeric PCFPs (mClavGR2, mKikGR) may be used. The use of monomeric PAFPs (PA-GFP, PA-mCherry, etc.) may provide data equivalent to the data obtained by using Dendra2. However, unlike PCFPs such as Dendra2, PAFPs have the disadvantage that they require photoactivation for visualization. Thus, it is not possible to verify the PAFP expression level just before the mobility analysis experiment. For the same reason, the use of monomeric PSFPs that are initially nonfluorescent ("off"-state, e.g., Padron) can be also tricky. Furthermore, PSFPs which are initially in "on"-state such as Dronpa may not be suitable to analyze the intercellular mobility. During observation of intercellular movement of Dronpa-type PSFPs, newly synthesized "on"-state Dronpa adds to the background. Moreover, in almost all of PSFPs, the excitation wavelength to image the

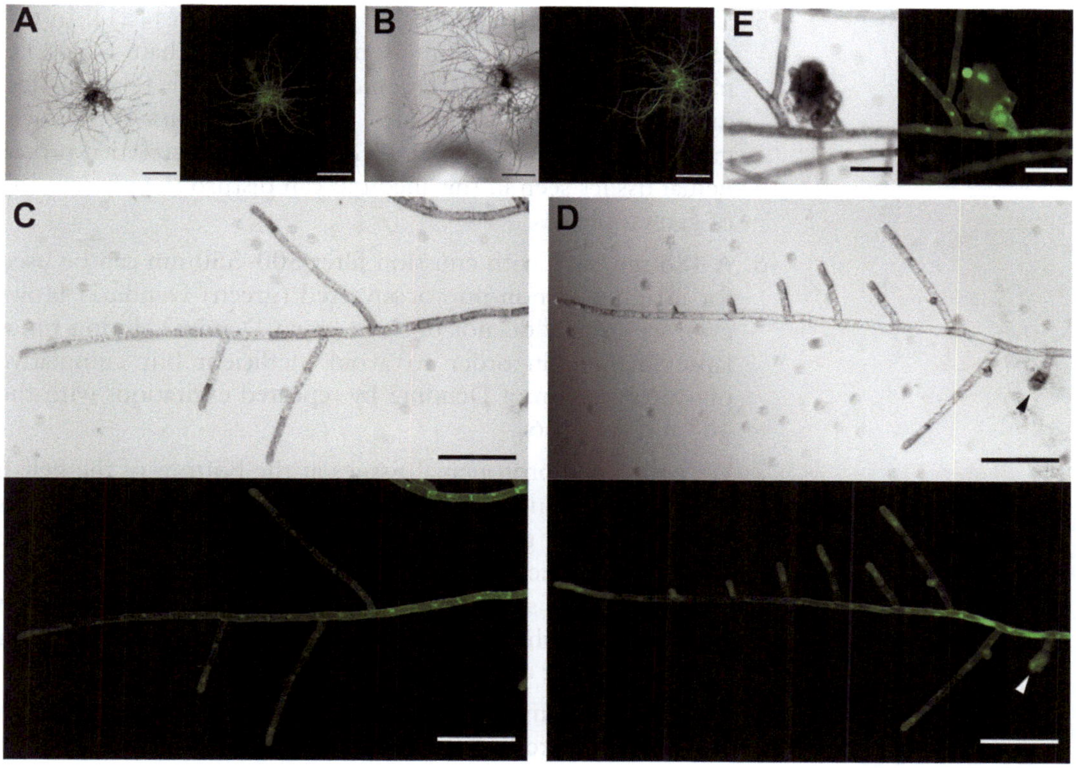

Fig. 3 The growth conditions of *ProEF1α:D2*. (**a**, **b**) Representative transmission and green Dendra2 fluorescence images of *ProEF1α:D2* protonemal tissues after 10 days of culture in BCDATG (**a**) and after 15-day culture in BCDG (**b**). (**c–e**) Representative transmission and green Dendra2 fluorescence images in a chloronema in BCDATG (**c**) and in a caulonema with a bud (**d**) and a young gametophore (**e**) cultured in BCDG. Scale bars = 1 mm (**a**, **b**), 200 μm (**c**, **d**), and 100 μm (**e**). An arrow head indicates a bud

"on"-state PSFP and the wavelength for "on" to "off" photoswitching are quite similar (e.g., there is only a 23 nm interval between the wavelength for excitation with the 488 nm laser and the wavelengths required for "on" to "off" photoswitching), resulting in decrease in fluorescence intensity of the tracked PSFPs. Such characteristics of Dronpa could be problematic especially in protonemal tissues of *P. patens* because intercellular mobility is low and excitations are repeatedly carried out to obtain images during a long period of time-lapse observation. Dronpa is, however, applicable for the analysis in, for example, root tissues of *A. thaliana* due to the relatively higher intercellular mobility compared to the mobility in protonemal tissues of *P. patens* and the absence of a requirement for many times of fluorescence excitation of Dronpa [17, 24].

4. Protonemal growth is reduced under red light conditions. Thus, glucose is supplemented to improve the growth. Instead of agar, gellan gum is used for solidification of medium to increase the translucence of medium which facilitates observation and imaging.

5. This protocol is also applicable to gametophores [21]. In this case, the BCDG medium is more suitable than BCDATG medium because the BCDG medium allows for relatively less growth of protonemal tissues. Although *P. patens* produces gametophores also in BCDATG medium, the crowded protonemal tissues seen in this medium can disturb the observation of gametophores.
6. A 488 nm laser with emission filter 500–550 nm can be used for imaging of non-photoconverted (green) Dendra2. However, we usually do not observe green Dendra2 during time-lapse imaging in order to avoid inefficient but cumulative photoconversion of Dendra2 by repeated excitations with the 488 nm laser [26].
7. We embedded protonemal tissues at the bottom of the gellan gum medium, onto the glass surface of the chamber slides. This allows them to grow along the glass surface soon after the inoculation. When protonemal tissues are embedded at multiple places in the same well of a four-well chamber slide, leave a space between the places to avoid overlapping growth during the 10-day long culture (Fig. 3a, b).
8. We wrap parafilm around the chamber slide once and half. In this case, gas exchange is maintained for protonemal growth while the medium is kept from drying out.
9. Culture under red light suppresses Dendra2 photoconversion as well as formation of protonemal side branches.
10. Fluorescent lamp intensity and the distance between the fluorescent lamp and sample determine the intensity of light and heat, affecting the growth of protonemal tissues. Thus, some optimization of culture conditions is required.
11. The amount of medium in the wells is adjustable for any purpose. Conditions using a small amount of medium described here are optimized to facilitate treatment with hormones [24]. However, while a small amount of medium allows more efficient treatment with hormones, this may also cause desiccation stress to the plants during a longer culture period. Using a larger amount of medium enables longer culture without desiccation. Thus, without pharmacological experiments, the use of more medium is advised.
12. We cannot rule out the possibility that 405 nm laser irradiation for photoconversion confers cellular damage. We tested the application of 0–100% laser power with 3 times irradiation at 8 s/frame (24 s in total), and the laser power applied in the protocol (10% output) did not arrest cellular growth following irradiation under our conditions (Table 1). One should pay attention to the laser power in Watt (mW) in addition to the

Table 1
Cellular damage by 405 nm laser irradiation. The growth of apical cells without and after 405 nm laser irradiation at different laser powers for 24 s (with 3 iterations of 8 s/frame). The absence or presence of cell division and cell elongation was monitored throughout time-lapse imaging at 20-min intervals for 12 h. Growth arrest was determined when the apical cells stopped growing during the time-lapse imaging

	The power of 405 nm diode laser (%)				
	0	10	25	50	100
Growth arrest	1.4% (2/147)	0% (0/12)	0% (0/12)	41.7% (5/12)	91.7% (11/12)

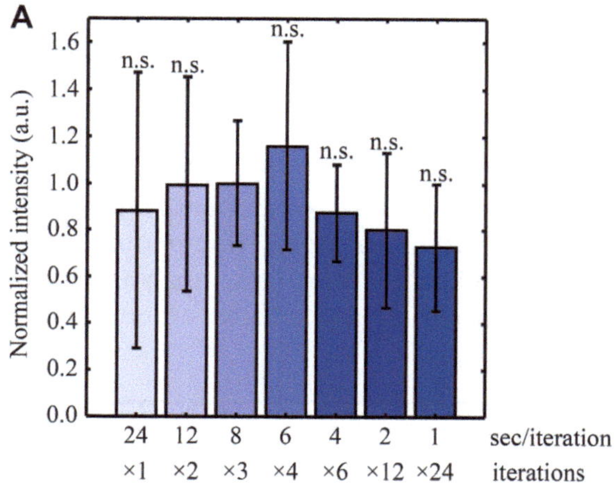

Fig. 4 Efficiency of Dendra2 photoconversion. (**a**, **b**) Red Dendra2 intensity measured (**a**) after applying different numbers of scan iterations during the 24-s long total laser irradiation time and (**b**) after applying different laser powers with 8-s long iterations during the 24-s long total irradiation time. Each value of red Dendra2 intensity was normalized in (**a**) to the average intensity obtained upon applying 3 iterations of irradiation for 8 s (normalized intensity set to 1.0) and in (**b**) to the value obtained with 10% laser power (normalized intensity set to 1.0). $N = 10$ (**a**) and $N = 6$ (**b**). The *P*-values were determined by the Welch's *t*-test and give the statistical significance between the values obtained for the different conditions and the value for the condition to which the data was normalized and set to 1.0. *n.s.* nonsignificant ($P \geq 0.05$); *, $P < 0.05$

laser power in % as well as to the duration of laser irradiation when checking the cellular growth. Under our conditions, 50% laser power clearly induced cellular damage (Table 1).

13. We checked the efficiency of photoconversion upon varying the number of scan iterations applied within a constant period of total irradiation time and dependent on laser power (10, 25, and 50%) (Fig. 4). The number of scan iterations did not significantly change the efficiency of photoconversion (Fig. 4a). However, when photoconversion was performed at higher laser power, the effect of photobleaching of Dendra2 also increased, resulting in the lower efficiency of photoconversion (Fig. 4b).

14. Despite the absence of photobleaching, proteolytic degradation of Dendra2 can possibly take place particularly in Dendra2 fused with an endogenous protein like ANGUSTIFOLIA 3 [27]. Since this leads to decrease in red Dendra2 intensity in photoconverted cells which can perturb the analysis of intercellular movement (a decrease in red Dendra2 intensity due to degradation can be falsely interpreted as indicative of molecules that moved from the photoactivated cells to neighboring cells), the contribution in the loss of red Dendra2 intensity due to proteolytic degradation should be quantified in isolated protoplasts using a time-lapse experiment. We tested proteolytic degradation of Dendra2 in protoplasts with the settings also used for photoconversion during the imaging of Dendra2 in protonemata [24] and found no significant proteolysis under these conditions.
15. Fluorescence intensity of FPs including Dendra2 is influenced by pH [28]. Thus, when applying hormones or other compounds that may alter pH, the effect on Dendra2 fluorescence and photoconversion must be monitored.
16. We found that photoconverted Dendra2 intensity continues to increase until 5–10 min after applying photoconversion conditions (Fig. 5). Thus, the time-lapse imaging to score the protein mobility should be started after around 10 min after the photoconversion.

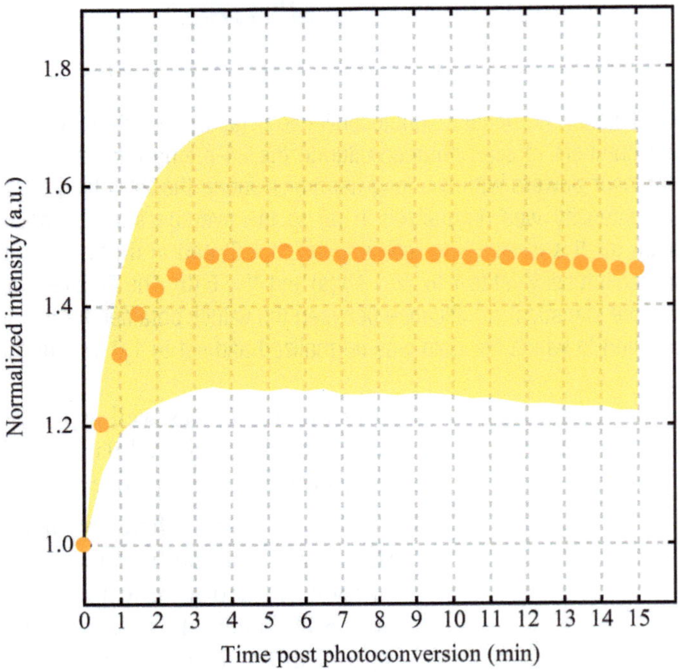

Fig. 5 Time to completion of Dendra2 photoconversion. Mean fluorescence intensity of red Dendra2 at 30-s intervals for 15 min of irradiation with the 561 nm laser in 18 protonemal cells. The standard deviation is indicated by the shaded area

17. Note that light conditions affect protonemal growth. Thus, the absence or presence of light during the time-lapse imaging can influence Dendra2 mobility [16, 24].
18. We performed the time-lapse imaging with the objective lens PlanApo 10×/0.45 NA and 561 nm solid laser (2.0% laser output, 10 mW) at 20-min intervals for 720 min and set the pinhole size to 16.60 μm in diameter (one airy unit), resulting in a 7.29-μm optical slice.
19. It is also advisable to acquire multiple z-stack images with optical sections of around 6 μm in each time interval for each ROI position to ensure the gaining of focused images of observed cells during the time-lapse experiment. In each time interval, we acquired 5 z-stack images (1024 × 1024 pixels in size) using 6 μm steps between optical sections.
20. The chamber slide and/or protonemal tissues can sometimes move during time-lapse imaging. In such a case, ROIs can be manually adjusted in the images of each time point.
21. We use Excel (Microsoft) to paste the "Result" into a table to extract the intensity of the image in focus, which empirically represents the maximum value in this setting, among the 5 z-stack images.

Acknowledgments

We thank Dr. Konstantin A. Lukyanov for providing plasmids containing *Dendra2* sequences and Drs. Minoru Kubo and Mitsuyasu Hasebe for the plasmid pT1OG possessing *EF1α* promoter. We thank the Spectrography and Bioimaging Facility at the NIBB Core Research Facilities for allowing access to a Leica M205FA stereomicroscope (Fig. 3) and a confocal microscope NikonA1 (Table 1 and Figs. 2, 4, and 5). We also thank the Model Plant Research Facility at the NIBB BioResource Center for technical support. The work was funded by KAKENHI (Grant-in-Aid for Challenging Exploratory Research (23657027 and 25650089) and Grant-in-Aid for Scientific Research on Innovative Areas (18H04829 and 20H04878)) to TF.

References

1. Crawford KM, Zambryski PC (2000) Subcellular localization determines the availability of non-targeted proteins to plasmodesmatal transport. Curr Biol 10:1032–1040
2. Crawford KM, Zambryski PC (2001) Non-targeted and targeted protein movement through plasmodesmata in leaves in different developmental and physiological states. Plant Physiol 125:1802–1812
3. Kim I, Cho E, Crawford K et al (2005) Cell-to-cell movement of GFP during embryogenesis and early seedling development in *Arabidopsis*. Proc Natl Acad Sci U S A 102:2227–2231

4. Martens HJ, Roberts AG, Oparka KJ, Schulz A (2006) Quantification of plasmodesmatal endoplasmic reticulum coupling between sieve elements and companion cells using fluorescence redistribution after photobleaching. Plant Physiol 142:471–480
5. Oparka KJ, Roberts AG, Boevink P et al (1999) Simple, but not branched, plasmodesmata allow the nonspecific trafficking of proteins in developing tobacco leaves. Cell 97:743–754
6. Lucas WJ, Bouché-Pillon S, Jackson DP et al (1995) Selective trafficking of KNOTTED1 homeodomain protein and its mRNA through plasmodesmata. Science 270:1980–1983
7. Kurata T, Ishida T, Kawabata-Awai C et al (2005) Cell-to-cell movement of the CAPRICE protein in *Arabidopsis* root epidermal cell differentiation. Development 132: 5387–5398
8. Koizumi K, Wu S, MacRae-Crerar A, Gallagher KL (2011) An essential protein that interacts with endosomes and promotes movement of the SHORT-ROOT transcription factor. Curr Biol 21:1559–1564
9. Xu XM, Wang J, Xuan Z et al (2011) Chaperonins facilitate KNOTTED1 cell-to-cell trafficking and stem cell function. Science 333: 1141–1144
10. Liu L, Liu C, Hou X et al (2012) FTIP1 is an essential regulator required for florigen transport. PLoS Biol 10:e1001313
11. Imlau A, Truernit E, Sauer N (1999) Cell-to-cell and long-distance trafficking of the green fluorescent protein in the phloem and symplastic unloading of the protein into sink tissues. Plant Cell 11:309–322
12. Liarzi O, Epel BL (2005) Development of a quantitative tool for measuring changes in the coefficient of conductivity of plasmodesmata induced by developmental, biotic, and abiotic signals. Protoplasma 225:67–76
13. Stadler R, Lauterbach C, Sauer N (2005) Cell-to-cell movement of green fluorescent protein reveals post-phloem transport in the outer integument and identifies symplastic domains in Arabidopsis seeds and embryos. Plant Physiol 139:701–712
14. Kawade K, Tanimoto H, Horiguchi G, Tsukaya H (2017) Spatially different tissue-scale diffusivity shapes ANGUSTIFOLIA3 gradient in growing leaves. Biophys J 113:1109–1120
15. Wu S, Koizumi K, MacRae-Crerar A, Gallagher KL (2011) Assessing the utility of photoswitchable fluorescent proteins for tracking intercellular protein movement in the *Arabidopsis* root. PLoS One 6:e27536
16. Kitagawa M, Fujita T (2013) Quantitative imaging of directional transport through plasmodesmata in moss protonemata via single-cell photoconversion of Dendra2. J Plant Res 126: 577–585
17. Gerlitz N, Gerum R, Sauer N, Stadler R (2018) Photoinducible DRONPA-s: a new tool for investigating cell–cell connectivity. Plant J 94: 751–766
18. Nicolas WJ, Grison MS, Trépout S et al (2017) Architecture and permeability of post-cytokinesis plasmodesmata lacking cytoplasmic sleeves. Nat Plants 3:1–11
19. Dempsey WP, Georgieva L, Helbling PM et al (2015) In vivo single-cell labeling by confined primed conversion. Nat Methods 12:645–648
20. Taniguchi A, Kimura Y, Mori I et al (2017) Axially-confined in vivo single-cell labeling by primed conversion using blue and red lasers with conventional confocal microscopes. Dev Growth Differ 59:741–748
21. Kitagawa M, Fujita T (2015) A model system for analyzing intercellular communication through plasmodesmata using moss protonemata and leaves. J Plant Res 128:63–72
22. Rensing SA, Goffinet B, Meyberg R et al (2020) The moss *Physcomitrium* (*Physcomitrella*) *patens*: a model organism for non-seed plants. Plant Cell 32:1361–1376
23. Kitagawa M, Tomoi T, Fukushima T et al (2019) Abscisic acid acts as a regulator of molecular trafficking through plasmodesmata in the moss *Physcomitrella patens*. Plant Cell Physiol 60:738–751
24. Tomoi T, Kawade K, Kitagawa M et al (2020) Quantitative imaging reveals distinct contributions of SnRK2 and ABI3 in plasmodesmatal permeability in *Physcomitrella patens*. Plant Cell Physiol 61:942–956
25. Nishiyama T, Hiwatashi Y, Sakakibara K et al (2000) Tagged mutagenesis and gene-trap in the moss, *Physcomitrella patens* by shuttle mutagenesis. DNA Res 7:9–17
26. Makarov NS, Cirloganu C, Perry JW et al (2014) Steady-state and time-resolved spectroscopic studies of green-to-red photoconversion of fluorescent protein Dendra2. J Photochem Photobiol A Chem 280:5–13
27. Kawade K, Horiguchi G, Hirose Y et al (2020) Metabolic control of gametophore shoot formation through arginine in the moss *Physcomitrium patens*. Cell Rep 32:108127
28. Pakhomov AA, Martynov VI, Orsa AN et al (2017) Fluorescent protein Dendra2 as a ratiometric genetically encoded pH-sensor. Biochem Biophys Res Commun 493:1518–1521

Chapter 23

Virus Genome-Based Reporter for Analyzing Viral Movement Proteins and Plasmodesmata Permeability

Ekaterina A. Lazareva, Anastasia K. Atabekova, Alexander A. Lezzhov, Sergey Y. Morozov, Manfred Heinlein, and Andrey G. Solovyev

Abstract

Plant virus movement proteins (MPs) mediate cell-to-cell movement of the virus genome through plasmodesmata (PD). MPs target PD to increase their size exclusion limit (SEL), and this MP function is essential for virus intercellular trafficking. In this chapter, we describe the use of a *Potato virus X* genome-derived reporter for agroinfiltration-based identification of virus genome-encoded MPs and analysis of the ability of individual viral MPs or plant proteins to increase the PD SEL.

Key words Plasmodesmata, Plant virus, Movement protein, Cell-to-cell movement, Size exclusion limit

1 Introduction

Cell-to-cell transport of plant viruses through plasmodesmata (PD) is mediated by virus-encoded movement proteins (MPs) [1]. In different virus groups, MPs can vary substantially in their sequence, predicted structure, and properties [2]. Therefore, the primary structure of a virus-encoded protein provides insufficient information for predicting a potential MP function, unless this candidate shows obvious structural homology to known MPs of other viruses. For candidate MPs lacking such homology, indirect indications of a MP function can be obtained by experiments determining their potential influence on PD function (see below) or showing the ability of the fluorescence-tagged candidate protein to move cell to cell upon transient expression or microinjection [3–9]. To determine more directly whether a protein under investigation is a viral MP functional in virus movement, a number of approaches have been used. These include techniques based on (a) mutational analysis of candidate MP gene(s) in an infectious

Yoselin Benitez-Alfonso and Manfred Heinlein (eds.), *Plasmodesmata: Methods and Protocols*,
Methods in Molecular Biology, vol. 2457, https://doi.org/10.1007/978-1-0716-2132-5_23,

cDNA copy of the viral genome [10–17], (b) *cis*-complementation tests involving insertion of a candidate MP gene into a movement-deficient heterologous virus genome with deleted MP gene [18, 19], and (c) a particle bombardment-based *trans*-complementation test, in which a cloned MP-deficient GFP-tagged reporter viral genome is delivered to individual epidermal cells together with a construct for expression of a candidate MP [7, 8, 20]. To set up these methods may be difficult as they require either infectious viral cDNA clones, which are often unavailable, or special equipment such as a particle gun. Therefore, a more straightforward approach that works without such specific needs would be very useful.

Typical and generally accepted properties of viral MPs include (a) the ability to modify the "size exclusion limit" (SEL) of PD and, thus, their conductivity for intercellular transport of macromolecules, (b) the ability of binding nucleic acids in a sequence-independent manner, and (c) interaction with the endoplasmic reticulum and the cytoskeleton for PD targeting [1]. Plant viruses may encode either a single MP responsible for all MP-specific functions [e.g., the MP of *Tobacco mosaic virus* (TMV)] or two, three, or more MPs that act together to facilitate virus movement [1]. Multicomponent transport systems have the advantage that different MP functions can be analyzed independently from each other. To determine which protein of such a multicomponent transport system has the ability to increase the PD SEL (the key MP function), a macromolecular diffusion test is often used, in which the candidate MP is co-expressed with green fluorescent protein (GFP) in individual leaf epidermal cells by particle bombardment. The ability to increase the SEL is evaluated based on the rate of GFP diffusion from these cells into adjacent cells and comparing this rate with control measurements in cells in which GFP is expressed alone [21]. In a variant of this approach, the GFP and the candidate MP are expressed by co-infiltration of plant leaf tissues with transformed agrobacteria carrying the respective protein-expressing binary constructs (co-agroinfiltration) [7, 22].

Here, we describe the use of the virus genome-derived reporter PVX-POL-GFP for agroinfiltration-based identification of viral MPs and the analysis of their ability to increase the PD SEL. PVX-POL-GFP represents a derivative of *Potato virus X* (PVX) from which the MP and capsid protein genes were deleted. Because this construct contains only the viral replicase gene and an inserted GFP gene (Fig. 1), it is able to replicate and express GFP in initially infected cells but is deficient in cell-to-cell movement [23]. In the described experimental setup, PVX-POL-GFP is agroinfiltrated into plant leaves at a high dilution to restrict the production of high expression levels of GFP to individual leaf cells that are distant from each other. The same leaves are co-agroinfiltrated with one or more constructs for expression of the candidate MP(s). The latter cultures are infiltrated in high concentration in order to express the

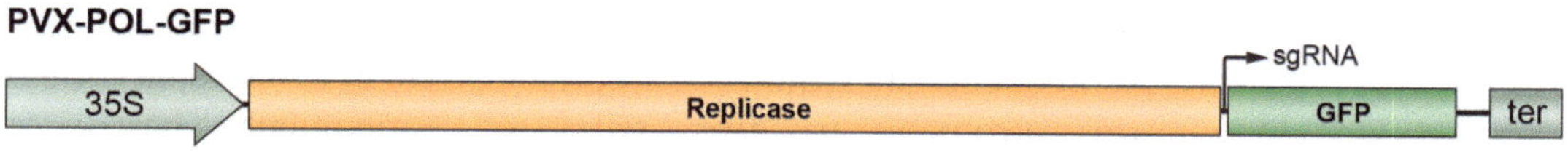

Fig. 1 Reporter construct PVX-POL-GFP. Protein-encoding open reading frames are shown by orange and green boxes. Positions of the *Cauliflower mosaic virus* 35S promoter and terminator are indicated. The position of the viral promoter driving the synthesis of GFP-encoding subgenomic RNA (sgRNA) is indicated

respective protein(s) in practically all cells of the infiltrated leaf area. This approach allows to analyze (a) the complementation of PVX-POL-GFP cell-to-cell movement and (b) GFP diffusion from the primary PVX-POL-GFP-infected cells into adjacent cells. Using this approach, PVX-POL-GFP has been successfully used for assigning the MP function to *Hibiscus green spot virus* (HGSV) proteins and for identification of the HGSV protein capable of increasing the PD SEL [23, 24].

Using PVX-POL-GFP as a reporter for the identification of virus-encoded MPs has several advantages over other systems. First, it employs an agroinfiltration method that does not require specific equipment and is easier to apply than particle bombardment. Second, the PVX-POL-GFP construct has all three PVX MP genes deleted. This eliminates the possibility of partial MP function complementation, which is theoretically possible when other PVX-based reporters are used, such as those with a frameshift mutation in one of the three PVX MP genes [25]. Moreover, in comparison to a recently described reporter system based on MP-deficient GFP-tagged genome of *Cucumber mosaic virus* (CMV) [26], which is also designed for agroinfiltration-mediated delivery into plant cells, PVX-POL-GFP is apparently preferable, as CMV has a tripartite genome. The CMV-based reporter system requires the co-infiltration of three agrobacterial cultures carrying derivatives of the three genomic RNAs of this virus, which may considerably decrease the efficiency of reporter system expression in individual leaf cells together with the gene of a candidate MP. Third, the PVX-POL-GFP replicon provides higher GFP expression levels than 35S promoter-driven vectors and thus provides a more sensitive system for detection of intercellular GFP diffusion and identification of MPs capable of increasing the PD SEL. The replication system for GFP expression provided by the reporter is critical in experiments with candidate MP proteins, which, under conditions of agroinfiltration, may reduce GFP expression by impairing the physiological status of the cells.

The protocol provided here can be used as a workflow for identification and characterization of MPs of previously uncharacterized viruses. In particular, the protocol includes methods for transformation of *Agrobacterium tumefaciens* cells with binary vectors containing candidate MP genes, preparation of agrobacterial cultures and infiltration of *Nicotiana benthamiana* plants, as well as subsequent data acquisition and analysis. We describe application of this protocol to two types of experiments based on co-agroinfiltration of PVX-POL-GFP and constructs for expression of candidate MP genes. The first type of experiment employs this protocol for a functional complementation assay to identify viral MPs. In this approach, a candidate protein having the MP function enables cell-to-cell transport of PVX-POL-GFP resulting in formation of multicellular infection sites, or infection foci, that are easily detectable under the UV light due to GFP expression in infected cells. In analyses of proteins encoded by multicomponent viral transport systems, where every single candidate protein is insufficient for virus movement, this assay allows simultaneous co-expression of two or three candidate MPs to define the set of viral proteins required for transport. The second type of experiment is a GFP diffusion test to determine which component of a multicomponent transport system is capable of increasing the PD SEL. In these experiments, components of a multicomponent transport system are co-expressed individually with PVX-Pol-GFP and, thus, are insufficient to cause PVX-POL-GFP movement. However, facilitated GFP diffusion from the PVX-POL-GFP-infected cells to neighboring cells indicates the ability of the expressed component to increases the PD SEL. Importantly, the GFP diffusion test can also be used for analysis of SEL modifications induced by non-viral plant proteins, therefore widening the application of the provided protocol toward the identification of cell-autonomous and cell-non-autonomous plant proteins capable of regulating PD conductivity.

2 Materials

2.1 Plants

1. 5–6-week-old (for GFP diffusion test) and 6–8-week-old (for complementation assay) *N. benthamiana* plants grown on commercial soil mixture (*see* **Note 1**) in small pots.

2.2 General Laboratory Equipment

1. Autoclaved 1.5 and 2 mL microcentrifuge tubes and pipette tips.
2. Sterile round-bottom 10 mL tubes with caps.
3. Sterile 50 mL plastic tubes with cap.
4. Plastic plates (Petri dishes), 90 mm diameter.
5. Membrane filters, 0.22 μm pore size.
6. pH meter.
7. Vortex.
8. Weighing scales.

9. Spectrophotometer and appropriate cuvettes.
10. Benchtop centrifuge.
11. Sterile hood.
12. Shaking incubator set to 28 °C.
13. Benchtop shaking platform.
14. Incubator set to 28 °C.
15. Heating block set to 37 °C.
16. Plant growth chamber (*see* **Note 2**).

2.3 Plasmid Vectors and Agrobacteria for Transient Expression in N. benthamiana *Plants*

1. Agrobacterial culture carrying the reporter construct PVX-POL-GFP [23] (*see* **Note 3**).
2. Agrobacterial cultures used as a negative control (agrobacteria carrying empty binary vector) and a positive control (*see* **Note 4**); a culture for expression of *Tomato bushy stunt virus* (TBSV) p19 RNA silencing suppressor [27, 28] (optional) (*see* **Note 5**).
3. Constructs for expression of candidate MPs or plant proteins under investigation. The respective genes should be cloned in advance into a 35S promoter-driven expression cassette of an *Agrobacterium*- and plant-compatible binary expression plasmid vector of choice, e.g., the pLH7000-based vector pLH* [29] (*see* **Note 6**).

2.4 Preparation of Agrobacterium Cultures and Agroinfiltration of N. benthamiana *Plants*

1. *Agrobacterium tumefaciens* strain C58C1 (*see* **Note 7**).
2. Liquid nitrogen.
3. 50% glycerol sterilized by autoclaving.
4. Antibiotics stock solutions (store at −20 °C) (*see* **Note 8**):
   (a) 50 mg/mL rifampicin: weigh 0.1 g of rifampicin, add 2 mL of 100% dimethyl sulfoxide (DMSO), and dissolve completely by vortexing.
   (b) 50 mg/mL kanamycin: dissolve 0.1 g of kanamycin in 2 mL of ultrapure, Milli-Q-purified water (Milli-Q water), and sterilize by filtering through 0.22 μm pore size filters.
   (c) 100 mg/mL streptomycin: dissolve 0.2 g of streptomycin in 2 mL of Milli-Q water, and then sterilize by filtering through 0.22 μm pore size filters.
   (d) 100 mg/mL spectinomycin: dissolve 0.2 g of spectinomycin in 2 mL of Milli-Q water, and then sterilize by filtering through 0.22 μm pore size filters (*see* **Note 9**).
5. 50 mM acetosyringone (AS) stock solution: dissolve 9.81 mg acetosyringone in 1 mL of DMSO. Store at −20 °C.
6. 1 M $MgCl_2$: dissolve 10.2 g ($MgCl_2 \cdot 6H_2O$) in 50 mL of Milli-Q water by careful mixing. Sterilize by filtering through

0.22 μm pore size filters and store at +4 °C. The solution can also be kept frozen in case of long breaks between experiments (*see* **Note 10**).

7. MES buffer solution (0.5 M 2-(*N*-morpholino)-ethanesulfonic acid (MES) buffer, pH 5.5): dissolve 5.3 g of MES, in about 40 mL Milli-Q water, mix and adjust pH with KOH (use 1 or 5 M stock solution), and add water to a total volume of 50 mL. Sterilize solution by filtration through 0.22 μm pore size filters. Store at +4 °C. The solution can also be kept frozen in case of longer interruptions between experiments.
8. Infiltration buffer (10 mM MES pH 5.5, 10 mM $MgCl_2$, 150 μM AS): add 1 mL of MES buffer solution, 0.5 mL of $MgCl_2$ stock, 150 μL of AS stock, and water to a total volume of 50 mL.
9. LB medium: 8 g tryptone, 5 g yeast extract, and 5 g NaCl in 1 L of Milli-Q water. Sterilize by autoclaving and store at room temperature (RT).
10. LB agar medium: 15 g agar in 1 L of LB. Sterilize by autoclaving and store at RT.
11. LB agar plates containing selective antibiotics (*see* Subheading 2.4, **item 4**). Melt LB agar medium in a microwave (or use it immediately after autoclaving). Pour the required amount of medium in a sterile container (e.g., 50 mL plastic tube); wait until the medium cools down to approximately 50 °C; add appropriate antibiotics from stock solutions (*see* Subheading 2.4, **item 4**), in amounts required to achieve final concentrations; carefully mix; and pour into plates. These operations should be done under a sterile hood.
12. Autoclaved toothpicks.
13. Needleless syringes (2–3 mL).

2.5 Equipment for Fluorescence Imaging

1. UV lamp (e.g., UVP™ Blak-Ray B-100).
2. Camera (*see* **Note 11**) and tripod.
3. Confocal laser scanning microscope (CLSM) with a 488 nm laser (we use LSM 780, Carl Zeiss AG) or an epifluorescence microscope with a 40× objective and a filter set for GFP detection (*see* **Note 12**).
4. Epifluorescence microscope with a low-magnification objective such as AxioZoom V16 (Carl Zeiss AG) equipped with a GFP filter set.
5. Glass slides (75 × 25 mm) and coverslips (50 × 24 mm and 24 × 24 mm).
6. Permanent water-proof marker.
7. Single-edged razor blades.

8. Adhesive tape.
9. Image processing software (e.g., ImageJ, http://rsbweb.nih.gov/ij/).

3 Methods

Perform all procedures at room temperature (RT) unless otherwise stated.

3.1 Transformation of *Agrobacterium tumefaciens* Cells with Binary Vectors Containing Candidate MP Genes

All manipulation with bacterial cultures should be performed under a sterile hood. Different methods can be used for transformation of *Agrobacterium tumefaciens* (C58C1 strain or other cells) with vectors carrying genes of candidate viral MPs or plant proteins under investigation [30]. We routinely use the following simple protocol, which is based on the freeze-thaw method [31] and requires neither transformation-competent cells nor electroporation.

1. Plate agrobacteria (C58C1) retrieved from a frozen glycerol stock on a LB plate with 100 μg/mL rifampicin (*see* **Note 13**). Incubate at 28 °C for 1–2 days.
2. Use a toothpick to transfer a bacterial colony into 5 mL of liquid LB medium supplemented with 50 μg/mL rifampicin. Grow the bacterial culture overnight in a shacking incubator at 28 °C and at 150–200 rpm (rotations per minute).
3. Transfer 500 μL of the bacterial culture to a sterile 2 mL tube. Add approximately 4 μg of a binary vector encoding the protein under investigation (usually 3–5 μL of a standard DNA miniprep) (*see* **Note 14**). Incubate on ice for 10 min.
4. Place the tube into liquid nitrogen for 2–3 min, and then transfer the frozen bacterial suspension to a heating block set to 37 °C for 15 min.
5. Add 800 μL of LB medium (without antibiotics at this stage), and incubate the tube containing the agrobacteria placed horizontally in a shacking incubator for 1.5–2 h with shaking (80 rpm) at 28 °C.
6. Centrifuge the tube in a benchtop centrifuge at 1100 × *g* for 5 min, discard 950 μL of the supernatant, and resuspend the agrobacteria in the 350 μL of remaining medium. Plate the cells on an LB plate with selective antibiotics (e.g., 100 μg/mL rifampicin, 100 μg/mL spectinomycin, and 300 μg/mL streptomycin for selection of C58C1 agrobacteria transformed with pLH*-based binary vector). Incubate at 28 °C for 2 days for colony formation.
7. Analyze individual colonies to validate the identity of contained binary vector (*see* **Note 15**).

8. Use a verified clone to prepare a glycerol stock by adding one volume of 50% glycerol to an overnight *Agrobacterium* liquid culture carrying the candidate MP gene, carefully mix, freeze in liquid nitrogen, and store at −70 °C.

3.2 Preparing Agrobacterial Cultures and Infiltration of N. benthamiana Plants

1. Retrieve individual agrobacterium cultures carrying (a) PVX-POL-GFP, (b) vectors for expression of candidate MPs or plant proteins under investigation, and (c) an empty binary vector (a negative control) from glycerol stocks onto LB agar plates containing selective antibiotics. For agrobacteria containing the PVX-POL-GFP construct, use 100 μg/mL rifampicin and 50 μg/mL kanamycin. Place the plates into an incubator for 2 days at 28 °C (*see* **Note 16**).
2. Use sterile toothpicks or sterile pipette tips to transfer single colonies of each agrobacterial culture into 2 mL of LB medium supplemented with appropriate antibiotics, 10 mM MES (40 μL of 0.5 M MES buffer solution per 2 mL), and 20 μM AS (0.8 μL of 50 μm AS stock solution per 2 mL). Use round-bottom 10 mL tubes. Incubate the liquid cultures overnight at 28 °C with vigorous shaking (150 rpm), preferably in an angle holder to increase aeration (*see* **Note 17**).
3. At the day of the experiment, prepare fresh infiltration buffer (*see* Subheading 2.4, **item 8**).
4. Transfer agrobacterial cultures (make sure that the suspension contains no visible clots or conglomerations) to 2 mL tubes and pellet cells by centrifugation at 1100 × *g* for 5 min at RT (*see* **Note 18**).
5. Resuspend the agrobacterial pellet in 1 mL of infiltration buffer by careful shaking. Do not vortex.
6. Incubate the bacterial suspension on a benchtop shaking platform for 2–3 h with shaking (80 rpm) at RT to induce virulence genes (*see* **Note 19**).
7. Measure the optical density of agrobacterial suspension with a spectrophotometer (OD_{600}) (*see* **Note 20**).
8. Immediately before infiltration of plant leaves, prepare a mixture of the agrobacteria containing the different constructs. For this, use the OD of each culture to calculate the amount required to achieve the desired dilution for each individual culture within the mixture. The total OD of all agrobacteria in the mixture should be equal between all mixtures used in the experiment to obtain equivalent bacterial loading of the infiltrated leaves. Depending on the experimental design, the relative amounts of agrobacterial cultures within mixtures can be adapted (*see* Subheading 3.3, **step 2**, and Subheading 3.4, **step 2**, for details).

9. Use a sterile 200 μL micropipette tip to puncture the plant leaf in the center of the anticipated infiltration area. Fill a needleless syringe with the bacterial mixture, and infiltrate the leaf by placing the opening of the syringe onto the puncture and by applying gentle pressure onto the syringe plunger. The leaf area acquiring the infiltration mixture changes color to dark green. Proceed with injection until the infiltrated leaf area is 2–2.5 cm in diameter.
10. Place the treated plants into a growth chamber, and keep at 16 h/8 h light/dark cycles with temperatures of 24 °C/20 °C, respectively.

3.3 Identification of Viral MPs in a Complementation-Assay

1. Prepare agrobacterial cultures for infiltration as described under Subheading 3.1. Prepare 5 mL cultures of specific agrobacteria carrying the following binary vector constructs: reporter PVX-POL-GFP, candidate MPs, and negative control (empty binary vector). A positive control (agrobacteria with a plasmid encoding the MP of TMV, e.g.) can also be included.
2. Set up infiltration mixtures (10 mL each) by mixing agrobacterial cultures according to Table 1. Note that the agrobacteria with the PVX-POL-GFP construct must be highly diluted to obtain individual transformed cells in the infiltrated leaf area, whereas cultures with constructs for candidate MPs and the control cultures are infiltrated at high density to transform as many cells as possible. The total optical density (OD_{600}) of the mixture (calculated as a sum of OD_{600} values of individual agrobacterial cultures) should be 0.3. Therefore, when cultures for expression of two individual MPs are added to the mixture, the OD_{600} of each of these should be 0.15; when three candidate MP cultures are added, the OD_{600} of each respective culture should be 0.1, thus keeping the total OD_{600} at 0.3 (*see* **Note 21**). As the amount of PVX-POL-GFP culture added to the mixture is several orders of magnitude less than the amounts of other cultures, the PVX-POL-GFP culture is not included in the calculation of final OD_{600}.
3. Infiltrate two fully expanded leaves of 6–8-week-old *N. benthamiana* plants (*see* **Note 22**). One large leaf can easily accommodate five infiltration areas as shown in Fig. 2. To avoid contamination, make sure that there are sufficient distances between infiltrated leaf areas; never allow these areas to contact or overlap each other. To identify infiltration zones at later steps, encircle them on the adaxial (upper) with a permanent water-proof marker immediately after infiltration.
4. After 4–5 days, use a UV lamp (such as Blak-Ray® UV) to visualize and compare the fluorescent foci in the infiltrated leaf areas (*see* **Note 23**) (Fig. 2).

Table 1
Preparation of agrobacterial mixtures for infiltration in movement complementation experiments[a]

Infiltration mixture	Individual agrobacterial cultures, OD_{600} in infiltration mixture				
	PVX-POL-GFP[b]	Empty binary vector	MP TMV	Candidate MP 1	Candidate MP 2
– Control	0.00003	0.3			
Experiment with single candidate MP					
+ Control	0.00003	0.3			
Candidate MP	0.00003			0.3	
Experiment with two candidate MPs					
+ Control	0.00003	0.15	0.15		
Candidate MP 1	0.00003	0.15		0.15	
Candidate MP 2	0.00003	0.15			0.15
Candidate MPs 1 + 2	0.00003			0.15	0.15

[a] Experimental setups for analyzing a single candidate MP and two candidate MPs, which comprise a virus transport system, are shown

[b] Agrobacterial culture carrying PVX-POL-GFP reporter should be first adjusted to the OD = 0.3 and then diluted 7000- to 10,000-fold (do it as a series of consecutive 10- to 100-fold dilutions). Plants grown in different climate chambers and greenhouses differ from each another; therefore, it is necessary, prior to experiment, to check optimal dilution for the reporter culture to get distantly located individual transformed cells

5. To quantitatively compare the sizes of fluorescent foci between the different infiltration areas (i.e., between different agrobacterium mixtures with specific constructs), prepare samples (*see* **Note 24**), and carry out microscopy using an AxioZoom or another epifluorescent microscope equipped with a low-magnification objective to take pictures for statistical analysis.
6. Use image processing software (ImageJ or similar) to measure the diameter or area of the fluorescent foci. For each infiltration mixture containing the specific constructs, a sufficient number of resulting foci (typically, more than 100 foci) should be measured to allow reliable statistical analysis with suitable software (e.g., Excel) (*see* **Note 25**).

3.4 GFP Diffusion Test to Determine Ability of Protein to Change Plasmodesmata Permeability

1. Prepare agrobacterium cultures for infiltration as in Subheading 3.1. Prepare 2 mL of the cultures carrying the following binary vector constructs: reporter PVX-POL-GFP, candidate MPs, and negative control (empty binary vector) (*see* **Note 21**).
2. Set up mixtures (5 mL each) of individual agrobacterial cultures. Each mixture consists of a highly diluted culture for the reporter construct (OD_{600} = 0.00038; prepare the reporter

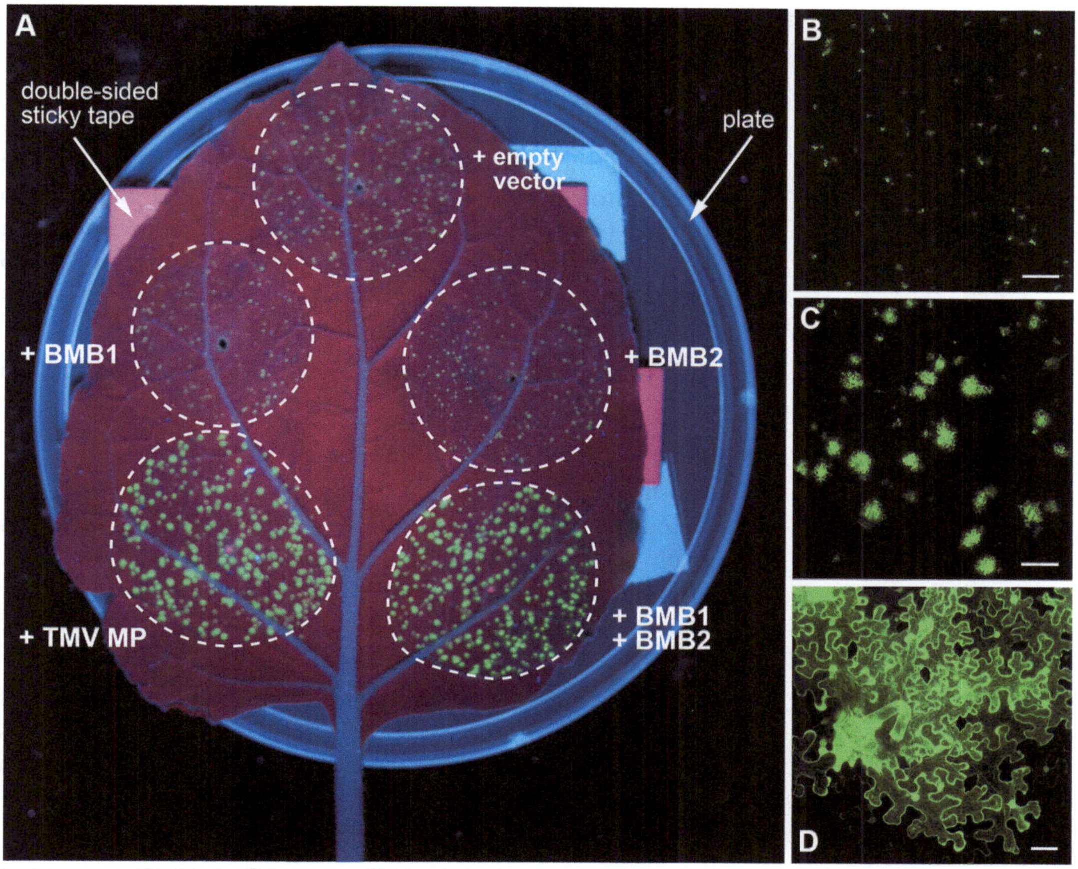

Fig. 2 Use of PVX-POL-GFP for identification of *Hibiscus green spot virus* (HGSV) MPs in a virus movement complementation assay. GFP fluorescence in a *N. benthamiana* leaf agroinfiltrated for infection of individual cells with diluted PVX-POL-GFP and locally co-agroinfiltrated with either an empty vector (negative control) or constructs for expression of TMV MP (positive control), HGSV BMB1 protein, HGSV BMB2 protein, or both BMB1 and BMB2 proteins. The HGSV BMB1 and BMB2 proteins are two virus-encoded MPs, which are, together, necessary and sufficient to mediate virus cell-to-cell transport [23]. (**a**) Imaging of infiltrated leaf 4 days post-agroinfiltration. To flatten the leaf surface and therefore facilitate imaging, the leaf was mounted on a plate with double-sided sticky tape as shown. Dashed lines indicate infiltrated leaf areas. Whereas the leaf areas expressing either BMB1 or BMB2 show fluorescence in only single infected cells, the leaf areas expressing either both BM1 and BMB2 or the MP of TMV show fluorescence in cell clusters, indicating expanding infection sites. (**b–d**) Fluorescence microscopy of infiltrated areas at low (**b**, **c**) and higher magnification (**d**). (**b**) Co-agroinfiltration of PVX-POL-GFP and BMB1 constructs leads to single cell fluorescence indicating a lack of complemented virus movement. (**c**, **d**) Co-agroinfiltration of PVX-POL-GFP with BMB1 and BMB2 constructs leads to the development of cell clusters showing GFP fluorescence indicating complementation of PVX-POL-GFP for movement. Scale bars, 1 mm in (**b**, **c**); 50 μm in (**d**)

culture with $OD_{600} = 0.3$ and add to the mixture the specific amount of culture to obtain a 1/800 dilution, i.e., 6.25 μL/ 5 mL) and either a culture carrying the empty binary vector (negative control) or a culture carrying a binary vector encoding the individual candidate MP or plant protein under investigation with an $OD_{600} = 0.3$.

3. Infiltrate upper leaves of 5–6-week-old *N. benthamiana* plants (avoid using tiny upper-most leaves) according to Subheading 3.1. Inject the agrobacterial culture encoding the candidate protein into one half of the leaf and the control culture into the other (*see* **Note 26**). To find the infiltration zones during sample preparation for microscopy (*see* Subheading 3.4, **step 5**), encircle these zones on the adaxial (upper) leaf side with a permanent marker immediately after infiltration.
4. Incubate plants in chamber for 2–2.5 days, but no more than 3 days.
5. Prepare leaf samples for microscopy (*see* **Note 24**).
6. Perform microscopy. Use a confocal microscope such as the Carl Zeiss LSM 780 or an epifluorescent microscope. For each initially transformed cell in the leaf specimen (*see* **Note 27**), count neighboring cells into which GFP has moved as seen by the green fluorescent nuclei as shown in Fig. 3c. Record the numbers of such cells (*see* **Note 28**).
7. Evaluate the statistical significance of the results. Application of a non-paired two-tailed Student's *t*-test is appropriate for this purpose. An example of the statistical analysis outcome for such experiments is shown in Fig. 3d.

4 Notes

1. Different brands of commercial soil mix can be used, for example, Compo Sana Universal (Germany).
2. To grow plants for described experiments, a growth chamber should be set to 16 h/8 h light/dark cycles and 24 °C/ 20 °C day/night temperatures and approximately 50% humidity.
3. The PVX-POL-GFP reporter can be obtained from A. Solovyev.
4. Store agrobacterial cultures at −70 °C as frozen glycerol stocks prepared as described under Subheading 3.1, **step 8**. As a positive control in cell-to-cell movement complementation assay, we used an agrobacterial culture for expression of TMV MP (pLH*-TMV-MP construct can be obtained from A. Solovyev).
5. The TBSV p19 suppressor of RNA silencing is often used to increase the accumulation level of target proteins in transient expression experiments [27, 28, 32].

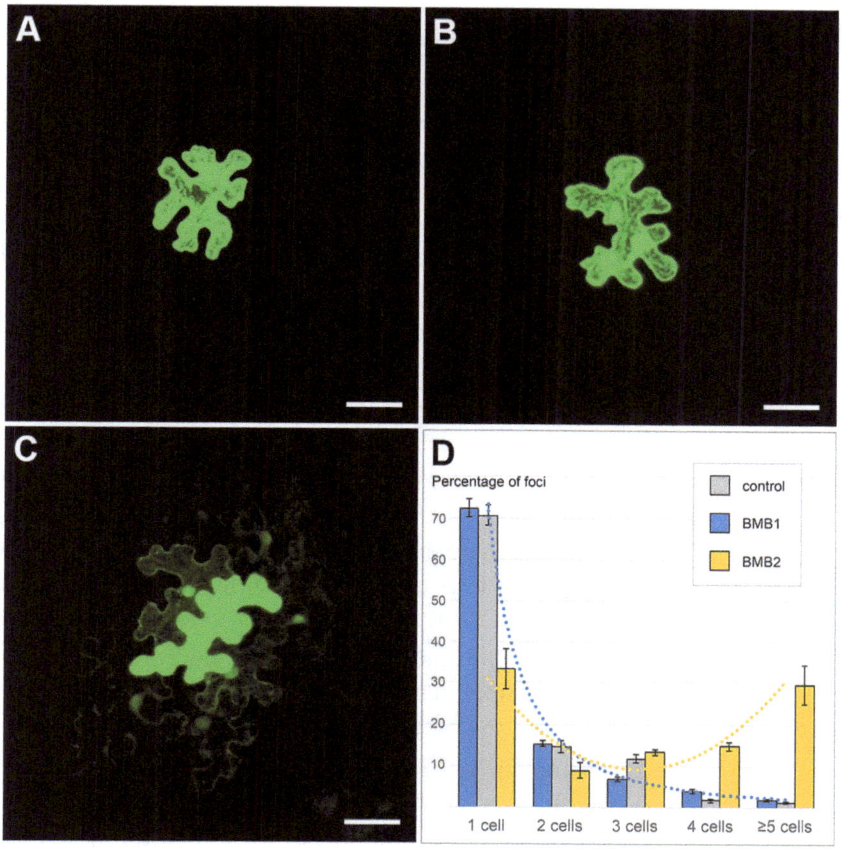

Fig. 3 Use of PVX-POL-GFP for testing the ability of BMB1 or BMB2, the two MPs encoded by *Hibiscus green spot virus*, to increase the PD SEL. PVX-POL-GFP was co-agroinfiltrated into *N. benthamiana* leaf with either an empty vector taken as a negative control (**a**) or constructs for expression of BMB1 (**b**) or BMB2 (**c**). Scale bars, 50 μm. (**d**) Data presentation showing that BMB2, and not BMB1, increases the PD SEL. Error bars indicate standard error

6. To monitor efficient expression of candidate MPs, they can be used in fusion to fluorescent proteins; however, it should be kept in mind that such fusions may fully or partially inhibit the functionality of the proteins. Therefore, for such experiments, the retainment of previously characterized activities should be verified before applying the fusion proteins in this test. As an alternative, shorter tags, such as FLAG or c-myc, that allow quantitative detection by Western blot analysis can be used to validate the expression of candidate MPs. If wild-type, non--fused protein candidates are addressed in the assay, the expression in the infiltrated tissues can be monitored by Western blot analysis using protein-specific antibodies or by reverse transcription-PCR with specific primers to detect expression at the RNA level.

7. Other agrobacterial strains may also be used. However, only strain C58C1 was tested for compatibility with the protocol described here.
8. The use of rifampicin selects for the agrobacterial strain C58C1 and the use of kanamycin selects for bacteria carrying the PVX--POL-GFP construct. Use both streptomycin and spectinomycin if candidate MP genes are cloned in pLH* or other vector-specific antibiotics if candidate MP genes are cloned in binary vectors other than pLH*.
9. Spectinomycin may form aggregates and precipitate. Dissolve it before use.
10. $MgCl_2$ salt is hygroscopic. Therefore, once the salt container is opened, the salt should be used within a short period of time. As an alternative to $MgCl_2$ salt, a ready-made commercial $MgCl_2$ solution can be used.
11. As an alternative to a dedicated camera, a mobile phone camera coupled to a special app for imaging under low light conditions (e.g., Slow Shutter Cam) can be used.
12. Both CLSM and epifluorescence microscopes can be used for virus complementation and GFP diffusion assays; however, a CLSM is preferred for high-quality imaging.
13. Rifampicin is used to select the C58C1 agrobacterial strain, which is resistant against this antibiotic.
14. Up to 20 μL of plasmid DNA can be used if its concentration is low.
15. To validate that the bacteria contain the plasmid encoding the expected candidate MP, perform colony PCR tests to verify amplification of a specific gene fragment with gene-specific primers. For further confirmation, the amplified DNA should be sequenced by Sanger sequencing.
16. Freshly plated cultures should be used for the experiments. In any case, avoid long cultivation of agrobacterial cultures on LB plates. Reproducible results can be obtained with agrobacteria kept on plates at 4 °C no longer than 10 days. Never re-plate agrobacteria from old plates. Retrieve new bacteria from frozen glycerol stocks instead.
17. If more bacteria are needed, inoculate 5 mL LB with antibiotics in a 50 mL tube.
18. This and all the following steps can be performed outside the sterile hood and do not require antibiotics for further selection.
19. The bacteria may be incubated also without shaking, and the incubation time can be as long as 10 h.
20. For more accurate measurements, dilute the agrobacterial cultures tenfold before measurement.

21. If the expression level of the candidate MP is too low, add a culture for expression of the p19 silencing suppressor to an optical density (OD_{600}) of 0.1. The optical density (OD_{600}) of the agrobacterium mixture used for agroinfiltration should be 0.4 in this case.
22. The fully expanded leaves of 6–8-week-old plants provide sufficient surface for movement complementation experiments. Younger, 5–6-week-old plants are used for other agroinfiltration applications, such as the GFP diffusion test described in Subheading 3.4.
23. Detached leaves can be kept in a humid chamber on wet paper towels for 1–2 days that makes it possible to observe the expansion of the fluorescent foci over time.
24. To prepare samples for microscopy, use a razor blade to excise leaf fragments from the infiltrated areas. In our experiments, leaf areas of approximately 0.8×0.8 cm have been usually examined. If necessary, use big coverslips to observe larger infiltration areas. Place a leaf sample with its abaxial side up on a microscope slide and cover with a coverslip. Fix the coverslip with an adhesive tape. Fill the space between the glasses with Milli-Q water. Place slides into a desiccator, and use a vacuum pump to evacuate and to remove air from the leaf intercellular spaces. Repeat cycles of filling with water/removing air several times until leaf samples turn fully dark green.
25. To obtain convincing data, several leaves should be infiltrated with the same constructs in independent experiments. Combine the sizes of multiple foci developed in such leaves from independent agroinfiltration experiments in the statistical analysis.
26. As the PD permeability depends on leaf developmental stage, young leaves can show strong GFP diffusion even in the absence of MP activity (negative control). Thus, use expanded leaves.
27. The initially transformed cell is brightly fluorescent and can be easily recognized (Fig. 3). Disregard foci with more than one strongly fluorescent cell. Although statistically unlikely (given the low total number of transformation events in the agroinfiltrated area), the latter may be caused by simultaneous transformation of two or more neighboring cells rather than by GFP movement. To unequivocally identify the initially transformed plant leaf cell in all cases, it is advisable to use agrobacteria that in addition to GFP encode a cell-autonomous marker (e.g., a fluorescent protein of different color targeted to the nucleus) [33]. If microscopy shows that many cells in a locus are equally bright such that an initially transformed cell cannot be discerned (Fig. 2d), the tested candidate MP protein

(or combination of several MP) may be able to support the cell-to-cell transport of viral RNA in addition to PD modification.

28. Because limited GFP diffusion may occur in the absence of a PD-modifying protein (especially in young tissues), identifying the activity of a PD-modifying protein may depend on determining the efficiency of GFP diffusion as compared to the control. Thus, to obtain clear-cut results, it is essential to count the number of cells into which GFP has diffused from the initially transformed cell in each case, including the negative control. The numbers obtained for the negative control must be included in the statistical analysis.

Acknowledgments

This work was supported by a CNRS-RFBR Joint Research Project (Russian Foundation for Basic Research grant 18-54-15003 to E.A.L. and PICS 2018-2020 grant 252520 to M.H.) and the Agence Nationale de la Recherche (grant ANR-13-KBBE-0005-01 to M.H.).

References

1. Heinlein M (2015) Plant virus replication and movement. Virology 479–480:657–671
2. Lucas WJ (2006) Plant viral movement proteins: agents for cell-to-cell trafficking of viral genomes. Virology 344:169–184
3. Ding B, Li Q, Nguyen L, Palukaitis P, Lucas WJ (1995) Cucumber mosaic virus 3a protein potentiates cell-to-cell trafficking of CMV RNA in tobacco plants. Virology 207:345–353
4. Fujiwara T, Giesman-Cookmeyer D, Ding B, Lommel SA, Lucas WJ (1993) Cell-to-cell trafficking of macromolecules through plasmodesmata potentiated by the red clover necrotic mosaic virus movement protein. Plant Cell 5: 1783–1794
5. Noueiry AO, Lucas WJ, Gilbertson RL (1994) Two proteins of a plant DNA virus coordinate nuclear and plasmodesmal transport. Cell 76: 925–932
6. Itaya A, Hickman H, Bao Y, Nelson R, Ding B (1997) Cell-to-cell trafficking of cucumber mosaic virus movement protein: green fluorescent protein fusion produced by biolistic gene bombardment in tobacco. Plant J 12:1223–1230
7. Robles Luna G, Peña EJ, Borniego MB, Heinlein M, Garcia ML (2013) Ophioviruses CPsV and MiLBVV movement protein is encoded in RNA 2 and interacts with the coat protein. Virology 441:152–161
8. Xiong R, Wu J, Zhou Y, Zhou X (2008) Identification of a movement protein of the tenuivirus rice stripe virus. J Virol 82:12304–12311
9. Vijayapalani P, Maeshima M, Nagasaki-Takekuchi N, Miller WA (2012) Interaction of the trans-frame potyvirus protein P3N-PIPO with host protein PCaP1 facilitates potyvirus movement. PLoS Pathog 8: e1002639
10. Petty IT, French R, Jones RW, Jackson AO (1990) Identification of barley stripe mosaic virus genes involved in viral RNA replication and systemic movement. EMBO J 9:3453–3457
11. Beck DL, Guilford PJ, Voot DM, Andersen MT, Forster RLS (1991) Triple gene block proteins of white clover mosaic potexvirus are required for transport. Virology 183:695–702
12. Gilmer D, Bouzoubaa S, Hehn A, Guilley H, Richards K, Jonard G (1992) Efficient cell-to-cell movement of beet necrotic yellow vein virus requires 3′ proximal genes located on RNA 2. Virology 189:40–47
13. Herzog E, Hemmer O, Hauser S, Meyer G, Bouzoubaa S, Fritsch C (1998) Identification of genes involved in replication and movement of Peanut clump virus. Virology 248:312–322

14. Carrington JC, Jensen PE, Schaad MC (1998) Genetic evidence for an essential role for potyvirus CI protein in cell-to-cell movement. Plant J 14:393–400
15. Peremyslov VV, Hagiwara Y, Dolja VV (1999) HSP70 homolog functions in cell-to-cell movement of a plant virus. Proc Natl Acad Sci U S A 96:14771–14776
16. Alzhanova DV, Hagiwara Y, Peremyslov VV, Dolja VV (2000) Genetic analysis of the cell-to-cell movement of beet yellows closterovirus. Virology 268:192–200
17. Hacker DL, Petty ITD, Wei N, Morris TJ (1992) Turnip crinkle virus genes required for RNA replication and virus movement. Virology 186:1–8
18. Wu Z, Wu J, Adkins S, Xie L, Li W (2010) Rice ragged stunt virus segment S6-encoded nonstructural protein Pns6 complements cell-to-cell movement of tobacco mosaic virus-based chimeric virus. Virus Res 152:176–179
19. Lewandowski DJ, Adkins S (2005) The tubule-forming NSm protein from tomato spotted wilt virus complements cell-to-cell and long-distance movement of tobacco mosaic virus hybrids. Virology 342:26–37
20. Hiraguri A, Ueki S, Kondo H, Nomiyama K, Shimizu T, Ichiki-Uehara T, Omura T, Sasaki N, Nyunoya H, Sasaya T (2013) Identification of a movement protein of Mirafiori lettuce big-vein ophiovirus. J Gen Virol 94:1145–1150
21. Tamai A, Meshi T (2001) Cell-to-cell movement of potato virus X: the role of p12 and p8 encoded by the second and third open reading frames of the triple gene block. Mol Plant-Microbe Interact 14:1158–1167
22. Wei Y, Shi Y, Han X, Chen S, Li H, Chen L, Sun B, Shi Y (2019) Identification of cucurbit chlorotic yellows virus P4.9 as a possible movement protein. Virol J 16(82)
23. Lazareva EA, Lezzhov AA, Komarova TV, Morozov SY, Heinlein M, Solovyev AG (2017) A novel block of plant virus movement genes. Mol Plant Pathol 18:611–624
24. Lazareva EA, Lezzhov AA, Chergintsev DA, Golyshev SA, Dolja VV, Morozov SY, Heinlein M, Solovyev AG (2021) Reticulon-like properties of a plant virus-encoded movement protein. New Phytol 229:1052–1066
25. Morozov SY, Fedorkin ON, Jüttner G, Schiemann J, Baulcombe DC, Atabekov JG (1997) Complementation of a potato virus X mutant mediated by bombardment of plant tissues with cloned viral movement protein genes. J Gen Virol 78:2077–2083
26. Shen Y, Zhao X, Yao M, Li C, Miriam K, Zhang X, Tao X (2014) A versatile complementation assay for cell-to-cell and long distance movements by cucumber mosaic virus based agro-infiltration. Virus Res 190:25–33
27. Voinnet O, Pinto YM, Baulcombe DC (1999) Suppression of gene silencing: a general strategy used by diverse DNA and RNA viruses of plants. Proc Natl Acad Sci U S A 96:14147–14152
28. Vargason JM, Szittya G, Burgyán J, Hall TMT (2003) Size selective recognition of siRNA by an RNA silencing suppressor. Cell 115:799–811
29. Solovyev AG, Minina EA, Makarova SS, Erokhina TN, Makarov VV, Kaplan IB, Kopertekh L, Schiemann J, Richert-Pöggeler KR, Morozov SY (2013) Subcellular localization and self-interaction of plant-specific Nt-4/1 protein. Biochimie 95:1360–1370
30. Wise AA, Liu Z, Binns AN (2006) Three methods for the introduction of foreign DNA into Agrobacterium. Methods Mol Biol 343:43–53
31. Weigel D, Glazebrook J (2006) Transformation of Agrobacterium using the freeze-thaw method. Cold Spring Harb Protoc 2006:pdb.prot4666
32. Voinnet O, Rivas S, Mestre P, Baulcombe D (2003) An enhanced transient expression system in plants based on suppression of gene silencing by the p19 protein of tomato bushy stunt virus. Plant J 33:949–956
33. Vogler H, Kwon M-O, Dang V, Sambade A, Fasler M, Ashby J, Heinlein M (2008) Tobacco mosaic virus movement protein enhances the spread of RNA silencing. PLoS Pathog 4:e1000038

Chapter 24

Analysis of the Distribution of Symplasmic Tracers During Zygotic and Somatic Embryogenesis

Justyna Wróbel-Marek, Kamila Godel-Jędrychowska, and Ewa Kurczyńska

Abstract

Plasmodesmata (PD) are membraneous channels that span cell walls of adjacent cells to establish the symplasm. These connections are unique to plants and enable the cell-to-cell exchange of information via the symplasm. However, not every plant cell is connected to its neighbor. Absence of PD and lack of communication (symplasmic isolation) are important regulators of cell differentiation. To determine cell-to-cell symplasmic connectivity, the distribution of fluorescent tracers can be analyzed. Here, we describe in detail the entire procedure for conducting such analysis using fluorescence and confocal microscopy to study molecular fluxes in fluorescence recovery after photobleaching (FRAP) experiments. Studies using fluorochromes and fluorescent-labeled dextrans successfully inform the degree of symplasmic connectivity between cells in zygotic and somatic embryos. Small molecules, such as water and ions, travel through PD but also transcription factors and different types of RNA. Studies of symplasmic communication are important to determine the spatio-temporal correlation between cell differentiation and the exchange of information between cells. This information is necessary to determine the role of symplasmic communication during embryogenesis, which is a very important stage in plant development and morphogenesis.

Key words Fluorochromes, FRAP, Plasmodesmata, Somatic embryos, Symplasmic communication, Zygotic embryos

1 Introduction

The role of symplasmic communication in plant development was inferred, primarily, from ultrastructural studies [1–3]. The presence and absence of plasmodesmata (PD, channels facilitating cytoplasm-to-cytoplasm connections forming the symplasm) have been analyzed using transmission electron microscopy (TEM). This method is static and does not provide a measure of transport rates thus PD functionality. In contrast, analyzing the distribution of symplasmic tracers, such as the low molecular weight fluorochromes [4, 5], F-dextran (fluorescein isothiocyanate-conjugated

Yoselin Benitez-Alfonso and Manfred Heinlein (eds.), *Plasmodesmata: Methods and Protocols*, Methods in Molecular Biology, vol. 2457, https://doi.org/10.1007/978-1-0716-2132-5_24,

dextran) [6], or the green fluorescent protein (GFP) [7], provides information on PD dynamics in living cells in real time enabling a complete profile of symplasmically connected, or isolated, areas in the plant body. In this chapter, we describe, step by step, how to use the fluorescent tracers to determine symplasmic connectivity in zygotic, androgenic, and somatic embryos. We recommend low molecular weight fluorochromes, especially their acetic derivatives and the CMNB-caged fluorochromes, because their application does not require breaking the continuity of cell membranes, avoiding mechanical damage that may affect the results. We describe the use of the following dyes: 8-hydroxypyrene-1,3,6-trisulfonic acid trisodium salt (HPTS); 8-acetoxypyrene-1,3,6-trisulfonic acid trisodium salt (PTSA); fluorescein diacetate (FDA); 5(6)-carboxyfluorescein diacetate mixed isomers (CFDA); fluorescein bis-(5-carboxymethoxy-2-nitrobenzyl) ether, dipotassium salt (CMNB-caged fluorescein); Lucifer Yellow CH (LYCH); and fluorescein isothiocyanate-dextran (F-dextran/FITC-dextran). We recommend different methods: using membrane-permeable fluorochromes (including uncaging CMNB-caged fluorescein [8]), and using membrane-non-permeable fluorochromes and FRAP, to precisely determine symplasmic communication in embryos. These procedures and methods can be also applied to study intercellular communication in different organs/tissues and plant species.

2 Materials

2.1 Fluorochromes and Solvents

1. **8-Hydroxypyrene-1,3,6-trisulfonic acid trisodium salt (HPTS) solution**: Dissolve 5 mg HPTS in 1 mL of 1/2 MS medium or PBS or demineralized water (*see* **Note 1**).
2. **8-Acetoxypyrene-1,3,6-trisulfonic acid trisodium salt (PTSA) solution**: Dissolve 5 mg PTSA in 1 mL of 1/2 MS medium or PBS or demineralized water (*see* **Note 1**).
3. **Fluorescein diacetate (FDA) stock solution**: Dissolve 5 mg of FDA in 1 mL of acetone. Store the stock solution at a temperature of −20 ° C.
4. **Fluorescein diacetate (FDA) working solution**: Add 0.5 mL of the stock solution to 24.5 mL of demineralized water (*see* **Note 2**) or PBS buffer or 1/2 MS medium (v/v; *see* **Note 1**).
5. **5(6)-Carboxyfluorescein diacetate mixed isomer (CFDA) stock solution**: Dissolve 5 mg of CFDA in 1 mL of acetone. Store the stock solution at a temperature of −20 ° C.
6. **5(6)-Carboxyfluorescein diacetate mixed isomer (CFDA) working solution**: Add 0.25 mL of stock solution to 5 mL of demineralized water, PBS buffer, or 1/2 MS medium (v/v; *see* **Note 1**).

7. **Fluorescein bis-(5-carboxymethoxy-2-nitrobenzyl) ether, dipotassium salt (CMNB-caged fluorescein) solution**: Dissolve 5 mg of CMNB-caged fluorescein in 2 mL of 0.2% DMSO (v/v). Then, add 48 mL of 0.2% DMSO, and divide the solution into smaller aliquots (*see* **Note 3**).
8. **Lucifer Yellow CH (LYCH) solution:** Dissolve 2.5 mg of LYCH in 1 mL of a 1/2 MS medium, PBS, or demineralized water (*see* **Note 1**).
9. **Fluorescein isothiocyanate-dextran (F-dextran) solution**: Dissolve 5 mg of F-dextran in 1 mL of a 1/2 MS medium, PBS, or demineralized water (*see* **Notes 1** and **4**).
10. **1/2 Murashige and Skoog (1/2 MS) medium**: Add about 500 mL demineralized water to a 1 L graduated cylinder (or glass beaker). Weigh *macronutrients*: 825 mg ammonium nitrate (NH_4NO_3), 220 mg calcium chloride ($CaCl_2{\cdot}2H_2O$), 185 mg magnesium sulfate ($MgSO_4{\cdot}7H_2O$), 85 mg monopotassium phosphate (KH_2PO_4), and 950 mg potassium nitrate (KNO_3); *micronutrients*: 3.1 mg boric acid (H_3BO_3), 0.0125 mg cobalt chloride ($CoCl_2{\cdot}6H_2O$), 13.9 mg ferrous sulfate ($FeSO_4{\cdot}7H_2O$), 18.65 mg ethylenediaminetetraacetic acid disodium salt (Na_2-EDTA) (*see* **Note 5**), 11.15 mg manganese(II) sulfate ($MnSO_4{\cdot}4H_2O$), 0.415 mg potassium iodide (KI), 0.125 mg sodium molybdate ($Na_2MoO_4{\cdot}2H_2O$), 4.3 mg zinc sulfate ($ZnSO_4{\cdot}4H_2O$), and 0.0125 mg copper sulfate ($CuSO_4{\cdot}5H_2O$); and *vitamins*: 50 mg myo-inositol, 0.25 mg nicotinic acid, 0.25 mg pyridoxine·HCl, 0.05 mg thiamine·HCl, 1 mg glycine, and 10 g sucrose, and transfer to the cylinder. Add demineralized water to a volume of 900 mL. Mix and adjust pH to 5.8 with 0.1 M KOH. Make up to 1 L with demineralized water (*see* **Note 6**). Sterilize by autoclaving.
11. **Phosphate-buffered saline (PBS)**: Dissolve 8 g NaCl, 0.2 g KCl, 1.44 g Na_2HPO_4, and 0.24 g KH_2PO_4 in 1 L of distilled water, and adjust the pH to 7.2 with 1 M HCl (if necessary).
12. **Dimethyl sulfoxide (DMSO) solution**: Mix 0.2 mL DMSO with 99.8 mL of demineralized water (v/v).

2.2 Other Reagents and Dyes

1. **2-Deoxy-D-glucose (DDG) 0.1 M stock solution**: Weigh 0.164 g of DDG, and dissolve in 10 mL of demineralized water.
2. **2-Deoxy-D-glucose (DDG) 0.1 mM working solution**: Add 0.01 mL of DDG stock solution to 9.99 mL of demineralized water (or 1/2 MS medium and adjust the pH to 5.8 with 0.1 M KOH).

3. **4′,6-Diamidino-2-phenylindole (DAPI) solution**: Dissolve 5 mg of DAPI in 1 mL of PBS to get the final concentration of 2 μg/mL.
4. **Evans Blue 0.05% solution (w/v)**: Dissolve 0.05 g of Evans Blue in 100 mL of 0.1 M $CaCl_2$ (w/v) at pH 5.6 (*see* **Note 7**).

2.3 Equipment

1. **Syringe filters**: sterile syringe filter with a 0.22 μm pore size.
2. **Filter papers**: commonly used cellulose filter paper for filtration.
3. **Epifluorescence microscope**: equipped with filters for UV and blue light illumination and barrier filters for green emission.
4. **Confocal laser scanning microscope (CLSM)**: equipped with UV laser (≤364 nm excitation wavelength) or LD lasers (405 nm excitation wavelength and 488 nm excitation wavelength).
5. **Microscope slides.**
6. **Coverslips.**
7. **Eppendorf tubes.**
8. **Microcapillaries**: borosilicate glass.
9. **Plan-Neofluar 5× objective.**
10. **Plan-Neofluar 10× objective.**
11. **LD-Achroplan 20× objective.**
12. **Chamber slides.**

3 Methods

The protocol describes testing connectivity in zygotic and somatic embryos of *Arabidopsis thaliana* (L.) and *Daucus carota* (L.) and androgenic embryos of *Hordeum vulgare* (L.), but other plant material could be also analyzed (*see* **Note 8**). Carry out all steps with the fluorochromes in the dark, and only perform the analysis on embryos where fluorochrome presence in cell walls is not detected.

3.1 Autofluorescence Determination

Autofluorescence has to be collected before the fluorochrome fluorescence analysis.

1. Place a non-treated (*see* **Note 9**) embryo in a drop of 1/2 MS medium (or PBS or demineralized water) on a microscope slide, and cover with a coverslip (*see* **Note 10**).
2. Analyze autofluorescence under CLSM: set the laser power to 10–20% (or more if no autofluorescence is visible), or increase a

voltage on a photomultiplier. Make a single scan of an embryo using CLSM to observe autofluorescence and save the image.

3. Decrease the values of microscope settings (e.g., laser power or voltage on a photomultiplier) so that the autofluorescence is not visible or almost invisible, and make a single scan. Save the image.
4. Make a scan of the embryo treaded with a fluorochrome. Make sure that fluorescence of the fluorochrome is well visible (*see* **Note 11**). Save the image.
5. Check the autofluorescence again. Make a scan of an embryo untreated with the fluorochrome. If autofluorescence is still visible, repeat all the steps starting from **step 1** until you find appropriate microscope settings (*see* **Note 12**).

3.2 Analysis of Membrane-Permeable Fluorochromes: PTSA, FDA, and CFDA

1. Prepare the fluorochromes: PTSA or FDA or CFDA (*see* **Notes 13** and **14**).
2. Incubate embryos in a 0.1 mM solution of DDG for 30–60 min (*see* **Note 15**) to prevent callose deposition.
3. Wash the embryos several times in demineralized water, PBS, or 1/2 MS.
4. Immerse embryos in the fluorochrome solution (either PTSA or FDA or CFDA), and incubate for 30 min to 1 h (*see* **Note 16**).
5. Rinse embryos several times (each for at least 5 min) with demineralized water, PBS, or 1/2 MS to wash out the excess of fluorochrome.
6. Determine embryo cells' viability before and after treatment with the fluorochrome using staining with 0.05% Evans Blue for 10–15 min (*see* **Note 17**).
7. Place stained undamaged embryos on a microscope slide, into a drop of water, PBS, or 1/2 MS, and cover with a coverslip (*see* **Note 10**).
8. Analyze PTSA/FDA/CFDA distribution within the embryo using a CLSM (*see* **Note 18**) (Fig. 1 shows an example) or an epifluorescence microscope (*see* **Note 19**) (Fig. 2 shows an example).

3.3 Analysis of CMNB-Caged Fluorescein

The protocol below is for CMNB-caged fluorescein, which is also a membrane-permeable fluorochrome, but where the release of fluorescein requires uncaging.

3.3.1 Uncaging CMNB-Caged Fluorescein with CLSM

1. Incubate the embryos in CMNB-caged fluorescein as described above (*see* Subheading 3.2) (*see* **Note 20**).
2. Make a single scan of an embryo using CLSM (*see* **Note 21**).

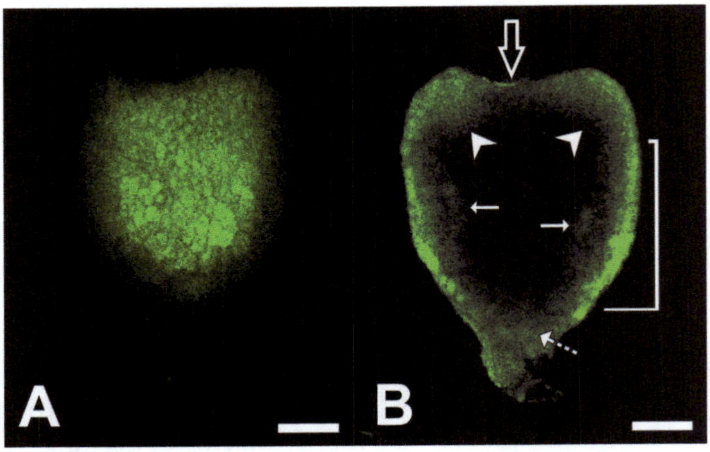

Fig. 1 Distribution of the PTSA within the *Daucus carota* L. somatic embryo. (**a**) The surface of the embryo in heart stage was imaged using a CLSM (excitation wavelength 488 nm, emission wavelength 510–530 nm). Fluorochrome is visible in all protodermal cells. (**b**) Optical section through the embryo (*Z*-stack) shown in (**a**). Fluorochrome is present in cells of protodermis throughout the embryo with stronger intensity in the hypocotyl (clamp), the proximal part of the cotyledons (arrowheads), the hypocotyl subprotodermis (small arrows), the shoot apical meristem (empty arrow), and the embryo root (dashed arrow). Scale bars = 100 μm

3. Mark a region of interest (ROI) on the image using the CLSM software.
4. Illuminate the ROI with a laser emitting a ≤364 nm wavelength or a LD laser emitting a 405 nm wavelength (laser power set at 60–70%) for 30–120 s (*see* **Note 22**).
5. Analyze fluorescein distribution within the embryos at different time points after uncaging (*see* **Note 18**).
6. Photodocument the findings (Fig. 3 shows an example).

3.3.2 Uncaging CMNB-Caged Fluorescein with Epifluorescence Microscope

Uncaging of CMNB-caged fluorescein can be also done using an epifluorescence microscope as described below.

1. Select an area of interest, and find an appropriate plane of focus (the image of the selected region should be sharp; *see* **Note 23**).
2. Close the numerical aperture of the objective lens, and illuminate the region with light at a 330–380 nm wavelength for variable time lengths depending on the tissue (*see* **Note 24**).
3. Open the numerical aperture of the objective lens immediately after uncaging.
4. Observe the embryos immediately under the 465–490 nm wavelength (the emission of the fluorescein is between 510 and 530 nm and produces a green-yellow fluorescence).

Fig. 2 Fluorochrome distribution in *Arabidopsis thaliana* (L.) samples treated with CFDA. Fluorochrome presence in the leaf pavement cells and trichome of the seedling. The image was taken using an epifluorescence microscope (excitation wavelength was 465–490 nm; emission wavelength was 510–530 nm). Scale bar = 10 μm

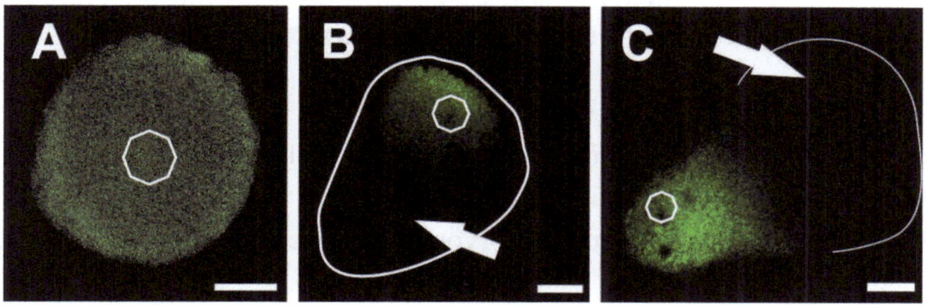

Fig. 3 Distribution of the fluorescein within the *Daucus carota* (L.) somatic embryos treated with CMNB-caged fluorescein. (**a**) At the globular stage, the fluorochrome was present in each embryo cell which indicates that fluorochrome moved freely throughout the embryo. (**b**) Presence of fluorochrome in the heart stage is restricted to the cells in the vicinity of uncaging. (**c**) At the mature embryo stage, fluorochrome was detected only in cells of the basal pole of the embryo. The white polygons represent the ROI selected for fluorescein uncaging; arrows indicate the embryo parts without fluorochrome. Results presented on (**b**, **c**) indicate the restriction in fluorochrome movement (images were taken in CLSM; the wavelength for CMNB-caged fluorescein uncaging was 405 nm, excitation wavelength was 488 nm, and emission was 510–530 nm; the power of LD laser for uncaging was 60%; the power of argon laser emitting 488 nm wavelength was set to 8%). Scale bars = 50 μm (**a**) and 100 μm (**b**, **c**)

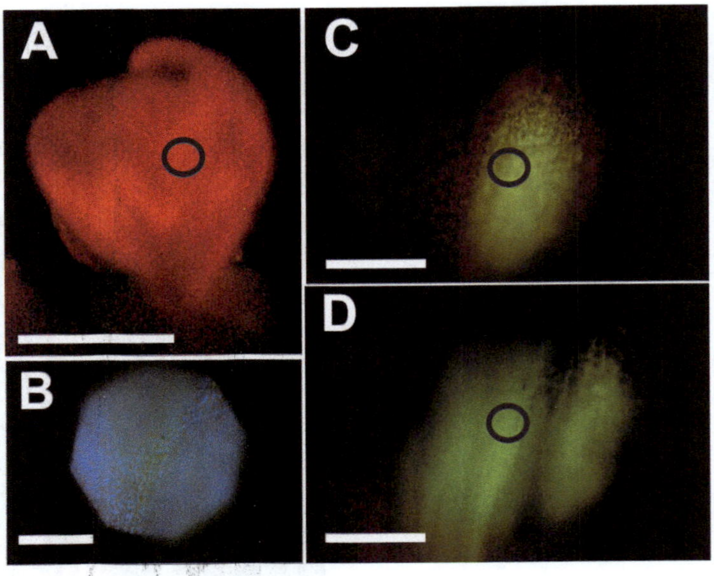

Fig. 4 Uncaging of the CMNB-caged fluorescein under the epifluorescence microscope. In the images, the marked area (the black circle) was selected for the uncaging (excitation wavelength for uncaging, 330–380 nm; excitation of fluorescein, 465–490 nm; emission of fluorescein, 510–530 nm). (**a**) Image of the somatic embryo before uncaging (red color—chlorophyll autofluorescence). (**b**) Image visible in the eyepiece after closing the numerical aperture of the objective and during UV illumination. (**c**) Fluorescein distribution 1 min after uncaging. (**d**) Fluorescein distribution 5 min after uncaging. The planes of visualization were focused to see sharp images of the surface of the analyzed somatic embryo. Scale bars = 200 μm (**a**) and 20 μm (**b–d**)

5. Analyze fluorescein distribution within the embryos at different time points after uncaging (*see* **Note 19**).
6. Photodocument the findings (example shown in Fig. 4).

3.4 Analysis of Non-membrane-Permeable Fluorochromes (HPTS, LYCH, FITC-Dextran)

1. Incubate embryos in a 0.1 mM solution of DDG for 30–60 min (*see* **Note 15**) to prevent callose deposition.
2. Wash the embryos in demineralized water, PBS, or 1/2 MS several times.
3. Place the embryos in an excess of fluorochrome (either HPTS, LYCH, or FITC-dextran solutions) on a microscopic slide or in a 35-mm-diameter Petri dish (which we recommend).
4. Using a stereomicroscope, injure the embryo in the area of interest with a microcapillary (with an outer diameter ranging between 3 and 7 μm; *see* Fig. 5).
5. Remove the microcapillary.
6. Incubate the embryos in the fluorochrome for 30–90 min.

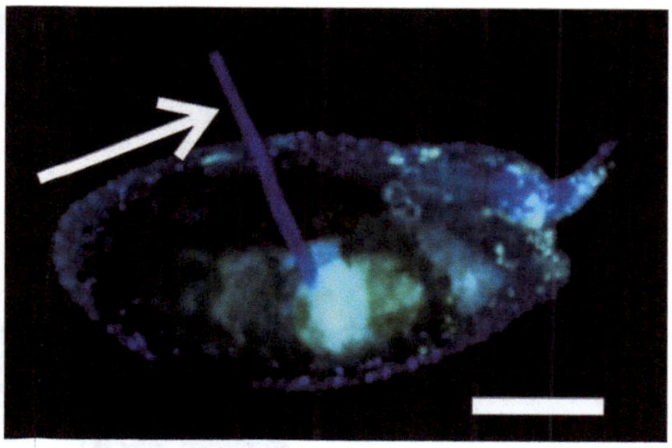

Fig. 5 Microcapillary injected into the *Sedum acre* seed. Microcapillary (arrow) is injected into the zygotic embryo through the integument. The plant material was incubated in DAPI. Observations were made under UV light (330–380 nm). Scale bar = 50 μm

7. Wash the embryos several times in demineralized water, PBS, or 1/2 MS (to remove any unincorporated dye).
8. Check the viability of embryos with Evans Blue staining (*see* **Note 17**).
9. Place undamaged embryos on a microscope slide and cover with a coverslip (*see* **Note 10**).
10. Analyze the fluorochrome distribution within the embryos using epifluorescence microscope (*see* **Notes 19** and **25**), and photodocument the findings (Figs. 6 and 7).

3.5 Analysis of Symplasmic Connectivity Using Fluorescence Recovery After Photobleaching (FRAP)

The FRAP method allows to trace the diffusion rate between cells and to determine plasmodesmal coupling. This method can be useful to analyze the spatio-temporal dynamics of PD connectivity. It is performed exactly as described for CMNB-caged fluorescein (Subheading 3.3), but, instead of photo-uncaging the fluorochrome, a bleaching of the fluorochrome occurs.

1. Incubate the embryos for 30–60 min in FDA in the dark.
2. Wash the embryo several times in demineralized water, PBS, or 1/2 MS medium.
3. Place the embryo on a microscope slide and cover with a coverslip (*see* **Note 10**).
4. Select the area of interest (ROI) using a CLSM software.
5. Irradiate the area with a 405 nm laser (laser power set at 60–90%) for 30–60 s in order to bleach the fluorescein.
6. Collect a series images to determine whether the fluorochrome had diffused into the bleached area/cells (excitation 488 nm, emission 510–530 nm). Take the first image just after photobleaching. Acquire subsequent images within the next 3–5 min (*see* **Note 26**) until the mean value of fluorescence increases to 20% (or more) (*see* Fig. 8).

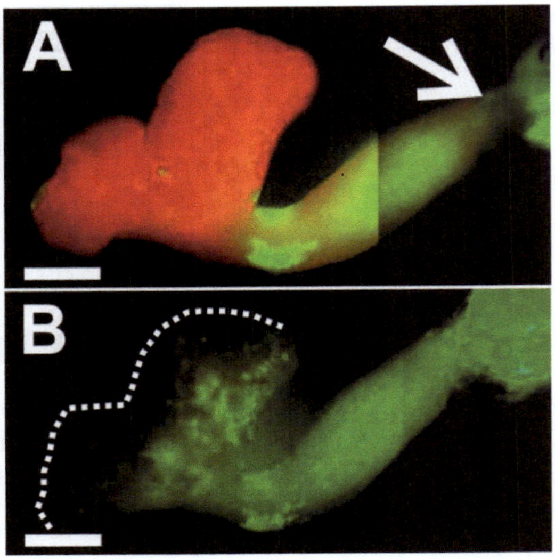

Fig. 6 Distribution of HPTS within the somatic embryos of *Arabidopsis thaliana* (L.). Image from epifluorescence microscope (at excitation wavelength of 465–490 nm and emission wavelength of 510–530 nm) after 5 min (**a**) and 30 min (**b**) after fluorochrome application. Arrow points to the site of the fluorochrome injection through the microcapillary puncture; red color shows the autofluorescence of chlorophyll and green color—HPTS; the vertical line in both photos was caused by the folding of two photos as the somatic embryo was too large to document in a single picture; white dotted line marks the cotyledons' borders. Scale bars = 100 μm

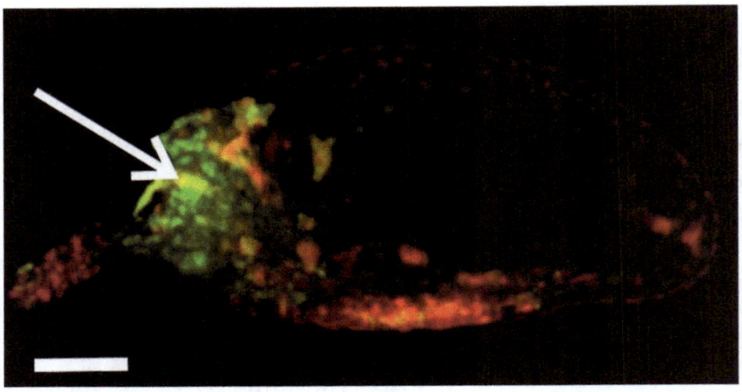

Fig. 7 Example of 3 kDa F-dextran distribution within the *Sedum acre* integument. The arrow points to the site of the F-dextran application with the microcapillary. The excitation wavelength was 465–490 nm and emission collection spectrum of 510–530 nm (image was taken in epifluorescence microscope). Scale bar = 50 μm

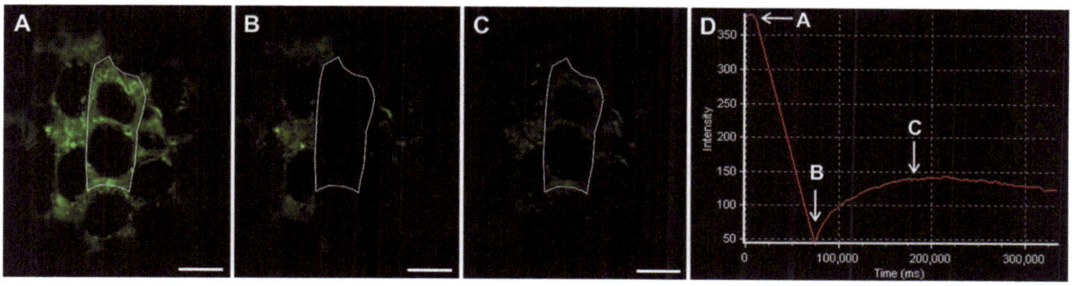

Fig. 8 Successive stages of the FRAP method performed on cotyledon cells of the *Arabidopsis thaliana* somatic embryo. Presence of fluorescein in the embryo cells was imaged using CLSM (it was irradiated with 405 nm laser with the power set at 60–90%) for 30 s to bleach fluorescein and analyzed at excitation 488 nm to check if the fluorochrome moved into the bleached area. The image of the cell before photobleaching (**a**), immediately after photobleaching (**b**), and 2.5 min of recovery after photobleaching (**c**). Intensity of fluorescein fluorescence during FRAP is indicated in a graph (**d**). Intensity of fluorochrome fluorescence observed in panels (**a–c**) is shown using arrows followed by the appropriate letters. The white outline indicates the region of the photobleaching (**a–c**). Scale bars = 10 μm

4 Notes

1. We recommend using the medium on which the zygotic, somatic, or androgenic embryos were cultured, but the solution of HPTS, LYCH, F-dextran, and working solution of FDA and CFDA can also be prepared in PBS or demineralized water.
2. Working solution of FDA prepared with demineralized water should be filtered using a syringe filter or filter paper prior to its use because the crystals of FDA precipitate in the working solution.
3. We recommend storing 1 mL of the solution in Eppendorf tubes. Keep the solution in the dark at −20 °C. Defreeze the solution directly before use.
4. F-dextran of different molecular weights (e.g., 3, 10, or 20 kDa) can be purchased. The decision of which dextran to use should be based on the literature. Three kilodalton dextran was used in our study because preliminary work showed that dextrans of higher molecular mass did not migrate from the site of application to other cells.
5. We recommend to prepare ethylenediaminetetraacetic acid ferric sodium (NaFe-EDTA) solution first. It is a stock solution containing 5.57 g $FeSO_4{\cdot}7H_2O$ and 7.45 g Na_2-EDTA per liter of water. Add 2.5 mL of stock solution per liter of 1/2 MS medium.
6. It is also possible to prepare 1/2 MS medium from commercial MS: Dissolve 2.2 g of commercial MS and 10 g of sucrose in 900 mL of demineralized water. Then mix the solution and adjust pH to 5.8 with 0.1 M KOH. Make up to 1 L with demineralized water. Sterilize by autoclaving.

7. Mix well until it is dissolved. The solution should always be prepared fresh!
8. Other plant material like roots, cotyledons, or leaves can be also analyzed, but the time of incubation will vary and must be determined experimentally. We can recommend for barley roots an incubation time of 15 min.
9. Autofluorescence has to be checked in an embryo untreated with the fluorochrome (a control embryo).
10. If using an inverted microscope, place the embryos in a chamber slide, and let embryos gravitate to the bottom of the chamber where they can be observed through the coverslip.
11. If you cannot see fluorochrome fluorescence, it means that microscope settings are not right. Change the settings, e.g., enhance the laser power or voltage on a photomultiplier, until you see fluorescence.
12. Microscope settings should be established to allow scanning the fluorescence of the fluorochrome only (without autofluorescence). Sometimes this is impossible to achieve, especially when fluorochrome fluorescence is weak. In such cases, changing the fluorochrome might help.
13. The procedure described applies to all the fluorochromes listed. The mentioned fluorochromes differ in the (a) molecular weight of the molecule (fluorescein = 332 Da; PTSA = 524 Da, CFDA = 460 Da), which is important to determine the size exclusion limit of PD; (b) potential sequestration to organelles, e.g., vacuoles (PTSA and CFDA have the least sequestration); and (c) sensitivity to UV light (PTSA is the most stable). The choice of the fluorochrome should be based on the above criteria, but we recommend PTSA. To prepare the fluorochrome solutions, we recommend 1/2 MS medium in the case of somatic/androgenic embryos, whereas, for zygotic embryos, demineralized water or PBS can be used.
14. PTSA, FDA, and CFDA are not fluorescent and easily diffuse into cells. After entering the cell cytoplasm, PTSA, FDA, and CFDA lose the acetic groups cleaved by intracellular nonspecific esterases releasing the membrane-impermeable, fluorescent forms of the fluorochromes [9]. The fluorescent form of PTSA is HPTS, the fluorescent form of FDA is fluorescein, and the fluorescent form of CFDA is carboxyfluorescein (CF). These fluorochromes move from one cell to another only via the PD and, therefore, are used in studies of symplasmic communication (Figs. 1 and 2) [9–11].
15. For somatic embryos, we recommend treatment with DDG for 60 min.

16. To study embryos with PTSA/FDA/CFDA, we recommend 15 min of incubation for globular embryos (zygotic, somatic, and androgenic), 30 min for the heart embryos of *Arabidopsis* or *Daucus* and transition stage of barley androgenic embryos, and 60 min for mature embryos. The entire procedure must be carried out in the dark. In the case of using CMNB-caged fluorescein, we recommend immersing the embryos in the solution for 1 h.
17. Evans Blue penetrates the cells only when the plasma membrane is damaged; thus, it is used in viability assays (dead cells are dark blue in a bright-field microscope).
18. CLSM settings for fluorochrome detection: observe the samples using a 5×, 10×, or 20× long-working-distance objective (Plan-Neofluar 5×, Plan-Neofluar 10×, LD-Achroplan 20×; other objectives/other magnifications are acceptable). Scan the sample at 488 nm excitation wavelength and 510–530 nm emission wavelength. To analyze the fluorochrome diffusion rate, scan the embryo in the *Z*-axis (i.e., multiple focus planes, *Z*-stacks) at multiple times within a given period of time. The time between successive *Z*-axis scans should be determined experimentally. For the somatic and zygotic embryos, the time gap between the successive scans can be 10 min. Save the images.
19. Epifluorescence settings for fluorochrome detection: observe the samples using a 5×, 10×, or 20× long-working-distance objective. Illuminate with a blue light (excitation wavelength 465–490 nm), and collect the green-yellow fluorescence (emission wavelength 510–530 nm). To analyze the fluorochrome diffusion rate, take several pictures of the samples within a given period of time. The time gap between the successive images should be determined experimentally. For the somatic and zygotic embryos, the time gap between the successive images can be 10 min. Save the images.
20. For globular and heart stage embryos, we recommend incubating the embryos in CMNB-caged fluorescein for 30 min. The somatic or androgenic embryos in the more advanced stages of development should be treated for a longer time.
21. CMNB-caged fluorescein is not fluorescent until uncaging with UV light [12]; thus, only autofluorescence of the sample is visible. The uncaged fluorescein is membrane impermeable. To determine autofluorescence (before uncaging), you cannot use UV lasers (excitation ≤405 nm). Instead, use 488 nm excitation wavelength and 500–530 nm emission wavelength.
22. ROI should be illuminated with a ≤364 nm wavelength for 30–45 s in order to detach the caging groups and release fluorescein. A LD laser emitting 405 nm wavelength can also

be used; however, the efficiency of fluorescein uncaging is lower than when using 364 nm wavelength; thus, the time of ROI illumination is longer lasting between 60 and 120 s. CMNB-caged fluorescein can be released in a single cell without causing any injuries to the plant material (unlike a microinjection) and then analyzed under an epifluorescence microscope or CLSM. CMNB-caged fluorescein is best used to study symplasmic communication during somatic embryogenesis/androgenesis or in zygotic embryos [12].

23. A plane of focus should be selected to obtain a sharp image of the region. If the embryo protodermis is the target of the observation, then protodermal cells have to be in focus. It is also necessary to determine how deep, within the embryo, the fluorescein is released. In somatic embryos, illuminating protodermis with UV light results in uncaging fluorescein not only in the protodermis but also in two to three layers of the ground pro-meristem beneath the protodermis.
24. The duration of UV illumination should be selected experimentally. The recommended time of illumination is usually 5–40 s (UV illumination should be kept as short as possible in order to prevent the harmful effects of UV light on cell viability). The duration of the uncaging can be prolonged to 60–120 s if no fluorescence is observed after the UV illumination.
25. In the case of LYCH, excite the fluorochrome with a 458 nm wavelength, and collect emission at a 530–600 nm wavelength to obtain a yellow fluorescence.
26. The time of image acquisition needs to be determined experimentally.

References

1. Canhoto JM, Mesquita JF, Cruz GS (1996) Ultrastructural changes in cotyledons of pineapple guava (Myrtaceae) during somatic embryogenesis. Ann Bot 78:513–521. https://doi.org/10.1006/anbo.1996.0149
2. Bilska A, Sowiński P (2010) Closure of plasmodesmata in maize (Zea mays) at low temperature: a new mechanism for inhibition of photosynthesis. Ann Bot 106:675–686. https://doi.org/10.1093/aob/mcq169
3. Nicolas WJ, Grison MS, Trépout S et al (2017) Architecture and permeability of post-cytokinesis plasmodesmata lacking cytoplasmic sleeves. Nat Plants 3:17082. https://doi.org/10.1038/nplants.2017.82
4. Wrobel J, Barlow PW, Gorka K et al (2011) Histology and symplasmic tracer distribution during development of barley androgenic embryos. Planta 233:873–881. https://doi.org/10.1007/s00425-010-1345-0
5. Grzyb M, Wróbel-Marek J, Kurczyńska E et al (2020) Symplasmic isolation contributes to somatic embryo induction and development in the tree Fern Cyathea delgadii Sternb. Plant Cell Physiol 61:1273–1284. https://doi.org/10.1093/pcp/pcaa058
6. Godel-Jedrychowska K, Kulinska-Lukaszek K, Horstman A et al (2020) Symplasmic isolation marks cell fate changes during somatic embryogenesis. J Exp Bot 71:2612–2628. https://doi.org/10.1093/jxb/eraa041
7. Kim I, Cho E, Crawford K et al (2005) Cell-to-cell movement of GFP during embryogenesis and early seedling development in Arabidopsis.

Proc Natl Acad Sci 102:2227–2231. https://doi.org/10.1073/pnas.0409193102

8. Martens HJ, Hansen M, Schulz A (2004) Caged probes: a novel tool in studying symplasmic transport in plant tissues. Protoplasma 223:63–66. https://doi.org/10.1007/s00709-003-0029-z
9. Wright KM, Oparka KJ (1996) The fluorescent probe HPTS as a phloem-mobile, symplastic tracer: an evaluation using confocal laser scanning microscopy. J Exp Bot 47:439–445. https://doi.org/10.1093/jxb/47.3.439
10. Gisel A, Hempel FD, Barella S, Zambryski P (2002) Leaf-to-shoot apex movement of symplastic tracer is restricted coincident with flowering in Arabidopsis. Proc Natl Acad Sci 99:1713–1717. https://doi.org/10.1073/pnas.251675698
11. Maule AJ, Gaudioso-Pedraza R, Benitez-Alfonso Y (2013) Callose deposition and symplastic connectivity are regulated prior to lateral root emergence. Commun Integr Biol 6: e26531. https://doi.org/10.4161/cib.26531
12. Wróbel-Marek J, Kurczyńska E, Płachno BJ, Kozieradzka-Kiszkurno M (2017) Identification of symplasmic domains in the embryo and seed of Sedum acre L. (Crassulaceae). Planta 245:491–505. https://doi.org/10.1007/s00425-016-2619-y

Chapter 25

Quantifying Intercellular Movement and Protein Stoichiometry for Computational Modeling

Lisa Van den Broeck, Mariah Gobble, and Rosangela Sozzani

Abstract

Analyzing protein movement dynamics and their regulation has shown to be important in the study of cell fate decisions. Such analyses can be performed with scanning fluorescence correlation spectroscopy (scanning FCS), a versatile imaging methodology that has been applied in the animal kingdom and recently adapted to the plant kingdom. Specifically, scanning FCS allows for qualitatively capturing protein movement across barriers, such as the active transport through plasmodesmata, the analysis of protein movement rates, and the quantification of the stoichiometry of protein complexes, composed of one or more different proteins. Importantly, the quantifiable data generated with scanning FCS can be used to inform computational models, enhancing model simulations of in vivo events, such as cell fate decisions, during plant development.

Key words Protein movement, Scanning fluorescence correlation spectroscopy (FCS), Intercellular movement, Computational modeling

1 Introduction

Growth and development in a multicellular organism are regulated by a plethora of factors, such as gene expression, cell-to-cell signaling, protein movement, protein post-translational modifications, and environmental cues. Together, these signals allow organisms to maintain homeostasis and to regulate growth and development in a changing environment. To ensure proper development, it is critical that these signals are spatially and temporally restricted to prevent overactivation of a response or, in the opposite case, the absence of responses. In plants, the spatiotemporal control of regulatory proteins is regulated by mechanisms that include their interactions with other proteins and their ability to move from cells of their expression to neighboring cells. Protein movement occurs through plasmodesmata, the cytoplasmic bridges between neighboring cells [1, 2]. Such intercellular movement of regulatory

Yoselin Benitez-Alfonso and Manfred Heinlein (eds.), *Plasmodesmata: Methods and Protocols*,
Methods in Molecular Biology, vol. 2457, https://doi.org/10.1007/978-1-0716-2132-5_25,

proteins and the formation of protein complexes play important roles in the regulation of many developmental processes, including cell division and cell differentiation. For example, movement of WUSCHEL (WUS), a stem cell fate regulator, in the shoot apical meristem of plants is regulated by two receptor kinases residing at the plasmodesmata [3]. Analyzing the dynamics of single proteins or protein complexes and their regulated movement is thus key in the study of cell fate decisions and plant development.

Over the past decade, a set of imaging methodologies have emerged that capture protein localization, quantify protein interactions, identify active protein transport through plasmodesmata, and analyze protein dynamics within and across cells. Specifically, scanning fluorescence correlation spectroscopy (FCS) methodologies can be used to measure temporal and spatial information of protein movement and interactions by tagging the proteins with a fluorophore and measuring fluctuations in fluorescence [4]. The movement or kinetics of fluorescent molecules can be quantified by analyzing fluorescence fluctuations over time, while the amplitude of the fluctuations allows for calculating protein concentration [5–7]. Several applications of scanning FCS are used for specific research questions. (a) Raster image correlation spectroscopy (RICS) quantifies the rate of protein diffusion/movement and has been used to calculate diffusion coefficients of proteins in various root cell types [8, 9]. (b) Pair correlation function (pCF) assesses restriction and directionality of protein movement through plasmodesmata and thus intercellular connectivity [5]. (c) Cross-RICS determines diffusion of a complex that can be composed of two proteins or a protein and another macromolecule, such as DNA [10, 11]. (d) Number and brightness (N&B) analysis measures the oligomerization state of single proteins and heterologous protein complexes [12]. With these tools, in vivo imaging of protein complex formation and trafficking can be implemented non-invasively and as a routine technology. The resulting image-derived quantifiable data, including diffusion rates and complex composition, can be combined with molecular datasets to enable systems-level computational modeling [9]. For example, kinetic data, such as the rate of protein diffusion, can be integrated into computational models by adding a linear term for protein diffusion. Computational analysis of spatiotemporal cellular activity allows for a deeper understanding of protein functions, the key components behind organismal growth and development [8, 9, 12, 13]. In this chapter, image acquisition and data analysis are described for several scanning FCS methods, as well as how the generated scanning FCS data can be used to inform computational models, enhancing model simulations of in vivo events.

2 Materials

1. Plant samples, for example, *Arabidopsis thaliana* seedlings.
2. Glass slides (25 × 75 mm).
3. Coverslips (22 × 40 mm).
4. Micropipette and tips.
5. Propidium iodide (PI): Dissolve 10 mg of PI powder in 10 mL of water. Store at 4 °C in the dark. For imaging, aliquot 100 μL of this stock solution into 10 mL of water.
6. Tissue wipers.
7. Forceps.
8. Heptane glue solution: Prepare by adding short pieces (+/−5 cm) of double-sided sticky tape and heptane until the pieces are covered to a 200 mL glass bottle. Gently shake the bottle overnight at room temperature.
9. SimFCS software (www.lfd.uci.edu/globals).
10. Commercial confocal fluorescence microscope, such as the Zeiss LSM 710, LSM 780, or LSM 880.

3 Methods

3.1 Sample Preparation

1. Apply 30 μL of the heptane glue solution to a glass slide to prevent the tissue from moving during imaging.
2. Allow to dry for 1 min.
3. Mount three to five plant samples very gently with forceps on top of the dried glue and add water. Use as little water as possible so that the plant tissue cannot move during imaging but is covered in water.
4. Carefully place a coverslip on top of the plant tissue. When imaging plant roots, the shoots should be outside of the coverslip.

3.2 Quantification of Intra- and Intercellular Movement

The diffusion of a fluorescent-tagged protein can be calculated from time series raster scan images with RICS (raster image correlation spectroscopy). Briefly, by leveraging high-resolution time series images to map fluorescence fluctuations over time, and removing fluorescence fluctuations as a result of the speed of the scan, an intrinsic setting of the confocal microscope [14], RICS allows to calculate the diffusion of the pixels across space and time. For example, one can measure pixel diffusion of a fluorescent signal as a result of plasmodesmata-regulated movement between two cells. To quantify the diffusion coefficient of a protein of interest fused to a GFP (green fluorescent protein), include the same GFP

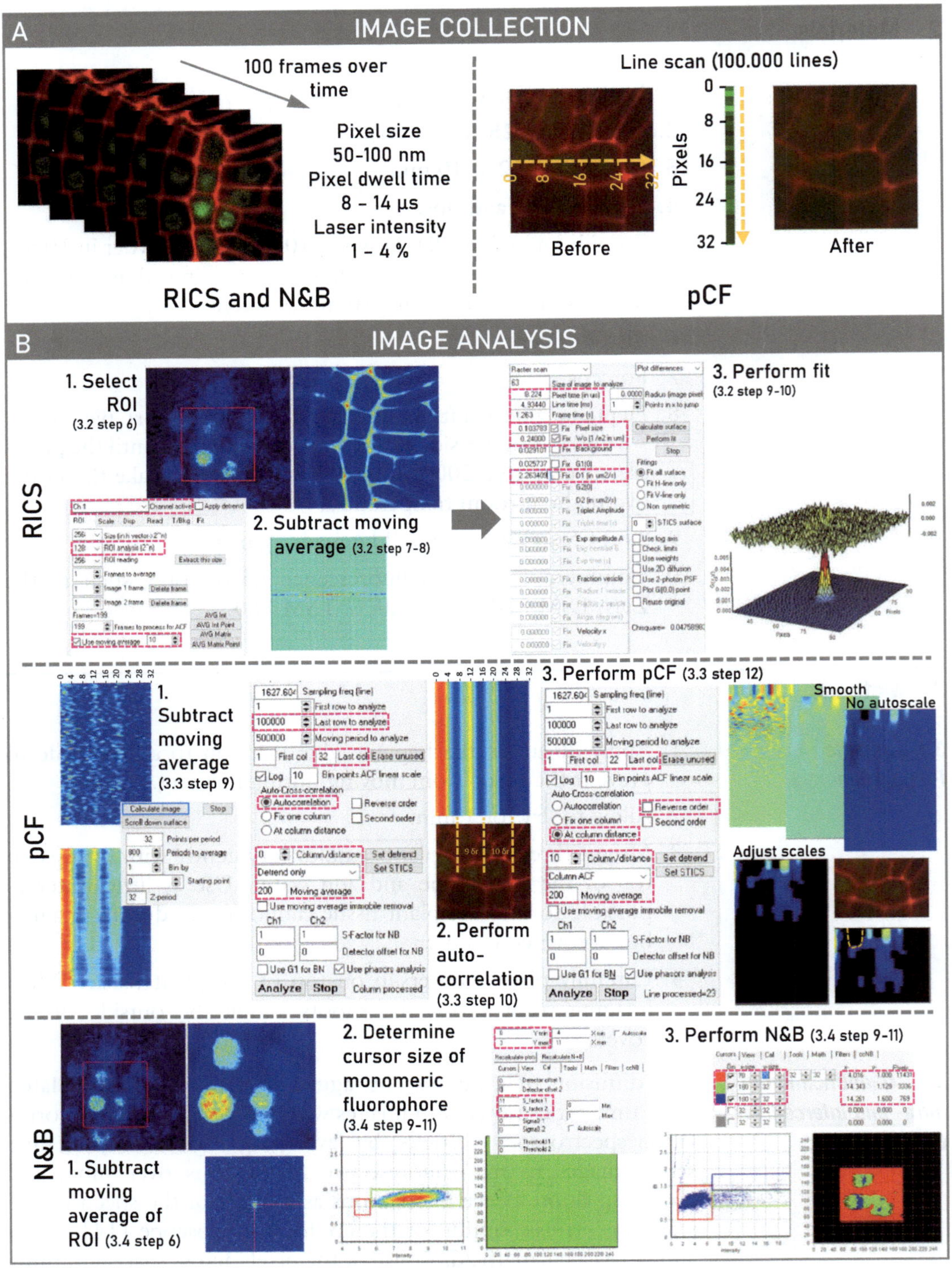

Fig. 1 Image acquisition and analysis to quantify protein movement (RICS) and protein oligomeric state (N&B) and to identify protein movement restrictions (pCF). (**a**) Image acquisition of a raster scan time series for RICS and N&B (left) and image acquisition of a line scan time series for pCF (right). (**b**) RICS (top), pCF (middle), and N&B (bottom) analysis using the SimFCS software. (Top) RICS analysis windows and representative displays

expressed in all cells under a constitutive promoter as your positive control (*see* **Note 1**). The same instructions apply for translational fusions with YFP (yellow fluorescent protein), mCherry, RFP (red fluorescent protein), or any other fluorophore. It is also strongly advised to include a cell-autonomous fluorophore, which does not move between cells, for example, 35S::3xGFP, as negative control.

1. Acquire raster images of plant tissues of interest as follows: Set the gain to 800, digital offset to 0, and digital gain to 1. The laser intensity should be set on a low percentage to avoid photobleaching (e.g., 1–4% laser intensity), but still high enough so that the fluorophore is visible. Zoom in on a region of interest (ROI) such that the pixel size is between 50 and 100 nm. Set frame size to 256 × 256 pixels and pixel dwell time between 8 and 14 μs (Fig. 1a—left panel) (*see* **Note 2**).
2. Collect a time series of 100 raster scans (*see* **Note 2**) (Fig. 1a—left panel).
3. After acquiring images, start the SimFCS software [15], and click on the RICS button.
4. Load an image stack by selecting File → Open multiple images and navigating to your file. A pop-up window will open with image information that should be closed to apply default settings.
5. Select the channel used to image your fluorescent protein, i.e., *Ch 1* should be selected if GFP was imaged in channel 1.
6. Optional: choose a ROI of 128 or 64 pixels to zoom on a region within the image, for example, a specific cell type (Fig. 1b—top panel, section 1).
7. Mark the "Use moving average" box and set the moving average to 10 (*see* **Note 3**).
8. Select Tools → RICS → Subtract moving average to calculate the autocorrelation function (ACF) (*see* **Note 4**) (Fig. 1b—top panel, section 2).
9. In the top menu, click "Fit" to fit a diffusion model to the ACF. SimFCS 4 automatically fills in the pixel time, line time (*see*

Fig. 1 (continued) are shown for ROI selection (Subheading 3.2, **step 6**), subtracting moving average (Subheading 3.2, **steps 7–8**), and fitting (Subheading 3.2, **steps 9–10**). (Middle) pCF analysis windows and representative carpets are shown for subtracting moving average (Subheading 3.3, **step 9**), performing autocorrelation (Subheading 3.3, **step 10**), and performing pCF (Subheading 3.3, **step 12**). (Bottom) N&B analysis windows and representative displays and plots are shown for subtracting the moving average (Subheading 3.4, **step 6**), the N&B analysis of the monomeric control (Subheading 3.4, **steps 9–11**), and the N&B analysis of the protein of interest (Subheading 3.4, **steps 9–11**). Residuals are shown at the top of the graph. Selection boxes, fill-in boxes, or parameters outlined with a dashed pink box are set during the analyses

Note 5), frame time, and pixel size. Please evaluate that these values are correct. Pixel time should be between 8 and 14 μs and pixel size between 0.05 and 0.1 μm. The Wo should be set to the PSF of the objective (*see* **Notes 6** and 7). Click the checkboxes of pixel size and Wo so that these are fixed (Fig. 1b—top panel, section 3).

10. Select Perform fit. A plot will appear that shows the diffusion model fit to the ACF (Fig. 1b—top panel, section 3). The value returned in the D1 box is the diffusion coefficient that best fits the data.

3.3 Qualification of Intra- and Intercellular Movement

To determine whether there is spatial variation in protein movement into or out of the nucleus (intracellular movement), or between different cell types through plasmodesmata (intercellular movement), pair correlation function (pCF) analysis can be used to measure any restriction of movement across the nuclear envelope or across cell walls. pCF uses a high-resolution time series of a single line scanned across the potential movement barrier to calculate the probability of fluorescence fluctuations in case of free movement and correlate this with the measured fluorescence [5].

1. pCF requires a line scan across the region of interest. Set the confocal settings (*see* **Note 2**) as follows:
   (a) Set gain to 800, digital offset to 0, and digital gain to 1.
   (b) The laser intensity should be set on a low percentage to avoid photobleaching (e.g., 1–4% laser intensity), but still high enough so that the fluorophore is visible.
   (c) Select a ROI of the root that contains three to five cells. The region of interest should be cropped such that the line scan crosses the barrier that is under investigation (Fig. 1a—right panel).
   (d) Set frame size to 256 × 256 pixels and pixel dwell time between 8 and 14 μs (Fig. 1a—right panel).
2. Take a snapshot of the region you are going to image and save the image. This image will serve as a reference to evaluate sample movement and photobleaching at **step 5**.
3. Change the image type from a frame to a line. Set the line to 32 pixels long (Fig. 1a—right panel).
4. Acquire a line scan time series of 100,000–200,000 lines.
5. After acquiring the line scan, reuse the settings from the reference image (**step 2**), and take a second snapshot of the region. Compare this second reference image to the first reference image to evaluate sample movement during imaging. If the sample moved, the line scan should not be used for further analysis, and **steps 2** to **5** should be repeated.

6. After acquiring images, open the SimFCS software and select SimFCS.
7. Select Analysis → Scanning FCS.
8. Open an image file by selecting File → Open .tif line file. A pop-up window will open with image information that should be closed to apply default settings. The image is loaded as an intensity carpet that represents the line scan over time (x-axis, pixels of the line scan; y-axis, time).
9. Apply an average across the intensity carpet to check for photobleaching or movement. Set the *Points per period* to 32 (i.e., number of pixels of the line scan) and the *Periods to average* to 800 (i.e., time points to average across). Then click "Calculate Image" to retrieve an averaged intensity carpet (Fig. 1b—middle panel, section 1) (*see* **Note 8**).
10. Calculate the autocorrelation of the carpet by selecting "Carpet analysis." In the pop-up window, set the *last row* to 100,000–200,000, depending on the number of line scans acquired (*see* **Note 8**) and the *last col* to 32. Mark "Autocorrelation", set *Column/distance* to 0, select "Detrend only", and set *Moving average* to 200. Then click "Analyze". The resulting image is 32 columns, where the color of each column represents the autocorrelation function (ACF) at that pixel (Fig. 1b—middle panel, section 2).
11. Align the carpet with the reference image to identify column distance (δr), i.e., the number of pixels that correspond approximately to the width of one cell (*see* **Note 9**) (Fig. 1b—middle panel, section 2). To align the carpet, select "Hold image 1" in the drop-down box on the bottom right to obtain a pop-up window of the carpet or take a screenshot of the carpet.
12. Perform pCF by selecting "Carpet analysis." In the pop-up window, set *last col.* to $32 - \delta r$. Mark "at column distance", set *column/distance* to δr, select "Column ACF", and set *moving average* to 200. Click "Analyze" to obtain the pCF carpet (Fig. 1b—middle panel, section 3).
13. Under the "Smooth" tab in the main window, click "Smooth columns" twice and then "Smooth surface" twice (Fig. 1b—middle panel, section 3).
14. Under the "Scale" tab, adjust the color scale by unchecking "Autoscale" and then drag the top slider to the right until the large area of blue near the bottom turns black (*see* **Note 10**).
15. Align the obtained carpet with the reference image. To align the carpet, select "Hold image 1" from the drop-down box on the bottom right to obtain a pop-up window of the carpet or take a screenshot. The carpet can be saved as an image by selecting File → Copy bitmap to clipboard, pasting the image

into Paint or a similar program, and subsequently saving the image.

16. Analyze the same line scan in the reverse direction by repeating **steps 10–15**. In addition to the guideline at **step 10**, also check "Reverse order", and set *last col* to 32 and the *first col* to 1 + δr (Fig. 1b—middle panel, section 3).
17. Analyze the same line scan with a column distance of $\delta r \pm 2$ in the forward and reverse direction by repeating **steps 10–16**. A total of six pCF analyses should be performed on one line scan.
18. Open an Excel file, and list for each image analysis the name of the image, the column distance, the direction of the pCF analysis (i.e., forward or reverse direction), and whether movement was observed (list as "1") or not (list as "0"). If there is an arch directly underneath a barrier, such as a cell wall, then autocorrelation is detected, and the protein can pass the barrier. There is delayed movement. If there is no arch and black background underneath the arch, there is no autocorrelation and thus no protein movement (Fig. 1b—middle panel, section 3).
19. To calculate the movement index (MI) for each image, average the six technical replicates/pCF analyses of each line scan. Then, calculate the average MI and standard deviation across all images from the same plant line.

3.4 Quantification of the Protein Oligomeric State

The acquired time series raster scan images taken for RICS (*see* Subheading 3.2) can, in addition to qualifying the diffusion coefficient, be used to identify the oligomeric state of proteins. Specifically, number and brightness (N&B) allows for the detection of different oligomeric states of a fluorescent-tagged protein while taking into account immobile fluorescent-tagged proteins, autofluorescence, and bleaching in a living cell [12, 16]. To quantify the oligomeric state of a fluorescent-tagged protein, include the same monomer fluorophore under a constitutive promoter to set parameters needed for the analysis of your protein of interest. The following instructions are specific for SimFCS 2.

1. Acquire images of plant tissues of interest as follows: Set the gain to 800, digital offset to 0, and digital gain to 1. The laser intensity should be set on a low percentage to avoid photobleaching (e.g., 1–4% laser intensity), but still high enough so that the fluorophore is visible. Zoom in on a ROI such that the pixel size is between 50 and 100 nm. Set frame size to 256 × 256 pixels and pixel dwell time between 8 and 14 μs (Fig. 1a—left panel). The images from the monomer control should contain a region of the plant expressing the fluorophore and a region that is outside of the plant or not expressing the fluorophore

(referred to as the background). The images of your protein of interest do not need to include a background region.

2. Collect a time series of 100 raster scans.
3. After acquiring images, start the SimFCS software [15] and click on the RICS button.
4. First analyze the images from the monomer control to identify the distribution of pixels representing monomers. Parameters identified during this analysis will be noted and used for the analysis of the images of the protein of interest.
5. Load an image stack by selecting File → Open multiple images and navigating to your file. A pop-up window will open with image information that should be closed to apply default settings.
6. Select Tools → RICS → Subtract moving average to calculate the autocorrelation function (ACF) (*see* **Note 3**) (Fig. 1b—bottom panel, section 1).
7. Select Tools → N and B → Detrend to perform N&B analysis.
8. Uncheck "autoscale," mark "Use white background," and uncheck "Change cursor position on refresh" under the view tab, and click "Recalculate plots." Each time a value is changed, "Recalculate plots" should be clicked to visualize the changes in the plot. Under the Filter tab, apply a filter to reduce noise by clicking *smooth* once.
9. Under the Cursors tab, change Cursor type to "Rectangle," and, in addition to the red cursor, add the blue and green cursor and adjust their dimensions:
   (a) The red cursor will capture the background pixels and should have *x*-size and *y*-size between 60 and 80.
   (b) The green cursor will capture the monomer pixels. The *x*-size should be larger than 120 so it captures all pixels. The *y*-size is determined with the images of the monomer control (**steps 9–11**) (*see* **Note 11**).
   (c) The blue cursor will capture the homodimeric pixels. The *x*-size and *y*-size are the same as for the green cursor.
   (d) Additional cursors can be added for higher-order oligomers with the same *x*-size and *y*-size as the green cursor.

   The area selected by the red, green, blue, or other cursors in the brightness vs. intensity plot is highlighted in the selection map in their respective colors (Fig. 1b—bottom panel, sections 2–3).
10. Position and potentially adjust the dimensions of the red cursor to cover all the background pixels in the brightness vs. intensity graph (*see* **Note 12**). The visualization in the selection map should help with positioning the red cursor (Fig. 1b—bottom panel, sections 2–3). The pixels should be distributed across

the plot such that the background pixels (captured by the red cursor) are centered around $\Upsilon = 1$ (brightness) and that all pixels are visible within the Υ min $= 0$ and Υ max $= 3$ range. The distribution of the pixels is adjusted by changing the *S*-factor under the Cal tab (Fig. 1b—bottom panel, section 2) (*see* **Note 13**). Each time the value is changed, click "Recalculate plots" to update the plot.

11. Once the *S*-factor is adjusted, position and adjust the *y*-size of the green cursor to cover all the monomer pixels in the brightness vs. intensity graph. The visualization in the selection map should help with positioning and setting the green cursor (Fig. 1b—bottom panel, section 2). Note the *y*-size of the cursor for the analysis of the images of the protein of interest (*see* **Note 11**).
12. Repeat **steps 5–10** for the images of the protein of interest.
13. After adjusting the *S*-factor, position the green cursor with a *y*-size identified in **step 11** to cover all the monomer pixels in the brightness vs. intensity graph (Fig. 1b—bottom panel, section 3).
14. The blue cursor of the same size as the green cursor can be placed on top of the green cursor (Fig. 1b—bottom panel, section 3). Pixels that fall inside this box are homodimers. Additional cursors can be added on top of each other to detect homotrimers and higher oligomeric states.
15. Under the Cursors tab, under "pixels", the number of pixels for each cursor and thus each oligomeric state can be found. To calculate the percentage of monomers, divide the number of monomer pixels by the sum of the pixels of all oligomeric forms (monomer, homodimer, etc.). This can be repeated to find the percentage of homodimer and other higher oligomeric forms (Fig. 1b—bottom panel, section 3).
16. Recheck the "Change cursor position on refresh" and "Autoscale" box each time an image analysis is finished to ensure the proper functioning of the software.

3.5 Integrating Experimental Parameters in Ordinary Differential Equations

To allow for a deeper understanding of protein function, systems-level computational modeling, such as ordinary differential equations (ODEs), can be used. ODEs measure how each of the variables, i.e., proteins, changes over time (*see* **Note 14**) and can incorporate the protein stoichiometry. As such, an ODE model can be used to model the dynamics of protein expression as a result of upstream regulatory interactions and/or protein movement.

Cell-autonomous protein expression over time can be modeled with ODEs consisting of two parts: (1) the first part is a production term that depends on upstream regulations; (2) the second part is a degradation term that is assumed to be linear. The promoter binding events of upstream regulators are modeled using Hill equations.

The numerator of the Hill fraction contains the terms that lead to the transcription and translation of the protein, while the denominator contains all possible binding events on the target's promoter. The induction of transcription by a transcriptional regulator is determined by the concentration of that regulator, while the absence of regulation is determined by a dissociation constant (K_D) that depends on the regulator. Taken together the ODEs should be written as follows:

- Considering the dynamics of a protein (Target1) of which the transcription is induced by a single activator (TFa), the ODE is:

$$\frac{\mathrm{dTarget1}}{\mathrm{d}t} = k_1 * \left(\frac{\mathrm{TFa}}{K_{\mathrm{Da}} + \mathrm{TFa}}\right) - d_1\mathrm{Target1}$$

- The expression dynamics of a protein (Target2) regulated by a single repressor (TFb) should have the following ODE:

$$\frac{\mathrm{dTarget2}}{\mathrm{d}t} = k_2 * \left(\frac{K_{\mathrm{Db}}}{K_{\mathrm{Db}} + \mathrm{TFb}}\right) - d_2\mathrm{Target2}$$

- To model the dynamics of a protein (Target3) regulated by two activators (TFa, TFc), of which their coordinated regulation operates in an OR gate, i.e., both transcriptional regulators can bind to the promoter and regulate the expression of the target gene, the following ODE should be used:

$$\frac{\mathrm{dTarget3}}{\mathrm{d}t} = k_3 * \left(\frac{K_{\mathrm{Da}}\mathrm{TFc} + K_{\mathrm{Dc}}\mathrm{TFa} + \mathrm{TFa\ TFc}}{K_{\mathrm{Da}}K_{\mathrm{Dc}} + K_{\mathrm{Da}}\mathrm{TFc} + K_{\mathrm{Dc}}\mathrm{TFa} + \mathrm{TFa\ TFc}}\right) - d_3\mathrm{Target3}$$

- Considering a protein (Target4) regulated by two activators (TFa, TFc) that operate through an AND gate, i.e., the expression of the target gene is only activated when both transcriptional regulators bind to the promoter, the ODE is:

$$\frac{\mathrm{dTarget4}}{\mathrm{d}t} = k_4 * \left(\frac{\mathrm{TFa\ TFc}}{K_{\mathrm{Da}}K_{\mathrm{Dc}} + K_{\mathrm{Da}}\mathrm{TFc} + K_{\mathrm{Dc}}\mathrm{TFa} + T\mathrm{Fa\ TFc}}\right) - d_4\mathrm{Target4}$$

To increase accuracy of the ODEs, use N&B to determine the oligomeric state of the upstream transcriptional regulators (*see* Subheading 3.4). The regulators' oligomeric states can be incorporated into the ODEs by including the degree of the regulator. The degree of the regulator depends on the oligomeric state: a dimer will be incorporated as the regulator to the second power, a trimer as the regulator to the third power, and so forth.

- The ODE of a protein (Target5) that is regulated by the dimerized regulator (TFd) is:

$$\frac{\mathrm{dTarget5}}{\mathrm{d}t} = k_5 * \left(\frac{\mathrm{TFd}^2}{K_{\mathrm{Dd}}^2 + K_{\mathrm{Dd}}\mathrm{TFd} + \mathrm{TFd}^2}\right) - d_5\mathrm{Target5}$$

– To model the dynamics of a protein (Target6) that is regulated by two activators, of which one operates as a dimer (TFd), the second operates as a monomer (TFa), and their coordinated regulation operates with on OR gate, the ODE is:

$$\frac{\mathrm{dTarget6}}{\mathrm{d}t} = k_6 * \left(\frac{K_{\mathrm{Dd}}^2\mathrm{TFa} + K_{\mathrm{Da}}K_{\mathrm{Dd}}\mathrm{TFd} + K_{\mathrm{Dd}}\mathrm{TFa\ TFd} + K_{\mathrm{Da}}\mathrm{TFd}^2 + \mathrm{TFa\ TFd}^2}{K_{\mathrm{Dd}}^2K_{\mathrm{Da}} + K_{\mathrm{Dd}}^2\mathrm{TFa} + K_{\mathrm{Da}}K_{\mathrm{Dd}}\mathrm{TFd} + K_{\mathrm{Dd}}\mathrm{TFa\ TFd} + K_{\mathrm{Da}}\mathrm{TFd}^2 + \mathrm{TFa\ TFd}^2}\right) - d_6\mathrm{Target6}$$

In addition to cell-autonomous regulation, some proteins are non-cell autonomously expressed, i.e., the protein is not transcribed and translated in the cell itself, but moves from neighboring cells into the cell. The mathematical equation for non-cell-autonomous proteins is constructed as follows:

1. Use pCF analysis to determine whether the movement is bidirectional (*see* Subheading 3.3).
2. Use RICS to experimentally determine the diffusion coefficient of the protein in different cells (*see* Subheading 3.2).
3. Measure the cell area of your cell type of interest for which RICS and pCF data was gathered with ImageJ as follows:
   (a) Open ImageJ.
   (b) Open the first image by selecting File → Open → browse to your images.
   (c) Select in the tool bar the Freehand Line tool by right-clicking on the Line selection tools icon.
   (d) Contour the cell with the Freehand Line tool.
   (e) Press M or select Results → Measure to record the area of the drawn contour.
   (f) Convert the measured areas from pixels to μm^2 or set in advance the scale in ImageJ.
   (g) Average the measured areas from at least ten different images.
4. Use a linear term for gradient-independent diffusion:

$$\frac{\mathrm{dTarget7}}{\mathrm{d}t} = a_1\mathrm{Target7} - a_2\mathrm{Target7} - d_7\mathrm{Target7}$$

$$a_1 = \frac{D_1}{A_1}$$

$$a_2 = \frac{D_2}{A_2}$$

with A_i = the area of the cell from which the protein moves and D_i = the experimentally calculated diffusion coefficient. In case the movement is unidirectional a_2 will be equal to zero.

5. Lastly, cell-autonomous and non-cell-autonomous regulation can be combined when your protein is found to be transcribed and translated within the cell of interest and is known to be mobile. After determining the diffusion coefficient and potential movement barriers, the ODE of a mobile protein regulated by one activator is constructed as follows:

$$\frac{\text{dTarget8}}{\text{d}t} = k_8 * \left(\frac{\text{TFa}}{K_{\text{Da}} + \text{TFa}}\right) + a_1\text{Target8} - a_2\text{Target8} - d_8\text{Target8}$$

$$a_1 = \frac{D_1}{A_1}, \quad a_2 = \frac{D_2}{A_2}$$

4 Notes

1. GFP has a known diffusion coefficient of 78 $\mu m^2/s$ [17].
2. To set the best imaging parameters such as pixel dwell time, pixel size, and laser intensity for the system, we recommend using a plant line with a known diffusion coefficient, such as 35S:GFP [12]. The imaging parameters vary depending on the confocal model and objective, so setting them using a known control can save time when imaging experimental samples. Suggested ranges of imaging parameters for the Zeiss 710 and 780 confocal models have been previously described [12]. For RICS, we recommend a smaller pixel size (typically 50 nm). In addition, while 100 frames are typically used, the number can range from 50 to 300 frames, where a higher number of frames minimize the error in the diffusion model fit to the ACF [18, 19].
3. The moving average is used to subtract the distribution of the immobile molecules and correct for sample movement and photobleaching during image acquisition [18]. Typically, the moving average is set on ten frames. The average of the first ten frames is calculated and subsequently subtracted from the fifth frame, since this frame is the center of the moving average window. These calculations are repeated for the next ten frames shifting one frame at a time.
4. If images within the stack show a high degree of photobleaching or have moved drastically, they should be eliminated from the analysis by changing the last frame (frames to process for ACF) or by deleting individual frames. For example, if the last

ten images for RICS are severely photobleached, change the last frame to 90, and recalculate the moving average.

5. To calculate the line time, divide the frame time at a specific pixel dwell time by the number of lines in one frame, and subsequently convert it from seconds to milliseconds. In this chapter, 256 × 256 images are taken, as such 1 frame consists of 256 lines.
6. The point spread function (PSF) measures the radius of the laser beam and depends on the type of confocal microscope and objective lens. PSF should be calculated by fitting the autocorrelation function (ACF) to the intensity fluctuations of free GFP in aqueous solution, obtained from the raster images, while fixing the known GFP diffusion coefficient of 78 $\mu m^2/s$ [17, 20].
7. The expected PSF should be between 0.2 and 0.4 μm [21].
8. The intensity carpet obtained after ACF should display vertical bands of similar colors. If these bands shift or change color along the vertical axis of the carpet, sample movement or photobleaching, respectively, has occurred during imaging. Lines showing movement or bleaching should be excluded from further analysis. For example, if the last 50,000 lines of a 200,000 lines scan appear to show photobleaching, they should be excluded by setting the *last row* in **step 10** to 150,000.
9. The *x*-coordinate of the carpet displays 32 pixels and can be used to estimate the number of pixels the width of one cell corresponds to. This number will be the column distance (δr) and should not exceed 14 pixels.
10. By adjusting the first slider, the bottom two thirds of the carpet should be colored black. By adjusting the second slider, areas of yellow or red that have high autocorrelation can be highlighted. It is advised to minimize the areas of red as these can indicate oversaturation.
11. The identified *y*-size can then be used for all future images taken with that specific fluorophore and under similar conditions (i.e., species, tissue, and objective).
12. Pixels that represent background fluorescence or immobile structures should not be used for the N&B analysis. Any background or immobile fraction of the image will have low intensity and a brightness close to 1. When positioning the cursor, only the region of the image with no fluorescence should be marked as background in the selection map.
13. When adjusting the *S*-factor, if the background is above 1, increase the *S*-factor. If the background is below 1, decrease the *S*-factor.

14. These models assume transcription and translation happen fast so that transcriptional regulation and protein expression can be modeled in the same equation.

Acknowledgments

Support for this work was provided to R.S. by the National Science Foundation (NSF) (CAREER MCB-1453130) and the Foundation for Food and Agriculture Research (FFAR CA18-SS-0000000026).

References

1. Gallagher KL, Sozzani R, Lee C (2014) Intercellular protein movement deciphering the language of development. Annu Rev Cell Dev Biol 30:207–233
2. Wu S, Gallagher K (2014) The movement of the non-cell-autonomous transcription factor SHORT-ROOT relies on the endomembrane system. Plant J 80:396–409
3. Stahl Y, Grabowski S, Bleckmann A, Kühnemuth R, Weidtkamp-Peters S et al (2013) Moderation of Arabidopsis root stemness by CLAVATA1 and ARABIDOPSIS CRINKLY4 receptor kinase complexes. Curr Biol 23:362–371
4. Dupuy L, Mimault M, Patko D, Ladmiral V, Ameduri B et al (2018) Micromechanics of root development in soil. Curr Opin Genet Dev 51:18–25
5. Digman MA, Dalal R, Horwitz AF, Gratton E (2008) Mapping the number of molecules and brightness in the laser scanning microscope. Biophys J 94:2320–2332
6. Park J, Kurihara D, Higashiyama T, Arata H (2014) Fabrication of microcage arrays to fix plant ovules for long-term live imaging and observation. Sens Actuat B Chem 191: 178–185
7. Reynaud EG, Peychl J, Huisken J, Tomancak P (2015) Guide to light-sheet microscopy for adventurous biologists. Nat Methods 12: 30–34
8. Clark NM, Fisher AP, Berckmans B et al (2020) Protein complex stoichiometry and expression dynamics of transcription factors modulate stem cell division. Proc Natl Acad Sci 11726:15332–15342
9. Van den Broeck L, Spurney R, Fisher A, Schwartz M, Clark N et al (2021) A hybrid model connecting regulatory interactions with stem cell divisions in the root. Quant Plant Biol 2:e2
10. Al Asafen H, Clark NM, Jacobsen T, Sozzani R, Reeves GT (2018) Dorsal/NF-κB exhibits a dorsal-to-ventral mobility gradient in the Drosophila embryo. bioRxiv. https://www.biorxiv.org/content/10.1101/320754v1.full
11. Digman MA, Gratton E (2009) Fluorescence correlation spectroscopy and fluorescence cross-correlation spectroscopy. Wiley Interdiscip Rev Syst Biol Med 1:273–282
12. Clark NM, Hinde E, Winter CM, Fisher AP, Crosti G et al (2016) Tracking transcription factor mobility and interaction in Arabidopsis roots with fluorescence correlation spectroscopy. elife 5:e14770
13. Akhunzada MJ, D'Autilia F, Chandramouli B, Bhattacharjee N, Catte A, Di Rienzo R et al (2019) Interplay between lipid lateral diffusion, dye concentration and membrane permeability unveiled by a combined spectroscopic and computational study of a model lipid bilayer. Sci Rep 9:1508
14. Denninger P, Bleckmann A, Lausser A, Vogler F, Ott T et al (2014) Male–female communication triggers calcium signatures during fertilization in Arabidopsis. Nat Commun 5: 4645
15. Gratton E. (2006) Globals software - laboratory for fluorescence dynamics. https://www.lfd.uci.edu/globals/
16. Digman MA, Brown CM, Sengupta P, Wiseman PW, Horwitz AR, Gratton E (2005) Measuring fast dynamics in solutions and cells with a laser scanning microscope. Biophys J 89: 1317–1327
17. Chen Y, Mu JD, Ruan Q, Gratton E (2002) Molecular brightness characterization of EGFP in vivo by fluorescence fluctuation spectroscopy. Biophys J 82:133–144

18. Digman MA, Brown CM, Sengupta P et al (2005) Measuring fast dynamics in solutions and cells with a laser scanning microscope. Biophys J 89:1317–1327
19. Brown CM, Dalal RB, Hebert B et al (2008) Raster image correlation spectroscopy RICS for measuring fast protein dynamics and concentrations with a commercial laser scanning confocal microscope. J Microsc 229:78–91
20. Clark NM, Hinde E, Winter CM et al (2016) Tracking transcription factor mobility and interaction in Arabidopsis roots with fluorescence correlation spectroscopy. eLife 5:e14770
21. Rossow MJ, Sasaki JM, Digman MA, Gratton E (2010) Raster image correlation spectroscopy in live cells. Nat Protoc 5:1761–1774

Chapter 26

Spatiotemporal Specific Blocking of Plasmodesmata by Callose Induction

Dawei Yan

Abstract

Plasmodesmata are nanoscale cell wall channels connecting neighboring cells in plants. Intercellular trafficking of molecules via plasmodesmata plays important roles in various developmental processes and stress responses. The turnover of callose, a β-1,3-glucan polysaccharide depositing in the cell wall around plasmodesmata, controls the plasmodesmal permeability and symplasmic transport. Here, we describe a protocol for the spatiotemporally controlled induction of callose synthesis and plasmodesmata closure using the *cals3m* system. In this system, *cals3m*, a mutant *CALLOSE SYNTHASE 3* (*CALS3*) gene, is driven by inducible tissue-specific promoters of interest. After appropriate induction by 17-β-estradiol, callose is overproduced within the corresponding specific domains, resulting in temporal closure of plasmodesmata at the cell-cell interfaces. This approach can be used to validate and dissect the function of plasmodesmata-mediated symplasmic communications.

Key words Plasmodesmata, Callose, cals3m, Tissue-specific promoter, Induction

1 Introduction

Callose is a linear polysaccharide that functions in plant development and defense. At plasmodesmata, callose accumulates in the cell walls surrounding the plasmodesmal neck regions, and its homeostasis controls the aperture and conductivity of the plasmodesmata pores [1]. Callose synthases and β-1,3-glucanases are responsible for the biosynthesis and degradation of callose, respectively [2]. In *Arabidopsis*, 12 putative callose synthases (CALS1-12) have been identified and characterized [3, 4]. Among them, CALS5, 9, 10, 11, and 12 are involved in pollen development, while CALS1 and 8 are key players in modulating plasmodesmal permeability in response to both pathogen infection and wounding stress [4, 5]. *CALS7* is specifically expressed in sieve elements and participates in sieve plate formation and phloem transport [6, 7].

Yoselin Benitez-Alfonso and Manfred Heinlein (eds.), *Plasmodesmata: Methods and Protocols*,
Methods in Molecular Biology, vol. 2457, https://doi.org/10.1007/978-1-0716-2132-5_26,

Three interesting semi-dominant mutants *cals3-1d*, *2d*, and *3d* were identified and analyzed by Vatén et al. [8]. Each of these mutants carries a single nucleotide mutation in the sequence encoding the cytosolic face of CALS3, a plasmodesmata-localized callose synthase. Elevated callose biosynthesis by these mutants leads to decreased plasmodesmal aperture and, thereby, reduced symplasmic trafficking and developmental defects in roots. By combining *cals3-1d* and *2d* mutations (named *cals3m*) into a LexA-VP16-ER(XVE)-based CALS3 expression vector driven by a tissue-specific promoter, an inducible system for ectopic callose production due to increased enzymatic activity was engineered [8–10]. By treating the transgenic plants carrying this system with 17-β-estradiol, induced expression of *cals3m* under the control of tissue-specific promoters leads to increased callose deposition at the plasmodesmata of the corresponding cells [8, 11]. These plasmodesmata are therefore obstructed temporally, thus interrupting symplasmic transport. For example, after induction, *cals3m* driven by the stele-specific CRE1promoter inhibits SHR movement from the stele to endodermis in *Arabidopsis* roots [8]. In transgenic plants carrying pCALS8::icals3m, callose overproduction in phloem pole pericycle blocks phloem unloading in roots [12]. This inducible *cals3m* system has been successfully used to validate mobile molecules and investigate their function during symplasmic communication [13–15]. Here, the protocol for generating an inducible cals3m system and performing callose induction and analysis in transgenic *Arabidopsis* plants is described using pPEAR1-XVE::cals3m and pAPL-XVE::cals3m as examples.

2 Materials

2.1 Plant Materials

1. 4- to 6-week-old plants of *Arabidopsis thaliana* ecotype Col-0 growing on soil in a growth chamber at 23 °C with 16/8 h light/dark.

2.2 Plasmid Vectors

1. Gateway vectors: p1R4-ML:XVE, pEN-cals3m, pEN-nosT, and pHm43GW [8, 9] (available upon request).

2.3 Oligonucleotides (PCR Primers)

1. pPEAR1-F: GGGGTACCGTGTTGCCTAACTCTTGA.
2. pPEAR1-R: CCCTCGAGTGGTTATTCTCTTTTGAT.
3. pAPL-F: GGGGTACCCGATCGATTTATCGAATTTAGTGTT.
4. pAPL-R: CCCTCGAGTCTCTCTCTCTCTCTCTGACTCTC.

2.4 Antibiotic Stock Solutions

1. 100 mg/ml spectinomycin. Dissolve 1 g spectinomycin in 10 ml of sterile H_2O. Sterilize through a 0.22 μm syringe filter and keep at −20 °C for up to 1 year.
2. 50 mg/ml rifampicin. Dissolve 0.5 g rifampicin in 10 ml of 100% methanol. Keep at −20 °C for up to 1 year.
3. 25 mg/ml gentamicin. Dissolve 0.25 g gentamicin in 10 ml of sterile H_2O. Sterilize by filtration through a 0.22 μm syringe filter and keep at −20 °C for up to 1 year.
4. 50 mg/ml hygromycin B. Dissolve 0.5 g hygromycin B in 10 ml of sterile H_2O. Sterilize by filtration through a 0.22 μm syringe filter and keep at −20 °C for up to 1 year.

2.5 Culture Media

1. LB medium: 10 g/l Bacto tryptone, 5 g/l yeast extract, and 10 g/l NaCl. Autoclave at 121 °C for 20 min. Add antibiotics for selection before use.
2. Selection LB agar medium plates (spec): LB medium supplemented with 10 g/l agar. Autoclave at 121 °C for 20 min, allow to cool to 60 °C, add 100 mg/l spectinomycin, distribute into round sterile petri dishes, and allow to solidify.
3. Selection LB agar medium plates (hyg/rif/genta): LB medium supplemented with 10 g/l agar. Autoclave at 121 °C for 20 min; allow to cool to 60 °C; add 100 mg/l hygromycin B, 50 mg/l rifampicin, and 25 mg/l gentamicin; distribute into round sterile petri dishes; and allow to solidify.
4. 1/2 MS medium plates: 2.2 g/l MS basal salt mixture, 10 g/l agar, 10 g/l sucrose, 0.5 g/l 4-morpholine ethanesulfonic acid (MES), pH 5.7–5.8. Autoclave at 121 °C for 20 min and pour into square sterile petri dishes (12 cm × 12 cm) after cooling to 60 °C.
5. Selection 1/2 MS medium plates: 1/2 MS medium plates containing 40 mg/l hygromycin B.
6. Induction 1/2 MS medium plates: 1/2 MS medium plates containing 5 μM 17-β-estradiol (*see* **Note 1**).
7. Non-induction medium plates: 1/2 MS medium plates containing 5 μM DMSO.
8. Round sterile petri dishes (diameter 9 cm).
9. Square sterile petri dishes (12 cm × 12 cm).

2.6 Plasmid Construction

1. Genomic DNA extraction kit.
2. KOD DNA polymerase and buffer kit.
3. Gel extraction kit.
4. PCR cleanup kit.

5. Endonucleases *KpnI* and *XhoI* with corresponding 10× buffers.
6. T4 DNA ligase with 10× buffer.
7. Gateway LR Clonase II Plus Enzyme Mix (Thermo Fisher Scientific).
8. Competent cells: *Escherichia coli* DH5α and *Agrobacterium tumefaciens* GV3101.

2.7 Plant Transformation

1. Transformation solution: 5% sucrose and 0.05% Silwet L-77.

2.8 Seed Sterilization

1. 70% ethanol.
2. Sterile distilled water.

2.9 Aniline Blue Staining

1. Aniline blue fluorochrome stock: 0.1 mg/ml in H_2O.
2. 67 mM K_3PO_4, pH 9.5.
3. Citifluor AF3 antifade mounting solution.
4. Forceps.
5. 6-well plastic plate.

2.10 Callose Imaging

1. Confocal laser scanning microscope.
2. ImageJ software [16].

3 Methods

3.1 Construction of Inducible Vectors

1. Extract genomic DNA from *Arabidopsis thaliana* Col-0 plants using the genomic DNA extraction kit.
2. Use PCR to amplify the 1.5 kb *PEAR1* promoter fragment and the 2.9 kb *APL* promoter fragment with KOD DNA polymerase using primer sets pPEAR1-F/R and pAPL-F/R, respectively. The resulting fragments are flanked by restriction endonuclease sites. Purify the PCR fragments using the gel extraction kit, and digest them and also the plasmid vector p1R4-ML-XVE [9] with *KpnI/XhoI.* After purification with the PCR cleanup kit, ligate each of the digested fragments to the linearized plasmid using T4 DNA ligase to produce plasmids p1R4-PEAR1-XVE and p1R4-APL-XVE (entry clone 1) (*see* Fig. 1).
3. Perform MultiSite Gateway LR reactions according to the MultiSite Gateway Three-Fragment Vector Construction Kit manual (Thermo Fisher Scientific; catalog no. 12537-023) with the respective entry clone 1. Combine each specific entry clone with pEN-cals3m [8], pEN-nosT [9], and the destination vector pHm43GW, and perform the recombination

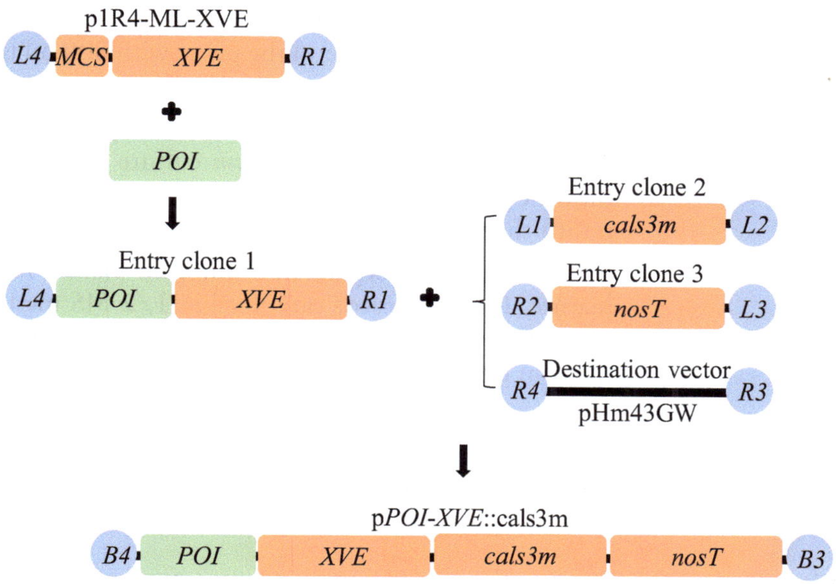

Fig. 1 Schematic representation of the inducible *cals3m* construction. Promoters of interest (POI) (promoters of PEAR1 and APL genes in this protocol) are cloned into the MCS site of p1R4-ML-XVE to generate the entry clone 1 using restriction enzyme digestion and ligation. pEN-cals3m, pEN-nosT, and destination vector pHm43GW are mixed together with entry clone 1 in a MultiSite Gateway Pro LR recombination reaction to create the final expression clone pPOI-XVE::cals3m. MCS, multiple cloning site. POI, promoter of interest

reaction with the Gateway LR Clonase II Plus enzyme mix to produce the expression vectors pPEAR1-XVE::cals3m and pAPL-XVE::cals3m (*see* Fig. 1).

4. Propagate pPEAR1-XVE::cals3m and pAPL-XVE::cals3m vectors by transforming DH5α competent cells and selecting them on Selection LB agar medium plates (spec).

3.2 Plant Transformation

1. Introduce pPEAR1-XVE::cals3m and pAPL-XVE::cals3m into *Agrobacterium tumefaciens* GV3101 competent cells using heat shock method. Shake the transformed cells in 1 ml LB medium without antibiotics at 28 °C for 2 h.
2. Spread 100 μL of the culture on Selection LB agar medium plates (hyg/rif/genta), and incubate at 28 °C for 48 h.
3. Inoculate 5 ml LB medium (supplemented with 100 mg/l hygromycin B, 50 mg/l rifampicin, and 25 mg/l gentamicin) with a single colony of *Agrobacterium* harboring pPEAR1-XVE::cals3m/pAPL-XVE::cals3m, and shake the culture (250 rpm) at 28 °C for 48 h until its optical density at 600 nm (OD 600) reaches 1.0.
4. Transfer 500 μl of the culture into 500 ml LB medium with antibiotics (100 mg/l hygromycin B, 50 mg/l rifampicin, and 25 mg/l gentamicin), and shake the culture (250 rpm) at 28 °C overnight.

5. Centrifuge the culture at 3000 × *g* for 1 min, discard the supernatant, and resuspend the pellet with 500 ml transformation buffer.
6. Dip the inflorescence shoots of Col-0 plants with siliques removed into the *Agrobacterium* culture for 20–30 s. Place the whole dipped plants into plastic bags for 24 h, and then return them to normal growth conditions.
7. Harvest the seeds separately from each plant when they get mature. Place the seeds on selection 1/2 MS medium plates, and select the transgenic lines with hygromycin B resistance that can germinate and grow.
8. Isolate and select transgenic lines, of which at least around 3/4 plants are hygromycin B-resistant in the T2 population. Of these, further select homozygous lines, of which all plants are fully hygromycin B-resistant in the T3 population (*see* **Note 1**).

3.3 Induction of Callose Biosynthesis

1. Sterilize the seeds of homozygous transgenic plants expressing pPEAR1-XVE::cals3m and pAPL-XVE::cals3m with 70% ethanol for 2–3 min, and then wash the seeds three times with sterile distilled water.
2. Keep the seeds in sterile water at 4 °C for 2–3 days.
3. Germinate the seeds on 1/2 MS medium plates, and keep the plates vertically in the growth chamber with 16/8 h light/dark at 23 °C for 5 days.
4. Use forceps to carefully transfer seedlings onto induction 1/2 MS medium plates. Transfer other seedlings also onto non-induction medium plates as control (*see* **Note 2**). Make sure that the roots of transferred seedlings contact the medium closely to allow efficient entry of estradiol. Return the plates into the growth chamber and keep them vertically for 12 h (*see* **Note 3**).

3.4 Aniline Blue Staining

1. Prepare aniline blue staining solution by diluting the aniline blue fluorochrome stock with 67 mM K_3PO_4 (pH 9.5) in the ratio 1:3.
2. Add 5–10 ml of aniline blue staining solution into the wells of a 6-well plate. Transfer the whole seedlings from the induction medium into one well filled with aniline blue staining solution and that from the non-induction medium plates into another well. 20–30 seedlings from the same treatment can be transferred into the same well. Keep the roots immersed into the solution, and place the 6-well plate on a rotary shaker (100 rpm) at room temperature for 2 h (*see* **Note 4**).
3. Add 50–100 μl Citifluor AF3 antifade solution on a glass slide. Transfer a seedling onto the glass slide, and place the root flatly

into the antifade solution. Place a coverslip over the root for confocal imaging. 5–10 roots can be examined per slide.

3.5 Imaging

1. Set up the confocal laser scanning microscope for using the ultraviolet laser for excitation at 405 nm and for appropriate emission light filtering at 425–525 nm.
2. Move the roots into focus (with a 20× or 40× objective) using bright-field illumination, and then use filtered light to scan the roots for aniline blue fluorescence. Image the fluorescence signal of aniline blue-stained callose in roots with same settings for each root sample (*see* **Note 5**). In the roots expressing pPEAR1-XVE::cals3m, callose accumulation is mainly induced in the early protophloem sieve elements in the meristem zone (*see* Fig. 2a), whereas in the roots expressing pAPL-XVE::cals3m, callose induction mainly occurs in the developing sieve elements (*see* Fig. 2b and **Note 6**).
3. Quantify the callose signal with ImageJ software [16].

4 Notes

1. Choose transgenic plants from different generations. T2 transgenic plants can already be used to determine if the inducible *cals3m* system is active, whereas T3 homozygous plants are required for quantitative analysis of the induced callose levels.
2. Transfer seedlings gently using forceps. Keep the roots from being damaged to prevent callose induction due to wounding.
3. The induction effects are not the same for different lines due to the different activities of promoters and plant growth conditions. The appropriate concentration of 17-β-estradiol and the appropriate duration of treatment for efficient induction must be empirically determined for each newly generated transgenic line.
4. Adjust the incubation time when analyzing different tissues. Vacuum infiltration can be applied in some cases to improve penetration of the dye.
5. Try to capture images with short exposure time to avoid the photobleaching of fluorescence signal. If available, choose the bidirectional laser acquisition mode to reduce bleaching and for faster image acquisition.
6. Specific cells in which the promoter is active due to induction by 17-β-estradiol should show a significantly strong punctate fluorescent signal associated with cell walls. This signal should be absent or significantly weaker in non-induced control plants.

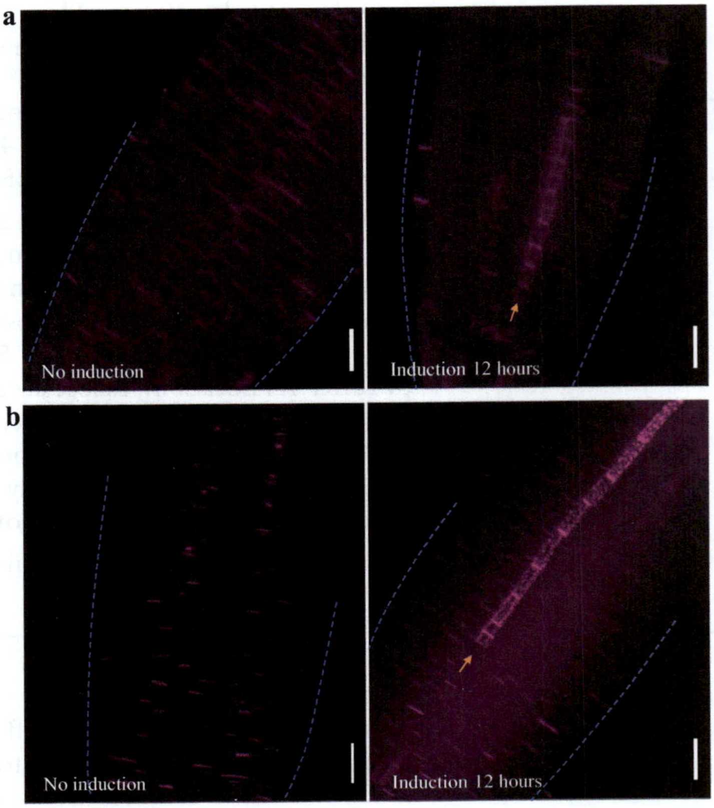

Fig. 2 Aniline blue staining for callose in the primary roots of transgenic plants. (**a**) Roots expressing pPEAR1-XVE::cals3m with and without induction. Orange arrow indicates the location of overproduced callose after 12 h of induction in the early protophloem sieve elements. (**b**) Roots expressing pAPL-XVE::cals3m with and without induction. Orange arrow indicates the location of overproduced callose after 12 h of induction in the developing protophloem sieve elements. Scale bars, 20 μm

Acknowledgments

We thank Ykä Helariutta and Ari Pekka Mähönen labs for the Gateway vectors. This work was supported by the National Natural Science Foundation of China (32070192) and the 111 Project#D16014.

References

1. De Storme N, Geelen D (2014) Callose homeostasis at plasmodesmata: molecular regulators and developmental relevance. Front Plant Sci 5:138. https://doi.org/10.3389/fpls.2014.00138
2. Zavaliev R, Ueki S, Epel BL, Citovsky V (2011) Biology of callose (beta-1,3-glucan) turnover at plasmodesmata. Protoplasma 248(1): 117–130. https://doi.org/10.1007/s00709-010-0247-0
3. Ellinger D, Voigt CA (2014) Callose biosynthesis in Arabidopsis with a focus on pathogen response: what we have learned within the last

decade. Ann Bot 114(6):1349–1358. https://doi.org/10.1093/aob/mcu120
4. Chen XY, Kim JY (2009) Callose synthesis in higher plants. Plant Signal Behav 4(6): 489–492. https://doi.org/10.4161/psb.4.6.8359
5. Cui W, Lee JY (2016) *Arabidopsis* callose synthases CalS1/8 regulate plasmodesmal permeability during stress. Nat Plants 2(5):16034. https://doi.org/10.1038/nplants.2016.34
6. Barratt DH, Kolling K, Graf A, Pike M, Calder G, Findlay K, Zeeman SC, Smith AM (2011) Callose synthase GSL7 is necessary for normal phloem transport and inflorescence growth in Arabidopsis. Plant Physiol 155(1): 328–341. https://doi.org/10.1104/pp.110.166330
7. Xie B, Wang X, Zhu M, Zhang Z, Hong Z (2011) *CalS7* encodes a callose synthase responsible for callose deposition in the phloem. Plant J 65(1):1–14. https://doi.org/10.1111/j.1365-313X.2010.04399.x
8. Vatén A, Dettmer J, Wu S, Stierhof YD, Miyashima S, Yadav SR, Roberts CJ, Campilho A, Bulone V, Lichtenberger R, Lehesranta S, Mahonen AP, Kim JY, Jokitalo E, Sauer N, Scheres B, Nakajima K, Carlsbecker A, Gallagher KL, Helariutta Y (2011) Callose biosynthesis regulates symplastic trafficking during root development. Dev Cell 21(6):1144–1155. https://doi.org/10.1016/j.devcel.2011.10.006
9. Siligato R, Wang X, Yadav SR, Lehesranta S, Ma GJ, Ursache R, Sevilem I, Zhang J, Gorte M, Prasad K, Wrzaczek M, Heidstra R, Murphy A, Scheres B, Mahonen AP (2016) MultiSite gateway-compatible cell type-specific gene-inducible system for plants. Plant Physiol 170(2):627–641. https://doi.org/10.1104/pp.15.01246
10. Zuo JR, Niu QW, Chua NH (2000) An estrogen receptor-based transactivator XVE mediates highly inducible gene expression in transgenic plants. Plant J 24(2):265–273. https://doi.org/10.1046/j.1365-313x.2000.00868.x
11. Yadav SR, Yan D, Sevilem I, Helariutta Y (2014) Plasmodesmata-mediated intercellular signaling during plant growth and development. Front Plant Sci 5:44. https://doi.org/10.3389/fpls.2014.00044
12. Ross-Elliott TJ, Jensen KH, Haaning KS, Wager BM, Knoblauch J, Howell AH, Mullendore DL, Monteith AG, Paultre D, Yan D, Otero S, Bourdon M, Sager R, Lee JY, Helariutta Y, Knoblauch M, Oparka KJ (2017) Phloem unloading in Arabidopsis roots is convective and regulated by the phloem-pole pericycle. elife 6:24125. https://doi.org/10.7554/eLife.24125
13. Daum G, Medzihradszky A, Suzaki T, Lohmann JU (2014) A mechanistic framework for noncell autonomous stem cell induction in *Arabidopsis*. Proc Natl Acad Sci U S A 111(40):14619–14624. https://doi.org/10.1073/pnas.1406446111
14. Gaudioso-Pedraza R, Beck M, Frances L, Kirk P, Ripodas C, Niebel A, Oldroyd GED, Benitez-Alfonso Y, de Carvalho-Niebel F (2018) Callose-regulated symplastic communication coordinates symbiotic root nodule development. Curr Biol 28(22):3562–3577.e6. https://doi.org/10.1016/j.cub.2018.09.031
15. Wu S, O'Lexy R, Xu M, Sang Y, Chen X, Yu Q, Gallagher KL (2016) Symplastic signaling instructs cell division, cell expansion, and cell polarity in the ground tissue of *Arabidopsis thaliana* roots. Proc Natl Acad Sci 113(41): 11621–11626. https://doi.org/10.1073/pnas.1610358113
16. Zavaliev R, Epel BL (2015) Imaging callose at plasmodesmata using aniline blue: quantitative confocal microscopy. Methods Mol Biol 1217: 105–119. https://doi.org/10.1007/978-1-4939-1523-1_7

Chapter 27

A Forward Genetic Approach to Identify Plasmodesmal Trafficking Regulators Based on Trichome Rescue

Munenori Kitagawa and David Jackson

Abstract

Plasmodesmata (PD) are channels in the walls of plant cells which enable cell-to-cell information transfer. This includes the selective transport of specific transcription factors that control cell fate during plant development. KNOTTED1 (KN1) homeobox (KNOX) family transcription factors that are essential for the maintenance and function of stem cells in shoot meristems use this trafficking pathway, but its mechanism is largely unknown. Here we describe a forward genetic approach to the identification of regulators of selective KN1 trafficking through PD, using a trichome rescue system that permits simple visual analysis in *Arabidopsis* leaves. A KN1 trafficking regulator identified in this approach had the capacity to regulate the transport not only of KN1 but also of another mobile regulatory protein, TRANSPARENT TESTA GLABRA1 (TTG1). Our system could be easily adapted to reveal the mechanism underlying the selective transport of additional mobile signals through PD.

Key words Plasmodesmata, Selective transport, Transcription factor, *Arabidopsis*, Maize, KNOTTED1 (KN1), SHOOT MERISTEMLESS (STM), Shoot apical meristem (SAM), Stem cell, Trichome rescue system

1 Introduction

Plasmodesmata (PD) are membrane-lined nano-channels that permit efficient exchange of information between plant cells across the cell wall, a physical barrier to cell-to-cell signaling [1]. PD mediate the cell-to-cell trafficking of signals that are essential for a broad range of biological processes and plant viability [2–8]. Signaling via PD involves the selective transport of specific transcription factors that control cell fates [9]. While smaller molecules and proteins can diffuse cell to cell, depending on PD aperture and density, larger proteins are thought to traffic actively, either by changing their structure or by interacting with other factors that increase PD permeability [1]. The first transcription factor found to use this pathway was the maize homeodomain protein KNOTTED1

Yoselin Benitez-Alfonso and Manfred Heinlein (eds.), *Plasmodesmata: Methods and Protocols*,
Methods in Molecular Biology, vol. 2457, https://doi.org/10.1007/978-1-0716-2132-5_27,

(KN1), which moves between cells and helps maintain stem cells identity in shoot apical meristems (SAMs) [10–12]. A KN1 homolog in *Arabidopsis*, SHOOT MERISTEMLESS (STM) [13], is also a mobile transcription factor and has an analogous function in SAM maintenance [12, 14]. Elucidation of the trafficking machinery of KNOX transcription factors could provide insights into mechanisms of SAM development [15] and permit manipulation of shoot architecture for agricultural benefits [16]. Previous studies identified several KNOX trafficking regulators [17–19], but a more complete picture of the mechanism requires the identification of additional factors and clarification of their interactions. Here, we describe a forward genetic approach for identifying KN1 trafficking regulators using the "trichome rescue system" that we previously developed to assay KN1 trafficking in the genetically tractable model plant, *Arabidopsis thaliana* [11, 18].

Maize KN1 can complement *Arabidopsis stm* mutants [12]. Thus, based on this conserved function of homologs, we developed a system to visually assess the cell-to-cell mobility of KN1 through a simple visual readout: trichome formation on the true leaves of *Arabidopsis* seedlings [11, 18]. *Arabidopsis* trichomes develop from the leaf epidermis and require the cell-autonomous activity of a MYB transcription factor, GLABROUS1 (GL1) [20]. In our system, a fusion protein of GL1 with the green fluorescent protein (GFP), and the KN1 C-terminal trafficking domain, $KN1^C$, is expressed in the mesophyll cell layer of *gl1* mutants using a *Rubisco small subunit 2b* promoter (*pRbcS::GFP ~ GL1 ~* $KN1^C$) (Fig. 1). The $KN1^C$ domain promotes PD trafficking of *GFP ~ GL1* fusion protein from the mesophyll layer to the epidermis. Once in the epidermal cells, GL1 activity from the GFP ~ GL1 ~ $KN1^C$ fusion protein rescues trichome formation; hence, trichome number is an output of KN1 trafficking. Using this system, we designed a forward genetic screen, as shown in Fig. 2. Putative KN1 trafficking mutants can be efficiently isolated by ethyl methanesulfonate (EMS) mutagenesis followed by screening in a dissecting microscope for mutants with absent or reduced trichomes. However, mutants in other genes required for trichome formation could lead to false positive results, as could mutations in the trichome rescue transgene itself, or silencing of the transgene. Therefore, the GFP signal from the fusion protein is used to verify the defect in KN1 trafficking in putative mutants and that trichome loss is not due to alternative causes. The parental trichome rescue lines display GFP signal in both the epidermis and mesophyll layers, as a result of GFP ~ GL1 ~ $KN1^C$ trafficking between the layers (Fig. 1). In contrast, GFP signal should be present in the mesophyll layer of KN1 trafficking mutants, but reduced or absent in the epidermis, due to the defective trafficking of the fusion protein (Fig. 1). After this confirmation, the mutant can be backcrossed to wild-type plants to reduce the load from mutagenesis and

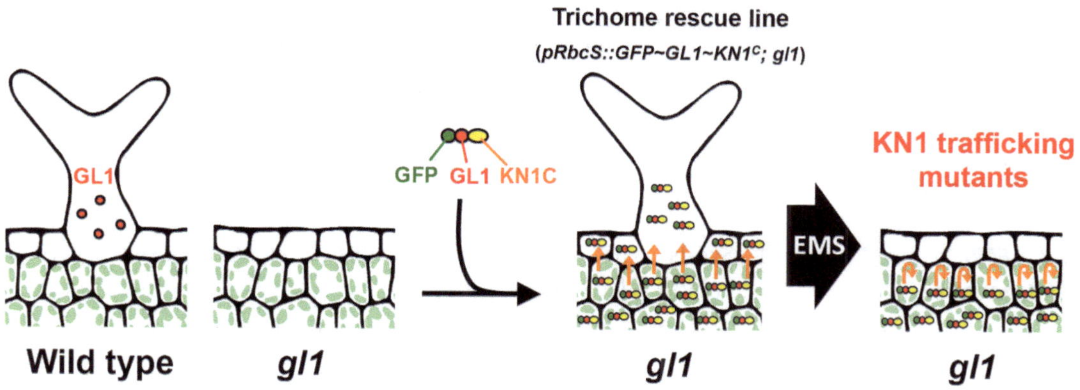

Fig. 1 The trichome rescue system enables efficient screening of KN1 trafficking mutants by an easy readout of trichome formation. In *Arabidopsis* leaves, trichomes develop from an epidermal cell by the cell-autonomous function of GL1, and thus *gl1* mutants lack trichomes. The trichome rescue system expresses the GFP ~ GL1 ~ $KN1^C$ fusion protein under a mesophyll-specific promoter (*pRbcS::GFP ~ GL1 ~ $KN1^C$*) in the *gl1* mutant background, and this fusion protein traffics to the epidermis where it rescues trichome formation. In KN1 trafficking mutants, the fusion protein cannot move to the epidermis, leading to a lack of rescue of trichome formation

confirm its genetic segregation. Subsequently, the mutated gene can be identified by a combination of genetic mapping and whole-genome sequencing to compare mutant and non-mutant genomes. Once a candidate gene is identified, it is then verified by complementing trichome rescue in the mutant background using a genomic expression construct.

An advantage of our screening system is that it can identify weak alleles of KN1 trafficking regulatory genes. This is important, because PD trafficking of transcription factors may be critical for plant development [14, 21]; therefore, null alleles of trafficking regulators may be lethal and not identified in forward screens. Indeed, a weak mutant allele of CHAPERONIN CONTAINING T-COMPLEX POLYPEPTIDE 1 (CCT) 8, a KN1 trafficking regulator isolated in our system, was defective in transport of KN1 and developed inflorescences and seeds, whereas *cct8* null alleles are lethal [18]. Thus, our system sheds light on the PD cell-to-cell signaling machinery, which is essential for plant viability. Another advantage of our system is that it could be used to study the trafficking of additional mobile signaling proteins. For example, we found that CCT8 also regulates the trafficking of TRANSPARENT TESTA GLABRA1 (TTG1), a mobile WD40 repeat protein that controls trichome spacing [18]. Thus, the KN1 trafficking regulators identified in this system may control the transport of additional signals. However, trafficking of SHORT ROOT, a mobile transcription factor that functions in ground tissue patterning in roots, was independent of CCT8 [18], suggesting that distinct machinery exists for different mobile proteins depending on the protein itself or the cellular context. Collectively, our system is a powerful tool to identify trafficking regulators of mobile proteins in plants. This chapter describes how KN1 trafficking mutants are isolated in our system and how to identify the responsible genes.

2 Materials

2.1 EMS Mutagenesis

1. Seeds of *Arabidopsis* trichome rescue line (*pRbcS::GFP ~ GL1 ~ KN1*^{C}*;gl1*).
2. 50–100 mM ethyl methanesulfonate (EMS) solution.
3. 5 M sodium hydroxide solution.
4. 0.1% agar solution.
5. Sterilized water.

2.2 Trichome Rescue Screen and GFP Rescreen

1. Murashige and Skoog (MS) medium agar plates (100 × 15 mm and 150 × 15 mm): 0.22% MS medium salts, 0.05% 2-(*N*-morpholino) ethanesulfonic acid (MES), and 0.8% agar; adjust to pH 5.7 with 1 M potassium hydroxide, and sterilize.
2. Bleach (6% sodium hypochlorite).
3. Hydrochloric acid (34–37%).
4. 0.1% agar solution.
5. Dissecting microscope (e.g., Nikon SMZ1500).
6. Confocal laser scanning upright or inverted microscope (e.g., ZEISS LSM900, upright microscope).
7. Fiji software (ImageJ; https://imagej.net/Fiji).

2.3 Whole-Genome Sequencing and Data Analysis

1. Liquid nitrogen.
2. Urea DNA extraction buffer: 7 M urea, 350 mM sodium chloride, 50 mM Tris, 20 mM ethylenediaminetetraacetic acid (EDTA), and 34 mM *N*-lauroylsarcosine.
3. Phenol/chloroform/isoamyl alcohol (25:24:1).
4. 3 M ammonium acetate solution, pH 5.2.
5. Isopropanol.
6. 70% ethanol.
7. Sterilized water.
8. Whole-genome sequencing library preparation kit (e.g., Illumina TruSeq Nano DNA Low Throughput Library Prep Kit).
9. Bioanalyzer (e.g., 2100 Bioanalyzer).
10. Next-generation sequencer (e.g., Illumina NextSeq Mid Output [2 × 75 bp]).
11. Pipeline for sequence data analysis. (e.g., MutMap+; https://genome-e.ibrc.or.jp/resource/mutmap, [22]).
12. *Arabidopsis* reference genome sequence (https://www.arabidopsis.org/, TAIR10_chr_all.fas).

3 Methods

3.1 Mutagenesis and Generating M2 Families for Trichome Rescue Screening

1. Prepare 20 ml of EMS solution (*see* **Note 1**).
2. Prepare the seeds of trichome rescue lines for mutagenesis (*see* **Note 2**).
3. Soak seeds of trichome rescue lines in the EMS solution in a disposable 50 ml plastic tube, and incubate them with rotating for 3–5 h.
4. Wash the seeds eight times with water. For each wash, incubate the seeds in 50 ml water with rotating for 1 min. For the last wash step, incubate the seeds in 50 ml water with rotating for an hour to remove any remaining EMS. Subsequently, suspend the seeds in 0.1% agar solution.
5. Incubate the tube containing mutagenized seeds for 2 days at 4 °C in the dark for stratification to synchronize germination. Then, drip the agar solution containing the seeds onto soil, and germinate them under long-day growth conditions. ~2 weeks later, transplant seedlings individually into pots with soil, and continue to grow in long-day conditions.
6. Harvest the seeds from each M1 plant separately to generate a collection of M2 families (Fig. 2).

3.2 Trichome Rescue Screening

1. Put ~100 seeds from each M2 family into separate disposable 15 ml plastic tubes, remove the tube lids, and sterilize the seeds with chlorine gas in a closed container in the hood (*see* **Note 3**).
2. Add ~7 ml of 0.1% agar solution to each tube, close the lid, and suspend the seeds by inverting tubes. Incubate these tubes for 2 days at 4 °C in the dark for stratification.
3. Pour the agar solution containing the seeds onto an MS agar plate (150 × 15 mm), and tilt the plate gently to spread the agar solution over the plate so that the seeds are evenly distributed (*see* **Note 4**).
4. Leave the plates with the lids open in a laminar flow hood to dry until the seeds do not move when the plates are tilted. Cover the plate with the lid, seal it with surgical tape, and incubate in a growth chamber under long-day conditions for 7–10 days.
5. Using a dissecting microscope, screen for plants that lack trichomes or have reduced trichome formation in the first or second true leaves, as candidates of KN1 trafficking defective mutants (Fig. 3a) (*see* **Notes 5–7**).
6. Transplant the candidate M2 mutants onto soil, grow them, and harvest the M3 seeds. Carefully number each M2 family, and keep seeds for each family that gave rise to candidate mutants (*see* **Note 8**).

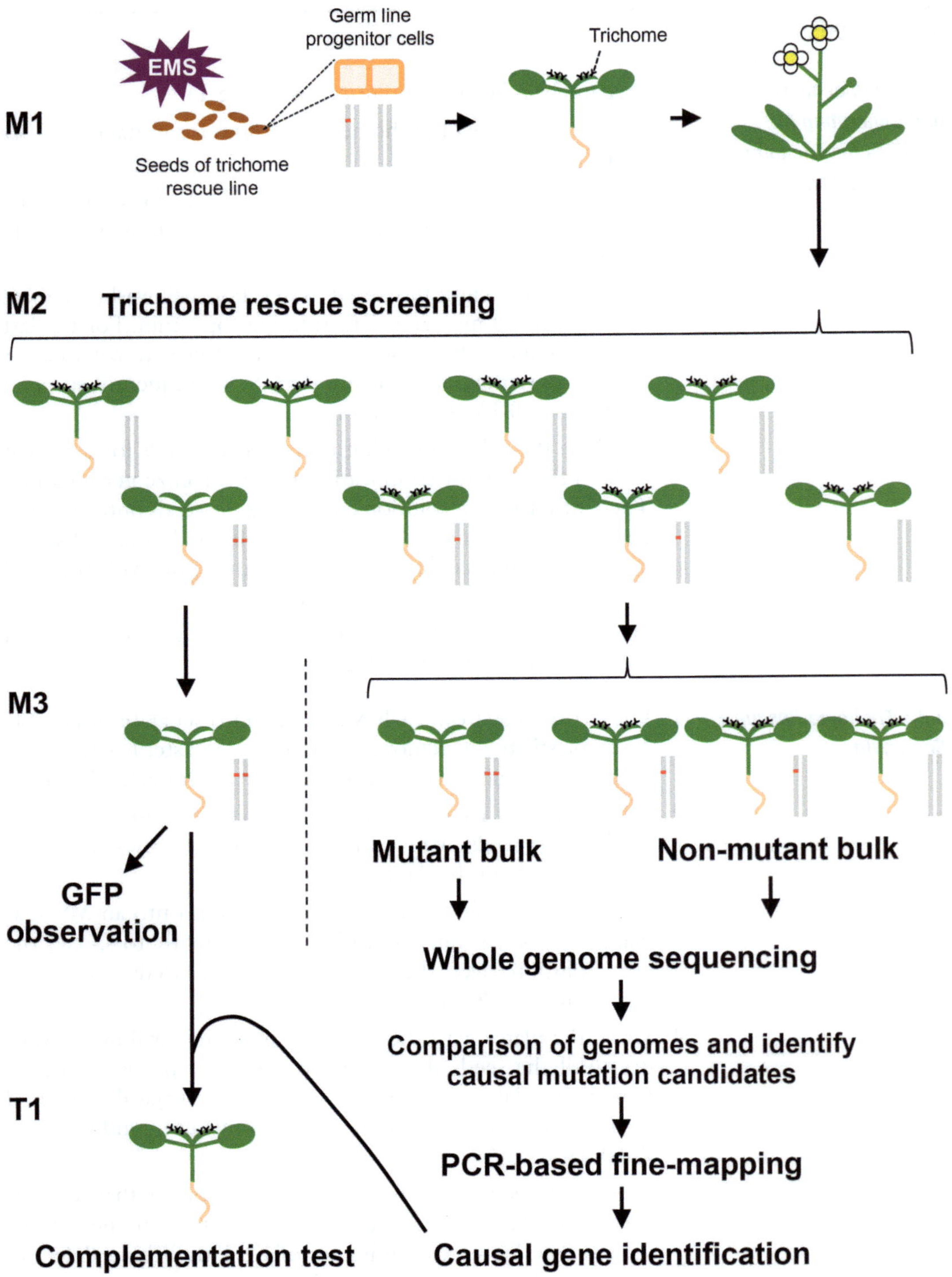

Fig. 2 Outline of the trichome rescue screening and gene identification. Seeds of trichome rescue lines (*pRbcS::GFP* ~ *GL1* ~ *KN1*^{C}; *gl1*) are mutagenized using EMS to generate M1 plants, and the M2 seeds are harvested from each M1 plant. The homozygous mutants of a recessive mutation that inhibits KN1 trafficking (red dot) segregate less than 3:1 in the M2 generation and display a trichomeless phenotype in seedling true leaves. The impaired KN1 trafficking in the mutants is verified by GFP observation in the M3 generation. The causal gene is determined by comparing genomes between mutants and non-mutants. First, the heterozygous

3.3 GFP Screening

1. Following the same procedure as in **steps 1–4** in Subheading 3.2, sow ~30 seeds of the candidate mutant M3 seeds on MS agar plates (100 × 15 mm).
2. Using confocal laser scanning microscopy, perform Z-stack imaging (e.g., pinhole = 1 airy unit [~0.6 μm section], interval = 1.5 μm) to scan the GFP signal (excitation wavelength, 488 nm; detection wavelength, 493–511 nm) and chloroplast autofluorescence (chlorophyll A, excitation wavelength, 640 nm; detection wavelength, 656–700 nm) from the adaxial epidermis to mesophyll cell layer at the apical region of the first or second true leaves (~1–2 mm long) of 7–10-day-old seedlings (*see* **Note 9**). Avoid imaging of leaf margin regions (Fig. 3b) (*see* **Note 10**).
3. Merge the GFP and chlorophyll A channels using Fiji software (*see* **Note 11**). The chloroplast autofluorescence can be detected in the GFP channel, which is distinguished from the true GFP signal by comparing it with the chlorophyll A signal (*see* **Note 12**).
4. Search for mutants that have close to normal levels of GFP ~ GL1 ~ $KN1^C$ fusion protein expression in the mesophyll cell layers, but no or very low GFP signal in the epidermis (Fig. 3c) (*see* **Note 13**).
5. Measure the GFP intensity in mesophyll cells of the non-mutant control lines (i.e., parental trichome rescue lines) and mutants, and compare them to confirm that mutant mesophyll cells have similar GFP ~ GL1 ~ $KN1^C$ expression to control lines (*see* **Note 14**).

3.4 Mapping by Whole-Genome Sequencing

Once a KN1 trafficking mutant is obtained, its heterozygotes should be segregating from the same M2 family (Fig. 2). Next, M3 bulks of mutants or non-mutant siblings are prepared for whole-genome sequencing.

1. Grow ~20 non-mutant siblings (normal trichomes) from each M2 family where a KN1 trafficking mutant was confirmed in the M3 (Subheading 3.3), and harvest M3 seeds from each of them.

Fig. 2 (continued) mutants are isolated from the M2 family from which KN1 trafficking mutants were isolated. The heterozygous mutants are confirmed by a 3:1 segregation of the homozygous mutants (trichomeless plants) in the next generation (M3), and the M3 bulks of the mutants (lacking trichomes) and non-mutants (normal trichomes) are obtained from this segregating population. Genomic DNA of each bulk is subjected to whole-genome sequencing and compared to identify candidate causal mutations. The true causal mutation is narrowed down by PCR-based fine-mapping. A genomic expression construct of the candidate gene is transformed into the M3 homozygous mutants, and the causal gene is verified by trichome rescue in the T1 plants

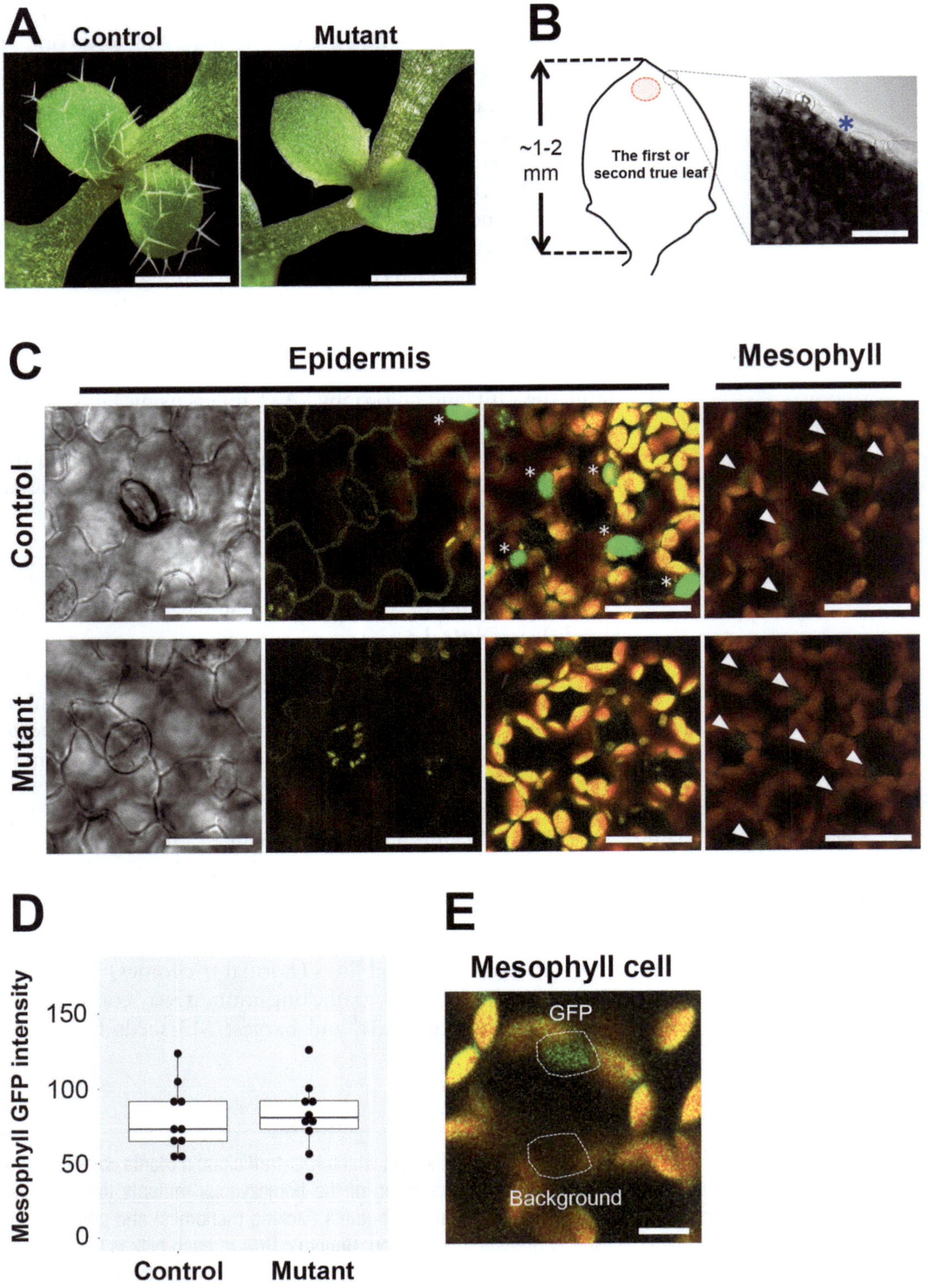

Fig. 3 KN1 trafficking mutants lack trichome rescue and are defective in KN1 protein trafficking. (**a**) The GFP ~ GL1 ~ KN1^{C} fusion protein expressed in the mesophyll layer traffics to the epidermis and rescues trichome formation in the non-mutant (control) plants, whereas defective trafficking in the mutants results in a lack of trichome development. Scale bars = 1 mm. (**b** and **c**) Defective KN1 trafficking in the mutants is

2. Grow ~20 M3 families on soil, and select the families that have ~3:1 segregation for non-mutant (normal trichomes) and mutant (lacking trichomes) (Fig. 2).
3. Collect leaves from ~100 plants each of M3 non-mutant or M3 mutant plants, and prepare two bulk DNA samples for sequencing (*see* **Notes 15** and **16**). As a reference, also prepare DNA of the parental trichome rescue line.
4. Prepare sequencing libraries of each DNA sample (non-mutant and mutant bulks and the parent trichome rescue plants), and then sequence to a depth of ~40–48-fold coverage for each sample (*see* **Note 17**).
5. Sample remaining tissues from the ~100 non-mutant or ~100 mutant plants individually, and extract DNA from each for PCR-based fine-mapping in a later step.

3.5 Sequence Data Analysis and PCR-Based Mapping

For identifying the causal mutation of a KN1 trafficking mutant, the single nucleotide polymorphisms (SNPs) in the genomes are compared between non-mutant and mutant bulks. We used the MutMap+ pipeline for this analysis [22].

1. Using MutMap+ pipeline, select the high-quality reads from the short reads (~75 bp) sequenced for each sample (non-mutant and mutant bulks and the parent trichome rescue line), align them to the *Arabidopsis* reference genome sequence to call SNPs, and exclude false-positive SNPs detected in the reference (parent trichome rescue line) genome from the following analysis. Subsequently, calculate a SNP-index for each position of all SNPs in each of mutant and non-mutant bulks (*see* **Note 18**), subtract the SNP-index of the non-mutant bulk from that of mutant bulk for each SNP to calculate Δ(SNP-index), and identify the SNPs with high Δ(SNP-index) as the candidates of the causal mutation of the KN1 trafficking mutant.

Fig. 3 (continued) verified by observing GFP fluorescence from the fusion protein in the epidermis and mesophyll layers. The apical region of a developing ~1–2 mm true leaf is used to reproducibly find defective KN1 trafficking (**b**, red circle), while the leaf margin and its neighboring cells should be avoided because of their unclear cell identities (**b**, gray circle, asterisk). Z-stack images from the epidermis to mesophyll layer indicate that the control lines have GFP signal in the cytoplasm at the periphery of epidermal cells and in nuclei of both epidermis (asterisks) and mesophyll layer cells (arrowheads). In contrast, the mutants have greatly reduced GFP signal in the epidermis, despite normal expression of the fusion protein in the mesophyll layer (arrowheads) (**c**). Scale bars = 50 μm (**b**) and 20 μm (**c**). (**d**) Quantification of nuclear GFP intensity in the mesophyll cells indicates that the mutants have similar transgene expression compared to the non-mutant control lines, suggesting that the loss or reduction of trichomes and epidermal GFP signal are not due to an overall decrease in GFP ~ GL1 ~ KN1^{C} expression in the mutants. (**e**) The net intensity of nuclear GFP is calculated by subtracting the fluorescence signal intensity in the empty region (background) from the actual measurement value of nuclear GFP (GFP). Scale bar = 5 μm

2. Design cleaved amplified polymorphic sequence (CAPS) or derived CAPS (dCAPS) primers for each SNP identified in **step 1**, Subheading 3.5, and perform PCR-based mapping using DNA extracted from individual plants in **step 5**, Subheading 3.4. A unique SNP corresponding to the causal gene in the KN1 trafficking mutant should show complete linkage.
3. To confirm the causal gene, transform a genomic clone containing promoter, coding, and 3′ regions of the identified gene (**step 2**, Subheading 3.5) into the M3 mutants (*pRbcS::GFP ~ GL1 ~ KN1*^{C}; *gl1* background) (Subheading 3.3) to test if the trichome formation is rescued in the T1 plants (*see* **Notes 19** and **20**).

4 Notes

1. Put 20 ml of water in a 50 ml disposable plastic tube, and then add EMS into the water using a filter tip (EMS should be added under the surface of the water to avoid aerosols). Then mix the solution by gently shaking the tube, being careful not to let the EMS solution stick to the tube lid. Up to 5000 seeds can be mutagenized with 20 ml of the EMS solution [23]. A concentration range of 50–100 mM EMS will introduce mutations at a high frequency while avoiding some lethality and sterility of M1 plants [23]. EMS is a dangerous mutagen. Handle with appropriate protective equipment in a fume hood. All waste plastics and solutions must be discarded into a beaker containing 1 L of 5 M sodium hydroxide solution to deactivate the EMS.
2. When screening in M2 families from individual M1 plants, a minimum of ~1870 M1 plants is required for saturation [24]. Due to space and time constraints, we aimed to produce ~500 M1 plants. First, the germination rate of the initial trichome rescue lines seeds should be tested by sowing 100 seeds on soil and growing in a growth chamber after stratification at 4 °C for 2 days to synchronize germination. After a week, count the number of seedlings, and calculate the germination rate. Based on this germination rate, calculate the number of seeds that will yield 500 M1 plants. Seeds can be lost during EMS treatment and washes, and the germination rate can decrease further by EMS treatment, so mutagenize ~1.5- to two-fold more seeds than the number of M1 plants required.
3. We generate chlorine gas by mixing 80 ml of bleach with 10 ml of hydrochloric acid (34–37%) in a beaker in a closed container in a fume hood and expose the seeds to the gas for ~5 h. After sterilization, leave the tubes containing the seeds with the lids

off in a sterile hood for several hours or overnight to release the remaining chlorine gas.

4. Do not shake the plates, as this causes the seeds to clump in the center of the plate.
5. The embryonic SAM in the mature *Arabidopsis* seed has ~ two cells that will form a lineage that differentiates into germ cells in the inflorescence [25]. Therefore, if a recessive mutation causing the KN1 trafficking defect occurs in one of these germ line progenitor cells in a mutagenized M1 seed, homozygous mutants (trichomeless plants) will segregate in the M2 generation at a ratio of less than 3:1; however, this is variable depending on the size of germ line sectors (Fig. 2). If the frequency of trichomeless plants is very different from this ratio, the defect may be due to a dominant mutation or factors unrelated to KN1 trafficking, such as *pRbcS::GFP ~ GL1 ~ KN1*^{C} transgene silencing.
6. Although dominant alleles can be found in this screening, we here focus on the isolation and analysis of recessive alleles, since we have so far successfully identified such mutants in our system.
7. The leaf margins sometimes develop trichomes even in trichomeless mutants, so they should be ignored in screening trichome number.
8. If the homozygous mutants are sterile, recover the heterozygous mutants from the seeds of the M2 family. Additionally, in Subheading 3.4, heterozygous mutants from the M2 family will be self-fertilized, and the mutant and non-mutant siblings segregated in the M3 generation will be used for mapping by whole-genome sequencing.
9. The GFP signal from the GFP ~ GL1 ~ KN1^C fusion protein is found in the nuclei of epidermal and mesophyll cells in non-mutant control lines (i.e., parental trichome rescue line) (Fig. 3c). In addition, the epidermal cells also have some GFP signal in the cytoplasm (Fig. 3c). In the immature true leaves used for screening, the nuclei of epidermal cells were often pushed to the bottom of the cell (near the interface between epidermal and mesophyll cells) by the enlarged vacuole. Therefore, it is difficult to focus on both cytoplasmic and nuclear GFP signals in a single plane image, and we suggest scanning the entire epidermal cell by Z-stack imaging.
10. For GFP screening, we recommend observing the apical region of true leaves (Fig. 3b), where we reproducibly find defects in GFP ~ GL1 ~ KN1^C trafficking in our mutants. In contrast, the leaf margins and their neighboring cells should be avoided from observation, because these cells sometimes make trichomes even in trichomeless mutants and also because leaf margins

can be made by periclinal divisions of L1 cells [26]; thus, the identities of these cells are unclear.

11. We export confocal Z-stack images in LSM or CZI format. The stack images of all channels (GFP, chlorophyll A, bright field) are integrated in these files. First, separate this integrated stack images by channel (image > color > split channels). Then, adjust the brightness and contrast of the stack images for each channel (image > adjust > Brightness/Contrast), and merge them again (image > color > merge channels).
12. The epidermal nuclei are usually located above the mesophyll chloroplasts (when using an inverted sample on an inverted microscope, they will be below). In contrast, the mesophyll nuclei are in the gap between their chloroplasts, and their size tends to be smaller than the epidermal nuclei (Fig. 3c). These observations allow us to distinguish the epidermal and mesophyll nuclei.
13. This step allows the selection of candidate KN1 trafficking defective mutants as ones that are impaired in trichome rescue and accumulate GFP ~ GL1 ~ KN1^{C} in the mesophyll layers but not in the epidermis. Non-mutant control lines (i.e., parental trichome rescue line) have GFP signal in both layers because of fusion protein trafficking, whereas KN1 trafficking mutants do not accumulate fusion protein in the epidermis, due to impaired trafficking (Figs. 1 and 3c). If both epidermis and mesophyll layers have strong GFP signal in a trichomeless mutant, this indicates that GFP ~ GL1 ~ KN1^{C} protein is trafficking normally from the mesophyll layer to the epidermis, and the mutant is not a KN1 trafficking defective mutant. These mutants may have mutations in genes such as *TRANSPARENT TESTA GLABRA* (*TTG*) associated with trichome initiation/development. *ttg* mutants have yellow seeds [27, 28], which allow us to exclude them from our candidate mutants. In contrast, lines where GFP signal is not detected in either mesophyll layer or epidermis are likely due to a mutated or silenced trichome rescue transgene.
14. The GFP intensity in the nuclei of mesophyll cells should be compared between the control lines and mutants, to confirm that the loss or reduction of trichomes and epidermal GFP signal in the mutants is not simply due to a decreased transgene expression (Fig. 3d). GFP intensity is measured using Fiji software. Set a region of interest (ROI) surrounding the nucleus as estimated by the GFP signal, and measure the mean fluorescence intensity (Analyze > Set measurements > Mean gray value, then Analyze > measure), and also measure the fluorescence signal (GFP channel) of a nearby empty region

as background (Fig. 3e). Subtract the background intensity from the GFP intensity to calculate the net intensity of GFP.

15. The optimal number of plants to use for mapping by sequencing can vary depending on the chromosomal position of the mutation, which affects recombination. In our previous mapping of KN1 trafficking mutants, whole-genome sequencing of DNA bulks from 74 non-mutant and 33 mutant plants identified 25 putative causal mutations in a ~7 Mb genomic region. In another case, DNA bulks from ~100 non-mutant or mutant plants identified 15 putative causal mutations in a ~5 Mb genomic region.
16. For DNA extraction, mix leaves from M3 plants in equal proportion for each bulk. We extract DNA using a urea-based extraction method [29]. Grind tissue in liquid nitrogen by mortar and pestle, and mix with urea extraction buffer, then add an equal volume of phenol/chloroform/isoamyl alcohol (25:24:1) in a tube, and vortex to mix well. After centrifugation (16,300 × *g* for 10 min), transfer the aqueous upper phase into another tube, and add a one-tenth volume of 3 M ammonium acetate pH 5.2 and 0.7 volumes of isopropanol to precipitate the DNA. After centrifugation (16,300 × *g* for 20 min), discard the supernatant liquid, and rinse the pellet two times using 70% ethanol. Air-dry the pellet and resuspend it in sterilized water.
17. We made DNA sequencing libraries by Illumina TruSeq Nano DNA Low Throughput Library Prep Kit and checked the quality using 2100 Bioanalyzer (the average library size was ~500 bp). Sequencing was performed by Illumina NextSeq Mid Output [2 × 75 bp, output = 16.25–19.5 Gb], and we obtained 5.4–6.5 Gb output per sample. Since the genome size of *Arabidopsis* is ~135 Mb, the sequence depth was ~40–48-fold coverage for each sample.
18. SNP-index represents the proportion of the short reads possessing nucleotides that are different from the reference sequence in a SNP position [22].
19. We perform *Arabidopsis* transformation by *Agrobacterium*-mediated floral dip [30].
20. The complementation of trichome rescue can be assayed in immature first or second true leaves of T1 seedlings.

Acknowledgments

The authors thank Drs. John Inglis, Damianos Skopelitis, Penelope Lindsay, and Thu Tran for assistance with editing the manuscript. This work is supported by the National Science Foundation

(IOS-1930101) and the Next-Generation BioGreen 21 Program (System & Synthetic Agro-biotech Center, Project No. PJ01322602), Rural Development Administration, Republic of Korea.

References

1. Li ZP, Paterlini A, Glavier M, Bayer EM (2020) Intercellular trafficking via plasmodesmata: molecular layers of complexity. Cell Mol Life Sci 78:799–816
2. Azim MF, Burch-Smith TM (2020) Organelles-nucleus-plasmodesmata signaling (ONPS): an update on its roles in plant physiology, metabolism and stress responses. Curr Opin Plant Biol 58:48–59
3. Gundu S, Tabassum N, Blilou I (2020) Moving with purpose and direction: transcription factor movement and cell fate determination revisited. Curr Opin Plant Biol 57:124–132
4. Hernández-Hernández V, Benítez M, Boudaoud A (2020) Interplay between turgor pressure and plasmodesmata during plant development. J Exp Bot 71:768–777
5. Maizel A, Markmann K, Timmermans M, Wachter A (2020) To move or not to move: roles and specificity of plant RNA mobility. Curr Opin Plant Biol 57:52–60
6. Paterlini A (2020) Uncharted routes: exploring the relevance of auxin movement via plasmodesmata. Biol Open 9:bio055541
7. Petit JD, Li ZP, Nicolas WJ, Grison MS, Bayer EM (2020) Dare to change, the dynamics behind plasmodesmata-mediated cell-to-cell communication. Curr Opin Plant Biol 53:80–89
8. Vu MH, Iswanto ABB, Lee J, Kim J-Y (2020) The role of plasmodesmata-associated receptor in plant development and environmental response. Plants 9:216
9. Kitagawa M, Jackson D (2017) Plasmodesmata-mediated cell-to-cell communication in the shoot apical meristem: how stem cells talk. Plants 6:12
10. Lucas WJ, Bouché-Pillon S, Jackson DP, Nguyen L, Baker L, Ding B, Hake S (1995) Selective trafficking of KNOTTED1 homeodomain protein and its mRNA through plasmodesmata. Science 270:1980–1983
11. Kim J-Y, Rim Y, Wang J, Jackson D (2005) A novel cell-to-cell trafficking assay indicates that the KNOX homeodomain is necessary and sufficient for intercellular protein and mRNA trafficking. Genes Dev 19:788–793
12. Kim J-Y, Yuan Z, Jackson D (2003) Developmental regulation and significance of KNOX protein trafficking in Arabidopsis. Development 130:4351–4362
13. Long JA, Moan EI, Medford JI, Barton MK (1996) A member of the KNOTTED class of homeodomain proteins encoded by the STM gene of Arabidopsis. Nature 379:66–69
14. Balkunde R, Kitagawa M, Xu XM, Wang J, Jackson D (2017) SHOOT MERISTEMLESS trafficking controls axillary meristem formation, meristem size and organ boundaries in Arabidopsis. Plant J 90:435–446
15. Fouracre JP, Poethig RS (2020) Lonely at the top? Regulation of shoot apical meristem activity by intrinsic and extrinsic factors. Curr Opin Plant Biol 58:17–24
16. Kitagawa M, Jackson D (2019) Control of meristem size. Annu Rev Plant Biol 70:269–291
17. Winter N, Kollwig G, Zhang S, Kragler F (2007) MPB2C, a microtubule-associated protein, regulates non-cell-autonomy of the homeodomain protein KNOTTED1. Plant Cell 19:3001–3018
18. Xu XM, Wang J, Xuan Z, Goldshmidt A, Borrill PG, Hariharan N, Kim JY, Jackson D (2011) Chaperonins facilitate KNOTTED1 cell-to-cell trafficking and stem cell function. Science 333:1141–1144
19. Liu L, Li C, Song S, Teo ZWN, Shen L, Wang Y, Jackson D, Yu H (2018) FTIP-dependent STM trafficking regulates shoot meristem development in Arabidopsis. Cell Rep 23:1879–1890
20. Oppenheimer DG, Herman PL, Sivakumaran S, Esch J, Marks MD (1991) A myb gene required for leaf trichome differentiation in Arabidopsis is expressed in stipules. Cell 67:483–493
21. Daum G, Medzihradszky A, Suzaki T, Lohmann JU (2014) A mechanistic framework for noncell autonomous stem cell induction in Arabidopsis. Proc Natl Acad Sci U S A 111: 14619–14624
22. Fekih R, Takagi H, Tamiru M, Abe A, Natsume S, Yaegashi H, Sharma S, Sharma S, Kanzaki H, Matsumura H (2013) MutMap+:

genetic mapping and mutant identification without crossing in rice. PLoS One 8:e68529
23. Leyser HO, Furner IJ (1993) EMS mutagenesis of Arabidopsis. Arabidopsis: the complete guide, Version 1, pp 9–10
24. Lightner J, Caspar T (1998) Seed mutagenesis of Arabidopsis. In: Martinez-Zapater J, Salinas J (eds) Arabidopsis protocols. Springer, New York
25. Page DR, Grossniklaus U (2002) The art and design of genetic screens: *Arabidopsis thaliana*. Nat Rev Genet 3:124–136
26. Dermen H (1969) Directional cell division in shoot apices. Cytologia 34:541–558
27. Johnson CS, Kolevski B, Smyth DR (2002) TRANSPARENT TESTA GLABRA2, a trichome and seed coat development gene of Arabidopsis, encodes a WRKY transcription factor. Plant Cell 14:1359–1375
28. Debeaujon I, Léon-Kloosterziel KM, Koornneef M (2000) Influence of the testa on seed dormancy, germination, and longevity in Arabidopsis. Plant Physiol 122:403–414
29. Chen J, Dellaporta S (1994) Urea-based plant DNA miniprep. In: Freeling M, Walbot V (eds) The maize handbook. Springer, New York
30. Clough SJ, Bent AF (1998) Floral dip: a simplified method for Agrobacterium-mediated transformation of *Arabidopsis thaliana*. Plant J 16:735–743

Part V

Other Approaches Contributing to Plasmodesmata Research

Chapter 28

In Vivo Visualization of Mobile mRNA Particles in Plants Using BglG

Eduardo J. Peña and Manfred Heinlein

Abstract

Cells have developed mechanisms for cytoplasmic RNA transport and localization that participate in the regulation and subcellular localization of protein synthesis. In addition, plants can exchange RNA molecules between cells through plasmodesmata and to distant tissues in the phloem. These mechanisms are hijacked by RNA viruses to establish their replication complexes and to disseminate their genomes throughout the plant organism with the help of virus-encoded movement proteins (MP). Live imaging of RNA molecules is a fundamental approach to understand the regulation and molecular basis of these processes. The most widely used experimental systems for the in vivo visualization of genetically encoded RNA molecules are based on fluorescently tagged RNA binding proteins that bind to specific motifs inserted into the RNA, thus allowing the tracking of the specific RNA molecule by fluorescent microscopy. Recently, we developed the use of the *E. coli* RNA binding protein BglG for the imaging of RNAs tagged with BglG-binding sites in planta. We describe here the detailed method by which we use this in vivo RNA tagging system for the real-time imaging of *Tobacco mosaic virus* (TMV) *MP* mRNA.

Key words RNA localization, RNA transport, RNA visualization, RNA labeling, RNA imaging, BglG, *Tobacco mosaic virus* (TMV), Movement protein (MP)

1 Introduction

The targeting of mRNA molecules to specific subcellular localizations is a widespread cellular mechanism to spatially control protein synthesis [1, 2]. Such localized mRNA sorting and protein synthesis occurs in a wide range of cell types and organisms including mammals, plants, yeasts, and even bacteria [3, 4]. Genome-wide transcriptome analyses have demonstrated that RNA localization is highly prevalent rather than exceptional [5, 6]. During transcription, maturation, and nuclear export, RNA structural elements called "zipcodes" are sequentially exposed and recognized by RNA-binding proteins (RBPs) that assemble the RNA into ribonucleoprotein complexes (RNPs) termed "RNA granules." RNA granules may contain single mRNA molecules [7] that are

Yoselin Benitez-Alfonso and Manfred Heinlein (eds.), *Plasmodesmata: Methods and Protocols*,
Methods in Molecular Biology, vol. 2457, https://doi.org/10.1007/978-1-0716-2132-5_28,

transported in a translationally repressed state by molecular motors along elements of the cytoskeleton [8, 9]. By determining the exact place at where a protein is synthesized, RNA localization influences the establishment of cell polarity and morphogenesis, asymmetric cell division, embryonic patterning during development, and cell motility [10]. In accordance to its prevalent role, the dysregulation of the RNA localization machinery is implicated in developmental diseases and genetic disorders [11, 12].

In addition to localize RNAs to specific subcellular compartments, plants transport RNA molecules between cells through plasmodesmata (PD) and to distant tissues via the phloem [13]. Intercellular and long-distance trafficking of mRNA molecules in plants plays important roles in the regulation of plant development, growth, and adaptation to stresses [14]. Although thousands of mobile mRNAs were identified by transcriptome analysis of grafted tissues in *Arabidopsis*, cucumber, and grapevine [15–17], mechanisms governing the selective targeting of transported mRNAs to PD, between cells and into the phloem stream, remain difficult to address.

As obligatory parasites, viruses depend on the host cell machinery to cause infection. Plant RNA viruses must interact with cellular mechanisms of RNA transport to spread their genomes between cells and, therefore, represent important probes for studying these mechanisms [18, 19]. Plant viruses encode specialized movement proteins (MPs) that bind RNA, localize to PD, modify PD regulation, and facilitate the cell-to-cell dissemination of the infection [19, 20]. *Tobacco mosaic virus* (TMV) is one of the most studied viruses [21] and has been the pioneer in studies related to virus movement through PD [22, 23]. The positive-sense RNA genome of this virus encodes two replicase subunits (126 and 183 kDa proteins) and also a MP (30 kDa) and a coat protein (CP, 17.5 kDa) produced from subgenomic RNAs upon virus replication [24]. Cell-to-cell movement is independent of the CP [25] during TMV infection; thus, this is an excellent model to study intra- and intercellular transport of RNPs in plants [19].

The use of TMV derivatives that express the MP fused to fluorescent reporters led to the discovery that this protein interacts with elements of the cortical microtubule (MT) and endoplasmic reticulum (ER)/actin networks [26–28]. The MP binds to the ER membrane surface [29] and the interaction of MP with MTs may occur through direct contact with the MT lattice [30]. In cells at the front of the infection, the ability of MP to interact with MTs facilitates the formation of distinct, ER-associated MP-containing particles that are anchored to cortical ER sites coinciding with underlying MTs (cortical MT-associated ER sites, cMERs [31]), and are tightly associated with MP function in virus movement [32–36]. At later stages when the MP is no longer active in virus movement and extracted from the ER by the CDC48 ATPase [37], the protein may accumulate to high levels along the length of the MTs before it is finally degraded [37–39]. While MP forms RNA particles with viral RNA and other viral proteins during infection [40], similar particles are also observed upon ectopic expression of MP and were shown to contain *MP* mRNA [41]. These mRNA-containing MP particles show stop-and-go movement along the ER-actin network and pause their trajectories at cMERs [31, 41]. Likewise, the MP targets PD and also interacts with MT independent of infection [42, 43].

Although RNA localization is a rational strategy for the spatio-temporal regulation of protein translation, the mechanisms through which RNA cis-elements and associated proteins regulate RNA trafficking and localization remain poorly understood. Methods to study the subcellular localization of RNA molecules are under continuous development since, in contrast to proteins, RNAs cannot be genetically encoded as fluorescent molecules. Initial evidence of uneven RNA distribution was obtained by in situ hybridization of *actin* mRNA on fixed ascidian eggs and early embryos [44]. Since then, new probe design as well as improved fluorophores and microscopy technologies allowed to dramatically increase sensitivity and spatial resolution. It is now possible to detect differently labeled mRNA molecules simultaneously or to map the subcellular locations of mRNAs in a transcriptome-wide scale [45–47]. Unlike in situ studies applied to chemically fixed biological specimens, in vivo studies of RNA localization in live cells and tissues rely on the mechanical delivery of in vitro synthesized, fluorescently labeled RNA using transfection or microinjection [48] or on labeling genetically encoded RNAs through binding of chemical ligands or florescence-tagged RNA-binding proteins (RBPs). RBPs like the Pumilio family of translational regulator proteins or a nuclease-inactivated CRISPR-Cas9 can be adapted to recognize and bind virtually any RNA sequence of interest [49, 50]. Other methods using chemical ligands (e.g., fluorogens) or RBPs depend on binding to specific RNA elements transcriptionally fused to the RNA under study [51–57]. One of the most universally applied strategies for the in vivo visualization of RNA molecules is the MS2 method [53, 58]. This method is based on the high-affinity binding of the bacteriophage MS2 coat protein (MS2-CP) to its cognate origin of assembly (OAS) and, thus, to an OAS-derived RNA structured element called MS2 stem loop (MS2-SL). The RNA under investigation is transcriptionally fused to a tandem array of the MS2-SL and co-expressed with MS2-CP translationally fused to a fluorescent reporter. To reduce the cytoplasmic background, the MS2-CP is targeted to the nucleus by the addition of a nuclear localization signal (NLS). Upon transcription, the tagged mRNA is decorated with a MS2-CP dimer at each transcribed MS2-SL, thus forming a fluorescent ribonucleoprotein (RNP) trackable by microscopy [53].

Attempts to visualize the TMV RNA genome by adding MS2-SLs to its genome interfered with the infectivity of this virus. However, the MP of this virus binds RNA in a sequence-nonspecific manner [59] and, as explained above, forms RNA particles also with its own mRNA upon ectopic expression [41, 60]. This allowed us to study *MP* mRNA transport and localization upon *Agrobacterium*-mediated transient expression of MP. Using the MS2 method, we showed that MP particles contain *MP* mRNA and that the RNA also accumulates with MP at PD [41, 60, 61]. In attempts to increase resolution, we recently compared the MS2 method with two other RNA labeling technologies and found that the application of the BglG method [55] led to improved mRNA imaging results [62]. Analogous to the MS2 method, the BglG method involves the addition of a tandem array of BglG-SL, a 29-nt-long RNA structural element, to the RNA under study. Each BglG-SL is recognized by a dimer of the 57-amino acid-long N-terminal portion (6.3 kDa) of the BglG transcriptional anti-terminator protein harboring the BglG RNA-binding domain. In its original design, the method involves the co-expression of the SL-tagged RNA together with the BglG peptide as a fluorescent fusion protein targeted to the nucleus [55]. However, in adapting the system to visualize *MP* mRNA, we removed the nuclear localization signal to increase the amount of BglG protein in the cytoplasm. This increased RNA labeling sensitivity and allowed us to achieve unprecedented resolution in *MP* mRNA detection. The mRNA was found to colocalize with MP in mobile MP particles and to be present at PD at much lower levels. Moreover, *MP* mRNA was shown to be absent from MP accumulating along MTs. In addition, the method allowed the detection of cell-to-cell transport of BglG-containing RNPs [62].

We describe here a detailed protocol for the application of the BglG system for the in vivo visualization and imaging of mRNA as developed for *MP*. The protocol includes the procedure for the addition of BglG-SLs to the RNA molecule of interest, the use of *Agrobacterium*-mediated transient expression of SL-tagged RNA and fluorescent protein-tagged BglG in *Nicotiana benthamiana* leaves, and the equipment and conditions to image and analyze the localization and trafficking of the fluorescently labeled RNA molecules in vivo.

2 Materials

2.1 DNA Cloning

1. PCR thermal cycler.
2. High-fidelity DNA polymerase for PCR.
3. Agarose gel electrophoresis system.
4. Reagents for DNA purification from agarose gels.

5. *E. coli* cells for standard DNA cloning (e.g., Top10 cells), competent for electro-transformation.
6. Electroporator system (e.g., Gene Pulser Xcell) and electroporation cuvettes.
7. LB liquid medium: 10 g/L Bacto tryptone, 5 g/L yeast extract, 85.6 mM NaCl, pH 7.0. Sterilize by autoclaving.
8. Sterile petri dishes.
9. LB-agar plates: Add 15 g Bacto agar to 1.0 L of LB liquid medium, sterilize by autoclaving, and pour plates while the solution is still warm and not solidified.
10. Static incubator for agar plates at 37 °C.
11. Shaker incubator for liquid cultures at 180–200 revolutions per minute (rpm) at 37 °C, compatible with 10–15 mL tubes.
12. Sterile 10–15 mL culture tubes (e.g., polypropylene screw cap tubes).
13. Antibiotics, stock solutions at 1000× concentration. Dissolve and sterilize by filtration through 0.22 μm pore membranes.
    (a) Zeocin: 20 mg/mL (in water).
    (b) Kanamycin: 50 mg/mL (in water).
    (c) Spectinomycin: 50 mg/mL (in water).
14. Plasmid DNA isolation kit.
15. Benchtop centrifuge for 1.5 mL tubes and providing 15,000 × *g*.
16. Gateway® BP Clonase™ mix and Gateway® LR Clonase™ mix (*see* **Note 1**).
17. Gateway® pDonor vector (e.g., pDONR®/Zeo).
18. Binary Gateway® destination vectors pMDC32 [63] and pMDC32-BglG-SL [62] (*see* **Note 2**).
19. Oligonucleotides (5′-3′):
    (a) M13fw: tgtaaaacgacggccagt.
    (b) M13rev: caggaaacagctatgac.

2.2 Transient Expression Assays

1. *Agrobacterium tumefaciens* strain GV3101, competent for electroporation.
2. Incubator for agar plates at 28 °C.
3. Shaker incubator for liquid cultures at 180–200 rpm at 28 °C, compatible with 50 mL tubes.
4. Sterile 50 mL culture tubes (e.g., 50 mL polypropylene screw cap tubes).
5. Antibiotics, stock solutions at 1000× concentration. Dissolve properly and sterilize by 0.22 μm pore membrane filtration.

(a) Rifampicin: 50 mg/mL (in DMSO).
(b) Gentamicin: 50 mg/mL (in water).

6. Spectrophotometer (e.g., SmartSpec Plus Spectrophotometer).
7. Spectrophotometer cuvettes.
8. Benchtop centrifuge for 50 mL tubes and providing 5000 × *g*.
9. Sterile distilled water.
10. Greenhouse or growth chambers for *N. benthamiana* plants with 16 h/8 h and 24 °C/22 °C day/night illumination and temperature conditions, respectively.
11. *N. benthamiana* plants, 4–5 weeks old with three to four fully expanded leaves.
12. pH 7-BglG:mRFP binary vector, for the expression of the RBP BglG fused to mRFP [62] (*see* **Note 3**).
13. 1.0–2.5 mL syringes without needle.
14. Hypodermic needles (e.g., 0.5-mm-diameter orange NN-2516R needles).

2.3 Fluorescence Microscopy Imaging

1. Epifluorescence, spinning disk, or laser scanning confocal microscope equipped with specific illumination and filters for GFP and RFP fluorescence, with low-magnification objective (e.g., 10× or 20×) and high-magnification, high-optical numerical aperture objective (e.g., 60× or 100×) and a sensitive and fast detection system compatible with high sample rates during time-lapse acquisition.
2. Slides and coverslips for microscopy (thickness compatible with available objectives).
3. Cork borer (5–10 mm diameter, for excision of samples from leaves; leaf disks).
4. Scotch tape.
5. Sterile distilled water.
6. Vacuum pump and vacuum desiccator.
7. Microscope image acquisition and analysis software (e.g., ImageJ, https://imagej.nih.gov/ij/).

3 Methods

3.1 Plasmid Constructs

RNA visualization with BglG depends on the co-expression of a BglG-SL-tagged RNA of interest with the fluorescent protein-fused BglG RBP in the same cells. We used BglG without nuclear localization signal, by which *MP* mRNA could be detected with high sensitivity [62] (*see* **Note 4**). BglG localized to MP particles when the *MP* mRNA was tagged with BglG-SLs. However, when

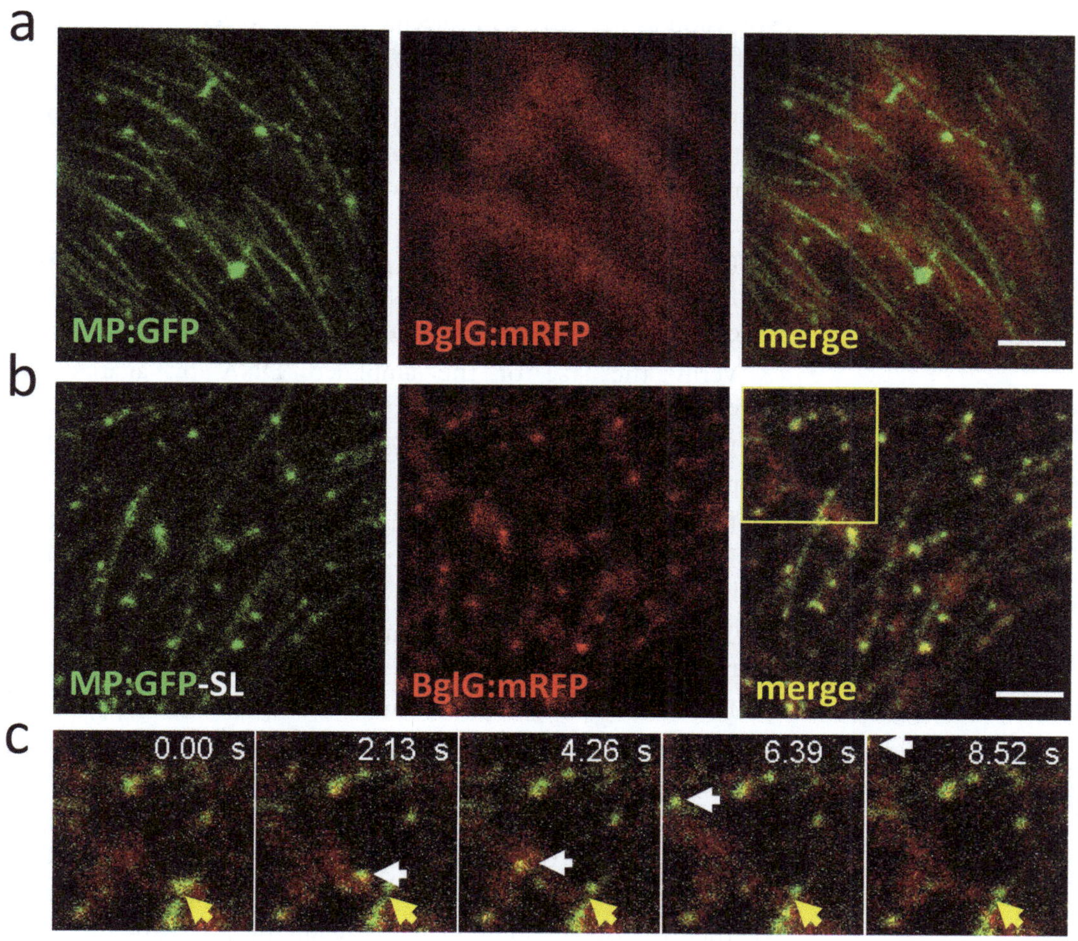

Fig. 1 Detection of *MP:GFP* mRNA in MP:GFP particles using the BglG system. Cortical views of *N. benthamiana* epidermal cells are shown. (**a**) Cell expressing MP:GFP (green channel, left) from non-tagged mRNA together with BglG:mRFP (red channel, center). An image showing the merged green and red channels is shown on the right. (**b**) Cell expressing MP:GFP from SL-tagged mRNA (MP:GFP-SL, green channel, left) together with BglG:mRFP (red channel, center). An image showing the merged green and red channels is shown on the right. *MP:GFP* mRNA colocalizes with MP:GFP in distinct particles. BglG-mRFP labels the *MP:GFP* mRNA only if tagged with SLs, thus demonstrating specific *MP:GFP* mRNA detection. Note that *MP:GFP* mRNA colocalizes only with MP:GFP in the particles but not with MP:GFP localizing to filaments (MTs). (**c**) The *MP:GFP* mRNA-containing MP:GFP particles are mobile. Time-lapse imaging of framed area of merged channel shown in (**b**). Yellow signal indicates colocalization of BglG:mRFP-labeled *MP:GFP* mRNA with MP: GFP. Arrows highlight the movement of a specific particle. Yellow arrows indicate the initial particle position seen in the first frame, and white arrows indicate the particle positions in subsequent frames. Images are acquired at 30–36 h post-agroinfiltration. *s* seconds. Scale bar, 5 μm

expressed together with non-tagged *MP* mRNA, BglG showed a diffuse nucleocytoplasmic distribution without localizing to the MP particles [62] (Fig. 1). Here, we describe a detailed procedure to obtain an entry vector containing the RNA of interest and the further recombination reactions to obtain binary vectors for its expression as free mRNA molecule or as a molecule transcriptionally fused with an array of BglG-SLs.

3.2 Cloning the RNA of Interest

1. Design oligonucleotides for the PCR amplification of the DNA fragment encoding your RNA of interest. Add *attB* recombination sequences for the Gateway® cloning by BP™ recombination (*see* **Note 1**).
2. Using a high-fidelity DNA polymerase and the appropriate DNA source (DNA or cDNA), amplify the DNA fragment encoding the RNA of your interest.
3. Verify the size of the amplified DNA by agarose gel electrophoresis.
4. Purify the DNA from the gel using the agarose gel DNA purification kit.
5. Insert the purified DNA fragment into pDONR/Zeo by a BP™ recombination reaction according to manufacturer's instructions.
6. Transform *E. coli*-competent cells with 1.0–3.0 μL of the recombination reaction mix.
7. Plate the transformed cells on LB-agar plates with appropriate antibiotics (20 mg/L Zeocin for pDONR®/Zeo).
8. Incubate overnight at 37 °C.
9. Grow liquid cultures from four to five isolated colonies in 15 mL sterile tubes containing 5 mL of LB liquid medium with antibiotics (20 mg/L Zeocin for pDONR®/Zeo).
10. Incubate overnight at 37 °C with agitation by 180–200 rpm.
11. Harvest bacteria by centrifugation, and isolate the plasmid DNA using the plasmid DNA isolation kit and following the manufacturer's instructions.
12. Confirm the correct DNA sequence by Sanger sequencing using M13fw and M13rev as primers, and select one of the verified plasmids, now termed "pZeo-RNA," to proceed with the following steps.
13. Perform two separate LR™ recombination reactions. (a) Recombine pZeo-RNA and pMDC32-BglG-SLs to obtain "pMDC32-RNA-BglG-SLs," a vector for the expression of your RNA transcriptionally fused to the BglG-SLs. (b) Recombine pZeo-RNA with pMDC32, to obtain "pMDC32-RNA," a vector for the expression of your RNA free of additional sequences. Follow LR™ manufacturer's instructions.
14. Transform independent aliquots of *E. coli*-competent cells with 1.0–3.0 μL of (a) pMDC32-RNA and (b) pMDC32-RNA-BglG-SLs.

15. Plate the transformed cells on LB-agar plates with appropriate antibiotics (50 mg/L kanamycin).
16. Incubate overnight at 37 °C.
17. Start liquid cultures from four to five isolated colonies in 15 mL sterile tubes containing 5 mL of LB liquid medium with antibiotics (50 mg/mL kanamycin).
18. Incubate overnight at 37 °C with agitation (180–200 rpm).
19. Harvest bacteria by centrifugation, and isolate plasmid DNA using the plasmid DNA isolation kit by following the manufacturer's instructions.
20. Confirm the insertion of the anticipated DNA fragment by DNA sequencing using internal primers. Select one of the verified clones for *Agrobacterium*-mediated transient expression as described in Subheading 3.3.

3.3 Agrobacterium-Mediated Transient Expression in Plant Leaf Tissues

Agrobacterium-mediated cell transformation is a fast and reliable method for the transient expression of high levels of proteins of interest in cells of intact leaves [64] and in vitro cultured cells [65, 66]. The combined and simultaneous expression of more than one protein can be achieved by combining *Agrobacterium* cultures harboring two different expression vectors. The vectors generated according to previous sections of this protocol and pH 7-BglG:mRFP [62] are binary vectors suitable for transformation of *Agrobacteria* and their use for transient protein expression upon agroinfiltration of plant tissues. We applied this method for the co-expression of BglG-SL-tagged or non-tagged RNA with fluorescent protein-fused BglG in the same cells.

1. Transform independent aliquots of electrocompetent *A. tumefaciens* cells with (a) pMDC32-RNA, (b) pMDC32-RNA-BglG-SLs, and (c) pH 7-BglG:mRFP.
2. Plate the transformed cells on LB-agar plates with appropriate antibiotics (rifampicin, gentamicin, and kanamycin for pMDC32-based vectors and rifampicin, gentamicin, and spectinomycin for pH 7-BglG:mRFP, all at 50 mg/L).
3. Incubate for 48 h at 28 °C.
4. Grow liquid cultures from isolated colonies from each plate in 50 mL sterile tubes containing 5 mL of LB liquid medium with antibiotics (rifampicin, gentamicin, and kanamycin for pMDC32-based vectors and rifampicin, gentamicin, and spectinomycin for pH 7-BglG:mRFP, all at 50 mg/L).
5. Incubate overnight at 28 °C with agitation (180–200 rpm).
6. Harvest bacteria by centrifugation (5 min at 5000 × *g*).
7. Discard the supernatant and gently resuspend the bacteria in sterile distilled water.

8. Determine the optical density (OD) of the suspension at 600 nm.
9. Mix water-resuspended cultures of (a) agrobacteria transformed with pH 7-BglG:mRFP and agrobacteria transformed with pMDC32-RNA-BglG-SLs for the expression BglG:mRFP together with BglG-SL-tagged RNA and (b) agrobacteria transformed with pH 7-BglG:mRFP and agrobacteria transformed with pMDC32-RNA for the expression BglG:mRFP together with non-tagged RNA. The concentration of each *Agrobacterium* strain in the infiltration mix needs to be empirically determined (*see* **Note 5**).
10. Select 4–5-week-old *N. benthamiana* plants with well-expanded and symmetric leaves. Using a sterile, hypodermic needle, apply a single hole on the abaxial side of each half leaf (choose the largest region without a major vein). Fill a 1.0–2.5 mL syringe without needle with the agrobacteria solution, place the syringe onto the small hole, and infiltrate the leaf half by applying gentle pressure to the plunger and by providing counterpressure by holding the leaf from the adaxial side with one finger. Infiltrate the agroinfiltration mixture with pMDC32-RNA into one half and the agroinfiltration mixture with pMDC32-RNA-BglG-SLs into the other half of the leaf.
11. Maintain the agroinfiltrated plants under normal growing conditions before analysis (*see* **Note 6**).

3.4 RNA Visualization by Fluorescence Microscopy

We adapted the BglG method for its application in plants and optimized the visualization of transiently expressed *MP:GFP* mRNA, i.e., the MP of TMV fused to GFP [62]. Figure 1 shows the localizations of co-expressed MP:GFP and BglG:mRFP in the cortical cytoplasm of epidermal cells. Whereas MP:GFP localizes to filaments (microtubules) and mobile particles, only a diffuse RFP signal is observed when MP:GFP is expressed from non-tagged mRNA (Fig. 1a). However, when MP:GFP is expressed from RNA tagged with BglG-SLs, the RFP signal localizes to MP:GFP in the distinct particles, indicating that the MP:GFP particles are RNA particles that also contain *MP:GFP* mRNA (Fig. 1b). Analysis of time-lapse movies provides information about the lateral displacement, the velocities, and the trajectories of the individual RNA particles (Fig. 1c).

1. Use a sharpened cork borer to excise leaf disk samples from the agroinfiltrated leaf areas.
2. To prepare the analysis, place leaf disks individually with their adaxial sides down onto microscope slides, and cover them with coverslips. Fix the coverslips to the slide with scotch tape.
3. Fill the airspace between slide and coverslip with sterile distilled water.

4. To replace the air present in intercellular spaces of the leaf tissue with water, place the sample into a vacuum desiccator, and evacuate the air until the pressure reaches −0.8 to −1.0 bar, and then slowly release the pressure to allow water entry. Upon water penetration, the leaf tissue assumes a darker green color.
5. Use a low-magnification objective of the microscope and appropriate filters for RFP imaging to examine tissue integrity and global expression levels of BglG:mRFP in the samples. Moreover, if the RNA under study encodes a fluorescent protein, the levels of its fluorescence seen with appropriate filters should be similar irrespective whether the RNA is tagged with BglG-SLs or not (*see* **Note 7**).
6. Use a high-magnification objective to explore the subcellular localization and accumulation pattern of BglG:mRFP in the abaxial pavement cells. Focus on the cytoplasm just beneath the plasma membrane facing the leaf surface. Here, the cortical ER and cytoskeleton are located. Also focus on central views of the cell, where the nucleus and the lateral cell walls are visible. The lateral cell walls have PD through which adjacent pavement cells are connected. Moreover, a narrow rim of cytoplasm and the plasma membrane should be seen. Acquire high-resolution images and time-lapse movies at high sampling rates from these cell areas in the samples expressing BglG:mRFP together with BglG-SLs tagged mRNA and in the samples expressing BglG: mRFP together with non-tagged mRNA as negative control. In samples with the non-tagged RNA, the red fluorescent BglG:mRFP signal should be homogenously distributed in the cytoplasm. However, in samples with the BglG-tagged RNA, the red fluorescent BglG:mRFP should be localized to discrete and mobile fluorescent spots, which are indicative of the binding of BglG:mRFP to the RNA, thus marking the location and movements of the tagged RNA. Observations should finish within 30 min after leaf disks were excised.
7. Use specialized software for image processing (e.g., ImageJ) to analyze the subcellular localization of the labeled RNA particles, their trajectories, and the velocity of their movements (*see* **Note 8**).

4 Notes

1. Gateway® technology allows the rapid and precise exchange of DNA fragments based on recombination reactions. While a BP Clonase™ reaction is used to create an entry vector from an *attB*-containing PCR product into a pDONR™ vector, the LR Clonase™ reaction is used to transfer the cloned fragment into

a "destination" vector to obtain the desired expression vector. Instructions for primer design and BP cloning into pDONR™ vectors are provided by the Gateway cloning systems provider (http://tools.thermofisher.com/content/sfs/manuals/gatewayman.pdf).

2. pMDC32-BglG-SL is a pMDC32 [63]-based vector that contains a tandem array of 18 copies of the 29-nt-long BglG-SL inserted between the Gateway® cassette and the polyadenylation signal [62]. The vector is available upon request.
3. This vector is available upon request.
4. In adapting BglG, we compared this method with the MS2 and λN RNA labeling methods [62]. All three methods were used with and without a NLS fused to the RBP. Although NLS-tagged versions of MS2 and λN have been used successfully in plants, the NLS-free version of BglG provided the best results for the in vivo imaging of *MP* mRNA. The imaging of other mRNA molecules or the use of other plant systems may depend on specific conditions for RNA imaging that should be optimized in each case. The identification of a compatible RNA-tagging system with and without NLS, the optimization of RNA and RBP expression levels, and the availability of appropriate imaging conditions are fundamental issues for achieving the successful visualization of the RNA of interest.
5. *Agrobacterium*-mediated transient expression in *N. benthamiana* leaves allows the fast accumulation of high levels of proteins since several bacteria can infect a single cell and thus transfer multiple T-DNA copies that can be simultaneously transcribed and subsequently translated. By varying the concentration of the tissue-infiltrated bacterial solution, the accumulation level of the desired expressed product can be adjusted. The optimal concentration of the transformed bacteria in the mixture for agroinfiltration must be empirically determined. For imaging of *MP* mRNA, the optical density (OD) of bacterial cultures were $OD_{600} = 0.5$ for the expression of *MP* mRNA (with or without BglG-SL tags) and $OD_{600} = 0.05–0.1$ for the expression of BglG:mRFP.

   Co-expression of RNA and RBP can also be achieved by using a single *Agrobacterium* culture sequential transformed with the respective plasmids and kept under selection using compatible antibiotics. However, while this approach can ensure that all transformed cells receive both T-DNAs, this approach has the disadvantage that the expression levels of RNA and RBP cannot be independently adjusted.
6. Upon agroinfiltration of *N. benthamiana* leaves, expression levels reach a maximum after 3–4 days and then slowly decline. Low or declining expression levels are in part due to the

induction of posttranscriptional gene silencing, which can be avoided by co-expression of a viral RNA silencing suppressor. To achieve sustained expression over longer time, we complement the agroinfiltration mix with agrobacteria ($OD_{600} = 0.2$) carrying a plasmid for the expression of the *Tomato bushy stunt virus* suppressor of RNA silencing P19 [67]. Enhanced protein expression can also be achieved by inducing the *Agrobacterium* virulence (*vir*) genes. To induce these genes, incubate the bacteria in infiltration medium containing acetosyringone (10 mM MES pH 5.6, 10 mM $NaCl_2$, 150 μM acetosyringone) for 2 h prior to tissue infiltration.

7. To determine the specificity of RNA labeling, the subcellular localization of the fluorescent RBP needs to be determined in the presence and absence of BglG-SL tags in the RNA. However, it is important that the presence or absence of the RNA tags does not influence the expression level of the RNA of interest or that of the RBP. The level of RNA expression can be monitored by using RNA encoding a fluorescent protein. In our experiments, we used RNA that encodes GFP-tagged MP (MP:GFP) [62].
8. To determine the trajectories and movement velocities of RNA particles, a high image sampling rate is needed. This can be achieved by reducing image size or resolution.

Acknowledgments

The authors thank the ANR (grant ANR-08-BLAN-0244–02 to MH), CONICET (grant PIP 2015-0847 to EP), and CNRS-CONICET (grant PICS 2014 to MH and EP) for their support.

References

1. St Johnston D (1995) The intracellular localization of messenger RNAs. Cell 81:161–170
2. Das S, Vera M, Gandin V, Singer RH, Tutucci E (2021) Intracellular mRNA transport and localized translation. Nat Rev Mol Cell Biol 7: 483–504
3. Eliscovich C, Singer RH (2017) RNP transport in cell biology: the long and winding road. Curr Opin Cell Biol 45:38–46
4. Nevo-Dinur K, Nussbaum-Shochat A, Ben-Yehuda S, Amster-Choder O (2011) Translation-independent localization of mRNA in E. coli. Science 331:1081–1084
5. Kannaiah S, Livny J, Amster-Choder O (2019) Spatiotemporal organization of the E. coli transcriptome: translation independence and engagement in regulation. Mol Cell 76: 574–589.e7
6. Lécuyer E, Yoshida H, Parthasarathy N, Alm C, Babak T, Cerovina T, Hughes TR, Tomancak P, Krause HM (2007) Global analysis of mRNA localization reveals a prominent role in organizing cellular architecture and function. Cell 131:174–187
7. Das S, Singer RH, Yoon YJ (2019) The travels of mRNAs in neurons: do they know where they are going? Curr Opin Neurobiol 57: 110–116
8. Buxbaum AR, Haimovich G, Singer RH (2015) In the right place at the right time: visualizing and understanding mRNA localization. Nat Rev Mol Cell Biol 16:95–109

9. Eliscovich C, Buxbaum AR, Katz ZB, Singer RH (2013) mRNA on the move: the road to its biological destiny. J Biol Chem 288: 20361–20368
10. Katz ZB, Wells AL, Park HY, Wu B, Shenoy SM, Singer RH (2012) β-actin mRNA compartmentalization enhances focal adhesion stability and directs cell migration. Genes Dev 26: 1885–1890
11. Bovaird S, Patel D, Padilla JA, Lécuyer E (2018) Biological functions, regulatory mechanisms, and disease relevance of RNA localization pathways. FEBS Lett 592: 2948–2972
12. Wang ET, Taliaferro JM, Lee JA, Sudhakaran IP, Rossoll W, Gross C, Moss KR, Bassell GJ (2016) Dysregulation of mRNA localization and translation in genetic disease. J Neurosci 36:11418–11426
13. Tian L, Chou HL, Fukuda M, Kumamaru T, Okita TW (2020) mRNA localization in plant cells. Plant Physiol 182:97–109
14. Ham B-K, Lucas WJ (2017) Phloem-mobile RNAs as systemic signaling agents. Annu Rev Plant Biol 68:173–195
15. Thieme CJ, Rojas-Triana M, Stecyk E, Schudoma C, Zhang W, Yang L, Minãmbres M, Walther D, Schulze WX, Paz-Ares J, Scheible WR, Kragler F (2015) Endogenous Arabidopsis messenger RNAs transported to distant tissues. Nat Plants 1:1–8
16. Zhang Z, Zheng Y, Ham B-K, Chen J, Yoshida A, Kochian LV, Fei Z, Lucas WJ (2016) Vascular-mediated signalling involved in early phosphate stress response in plants. Nat Plants 2:16033
17. Yang Y, Mao L, Jittayasothorn Y, Kang Y, Jiao C, Fei Z (2015) Messenger RNA exchange between scions and rootstocks in grafted grapevines. BMC Plant Biol 15:251
18. Heinlein M (2015) Plasmodesmata: channels for viruses on the move. Methods Mol Biol 1217:25–52
19. Peña EJ, Heinlein M (2012) RNA transport during TMV cell-to-cell movement. Front Plant Sci 3:1–10
20. Heinlein M (2015) Plant virus replication and movement. Virology 479–480:657–671
21. Scholthof K-BG, Adkins S, Czosnek H, Palukaitis P, Jacquot E, Hohn T, Hohn B, Saunders K, Candresse T, Ahlquist P, Hemenway C, Foster GD (2011) Top 10 plant viruses in molecular plant pathology. Mol Plant Pathol 12:938–954
22. Deom CM, Oliver MJ, Beachy RN (1987) The 30-kilodalton gene product of Tobacco mosaic virus potentiates virus movement. Science 237: 389–394
23. Citovsky V (1999) Tobacco mosaic virus: a pioneer of cell-to-cell movement. Philos Trans R Soc Lond Ser B Biol Sci 354:637–643
24. Goelet P, Lomonossoff GP, Butler PJ, Akam ME, Gait MJ, Karn J (1982) Nucleotide sequence of Tobacco mosaic virus RNA. Proc Natl Acad Sci U S A 79:5818–5822
25. Holt CA, Beachy RN (1991) In vivo complementation of infectious transcripts from mutant Tobacco mosaic virus cDNAs in transgenic plants. Virology 181:109–117
26. Heinlein M, Epel BL, Padgett HS, Beachy RN (1995) Interaction of tobamovirus movement proteins with the plant cytoskeleton. Science 270:1983–1985
27. Heinlein M, Padgett HS, Gens JS, Pickard BG, Casper SJ, Epel BL, Beachy RN (1998) Changing patterns of localization of the Tobacco mosaic virus movement protein and replicase to the endoplasmic reticulum and microtubules during infection. Plant Cell 10:1107–1120
28. Amari K, Di Donato M, Dolja VV, Heinlein M (2014) Myosins VIII and XI play distinct roles in reproduction and transport of Tobacco mosaic virus. PLoS Pathog 10:e1004448
29. Peiro A, Martinez-Gil L, Tamborero S, Pallas V, Sanchez-Navarro JA, Mingarro I, Simon A (2014) The Tobacco mosaic virus movement protein associates with but does not integrate into biological membranes. J Virol 88:3016–3026
30. Ashby J, Boutant E, Seemanpillai M, Sambade A, Ritzenthaler C, Heinlein M (2006) Tobacco mosaic virus movement protein functions as a structural microtubule-associated protein. J Virol 80:8329–8344
31. Peña EJ, Heinlein M (2013) Cortical microtubule-associated ER sites: organization centers of cell polarity and communication. Curr Opin Plant Biol 16:764–773
32. Boyko V, Hu Q, Seemanpillai M, Ashby J, Heinlein M (2007) Validation of microtubule-associated Tobacco mosaic virus RNA movement and involvement of microtubule-aligned particle trafficking. Plant J 51:589–603
33. Boyko V, Ashby JA, Suslova E, Ferralli J, Sterthaus O, Deom CM, Heinlein M (2002) Intramolecular complementing mutations in Tobacco mosaic virus movement protein confirm a role for microtubule association in viral RNA transport. J Virol 76:3974–3980
34. Boyko V, van der Laak J, Ferralli J, Suslova E, Kwon MO, Heinlein M (2000) Cellular targets of functional and dysfunctional mutants of Tobacco mosaic virus movement protein

fused to green fluorescent protein. J Virol 74: 11339–11346

35. Boyko V, Ferralli J, Ashby J, Schellenbaum P, Heinlein M (2000) Function of microtubules in intercellular transport of plant virus RNA. Nat Cell Biol 2:826–832
36. Ouko MO, Sambade A, Brandner K, Niehl A, Peña E, Ahad A, Heinlein M, Nick P (2010) Tobacco mutants with reduced microtubule dynamics are less susceptible to TMV. Plant J 62:829–839
37. Niehl A, Amari K, Gereige D, Brandner K, Mely Y, Heinlein M (2012) Control of Tobacco mosaic virus movement protein fate by CELL-DIVISION-CYCLE protein48. Plant Physiol 160:2093–2108
38. Curin M, Ojangu E-L, Trutnyeva K, Ilau B, Truve E, Waigmann E (2006) MPB2C, a microtubule-associated plant factor, is required for microtubular accumulation of Tobacco mosaic virus movement protein in plants. Plant Physiol 143:801–811
39. Niehl A, Peña EJ, Amari K, Heinlein M (2013) Microtubules in viral replication and transport. Plant J 75:290–308
40. Kawakami S, Watanabe Y, Beachy RN (2004) Tobacco mosaic virus infection spreads cell to cell as intact replication complexes. Proc Natl Acad Sci U S A 101:6291–6296
41. Sambade A, Brandner K, Hofmann C, Seemanpillai M, Mutterer J, Heinlein M (2008) Transport of TMV movement protein particles associated with the targeting of RNA to plasmodesmata. Traffic 9:2073–2088
42. Ferralli J, Ashby J, Fasler M, Boyko V, Heinlein M (2006) Disruption of microtubule organization and centrosome function by expression of Tobacco mosaic virus movement protein. J Virol 80:5807–5821
43. Boutant E, Didier P, Niehl A, Mély Y, Ritzenthaler C, Heinlein M (2010) Fluorescent protein recruitment assay for demonstration and analysis of *in vivo* protein interactions in plant cells and its application to *Tobacco mosaic virus* movement protein. Plant J 62: 171–177
44. Jeffery WR, Tomlinson CR, Brodeur RD (1983) Localization of actin messenger RNA during early ascidian development. Dev Biol 99:408–417
45. Femino AM, Fay FS, Fogarty K, Singer RH (1998) Visualization of single RNA transcripts in situ. Science 280:585–590
46. Moffitt JR, Pandey S, Boettiger AN, Wang S, Zhuang X (2016) Spatial organization shapes the turnover of a bacterial transcriptome. elife 5:e13065
47. Raj A, van den Bogaard P, Rifkin SA, van Oudenaarden A, Tyagi S (2008) Imaging individual mRNA molecules using multiple singly labeled probes. Nat Methods 5:877–879
48. Sokol DL, Zhang X, Lu P, Gewirtz AM (1998) Real time detection of DNA·RNA hybridization in living cells. Proc Natl Acad Sci U S A 95: 11538–11543
49. Adamala KP, Martin-Alarcon DA, Boyden ES (2016) Programmable RNA-binding protein composed of repeats of a single modular unit. Proc Natl Acad Sci U S A 113:2579–2588
50. Nelles DA, Fang MY, O'Connell MR, Xu JL, Markmiller SJ, Doudna JA, Yeo GW (2016) Programmable RNA tracking in live cells with CRISPR/Cas9. Cell 165:488–496
51. Paige JS, Wu KY, Jaffrey SR (2011) RNA mimics of green fluorescent protein. Science 333:642–646
52. Babendure JR, Adams SR, Tsien RY (2003) Aptamers switch on fluorescence of triphenylmethane dyes. J Am Chem Soc 125: 14716–14717
53. Bertrand E, Chartrand P, Schaefer M, Shenoy SM, Singer RH, Long RM (1998) Localization of ASH1 mRNA particles in living yeast. Mol Cell 2:437–445
54. Daigle N, Ellenberg J (2007) λN-GFP: an RNA reporter system for live-cell imaging. Nat Methods 4:633–636
55. Chen J, Nikolaitchik O, Singh J, Wright A, Bencsics CE, Coffin JM, Ni N, Lockett S, Pathak VK, Hu W-S (2009) High efficiency of HIV-1 genomic RNA packaging and heterozygote formation revealed by single virion analysis. Proc Natl Acad Sci U S A 106: 13535–13540
56. Larson DR, Zenklusen D, Wu B, Chao JA, Singer RH (2011) Real-time observation of transcription initiation and elongation on an endogenous yeast gene. Science 332:475–478
57. Ryckelynck M (2020) Development and applications of fluorogen/light-up RNA aptamer pairs for RNA detection and more. Methods Mol Biol 2166:73–102
58. Pichon X, Robert MC, Bertrand E, Singer RH, Tutucci E (2020) New generations of MS2 variants and MCP fusions to detect single mRNAs in living eukaryotic cells. Methods Mol Biol 2166:121–144
59. Citovsky V, Knorr D, Schuster G, Zambryski P (1990) The P30 movement protein of Tobacco mosaic virus single-strand nucleic acid binding protein. Cell 60:637–647
60. Peña E, Heinlein M, Sambade A (2015) In vivo RNA labeling using MS2. Methods Mol Biol 1217:329–341

61. Peña EJ, Heinlein M (2020) Visualization of transiently expressed mRNA in plants using MS2. Methods Mol Biol 2166:103–120
62. Peña EJ, Robles Luna G, Heinlein M (2021) *In vivo* imaging of tagged mRNA in plant tissues using the bacterial transcriptional antiterminator BglG. Plant J 105:271–282
63. Curtis MD, Grossniklaus U (2003) A gateway cloning vector set for high-throughput functional analysis of genes in planta. Plant Physiol 133:462–469
64. Kapila J, De Rycke R, Van Montagu M, Angenon G (1997) An Agrobacterium-mediated transient gene expression system for intact leaves. Plant Sci 122:101–108
65. Buschmann H, Green P, Sambade A, Doonan JH, Lloyd CW (2011) Cytoskeletal dynamics in interphase, mitosis and cytokinesis analysed through Agrobacterium-mediated transient transformation of tobacco BY-2 cells. New Phytol 190:258–267
66. Rademacher T, Sack M, Blessing D, Fischer R, Holland T, Buyel J (2019) Plant cell packs: a scalable platform for recombinant protein production and metabolic engineering. Plant Biotechnol J 17:1560–1566
67. Silhavy D, Molnár A, Lucioli A, Szittya G, Hornyik C, Tavazza M, Burgyán J (2002) A viral protein suppresses RNA silencing and binds silencing-generated, 21- to 25-nucleotide double-stranded RNAs. EMBO J 21:3070–3080

Chapter 29

Multi-Angle In Vivo Imaging of the *Arabidopsis thaliana* Shoot Apical Meristem (SAM)

Michael Fuchs and Jan U. Lohmann

Abstract

Fluorescent imaging, especially in living tissue, has become a key method in modern life sciences, with the development of new tools for sample preparation, imaging, and data analysis continuously advancing our understanding of biological principles. Here, we present our strategy for in vivo imaging of the *Arabidopsis* shoot apical meristem (SAM), a central structure in plant development. We implement simplifications to previously published workflows and present a novel approach to subsequentially image the meristem from multiple angles at high resolution. This tool may represent a valuable resource for shoot meristem-centered research in general, but also for studies on plasmodesmata or intercellular connectivity within the SAM: via the analysis of fluorescently labeled plasmodesmata-localized proteins, via the tracing of fluorescent dyes, via analyzing the cell-to-cell mobility of fluorescently labeled proteins, but also via the analysis of morphological features of meristematic cells in mutants or upon perturbation of symplastic connectivity.

Key words *Arabidopsis*, Shoot apical meristem, Dissection, Sample preparation, Multi-angle, Life imaging

1 Introduction

All aerial tissue of a plant is, directly or indirectly, generated by a small subset of stem cells located at its shoot apex. Even more importantly, the same stem cells also give rise to the plant's gametes and with that are indispensable for the next generation. Given their importance for the continuous tissue homeostasis of the whole plant, it is only reasonable that the number of these stem cells is heavily regulated on all levels. To this end, they are embedded in a stem cell niche called the shoot apical meristem (SAM), which serves as a nurturing environment in which stem cells and niche cells are in constant exchange of information via signaling molecules. Such intercellular communication follows two routes: first, through the apoplast via secretion (e.g., the CLAVATA3 (CLV3) peptide [1]) or via specific export proteins (e.g., the phytohormone

Yoselin Benitez-Alfonso and Manfred Heinlein (eds.), *Plasmodesmata: Methods and Protocols*,
Methods in Molecular Biology, vol. 2457, https://doi.org/10.1007/978-1-0716-2132-5_29,

auxin [2]) and, second, in a symplastic manner through the plasmodesmata, cytoplasmic bridges that connect neighboring cells, as is the case for the mobile transcription factors SHOOTMERISTEMLESS (STM) [3] and WUSCHEL (WUS) [4, 5] (for an overview of this topic, *see* ref. [6]). WUS is a key stem cell maintenance factor and the disruption of its non-cell-autonomous function via the blocking of plasmodesmata-mediated cell-to-cell movement leads to the loss of stem cells and termination of the stem [5]. While the symplastic connection of cells within the SAM is necessary for its function as a stem cell niche, plasmodesmata also allow the trafficking of pathogens or of pathogen effectors [7]. It follows that symplastic connectivity must be dynamically regulated in vivo in the shoot apical meristem, but also in general, and consequently the topic continues to raise the interest of plant biologists.

Plasmodesmata were first described in the late nineteenth century using light microscopy. Their study did not intensify before the advent of electron microscopy in the 1950s, and indeed, to this day, most insights on plasmodesmata structure have been gained by using a variety of different electron microscopy techniques [8]. Light microscopy, at the time, was limited to using aniline blue to stain the callose associated with plasmodesmata [8, 9]. This changed when fluorescent reporters, such as the green fluorescent protein (GFP) from *Aequorea victoria* [10], became available, and it did not take long until GFP was fused to viral movement proteins both to serve as an in vivo tag for plasmodesmata and to study the intercellular trafficking of these proteins [8, 11, 12]. Since then, fluorescent proteins and fluorescent dyes have been used extensively to study plasmodesmata-localized proteins, proteins that exhibit cell-to-cell mobility and symplastic domains.

Fluorescence microscopy has proven a great tool not only for research related to plasmodesmata but also to study the shoot apical meristem in general: In 2014, Heisler and Ohno have published a detailed collection of protocols for live imaging of the *Arabidopsis* inflorescence meristem [13]. Their workflows require the use of an upright confocal microscope with a water-dipping objective and enable the imaging of the shoot apex in its natural orientation along its apical-basal axis from a top-view perspective. Other protocols have explored methods to utilize inverted microscopes instead, such as encasing the meristem in low-melting agarose [14], custom-made objective attachments [15], commercially available auxiliary equipment (objective lens inverter attachment) [16], or custom-made coverslips [17]. In the following, we present three different strategies for live imaging of the *Arabidopsis* shoot apical meristem, all of which are primarily designed to be used with a water-dipping objective on an upright confocal microscope.

Our first strategy (described in Subheading 3.1) is based on the previous protocol from Heisler and Ohno for one-time snapshots of dissected shoot apices [13] from a top-view perspective, introducing a number of modifications with regard to fine dissection of primordia. Imaging from a top-view perspective allows to make use of the high *XY*-resolution (in comparison to *Z*-resolution) in the plane perpendicular to the apical-basal axis of the shoot. As a result, this approach is ideal to generate an overview of the whole meristem, including even young primordia, and to visualize differences or differential processes between its central zone and the periphery.

Our second strategy (described in Subheading 3.2) is a novel approach to prepare shoot apices for imaging from a side-view perspective that is perpendicular to their apical-basal axis. Consequently, imaging resolution is maximized in the plane parallel to the apical-basal axis of the shoot, which in turn allows to visualize differences or differential processes between meristematic cell layers in great detail. Since dissection of shoot apices for side-view imaging requires the complete removal of young primordia at the side facing the objective, as these would otherwise obstruct the clear view onto the meristem, this approach is both more time-consuming and more prone to causing accidental damage compared to imaging from a top-view perspective.

Our third strategy (described in Subheading 3.3) combines the two approaches described above: Here, a single shoot apex is dissected, mounted, and imaged from a top-view perspective, resulting in an image with resolution maximized in the plane perpendicular to the apical-basal axis of the shoot. Afterward, the same shoot apex is re-mounted and subsequently imaged from a side-view perspective (at a 90° angle to the first image) generating another image with its maximal resolution in the plane parallel to the apical-basal axis of the shoot. The combination of these two images maximizes image resolution along all axes (Fig. 1) and consequently allows for a comprehensive analysis of the meristem as a three-dimensional structure. It is noteworthy though, that, since one and the same shoot apex is imaged two consecutive times—essentially doubling light exposure and with that increasing the effects of phototoxicity and bleaching—this approach may be of limited use for weak signal requiring high laser intensity.

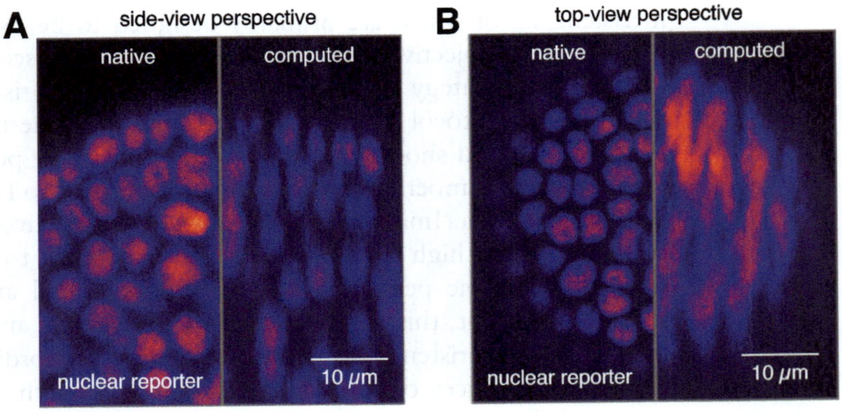

Fig. 1 Multi-angle in vivo imaging of the shoot apical meristem (SAM) can maximize image resolution along all axes. (**a**) Confocal image of an *Arabidopsis* SAM with ubiquitous expression of a nuclear reporter construct (*pUBQ10::3xmCherry-NLS::tUBQ10*), imaged from a native side-view perspective in comparison to a top-view perspective that was computed from the original side-view image. (**b**) Confocal image of the same meristem shown in (**a**), imaged from a native top-view perspective in comparison to a side-view perspective that was computed from the original top-view image

2 Materials

General Remark: Materials are listed in the order in which they appear in the methods sub-sections. A number of relevant tools for dissection and mounting of shoot apices are detailed in Fig. 2.

Materials necessary for all three Subheadings 3.1, 3.2, and 3.3:

1. Plastic petri dishes, 60 × 15 mm (diameter × height).
2. Molten agarose, 2%, in water (*see* **Note 1**).
3. P10 pipet tips.
4. Small, sharp scissors.
5. Turgid plants (*see* **Note 2**), length of the main shoot at least 2–3 cm.
6. Sturdy, sharp tweezers (e.g., re-sharpened precision tweezers).
7. Sterile, desalted water.
8. Stereomicroscope with a long working distance (e.g., 115 mm) and a zoom of 0.67–5× (10× eyepiece, 1× objective for 6.7–50× magnification) and equipped for sample illumination from the top.
9. Flat tweezers (e.g., "stamp tongs" or coverslip tweezers) (*see* **Note 3**).
10. Fine, sharp precision tweezers (e.g., Dumont No5 or equivalent).
11. Fine hypodermic injection needle (e.g., Sterican®, Ø 0.30 × 12 mm 30 G × ½″ or equivalent) (*see* **Note 4**).

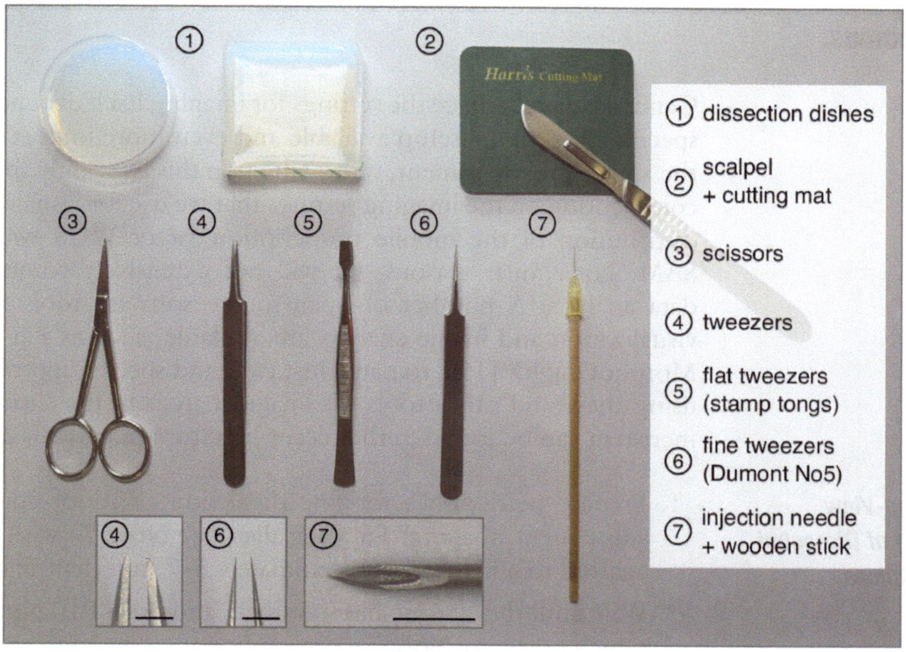

Fig. 2 Essential materials for dissection of the shoot apex. (1) Dissection dishes filled with solidified agarose. (2) Scalpel and cutting mat for the preparation of meristems for side-view imaging. (3) Scissors for cutting the main stem. (4) Sturdy, sharp tweezers for rough dissection. Scale bar = 1 mm. (5) Tweezers for holding the sample during dissection. For this, tweezers with flat, smooth, spade-like ends (e.g., "stamp tongs" or coverslip tweezers) are ideal. (6) Precision tweezers for fine dissection (e.g., Dumont No5). Scale bar = 1 mm. Note the difference in tip size between (4) and (6). (7) Hypodermic injection needle for fine dissection. Scale bar = 1 mm. Note the fine tip in combination with an angled bevel that has sharp edges at the inside but is blunt at the outside. A wooden stick that fits the size of the needle gauge can be helpful for handling the needle during dissection

12. Wooden stick; diameter needs to fit the size of the needle gauge.
13. Confocal fluorescence microscope with an upright stand and a water immersion objective with high numerical aperture (1.10), chromatic aberration correction (visible to near infrared), and a long working distance (2 mm).

Additional materials only necessary for Subheadings 3.2 and 3.3:

14. Rectangular dishes, e.g., 40 × 40 × 15 mm (length × width × height) in size, cut open at one side (*see* **Note 5**).
15. Sticky tape.
16. Cutting mat (*see* **Note 6**).
17. Scalpel.

3 Methods

General Remark: Since the settings for imaging itself depend on the specific microscope setup available and even more importantly on the aim of the experiment, we touch upon this issue only shortly by commenting on the imaging settings that we use for visualizing the distribution of the mobile transcription factor WUS within the SAM. For similar reasons, we will not extensively comment on data analysis: A number of open-source software tools for data visualization and image analysis are available, such as Fiji [18] or MorphoGraphX [19], to name just two, and specific approaches in using these and other tools for image analysis in the shoot apical meristem can be found in the recent literature (e.g., [20, 21]).

3.1 Top-View Imaging of Dissected Shoot Apices

1. Prepare petri dishes for fine dissection of shoot apices and subsequent imaging: Fill each dish (*see* **Note 7**) with molten agarose to a height of approximately 5–7 mm (half-filled).
2. Wait until the agarose has solidified, and use a P10 pipet tip to punch holes (*see* **Note 7**), perpendicular to the agarose surface, for holding stem pieces.
3. Start with rough dissection of a shoot apex: Use scissors to cut the main inflorescence (Fig. 3a) from the plant and remove all larger organs (emerging side shoots, leaves, siliques, and larger flowers) starting a few centimeters below the apex top (Fig. 3a). For this, gently hold the shoot with one hand, and use tweezers (*see* **Note 8**) to pinch off organs at the node, as close to the main stem as possible (*see* **Note 9**).
4. For further dissection of smaller floral organs, cut the stem 2–3 cm below the apex top, and transfer the stem piece including the shoot apex to a petri dish filled with sterile, desalted water (*see* **Note 10**).

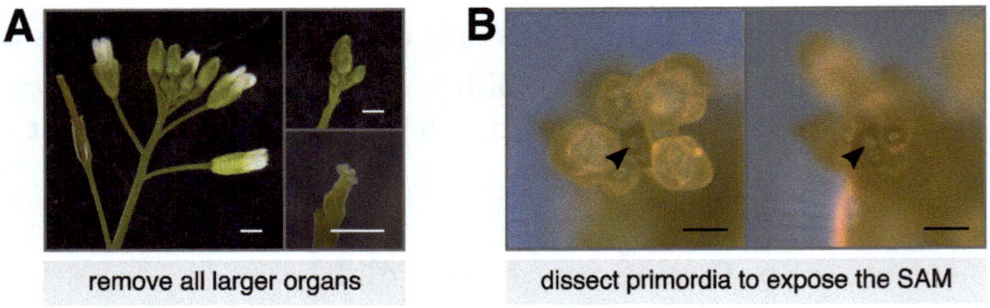

Fig. 3 Dissection of floral organs to expose the shoot apical meristem for imaging. The shoot apex dissected here is of the *Arabidopsis thaliana* ecotype Columbia-0 (Col-0). (**a**) Larger floral organs are removed using tweezers. Scale bars = 1 mm. (**b**) Primordia are removed using a hypodermic injection needle. The SAM is visible in the middle of the shoot apex and indicated by the black arrows. Scale bars = 0.1 mm (100 μm)

5. Continue dissection under the stereomicroscope (*see* **Note 11**) at 20–30× magnification: Hold the sample with flat tweezers at the basal end (*see* **Note 12**), and remove organs using fine tweezers (*see* **Note 13**). For this, place one lever of the unclosed fine tweezers directly at the node, between the stem and the petiole of the organ to be removed. Do not close the tweezers, but apply pressure in a basal direction onto the petiole, which will bend and, in many cases, break off easily. Alternatively, carefully grab the petiole, and pull it away from the apex top in a basal direction (*see* **Note 14**) until the organ breaks off. If the organ still cannot be removed, close the tweezers to pinch it off, as already done during initial rough dissection (*see* **Note 9**). To remove younger floral organs with only a small petiole, place one lever of the unclosed fine tweezers at the apex-facing side of the organ itself, and apply pressure in a basal direction until it breaks off.
6. Rotate the sample, whenever necessary, to access all sides of the stem piece: Carefully grab the basal end of the stem piece with the fine tweezers, and release the hold of the flat tweezers. Rotate the sample around its apical-basal axis, grab the stem piece again using the flat tweezers, and continue dissection until floral organs become too small to be conveniently removed (Fig. 3a, b) with the methods described in **step 5**.
7. For fine dissection of flower primordia, transfer the sample to a previously prepared dish (*see* **steps 1–2**): Use fine tweezers to separate the topmost 7–8 mm of the stem piece, including the shoot apex with the shoot apical meristem, from the rest of the stem. Insert the basal part of the stem piece into a hole in the prepared dish containing solidified agarose, and push down until only the top 1–2 mm remain above the surface. Proceed to fine dissection right away to avoid excessive dehydration of the sample (*see* **Note 15**).
8. Continue fine dissection under the stereomicroscope at 50× magnification: To remove older primordia, place the fine tip of a hypodermic injection needle at the apex-facing side of the primordium, and apply pressure in a basal direction until it breaks off. To remove younger primordia, place the needle between a primordium and the center of the shoot apex (*see* **Note 16**), and use the sharp edge to cut away the primordium by applying slight pressure in a basal direction. Continue fine dissection until the shoot apical meristem in the middle of the shoot apex is clearly visible and is no longer obstructed by organ primordia (Fig. 3b).
9. Pipet a drop of sterile, desalted water onto the dissected shoot apex to prevent dehydration of the sample (*see* **Note 15**). Go back to **step 3**, and proceed to dissect additional shoot apices

until either all holes in the prepared dish are filled or no additional samples are left.

10. Optional step: Perform sample staining or chemical treatment (*see* **Note 17**), if required for your experiment.
11. Fill up the petri dish with sterile, desalted water, completely submerging the sample(s) with liquid. Let the dish rest for 30 min to allow hydration of the agarose and thereby avoid movement of the sample(s) during image acquisition.
12. Place the petri dish, containing the sample(s) submerged in water, on the microscope stage. Lower the objective until it touches the water surface, make sure there is no air bubble trapped at the lens (*see* **Note 18**), and use the XY-controls of the stage to center the shoot apex to be imaged under the objective. Look through the eye-piece using regular bright-field settings, and slowly continue lowering the objective until the sample appears within the field of view. Use the XY-controls of the stage to center the sample within the field of view and lower the objective until the shoot apex with the SAM is clearly visible. Switch from bright-field to confocal settings and zoom onto the shoot apical meristem (*see* **Note 19**).
13. Proceed with imaging of the sample(s). Specific imaging settings will depend on the microscope setup available as well as on the aim of the experiment (*see* **Note 20**).
14. After all samples have been imaged, discard the agarose sheet containing the shoot apices, and clean the petri dish for later reuse.

3.2 Side-View Imaging of Dissected Shoot Apices

1. Prepare rectangular dishes for fine dissection of shoot apices: Temporarily seal the open side of each dish with tape, and fill each dish with molten agarose to a height of approximately 5–7 mm (half-filled). Make sure to have a small quantity of liquid molten agarose left over and store it at >55 °C for later usage.
2. Wait until the agarose within the dishes has solidified, and use a P10 pipet tip to punch holes (*see* **Note 21**), perpendicular to the agarose surface, for holding stem pieces.
3. Proceed with dissection of the shoot apex as has been described in the previous Subheading 3.1, **steps 3–9** (*see* **Note 22**).
4. Remove the sealing tape from the rectangular dish (Fig. 4a), carefully slide out the agarose sheet containing the sample(s), and place it on a cutting mat.
5. Use a scalpel to cut small asymmetric blocks of agarose (Fig. 4b), each of which contains a single shoot apex: For this, make a first straight cut, perpendicular to the agarose surface, at a distance of no further than 1 mm from the sample

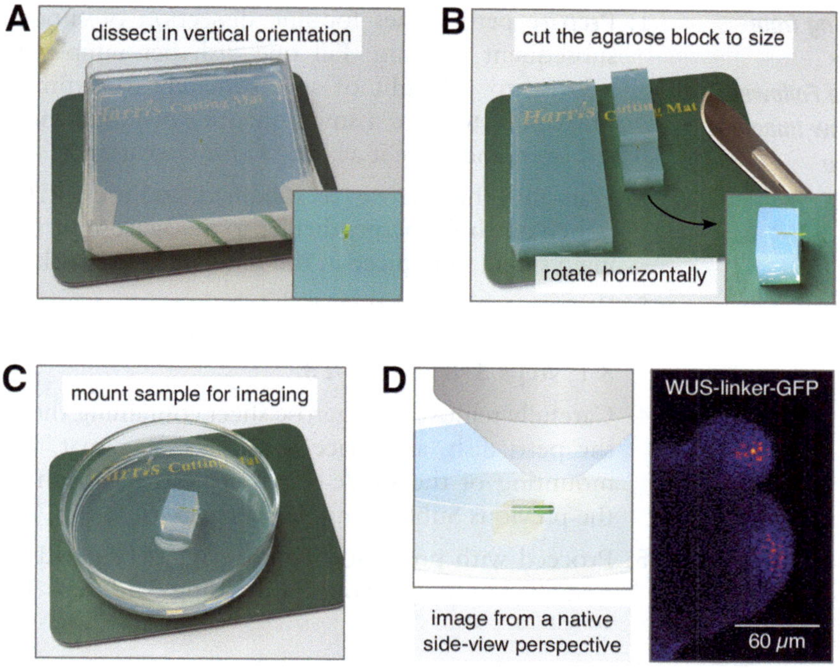

Fig. 4 Sample mounting for imaging from a native side-view perspective. (**a**) Fine dissection of the shoot apex is done with the sample oriented vertically. (**b**) A small agarose block with the shoot apex placed asymmetrically at one side is cut from the agarose sheet. The block is then rotated horizontally, so that the shoot apex sits in a horizontal orientation relative to its apical-basal axis. (**c**) The agarose block is placed in liquid agarose within a round petri dish. Make sure the shoot apex is elevated above the liquid agarose. (**d**) The sample is submerged in water and imaged using a water-dipping objective on an upright confocal microscope

to create the surface that is going to face the objective. Make a second cut, perpendicular to the agarose surface, parallel to the first cut and 7–8 mm away from it, to create the surface on which the agarose block will rest once it has been turned by 90° for later imaging. Make two more cuts, perpendicular to the first two cuts as well as to the agarose surface, each at a distance of 3–4 mm from the shoot apex, to create an agarose block of a suitable size.

6. Pour molten agarose into an empty petri dish to a height of 1–2 mm. Place the agarose block(s) within the liquid agarose in such a way, so that the sample is elevated above the agarose and its apical-basal axis is parallel to the surface of the petri dish (Fig. 4b, c) (*see* **Note 23**). Wait until the liquid agarose has solidified.

7. Proceed with preparation and imaging (Fig. 4d) as has been described in the previous Subheading 3.1, **steps 10–14**.

3.3 Imaging from a Top-View Perspective Followed by Side-View Imaging of the Same Shoot Apex

1. Prepare petri dishes for fine dissection of shoot apices and subsequent imaging: Fill one dish per sample with molten agarose to a height of approximately 5–7 mm (half-filled). Make sure to have a small quantity of liquid molten agarose left over, and store it at >55 °C for later usage.
2. Wait until the agarose has solidified, and use a P10 pipet tip to punch one hole, perpendicular to the agarose surface, for holding a single stem piece in the middle of each dish.
3. Proceed with dissection of the shoot apex, staining, mounting, and imaging as has been described in the previous Subheading 3.1, **steps 3–8** and **10–13**.
4. Carefully remove the agarose sheet containing the sample from the petri dish, and place it on a cutting mat. Proceed with mounting of the single shoot apex as has been described in the previous Subheading 3.2, **steps 5–6**.
5. Proceed with preparation and imaging (Fig. 4d) as has been described in the previous Subheading 3.1, **steps 10–14**.

4 Notes

1. After solidification, agarose provides physical support for the sample during sample dissection as well as imaging. Several published protocols use different agarose concentrations, ranging from 1% [13] to 2% [15] and up to 3% [17], suggesting that the exact percentage is not critical. Liquid agarose can be prepared in small or large quantities via boiling in a microwave oven until it is molten: 50 ml of molten agarose are sufficient for three to four petri dishes. Petri dishes with solidified agarose can be stored in the fridge for later usage; however, freshly prepared dishes are preferable. Molten agarose can be stored at >55 °C for several days, but changes in the texture of the agarose after solidifying may occur eventually.
2. Make sure your plants have been sufficiently watered the day before or at least several hours before starting with dissection. Turgid plants are much easier to dissect and will be more resilient to dehydration stress that may occur during the dissection procedure.
3. Tweezers with flat, spade-like ends, rather than a sharpened tip, are ideal for holding the sample during dissection. Accidentally using too much force while grabbing the stem piece with flat tweezers will only result in squeeze damage to the tissue in an area that is removed later. In contrast, applying excessive force with sharp tweezers will result in cutting through the stem, leaving a shortened stem piece that, depending on its length, can be difficult to handle during dissection.

4. A hypodermic needle of suitably small size has an ideal geometry for fine dissection: It combines a fine tip with an angled bevel that is sharpened at the edges and can be used for cutting, while the outside of the cannula is not sharp. In addition, hypodermic needles can be replaced easily and at a much lower price than fine tweezers, once they are bent or are no longer sharp.
5. For samples that are going to be imaged from a side-view perspective, we use 40 × 40 × 15 mm (length × width × height) rectangular dishes that have been cut open at one side, so that the sheet of solidified agarose can be conveniently removed. The exact length and width of these dishes, however, are not critical, as the agarose sheet is cut to size later and samples are mounted for imaging within 60 × 15 mm petri dishes.
6. While not strictly necessary, a cutting mat or equivalent surface protection can be used to prevent scratches on the tabletop surface when cutting the agarose sheet to size in sample preparation for side-view imaging.
7. The number of petri dishes that need to be prepared depends on the total number of samples as well as on the number of samples placed within the same dish: We have found that working with no more than four to six samples per petri dish, evenly spaced out within a circle with at least 1 cm distance to the rim of the dish, creates sufficient spacing between samples as well as ensures accessibility of each sample with the objective. In addition, the circular arrangement allows to label samples on the (outer) side of the petri dish, rather than at the bottom, making labels easily recognizable even when the dish is already mounted on the microscope.
8. Do not use your best precision tweezers during rough dissection. Due to large organ sizes and internode spacing, such a fine tip is not necessary, and in fact, it might suffer from the pressure required to pinch off thicker petioles. Use sharp tweezers with a sturdier tip or use older, potentially re-sharpened tweezers that are no longer good enough for fine dissection.
9. Pinching off organs will in many cases leave a piece of the petiole attached to the stem. During rough dissection, this can be neglected, because internodal spaces are large, and therefore petiole fragments of basal organs do not interfere with the removal of more apical organs. During subsequent dissection and especially during fine dissection however, as the internodes become smaller, petiole fragments that remain at the stem will be a hindrance, and therefore pinching off organs should be avoided here.
10. Due to surface tension, pockets of residual air can remain trapped between the petioles of larger organs or between

organs and the stem after transfer of the sample to water. These do, however, usually not hinder the dissection process and disappear as dissection proceeds.

11. Different stereomicroscopes have different options for illuminating the sample. The stereomicroscope we use features lighting from the top and, through a removable glass disc within the stand, also from below the sample. We have found that in our setup, illumination only from the top with a black acrylic glass disc replacing the opaque glass disc provides good contrast for dissection.
12. Grab the sample carefully, so that its apical-basal axis is roughly perpendicular to the axis of the tweezers. Use enough force for the sample to remain fixed during dissection, but do not squeeze too much as this will result in tissue damage. It is advantageous if the flat, spade-like ends are not entirely parallel to each other, but close at an acute angle. This will allow to completely close the tweezers, while at the same time allows to adjust the pressure on the sample by placing it closer to the front or further back. To stabilize the sample during dissection, it is helpful to let the tip of the closed tweezers rest on the bottom of the petri dish.
13. Dissection requires concentration and fine motor manipulations: movements over minimal distances, but with maximal precision. Assume a comfortable and relaxed position that can be hold even over an extended period of time. To avoid inadvertent movements of your arm, of your hand, or of the equipment therein, create contact sites with stabile surfaces (e.g., the table) to support your arm/hand and to reduce the amount of muscle tension necessary. Consider avoiding other factors that may lead to "shaky hands", such as hypoglycemia, dehydration, but also excessive consumption of caffeine.
14. Pulling away an organ in a basal direction from the stem can help to facilitate a clean break off point at the node. Keep in mind, however, that the stem itself can bend as well and is in danger of breaking off if too much force is applied.
15. Excessive dehydration of the sample will complicate the dissection procedure due to a loss of turgor pressure and in addition may lead to tissue damage, which will negatively impact on imaging quality.
16. To protect the meristem from the sharp inside of the cannula, make sure that when placing the needle between the meristem and the primordium to be removed, the blunt outside of the cannula is facing the meristem and the sharpened edges of the bevel are facing the primordium.
17. The sample(s) can be stained using fluorescent dyes or can be subjected to any other kind of chemical treatment: For this,

carefully remove the drop of water covering the sample(s), and pipet a drop of treatment solution directly on top. After incubation, fill up the petri dish with sterile, desalted water and decant again to dilute and wash away the treatment solution. In case an ongoing treatment is required and the solution used for treatment has no (optical) properties that interfere with imaging, the treatment can instead be performed by simply filling up the petri dish with a larger volume of treatment solution. In this case, washing is not required, and subsequent imaging takes place with the sample(s) submerged in the treatment solution. Be careful to clean the lens afterward to prevent the accumulation of potential precipitates.

18. An air bubble trapped at the lens will drastically reduce imaging quality due to light refraction. Air bubbles do not form every time and usually only occur when the dry objective gets in contact with water for the first time. Therefore, after having immersed the objective, raise it again, and check whether an air bubble has formed. If that is the case and if the air bubble does not pop on its own after a short time, dry the objective, and repeat the whole process until the lens is both coated with water and free of air bubbles.

19. Finding the shoot apex within the field of view can be cumbersome: Look from the side and try to center the sample under the objective. Carefully start lowering the objective while looking through the eye-piece using regular bright-field settings. If the sample has been properly centered, a dark green shape will eventually appear in the field of view, and the shoot apex will come in focus shortly afterward. If the sample is not centered, the shoot apex will not become visible, and eventually the agarose surface will come in focus instead. If this is the case, move the objective up a little bit again, and slowly start moving the stage in *XY*-direction until you find either the shoot apex already in focus or a dark green shade hinting at its presence.

20. To visualize the shoot apical meristem and to analyze the distribution of the mobile transcription factor WUS, we usually generate images with a scan size of 512 × 512 pixels and a zoom factor of 4, resulting in a final resolution of 0.25 μm per pixel in *XY*-direction. In *z*-direction, we scan with a step size of 0.5 μm, moving into the tissue as the scan progresses. For imaging from a top-view perspective, we usually generate a *z*-stack of about 50 μm, starting from 1 to 2 μm above the L1 surface. For imaging from a side-view perspective, we usually generate *z*-stacks of 60–70 μm, starting from the periphery in the direction of the central zone. Scanning is done at a speed of 0.5 frames per second (pixel dwell of 4.8 μs) and without averaging to reduce scanning time, which minimizes phototoxicity and bleaching as well as maximizes temporal resolution

between the frames within the *z*-stack. For the same reason, we use sequential scanning of multiple channels only when strictly necessary: This is the case when bleed-through from a channel for visualization of cell outlines (e.g., DAPI) occurs in a channel of interest (e.g., GFP), but not when bleed-through from the channel of interest (e.g., GFP) occurs in a channel for visualization of cell outlines (e.g., FM 4-64). We image with the pinhole set to 16.6 μm and balance the laser power and the gain to minimize noise (high laser power, low gain) while at the same time also trying to reduce phototoxicity and bleaching during image acquisition (low laser power, high gain). In general, we choose our settings so that in the structure of interest only a few pixels are saturated and try to avoid large clusters of over-saturated pixels.

21. Holes for holding shoot apices have to be created with sufficient distance to the rim of the dish as well as to each other. For example, in a 40 × 40 mm rectangular dish filled with solidified agarose, we would punch four holes, each with a distance of 10–15 mm to the rim of the dish and to all adjacent holes, ensuring that each block of agarose (containing a single shoot apex) later cut from the agarose sheet has an appropriate size.
22. Keep in mind that even though during dissection the meristem is viewed from a top-view perspective, imaging will take place from a side-view perspective. Therefore, it is essential to remove all primordia at the side that is going to be oriented toward the objective during later imaging, as these may obstruct a clear view onto the meristem.
23. If more than one agarose block per petri dish is placed, it is advantageous to arrange the blocks with shoot apices facing to the inside for better accessibility with the objective.

References

1. Rojo E, Sharma VK, Kovaleva V, Raikhel NV, Fletcher JC (2002) CLV3 is localized to the extracellular space, where it activates the Arabidopsis CLAVATA stem cell signaling pathway. Plant Cell 14:969–977. https://doi.org/10.1105/tpc.002196
2. Gälweiler L, Guan C, Müller A, Wisman E, Mendgen K, Yephremov A, Palme K (1998) Regulation of polar auxin transport by AtPIN1 in Arabidopsis vascular tissue. Science 282:2226–2230. https://doi.org/10.1126/science.282.5397.2226
3. Kim JY, Yuan Z, Jackson D (2003) Developmental regulation and significance of KNOX protein trafficking in Arabidopsis. Development 130:4351–4362. https://doi.org/10.1242/dev.00618
4. Yadav RK, Perales M, Gruel J, Girke T, Jönsson H, Venugopala Reddy G (2011) WUSCHEL protein movement mediates stem cell homeostasis in the Arabidopsis shoot apex. Genes Dev 25:2025–2030. https://doi.org/10.1101/gad.17258511
5. Daum G, Medzihradszky A, Suzaki T, Lohmann JU (2014) A mechanistic framework for noncell autonomous stem cell induction in Arabidopsis. Proc Natl Acad Sci 111:14619–14624. https://doi.org/10.1073/pnas.1406446111
6. Fuchs M, Lohmann JU (2020) Aiming for the top : non-cell autonomous control of shoot stem cells. J Plant Res 133:297–309. https://doi.org/10.1007/s10265-020-01174-3

7. Cheval C, Faulkner C (2018) Plasmodesmal regulation during plant–pathogen interactions. New Phytol 217:62–67. https://doi.org/10.1111/nph.14857
8. Bell K, Oparka K (2011) Imaging plasmodesmata. Protoplasma 248:9–25. https://doi.org/10.1007/s00709-010-0233-6
9. Currier HB (1957) Callose substance in plant cells. Am J Bot 44:478–488. https://doi.org/10.2307/2438916
10. Chalfie M, Tu Y, Euskirchen G, Ward WW, Prasher DC (1994) Green fluorescent protein as a marker for gene expression. Science 263:802–805. https://doi.org/10.1126/science.8303295
11. Heinlein M, Epel BL, Padgett HS, Beachyt RN, Beachy RN (1995) Interaction of Tobamovirus movement proteins with the plant cytoskeleton. Science 270:1983–1985. https://doi.org/10.1126/science.270.5244.1983
12. Epel BL, Padgett HS, Heinlein M, Beachy RN (1996) Plant virus movement protein dynamics probed with a GFP-protein fusion. Gene 173:75–79. https://doi.org/10.1016/0378-1119(95)00678-8
13. Heisler MG, Ohno C (2014) Live-imaging of the Arabidopsis inflorescence meristem. In: Riechmann JL, Wellmer F (eds) Flower development: methods and protocols. Springer, New York, pp 431–440
14. von Wangenheim D, Daum G, Lohmann JU, Stelzer EK, Maizel A (2014) Live imaging of Arabidopsis development. In: Sanchez-Serrano JJ, Salinas J (eds) Arabidopsis protocols. Humana Press, Totowa, pp 539–550
15. Prunet N, Jack TP, Meyerowitz EM (2016) Live confocal imaging of Arabidopsis flower buds. Dev Biol 419:114–120. https://doi.org/10.1016/j.ydbio.2016.03.018
16. Nimchuk ZL, Perdue TD (2017) Live imaging of shoot meristems on an inverted confocal microscope using an objective lens inverter attachment. Front Plant Sci 8:1–10. https://doi.org/10.3389/fpls.2017.00773
17. Wei J, Qi Y, Li M, Li R, Yan M, Shen H, Tian L, Liu Y, Tian S, Liu L, Zhang Y, Sun H, Bai Z, Zhang K, Li C (2020) Low-cost and efficient confocal imaging method for Arabidopsis flower. Dev Biol 466:73–76. https://doi.org/10.1016/j.ydbio.2020.07.012
18. Schindelin J, Arganda-Carreras I, Frise E, Kaynig V, Longair M, Pietzsch T, Preibisch S, Rueden C, Saalfeld S, Schmid B, Tinevez J-Y, White DJ, Hartenstein V, Eliceiri K, Tomancak P, Cardona A (2012) Fiji: an open-source platform for biological-image analysis. Nat Methods 9:676–682. https://doi.org/10.1038/nmeth.2019
19. de Reuille PB, Robinson S, Smith RS (2014) Quantifying cell shape and gene expression in the shoot apical meristem using MorphoGraphX. Methods Mol Biol 1080:121–134. https://doi.org/10.1007/978-1-62703-643-6_10
20. Gaillochet C, Stiehl T, Wenzl C, Ripoll JJ, Bailey-Steinitz LJ, Li L, Pfeiffer A, Miotk A, Hakenjos JP, Forner J, Yanofsky MF, Marciniak-Czochra A, Lohmann JU (2017) Control of plant cell fate transitions by transcriptional and hormonal signals. elife 6:1–30. https://doi.org/10.7554/eLife.30135
21. Ma Y, Miotk A, Šutiković Z, Ermakova O, Wenzl C, Medzihradszky A, Gaillochet C, Forner J, Utan G, Brackmann K, Galván-Ampudia CS, Vernoux T, Greb T, Lohmann JU (2019) WUSCHEL acts as an auxin response rheostat to maintain apical stem cells in Arabidopsis. Nat Commun 10:5093. https://doi.org/10.1038/s41467-019-13074-9

Chapter 30

More Insights from Ultrastructural and Functional Plasmodesmata Data Using PDinsight

Eva E. Deinum

Abstract

PDinsight is a Python-based tool for computing effective wall permeability for symplasmic transport based on plasmodesma (PD) size and distribution data. PDinsight can be used for direct computation of such permeabilities if full data is available, as well as in an explorative way if some data is either not available or considered unreliable. In this chapter, we briefly describe the basic model underlying the PDinsight calculations and discuss how the different modes of PDinsight can be used in relation to typical research questions. We also offer advice on choosing appropriate values for diffusion coefficients and particle size based on the currently most used experimental probes.

Key words Modeling plasmodesmata, Effective wall permeability, Symplasmic transport predictions, Plasmodesmata geometries

1 Introduction

Plasmodesmata (PDs) and their proper regulation are critical for normal plant development. Mutants affected in PD regulation are often embryo or seedling lethal and show severe patterning defects [9]. It is experimentally challenging, however, to probe the native state of PDs and, through that, estimate their aperture and individual and collective transport properties for all relevant substances. Estimates from functional (e.g., carboxyfluorescein (CF) mobility) and ultrastructural (EM) measurements often do not match well [4, 6]. Moreover, recent observations of the so-called type I PDs—straight PDs found in meristematic tissue that support high symplasmic transport of GFP-like molecules despite the apparent lack of a cytoplasmic sleeve [15, 27]—demonstrate that our current knowledge of and tools for assessing the native state of PDs are limited. Computational models can improve our understanding in two ways. First, an accessible tool for predicting effective permeability for symplasmic transport facilitates comparative studies: are

Yoselin Benitez-Alfonso and Manfred Heinlein (eds.), *Plasmodesmata: Methods and Protocols*,
Methods in Molecular Biology, vol. 2457, https://doi.org/10.1007/978-1-0716-2132-5_30,

predictions better for some cell types/interfaces than others, or are there large differences between species? If so, is it possible to identify, for example, certain PD or cell wall properties that correlate with good or bad predictions? Second, a computational model could be used to explore "what if" type hypotheses in a quantitative way. If the PDs appear to be too narrow to support the functionally measured effective permeability, how wide or with how many would they have to be? How do these numbers compare to the uncertainties in both the functional and ultrastructural assays, as well as potential fluctuations in native PD dimensions vs. the fixed samples used in electron microscopy [1, 27]? To these ends, the program PDinsight has been developed.

In this chapter, we will first briefly introduce the computational model that is used in PDinsight using the parameter names as used in the software. We then address which data are required to use the program. Most of these requirements would also hold for other models. We will then discuss how the different modes of PDinsight can be used for addressing different questions and how the available options depend on the experimental data.

This chapter is written about PDinsight version 2.0, which has more options for data analysis than the original published model and allows for different neck radii on either side of the PD [4].

2 The Computational Model

The computational model used in PDinsight is extensively described in [4]. In essence, effective wall permeability for symplasmic transport (`Peff`) is computed as the "unit permeability" (`Punit`) of a single channel, multiplied by PD density (`dens`) and a correction factor `Fih` that accounts for the fact that the wall is only permeable where the PDs are, and not uniformly.

$$\texttt{Peff} = \texttt{dens} * \texttt{Fih} * \texttt{Punit}. \quad (1)$$

As `Fih` varies only mildly for typical values of PD density and dimensions, this factor is not discussed here. See [4] for details.

Figure 1 shows the architecture and parameter names of the model PD used in PDinsight. A single PD is modelled using three cylinders for the central region (radius `Rc`) and possible neck constrictions (radius `Rn` or `Rn1` and `Rn2`) at either end surrounding a fourth cylinder, the desmotubule (DT in Fig. 1, radius `Rdt`). The total PD has length `Lpd`. The necks have length `Lneck` or `Lneck1` and `Lneck2` and the central region has length `Lc` = `Lpd` - `Lneck1` - `Lneck2`. The space between the desmotubule and the larger cylinders is the cytosolic sleeve and considered the sole compartment for symplasmic transport. The mobile solute is modelled as (non-additive) spherical particles with radius `x`. The walls have a so-called

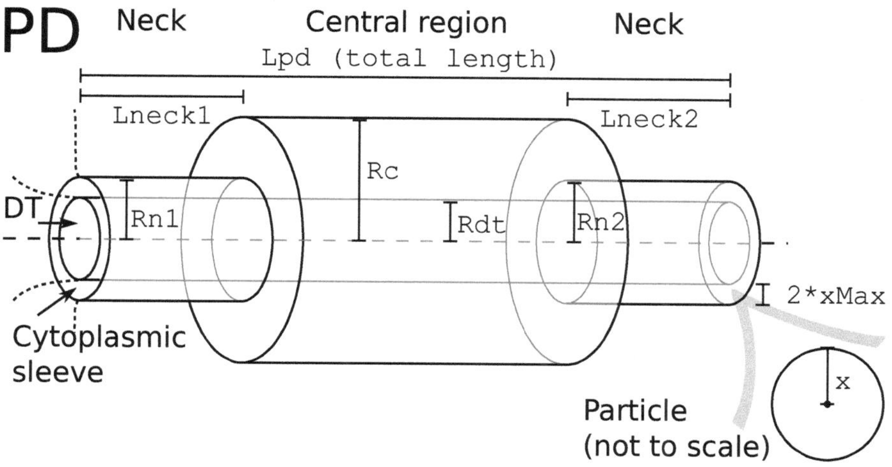

Fig. 1 Model PD used in PDinsight with parameter names. The necks at both ends can be set independently (as shown) or both at the same time using `Rn` and `Lneck`. Particles have radius `x`. The maximum radius that would fit through the channel is `xMax` = (`Rn` -`Rdt`)/2, with `Rn` the smallest of `Rn1` and `Rn2`. If `Rc` < `Rn`, PDinsight will use `Rc` = `Rn`. The available PD geometries per mode are detailed in Table 3

hard interaction with solutes. Due to these interactions, only part of the cytosolic sleeve is available to the particle center. This reduction of available volume is called steric hindrance. Additionally, diffusion close to the walls is slowed down by viscous drag. Both these hindrance effects are taken into consideration by PDinsight.

The unit permeability is then computed as

$$\texttt{Punit} = \texttt{diff}/(\texttt{Lneck1}/\texttt{An1} + \texttt{Lneck2}/\texttt{An2} + \texttt{Lc}/\texttt{Ac}), \tag{2}$$

where `diff` is the cytosolic particle diffusion coefficient and `An1`, `An2`, `Ac` are the respective hindrance corrected available cross-section areas. For advice on the diffusion coefficient, see Sect. 7.2. The cross-section areas are not used as input parameters but computed from PD and particle dimensions.

The PD density `dens` is inputted as the number of PDs per μm^2. In the model, PDs are considered to be spatially arranged on a regular grid (or random), see Fig. 2. The default is a hexagonal grid. The type of grid has very little impact on `Peff`, so using this default is a good choice. Clustering can reduce `Peff` compared to the same number of PDs more uniformly distributed over the cell wall. For investigating the impact of clustering, cluster size (`nPit`) and center-to-center distance of neighboring PDs within a cluster (`dPit`) can be controlled. All these effects are included in the computation of inhomogeneity factor `Fih`. For advice on clustering parameters, see Sect. 7.3.

For more details on the model, see [4].

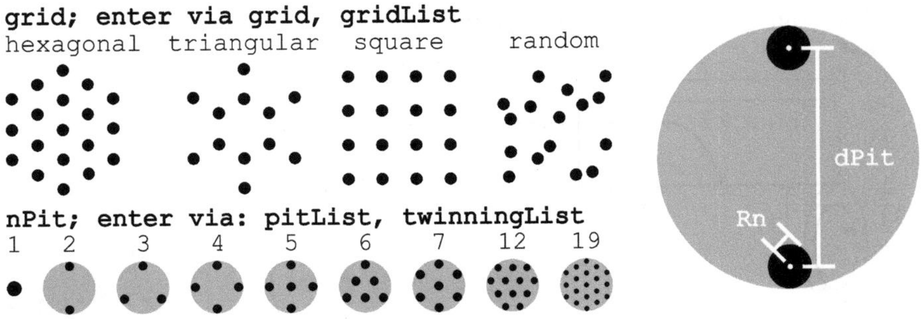

Fig. 2 Parameters describing PD distribution in the cell wall. PDs have a total density of `dens` PDs per μm^2. Individual PDs (if `nPit` = 1) or clusters (`nPit` > 1) are placed on a grid with structure `grid`. The figure shows small samples of the grids. A similar structure would cover the entire cell wall. In case of clustering, the PDs are placed on a (possibly different) regular grid inside the smallest possible replacement circle (gray). The arrangements used for each `nPit` are shown with black dots for PDs. Note that the subsequent images are not to scale. The distance between the centers of nearest PDs within a cluster is `dPit`

3 Which Data Are Required to Predict Effective Wall Permeability?

PDinsight has multiple computation "modes" and some require more data than others (see Table 1). In more general terms: the available options for predicting and exploring effective wall permeability (simply `Peff` here) depend on the available data. Below, the parameters are described per category. Which parameters belong to each category is summarized in Table 2.

3.1 Particle Information

Because the space for transport, the cytosolic sleeve, is so narrow relative to typical solute sizes, nothing can be computed without knowing particle size. For commonly used fluorescent proteins and carboxyfluorescein and similar dyes, good literature estimates are available. Alternatively, these can be estimated from chemical structures (e.g., see [12]). Related to particle size, the diffusion coefficient is also an essential parameter for computing `Peff`. As this parameter is temperature dependent, it may be hard to obtain from the literature. See Sect. 7.2 for a more extensive discussion of this parameter and useful references for a few commonly used probes.

3.2 PD Geometry

All modes require information about PD geometry: length and radius of all parts of the PD. Figure 1 shows the most elaborate PD geometry currently available in PDinsight, the "3-cylinder" model with separate parameters for the neck regions at either side, but depending on data quality and the interface of interest, and the PDs may be described equally well by a simple "1-cylinder" (straight) or "2-cylinder" (identical necks) model. The more advanced geometry models are only available in part of the modes, see Table 3. In all cases, an estimate of the desmotubule

Table 1
Required input parameters for PDinsight per mode. As some exploratory modes predict neck radius, `Rn` is included separately from the other PD geometry parameters. S: at least one value required per parameter. N: multiple values required per parameter

Mode	Required parameter categories			
	PD geometry - `Rn`	`Rn`	Density and distribution	Particle information
`computeVals`[a]	S	S	S	S
`computeUnitVals`[b]	S	S		S
`sensitivityAnalysis`[c]	S	S	S	S
`computeDens`[d]	S	S		S
`computeAperture`[e]	S		S	S
`computeRnDensityGraph`[f]	S			S
`singleInterface`[g]	N	N	S	S
`bootstrapInterface`[h]	N	N	S	S

[a]Compute `Peff` for one or more parameter combinations
[b]Compute `Punit` for one or more parameter combinations
[c]Perform a sensitivity analysis around one or more parameter combinations
[d]Compute required density (`dens`) for obtaining a given target `Peff`-value (given `Rn`)
[e]Compute required channel radius (`Rn`) for obtaining a given target `Peff`-value (given `dens`)
[f]Compute a graph (a list of `Rn`, `dens` pairs) required for obtaining a given target `Peff`-value
[g]Compute `Peff` from raw experimental data. Also computes the relative contribution of individual PDs (based on `Punit`) to `Peff` if the data is structured per PD
[h]Compute confidence intervals on `Peff` from raw experimental data using a bootstrapping approach

radius is required. It may be necessary to resort to literature values, particularly in case of "type I" PDs [15]. See also Sect. 7.1. For the most accurate predictions, it is recommended to use whole data sets rather than sample averages whenever possible. This also allows you to predict confidence intervals on `Peff` via a bootstrap. Typically, structured data, i.e., with all PD geometry data (except desmotubule) grouped per PD, will yield somewhat more accurate predictions of `Peff` and narrower confidence intervals than unstructured geometry data, i.e., when the measured values for the parameters are treated independently.

3.3 Density and Distribution

By far the most important parameter of this category is density, the number of PDs per μm^2. All other parameters in this category describe how the PDs are assumed to be distributed over the interface in the computation of inhomogeneity factor `Fih`. This is visualized in Fig. 2. If the PDs are not clustered, their exact distribution has little effect on `Peff`. See also Sect. 7.3.

Table 2
Parameter categories (Table 1) in detail. Which parameters are required depends on the mode as well as options chosen. For meaning of the parameters, see Figs. 1 and 2

PD model	Class: PD geometry ("Channel parameters")						
	`Lpd`	`Rn`	`Rn2`	`Lneck`	`Lneck2`	`Rc`	`Rdt`
Straight (`1cyl`)	X	X					X
2-cylinder (`2cyl`)	X	X		X		X	X
3-cylinder (`3cyl`)	X	X	X	X	X	X	X

Situation	Class: Density and distribution			
	`dens`	`grid`	`nPit`	`dPit`
Isolated PDs	X	X	(write 1)	
Clustered PDs	X	X	X	X

Situation	Class: Particle information	
	`diff`	`x`
Always	X	X

Table 3
Overview of available PD geometries per mode. 1-cylinder is a straight channel with desmotubule, 2-cylinder has identical neck regions on both sides, 3-cylinder has all features and parameters shown in Fig. 1. For a short description of the modes, see Table 1

Mode	Available PD geometries		
	1-cylinder	**2-cylinder**	**3-cylinder**
`computeVals`	X	X	X
`computeUnitVals`	X	X	X
`sensitivityAnalysis`	X	X	
`computeDens`	X		
`computeAperture`	X		
`computeRnDensityGraph`	X		
`singleInterface`	X	X	X
`bootstrapInterface`	X	X	X

4 Starting Point: Calculation of Effective Wall Permeability from Complete Data

Whether comparing ultrastructural measurements with functionally assessed permeability measures (e.g., CFDA photobleaching-based assays) or searching for relevant effective permeability values to use in a tissue level model, the first step will often be to compute `Peff` from data. This requires information about PD geometry and their density/distribution on the interface. In the simplest case, there is a single value for each model parameter, either a literature-based estimate or the average of a data set. In this case, the mode `computeVals` calculates `Peff` based on these parameters. It is also possible to enter multiple values in a single run, e.g., to include upper/lower bounds or compute multiple values in a single run. When you have access to the full experimental data underlying an average, you can obtain a more accurate estimate using the mode `analyzeInterface` and compute confidence intervals of `Peff` using the mode `bootstrapInterface` that are described in more detail under Sect. 6.1, "Impact of variability within data."

Ideally, calculated `Peff` matches with functionally obtained values within the error margins related to both approaches. Sometimes the calculation is "spot on," but this may not be the case. In the latter case, PDinsight can be used to generate quantitative hypotheses about native PD dimensions and density (see below under Sect. 5 "What if. . .").

5 What If: Using Exploratory Modes to Produce Quantitative Hypotheses About PD Geometry and Density

Besides calculating the full effective permeability, PDinsight can be used to ask "what if" type questions: what kind of parameters would be required to obtain a given (measured) effective permeability? This can be useful in two cases: first, when no complete data is available and second, when calculations based on full data do not match—which may indicate, for example, that native dimensions differ from those after sample preparation. Particularly for the type I PDs "without apparent cytosolic sleeve" [15], such discrepancy may be expected and it could prove very insightful to know which effective radius would be required to reproduce the functionally measured effective permeability by diffusion alone.

Depending on how density is estimated, this may also be a poorly known parameter. The "classical" method based on counting PDs in cross sections of the cell wall [7] has the advantage that the same images can be used as for measuring PD geometry but depends on the chance events of hitting PDs and hence introduces uncertain estimates if only a small number of good images are available and/or the PDs are highly clustered. In the latter case, a very large number of sections would be required for averaging out the effects of hitting or not hitting the larger clusters.

For these reasons, the least well-known parameters likely are PD (neck) radius and/or density. These can be addressed in the exploratory modes `computeAperture`, `computeDens`, and `computeRnDensityGraph`, which compute the required (neck) radius, density, and the combinations of both, respectively, that yield a chosen target `Peff`. For simplicity, these modes only use straight channels. These values then have to be interpreted in the light of relevant literature values. Based on current knowledge, radii measured in the central region of thick (> 500 nm) cell walls could provide a reasonable soft upper bound for the (neck) radius. Reported densities, however, vary by at least two orders of magnitude, so the literature-based reasonable bounds would have to be obtained for each context specifically.

The exploratory modes only exist for (neck) radius and density, but a similar approach centered around other parameters could be carried out "by hand" by entering multiple values into `computeVals`. To do this selectively, the most relevant parameters can be identified with a sensitivity analysis. For a general impression of how Peff and Punit depend on various parameters, we refer to the explorations presented in [4].

6 Gaining Deeper Insight: Tools for In-depth Exploration

Measurement uncertainty and natural variation in the data are a fact of life. Besides potential artifacts from sample preparation, the main reasons are that the variation among PDs can be substantial [13, 27] and clustering of PDs [17] could have a big impact on density estimates based on independent cross sections of the cell wall. This raises the question: how much does that matter?

The modes `singleInterface` and `bootstrapInterface` can be used to address the impact of natural variation within a given data set, whereas a sensitivity analysis addresses the potential impact of unknown or not-yet-quantified variation.

6.1 Impact of Variability Within an Experimental Data Set

6.1.1 `singleInterface`

The mode `singleInterface` computes `Peff` based on raw data and, if the data is structured per PD, also outputs the relative contributions of individual PDs to `Peff`. For structured data, this mode computes the unit permeability `Punit` of all individual PDs with complete data (i.e., no missing parameters for the chosen PD model) and computes their relative contribution to the total permeability. This shows whether all PDs contribute approximately equally to permeability, or most can be attributed to a small subset of PDs.

Next, PDinsight computes `Peff` in three different ways: based on the average `Punit` found before, based on the per parameter averages of all complete PDs, and based on the per parameter averages of all data. The difference between the first and second calculations can be more than 10% and go in either direction. To understand why different values occur, consider the hypothetical example where the particle diameter exactly matches the minimal sleeve width. Using parameter averages, the predicted `Peff` would be 0. With variation among PDs, however, the wider sleeves could still sustain some permeability.

For unstructured data, `singleInterface` yields the same information as `computeVals`, but with different inputs (and the option to predict desmotubule radius with a statistical model). If desired, `Punit` can also be computed directly using `computeUnitVals`.

6.1.2 *bootstrapInterface*

For all modes discussed so far, computed values for `Peff` etc. are single values. With a bootstrapping approach, it is possible to determine confidence intervals around these values. As a rule of thumb, at least ten measurements per parameter are required for meaningful results. The mode `bootstrapInterface` computes these confidence intervals for both structured and unstructured data by resampling many (e.g., n = 10,000) times and computing `Peff` based on average `Punit` of the resampled PDs. This resampled set has the same size as either the number of PDs (structured data) or the largest number of measurements (unstructured data). It is possible to add noise while resampling. Finally, the median `Peff` and upper and lower bounds of the (95%) confidence interval are reported.

When analyzing a structured data set without missing values both in a structured and unstructured way, median bootstrap value will come very close to the value of using all PDs exactly once with structure (for large enough n), whereas the unstructured value may deviate a few percent from this. Depending on the specific data set, the unstructured value may be closer to either of the two options of computing `Peff` that are used by `singleInterface`. Structured resampling, moreover, produces narrower confidence intervals. So, if the information is available, it is best to maintain and make use of the per PD structure in the data for bootstrapping.

6.2 Sensitivity Analysis

A sensitivity analysis is an exploratory method for identifying which parameters have the strongest impact on effective permeability. This question can be addressed for its own sake, or as a starting point for a more in-depth exploration of a smaller number of parameters—whether theoretically or experimentally. This does not require more data than computing `Peff` or `Punit` from single parameter estimates/averages, but knowing the relevant range of the parameters makes the interpretation of the results more straightforward.

The `sensitivityAnalysis` tool in PDinsight computes the so-called elasticities of all continuous model parameters around a certain point, i.e., a parameter set or "measurement." The elasticity ϵ for a given parameter, say density, and `Peff` is defined as

$$\epsilon = \texttt{dens/Peff} * \partial\texttt{Peff}/\partial\texttt{dens}. \quad (3)$$

In mathematical terms, this is the partial derivative of `Peff` to `dens`, normalized by the actual values of `Peff` and `dens`. This estimates the relative change of `Peff` if `dens` changes. A value of $\epsilon = 1$ means that a change of 1% in `dens` results in an (approximately) 1% increase of `Peff`. Note that the actual range of the parameter also determines how much it contributes to the variation in `Peff`. A parameter with $\epsilon = 2$ that varies by a few percent only contributes less to the variation in `Peff` than one with $\epsilon = 0.2$ that varies several fold.

PDinsight also outputs elasticities of `Punit`, `Fih` and the hindrance factor that is used internally in calculating `An` and `Ac`. These are defined analogously. In the output file, parameter names are followed by `_0.001` or similar. This indicates the stepsize that was used in the numerical calculations. The stepsize should be so small that the outcome hardly changes with further reduction. To allow for this quality check, therefore, two values are used by default.

For a general impression of how `Peff` and `Punit` depend on various parameters, we refer to the explorations presented in [4].

7 Advice on Parameter Choices and Data Analysis

Depending on the experimental strategy, it may not be possible to determine all required parameters in a single experiment. The following advice is intended as a starting point for filling in the blanks.

7.1 PD Dimensions

PDinsight is designed with the idea in mind that the user has access to some type of electron microscopy data. This data may perhaps be unreliable due to the impact of sample preparation procedures, suggesting a more exploratory use of PDinsight, but in general most parameters regarding PD geometry will be available.

A notable exception is the desmotubule radius, which may be hard to determine for reasons of contrast/resolution if the cytosolic sleeve is narrow. Moreover, when using data like PD diameter/radius from a literature source, the corresponding PD length may not be readily available.

7.1.1 Desmotubule Radius `Rdt`

Desmotubule radius has been recently quantified in [15] in the *A. thaliana* root tip (9.5 ± 1.3 nm) and cultured cells (8.5 ± 1.2 nm). Theoretical studies use values ranging from 7.5 to 10 nm [2, 4, 16]. Such values are supported by images presented in [5].

7.1.2 PD Length `Lpd`

When both ends of the PD appear in the same image, measuring PD length is straightforward. If part of the PD is missing from the image, local cell wall thickness could provide a reasonable alternative estimate. Note that cell wall thickness varies substantially within a single interface (see below).

When using literature values, cell wall thickness measured in the most relevant system would be a good candidate. This should be chosen with care, however, as cell wall thickness varies by about an order of magnitude, from < 100 nm in the meristem [15, 17] to well over 500 nm in older cell walls [15]. Common sense has it that a PD cannot be shorter than the local cell wall thickness, but it may differ from average cell wall thickness. Notably, PDs may be located in thinner sections of the cell wall, as reported for the pericycle phloem pole—endodermis interface (53 vs. 132 nm median thickness near/away from PDs in *A. thaliana* Col-0, respectively [17]), as well as in markedly thicker sections as observed for clustered PDs in various mature cells/tissues, e.g., [10, 19].

7.1.3 Neck Length `Lneck`

On certain published images, the neck appears as a nicely distinct and narrowed region. In that case, `Lneck` can be measured as the narrow part until widening for the central region starts. On other images, however, the neck region may be harder to delineate. In that case, perform some sensitivity analysis and/or compute effective permeability with your lower, median, and upper estimate of `Lneck`.

In case of a high-resolution reconstruction of the PDs from electron tomography, an alternative approach could be to use a more detailed PD model based on this data. Such could be implemented in a future version of PDinsight.

7.2 Diffusion Coefficients and Particle Sizes

Finding the correct diffusion coefficient from the literature is typically not easy. Diffusion coefficients are not intrinsic material properties but depend on temperature and viscosity and are therefore not reported on chemical fact sheets. In addition, the viscosity of the cytoplasm is higher than that of water and varies with location in the cell [23, 24]. These differences, moreover, are stronger for larger molecules [14, 26]. In the absence of data for the correct probe, a good approach would, therefore, be to search for a similarly sized measured reference value and use the Stokes–Einstein relation for spherical particles (below) to adjust:

$$D = \frac{k_B T}{6\pi\eta r}, \tag{4}$$

where D is the diffusion coefficient, $k_B = 1.380649 \cdot 10^{-23}$ J K^{-1} the Boltzmann constant, T the (absolute) temperature, η the fluid's viscosity, and r the particle's radius. The "Stokes radius" or "hydrodynamic radius"[1] is the radius of a replacement particle (hard sphere) that yields the same diffusion coefficient as the real, non-spherical particle with its hydration shell.

The diffusion coefficient D is temperature dependent in two ways: linearly with absolute temperature T and, additionally, the viscosity η is itself temperature dependent. Therefore, the Stokes radius or hydrodynamic radius is more often reported.

As a reference, [8] contain a nice overview of diffusion coefficients and Stokes radii of multiple GFP-likes (GFP, Dendra2, EGFP) measured with various methods, both in water and (mammalian) cells. Notably, reported aqueous diffusion coefficients differ by over 30%, which may partially be due to different measurement technologies [8]. The most cited Stokes radius for GFP is 2.82 nm [25], 2.42 nm for EGFP [18, 22], 2.4 ± 0.2 nm for Dendra2 [8], and 2.27 nm for YFP [21].

Diffusion coefficients for (carboxy)fluorescein (CF or F) that are used in the PD literature are 1/3 [20] or 1/2 [12] of the water value, respectively, based on measured values of 487 ± 22 $\mu m^2 s^{-1}$ for CF [11] and 425 ± 1 $\mu m^2 s^{-1}$ for F [3]. The default diffusion coefficient in PDinsight, 162 $\mu m^2 s^{-1}$, is the CF value used in [20]. Note that although these substances are added to the experiment as (C)FDA, upon entering the symplast, both acetyl groups of (C)FDA are enzymatically removed by nonspecific intracellular esterases, converting it into (C)F. Therefore, the relevant diffusion coefficient is that of (C)F.

7.3 Parameters Describing PD Distribution

Investigating clustering requires a front view on the cell wall, whereas investigating PD dimensions requires cross sections. Therefore, it may be hard to accurately obtain channel parameters and distribution parameters from a single sample, except perhaps using an extensive serial block electron microscopy approach [17].

Without clustering information, we suggest doing a first calculation with settings for a homogeneous wall: hexagonal grid and `nPit` = 1. Note that without clustering, the distribution in the wall has little impact, so the grid parameter is not very important [4]. Subsequently, it may be informative to explore cluster sizes (1), 3, 7, 12, and 19 to investigate the potential impact of clustering. These are the cluster sizes with the most efficient packing of PDs on a regular grid fitting inside a circle.

Another challenge is that (currently) PDinsight only accepts a few predefined cluster sizes (for geometrical reasons: see Fig. 2). When you expect or observe strong clustering, choose the cluster size that is closest to the typical (median) cluster size in the sample. If this happens to be 4, 5, or 6, also compare with 3 and 7, because those are more efficiently packed. Figure 7 in [4] shows that the impact of clustering is largest for low PD density.

[1] In polymer physics, hydrodynamic radius has a somewhat different meaning. In the PD literature, hydrodynamic radius typically refers to the Stokes radius.

Acknowledgements

I would like to thank Johannes Liesche for valuable comments on the manuscript.

References

1. Abou-Saleh RH, Hernandez-Gomez MC, Amsbury S, Paniagua C, Bourdon M, Miyashima S, Helariutta Y, Fuller M, Budtova T, Connell SD et al (2018) Interactions between callose and cellulose revealed through the analysis of biopolymer mixtures. Nat Commun 9:1–13
2. Blake JR (1978) On the hydrodynamics of plasmodesmata. J Theor Biol 74(1):33–47
3. Culbertson CT, Jacobson SC, Ramsey JM (2002) Diffusion coefficient measurements in microfluidic devices. Talanta 56(2):365–373
4. Deinum EE, Mulder BM, Benitez-Alfonso Y (2019) From plasmodesma geometry to effective symplasmic permeability through biophysical modelling. Elife 8:e49000. https://doi.org/10.7554/eLife.49000
5. Ding B, Turgeon R, Parthasarathy M (1992) Substructure of freeze-substituted plasmodesmata. Protoplasma 169(1):28–41
6. Goodwin PB, Shepherd V, Erwee MG (1990) Compartmentation of fluorescent tracers injected into the epidermal cells of *Egeria densa* leaves. Planta 181:129–136. https://doi.org/10.1007/BF00202335
7. Gunning BES (1978) Age-related and origin-related control of the numbers of plasmodesmata in cell walls of developing *Azolla* roots. Planta 143:181–190. https://doi.org/10.1007/BF00387789
8. Gura Sadovsky R, Brielle S, Kaganovich D, England JL (2017) Measurement of rapid protein diffusion in the cytoplasm by photo-converted intensity profile expansion. Cell Rep 18(11):2795–2806. https://doi.org/10.1016/j.celrep.2017.02.063. http://www.sciencedirect.com/science/article/pii/S2211124717302784
9. Kim I, Hempel FD, Sha K, Pfluger J, Zambryski PC (2002) Identification of a developmental transition in plasmodesmatal function during embryogenesis in Arabidopsis thaliana. Development 129(5):1261–1272
10. Kollmann R, Glockmann C (1999) Multimorphology and nomenclature of plasmodesmata in higher plants. In: Plasmodesmata. Springer, pp 149–172
11. Kramer EM, Frazer NL, Baskin TI (2007) Measurement of diffusion within the cell wall in living roots of Arabidopsis thaliana. J Exp Bot 58(11):3005–3015. https://doi.org/10.1093/jxb/erm155
12. Liesche J, Schulz A (2013) Modeling the parameters for plasmodesmal sugar filtering in active symplasmic phloem loaders. Front Plant Sci 4:207. https://doi.org/10.3389/fpls.2013.00207
13. Liesche J, Gao C, Binczycki P, Andersen SR, Rademaker H, Schulz A, Martens HJ (2019) Direct comparison of leaf plasmodesma structure and function in relation to phloem-loading type. Plant Physiol 179(4):1768–1778. https://doi.org/10.1104/pp.18.01353, http://www.plantphysiol.org/content/179/4/1768, http://www.plantphysiol.org/content/179/4/1768.full.pdf
14. Luby-Phelps K (1999) Cytoarchitecture and physical properties of cytoplasm: volume, viscosity, diffusion, intracellular surface area. In: International review of cytology, vol 192. Elsevier, pp 189–221
15. Nicolas WJ, Grison MS, Trépout S, Gaston A, Fouché M, Cordelières FP, Oparka K, Tilsner J, Brocard L, Bayer EM (2017) Architecture and permeability of post-cytokinesis plasmodesmata lacking cytoplasmic sleeves. Nat Plants 3(7):17082
16. Park K, Knoblauch J, Oparka K, Jensen KH (2019) Controlling intercellular flow through mechanosensitive plasmodesmata nanopores. Nat Commun 10(1):1–7
17. Paterlini A, Belevich I, Jokitalo E, Helariutta Y (2020) Computational tools for serial block electron microscopy reveal plasmodesmata distributions and wall environments. Plant Physiol 184(1):53–64
18. Petrášek Z, Schwille P (2008) Precise measurement of diffusion coefficients using scanning fluorescence correlation spectroscopy. Biophys J 94(4):1437–1448
19. Robinson-Beers K, Evert RF (1991) Fine structure of plasmodesmata in mature leaves of sugarcane. Planta 184:307–318. https://doi.org/10.1007/BF00195331
20. Rutschow HL, Baskin TI, Kramer EM (2011) Regulation of solute flux through

plasmodesmata in the root meristem. Plant Physiol 155(4):1817–1826. https://doi.org/10.1104/pp.110.168187

21. Schaefer M, Albrecht N, Hofmann T, Gudermann T, Schultz G (2001) Diffusion-limited translocation mechanism of protein kinase c isotypes. The FASEB J 15(9): 1634–1636, contains YFP diffusion
22. Schenk A, Ivanchenko S, Röcker C, Wiedenmann J, Nienhaus GU (2004) Photodynamics of red fluorescent proteins studied by fluorescence correlation spectroscopy. Biophys J 86(1):384–394
23. Scherp P, Hasenstein KH (2007) Anisotropic viscosity of the Chara (Characeae) rhizoid cytoplasm. Am J Botany 94(12):1930–1934
24. Srivastava A, Krishnamoorthy G (1997) Cell type and spatial location dependence of cytoplasmic viscosity measured by time-resolved fluorescence microscopy. Arch Biochem Biophys 340(2):159–167
25. Terry B, Matthews E, Haseloff J (1995) Molecular characterization of recombinant green fluorescent protein by fluorescence correlation microscopy. Biochem Biophys Res Commun 217(1):21–27
26. Verkman AS (2002) Solute and macromolecule diffusion in cellular aqueous compartments. Trends Biochem Sci 27(1):27–33
27. Yan D, Yadav SR, Paterlini A, Nicolas WJ, Petit JD, Brocard L, Belevich I, Grison MS, Vaten A, Karami L et al (2019) Sphingolipid biosynthesis modulates plasmodesmal ultrastructure and phloem unloading. Nat Plants 5(6):604–615

Chapter 31

Measuring Intercellular Interface Area in Plant Tissues Using Quantitative 3D Image Analysis

Gwendolyn V. Davis, Richard S. Smith, and George W. Bassel

Abstract

The cells which make up plant tissues remain fixed together through shared cell walls. Cell-to-cell communication principally takes place through these shared interfaces through a combination of plasmodesmata, transporters, and the apoplastic space. To better understand the capacity for intercellular communication in plant tissues, this chapter outlines a method which can be used to quantify the surface area of shared intercellular interfaces using whole mount imaging and quantitative 3D image analysis. This method allows the potential for intercellular communication as prescribed by cellular architecture to be measured at single cell resolution.

Key words 3D image analysis, Plant organ, Intercellular communication, Tissue architecture, MorphoGraphX

1 Introduction

The cells which make up plant tissues are glued together through shared cell walls [1]. Cellular patterning in plants is therefore fixed, unlike in animal tissues where cells can be mobile [2]. Highly stereotyped patterns of cells give rise to organ function following structure-function relationships [3]. Organs consisting of coherent assemblies of cells therefore perform functions which are greater than the sum of their parts [4].

Both pattern formation and organ function emerge as a result of cell-to-cell communication [5]. This communication occurs principally at intercellular interfaces through different means, including cytoplasmic channels termed plasmodesmata, transporters, and intercellular space called the apoplasm [6]. While detailed knowledge of the molecular and cellular basis for intercellular communication is emerging, less is understood about the higher-order intercellular communication in plant tissues. In order to address this, a computational technique to capture and quantify

Yoselin Benitez-Alfonso and Manfred Heinlein (eds.), *Plasmodesmata: Methods and Protocols*,
Methods in Molecular Biology, vol. 2457, https://doi.org/10.1007/978-1-0716-2132-5_31,

global cellular organization (or patterning) has been described previously [7–9]. This involves both the identification of cell types within 3D imaged plant organs [9] and their physical connectivity [7] using the software MorphoGraphX [10].

In the process of identifying physical connectivity, the size of shared cell interfaces can also be measured using this computational pipeline [7]. This represents the measurement of the intercellular junctions through which cell communication can take place.

This chapter outlines how to capture cell interface sizes and compare how these change under different conditions.

2 Materials

2.1 Sample Preparation

1. 10% bleach.
2. 1/2 MS plates with 1% agar (w/v).
3. Fixative solution: 3:1 (v/v) ethanol/acetic acid.
4. Clearing solution: 1% SDS and 0.2 N NaOH.
5. Amylase to degrade starch.
6. mPA-PI solution prepared as previously described [11, 12].
7. Chloral hydrate mounting media: 4 g chloral hydrate, 1 mL glycerol, and 2 mL water.

2.2 Image Acquisition

1. Glass bottom Greiner Bio-One CELLView™ cell culture dishes.
2. Inverted confocal microscope.

2.3 Image Analysis Software

1. Download Fiji (https://imagej.net/software/fiji/) [13] as this is needed to convert confocal images into TIFF stacks for further analysis in MorphoGraphX.
2. Download MorphoGraphX [10] as this is used for the segmentation of cells, identification of cell types, and quantitative measurement of cell interface area.

3 Methods

3.1 Sample Preparation

1. Sterilize *Arabidopsis* seeds using 10% bleach for 5 min.
2. Rinse three times in sterile water before plating onto 1/2 MS plates (1% agar).
3. Grow seedlings vertically for 4 days in the cabinets using 16 h light (light intensity 150–175 mmol m^2 s^{-1}).
4. Fix tissue overnight using fixative solution.
5. Clear in clearing solution overnight [11].

6. Wash tissue in distilled water three times.
7. Degrade starch granules using α-amylase as previously described [14].
8. Stain tissue using the mPA-PI method [11, 12] and clear with chloral hydrate mounting media.

3.2 Image Acquisition

1. Place plant tissue samples onto a glass bottom Greiner Bio-One CELLView™ cell culture dish, and mount using chloral hydrate clearing solution.
2. Image the tissue using an inverted confocal microscope as described previously [15]. Z-stacks need to be acquired such that the z-step size is 0.4 the optical slice thickness, pinhole size is kept to a minimum, and images are 16-bit depth. Using a 25× oil immersion lens, the resolution of acquired images should be 2048 × 2048.

3.3 Importing Images into MorphoGraphX, 3D Cell Segmentation, and Editing

1. Export images into TIFF stacks using the BioFormats Importer within Fiji [13], and then import them into MorphoGraphX for further analysis via Stack > Stack 1 > Main > Open (Fig. 1a).
2. Segment cells in MorphoGraphX as described previously using an autoseeded watershed algorithm [10, 12, 16]. Briefly, the TIFF confocal stack is blurred using Gaussian blur with a radius of 0.5. Cells are then segmented using ITK Autoseeded Watershed, and over-segmented cell edited using the tools available in MorphoGraphX described previously [10, 15] (Fig. 1b) (*see* **Notes 1–3**).

3.4 Creating Cell Meshes

1. Create meshes on segmented cell surfaces by running the Marching Cubes 3D algorithm Process > Mesh > Creation with a cube size of 0.5 and with 20 smooth passes at the time of mesh creation (Fig. 1c) (*see* **Note 4**). Running smoothing after mesh creation leads to the shrinking of the mesh and alteration in the geometric properties of the object in question [17] (*see* **Note 5**). This shrinking of the mesh also removes cell-cell connectivity making it impossible to extract networks later in the protocol.

3.5 Identification of Cell Types and Cell Interfaces Using 3DCellAtlas

1. Identify cell types using the 3DCellAtlas pipeline as described previously [9, 15]. This method uses a combination of 3D cell geometry and position within the organ to identify cell types and quantify cell anisotropy. Briefly, a Bezier curve is placed along the center of the axis of the organ and a mesh cast across the surface of the organ. The principal directions of cell length along each the longitudinal, radial, and circumferential axes are identified using these as guides. Cell types can then be

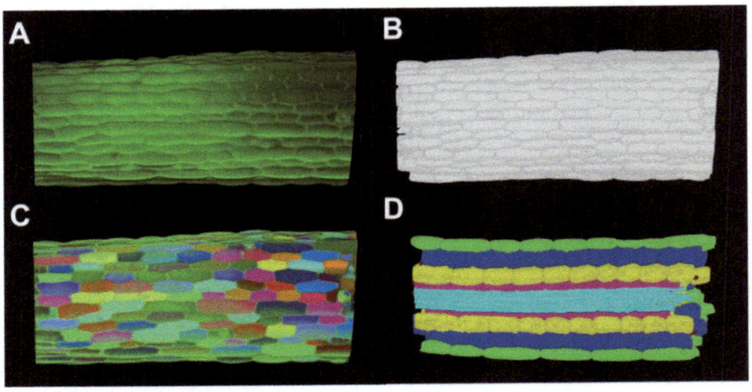

Fig. 1 Image analysis procedure to segment and annotate cells in the *Arabidopsis* hypocotyl. (**a**) The original 4-day-old light-grown *Arabidopsis* hypocotyl TIFF confocal image stack loaded into MorphoGraphX. (**b**) 3D cellular segmentation of the image stack from (**a**). (**c**) Creation of the 3D cell mesh using 3D Marching Cubes. (**d**) Annotation of cell types of the hypocotyl sample using 3DCellAtlas

identified by clustering cells based on their geometric properties and distance from the radial center of the organ (Fig. 1d).

2. Export the cell connectivity network. Once cells have been segmented and identified, their connectivity in the form of a network can be exported. This can be achieved using Process > Mesh > Export > Cell Graph 3D (Fig. 2). The result of this export process is a .CSV file which contains two sections of information: (a) the cell surface area, volume, and position in space and (b) the list of associations between cells expressed as pairwise interactions. In the third column of this section (after the first two describing the cell IDs which are connected) is the area in mm^2 of the size of the cell interface. It is these data which will be used for the further analysis of cell interface size.

3.6 Data Analysis

1. Analyze the data in the exported CSV file output from Cell Graph 3D using either a statistical analysis software or a network analysis software. The CSV file contains both the IDs of cells which are connected and the size of the shared interface between adjacent cells. Coupled with the identification of cell types, this raises the possibility of looking at the size of cell type-specific intercellular interfaces within whole tissues.
2. Using statistical analysis software such as R or Excel, import data and analyze the sizes of the interfaces. For example, the size and distribution of cell interfaces within the epidermis of a light-grown *Arabidopsis* hypocotyl after 4 days can be examined in a histogram (Fig. 3).
3. Use a network analysis software such as Cytoscape [18] to represent the nodes as the cells and the edges as their physical

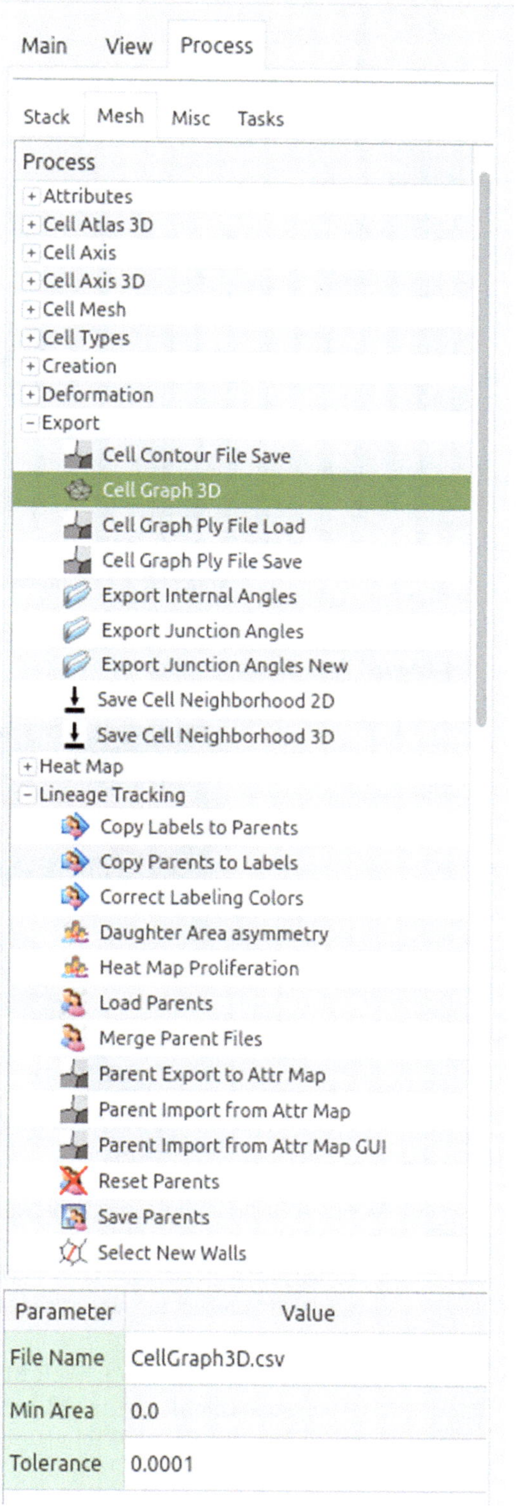

Fig. 2 Exporting the network and cell interface data. Screenshot showing where to export the network and cell interface data using Cell Graph 3D under the Mesh > Export section of MorphoGraphX

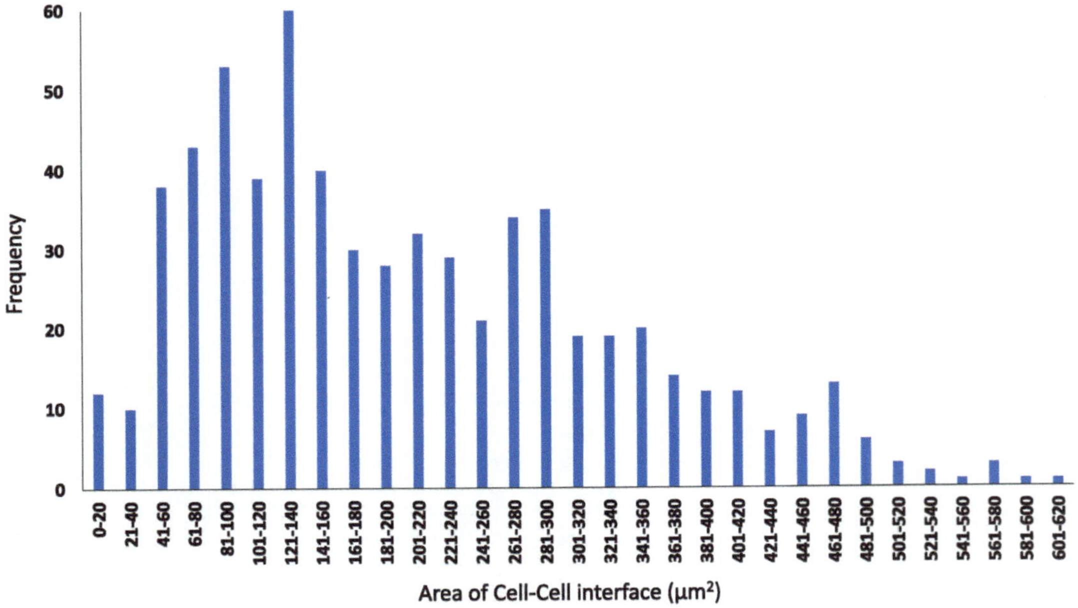

Fig. 3 Cell interfaces in *Arabidopsis* hypocotyls. Histogram showing the distribution of cell interface sizes in the epidermis of a 4-day-old light-grown *Arabidopsis* hypocotyl

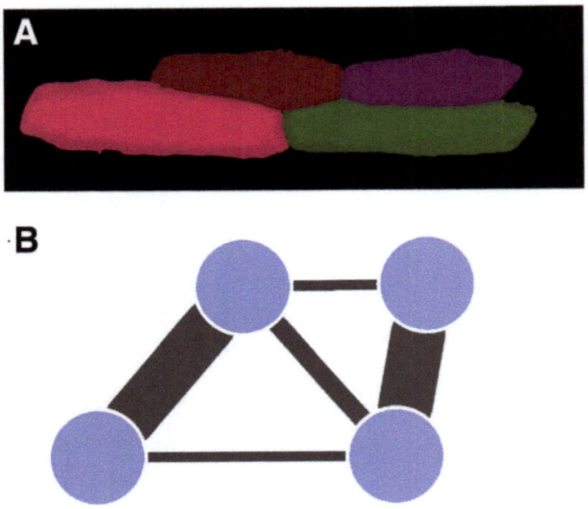

Fig. 4 Going from images to networks. (**a**) Subselection of four epidermal cells from a light-grown *Arabidopsis* hypocotyl. (**b**) Graphical representation of the connectivity of the four cells from (**a**) in Cytoscape. Edge line thickness is adjusted to reflect the size of shared interfaces between cells

interactions [19]. Weight the edges using the shared interface sizes calculated in MorphoGraphX. These can be visualized (Fig. 4) or used when performing centrality-based analysis of tissue topology [7].

4 Notes

1. The segmentation can be edited if it is over-segmented, but not if it is under-segmented. Labels can be merged where a single cell has been divided; but two cells cannot be split which have been labeled as one cell. Therefore, it is important to try different threshold values for the ITK Autoseeded Watershed in order to select the most suitable threshold.
2. Clipping planes are used for the visual inspection of a segmentation and editing. Clip 3 can be enabled, and the stack scrolled through in order to assess the success of a segmentation. The "dropper tool" can be used to identify individual cell labels during this inspection. Regular saving of the work stack is important when editing as a change to the segmentation once made cannot be undone.
3. In order to make visualization of the segmentation easier, the number of colors used for cell labels can be increased using the Label Editor. A greater number of colors than the default 16 can be set and randomized making visualization easier to interpret when examining a large number of cells.
4. Choice of cube size is important for defining what level of detail of the cell morphology is captured by the mesh. For capturing cell curvature, the cube size will need to be several times smaller than the size of the cells, while for the general shape of an organ or tissue, a larger cube size of roughly the same size as the cells may be more suitable. Mesh creation will take a longer amount of time with smaller cube sizes. Therefore, consideration should be put into what cube size is most appropriate for each study and for the tissue being examined. It may be useful to generate several meshes with different cube sizes in order to select the most appropriate.
5. Smoothing can also cause shrinking of the mesh during mesh creation; the greater the number of smooth passes, the more shrinking may occur. Smaller cube sizes are less affected by a larger number of smooth passes. A visual inspection of the mesh post-creation (using clipping planes) is important for ensuring that the mesh has not shrunk.

Acknowledgments

This work was supported by BBSRC grant BB/S002804/1, Leverhulme Grant RPG-2019-267, and HFSP Grant RGP0002/2020.

References

1. Coen E, Rolland-Lagan A-G, Matthews M, Bangham JA, Prusinkiewicz P (2004) The genetics of geometry. Proc Natl Acad Sci U S A 101(14):4728–4735
2. Gibson WT, Veldhuis JH, Rubinstein B, Cartwright HN, Perrimon N, Brodland GW, Nagpal R, Gibson MC (2011) Control of the mitotic cleavage plane by local epithelial topology. Cell 144(3):427–438
3. Thompson DW, Thompson DAW (1942) On growth and form, vol 2. Cambridge University Press, Cambridge
4. Ollé-Vila A, Duran-Nebreda S, Conde-Pueyo-N, Montañez R, Solé R (2016) A morphospace for synthetic organs and organoids: the possible and the actual. Integr Biol 8(4):485–503
5. Gordon SP, Heisler MG, Reddy GV, Ohno C, Das P, Meyerowitz EM (2007) Pattern formation during de novo assembly of the Arabidopsis shoot meristem. Development 134(19):3539–3548
6. Van Norman JM, Breakfield NW, Benfey PN (2011) Intercellular communication during plant development. Plant Cell 23(3):855–864
7. Jackson MD, Xu H, Duran-Nebreda S, Stamm P, Bassel GW (2017) Topological analysis of multicellular complexity in the plant hypocotyl. elife 6:e26023
8. Jackson MD, Duran-Nebreda S, Bassel GW (2017) Network-based approaches to quantify multicellular development. J R Soc Interface 14(135):20170484
9. Montenegro-Johnson TD, Stamm P, Strauss S, Topham AT, Tsagris M, Wood AT, Smith RS, Bassel GW (2015) Digital single-cell analysis of plant organ development using 3DCellAtlas. Plant Cell 27(4):1018–1033
10. de Reuille PB, Routier-Kierzkowska A-L, Kierzkowski D, Bassel GW, Schüpbach T, Tauriello G, Bajpai N, Strauss S, Weber A, Kiss A (2015) MorphoGraphX: a platform for quantifying morphogenesis in 4D. eLife 4: e05864
11. Truernit E, Bauby H, Dubreucq B, Grandjean O, Runions J, Barthélémy J, Palauqui J-C (2008) High-resolution whole-mount imaging of three-dimensional tissue organization and gene expression enables the study of phloem development and structure in Arabidopsis. Plant Cell 20(6):1494–1503
12. Bassel GW, Stamm P, Mosca G, de Reuille PB, Gibbs DJ, Winter R, Janka A, Holdsworth MJ, Smith RS (2014) Mechanical constraints imposed by 3D cellular geometry and arrangement modulate growth patterns in the Arabidopsis embryo. Proc Natl Acad Sci U S A 111(23):8685–8690
13. Schindelin J, Arganda-Carreras I, Frise E, Kaynig V, Longair M, Pietzsch T, Preibisch S, Rueden C, Saalfeld S, Schmid B (2012) Fiji: an open-source platform for biological-image analysis. Nat Methods 9(7):676–682
14. Wuyts N, Palauqui J-C, Conejero G, Verdeil J-L, Granier C, Massonnet C (2010) High-contrast three-dimensional imaging of the Arabidopsis leaf enables the analysis of cell dimensions in the epidermis and mesophyll. Plant Methods 6(1):1–14
15. Stamm P, Strauss S, Montenegro-Johnson TD, Smith R, Bassel GW (2017) In silico methods for cell annotation, quantification of gene expression, and cell geometry at single-cell resolution using 3DCellAtlas. In: Plant hormones. Springer, New York, pp 99–123
16. Yoshida S, de Reuille PB, Lane B, Bassel GW, Prusinkiewicz P, Smith RS, Weijers D (2014) Genetic control of plant development by overriding a geometric division rule. Dev Cell 29(1):75–87
17. Bassel GW (2015) Accuracy in quantitative 3D image analysis. Plant Cell 27(4):950–953
18. Shannon P, Markiel A, Ozier O, Baliga NS, Wang JT, Ramage D, Amin N, Schwikowski B, Ideker T (2003) Cytoscape: a software environment for integrated models of biomolecular interaction networks. Genome Res 13(11):2498–2504
19. Bassel GW (2019) Multicellular systems biology: quantifying cellular patterning and function in plant organs using network science. Mol Plant 12(6):731–742

INDEX

Yoselin Benitez-Alfonso and Manfred Heinlein (eds.), *Plasmodesmata: Methods and Protocols*,
Methods in Molecular Biology, vol. 2457, https://doi.org/10.1007/978-1-0716-2132-5,

D

E

F

Q

R

S

MIX
Papier aus verantwortungsvollen Quellen
Paper from responsible sources
FSC® C105338

If you have any concerns about our products,
you can contact us on
ProductSafety@springernature.com

In case Publisher is established outside the EU,
the EU authorized representative is:
Springer Nature Customer Service Center GmbH
Europaplatz 3, 69115 Heidelberg, Germany

Printed by Libri Plureos GmbH
in Hamburg, Germany